Springer Series in Reliability Engineering

For further volumes:
http://www.springer.com/series/6917

Krzysztof Kołowrocki
Joanna Soszyńska-Budny

Reliability and Safety of Complex Technical Systems and Processes

Modeling—Identification—Prediction—Optimization

Springer

Krzysztof Kołowrocki
Maritime University
ul. Morska 81-87
81-225 Gdynia
Poland
e-mail: katmatkk@am.gdynia.pl

Joanna Soszyńska-Budny
Maritime University
ul. Morska 81-87
81-225 Gdynia
Poland
e-mail: joannas@am.gdynia.pl

ISSN 1614-7839
ISBN 978-1-4471-2712-3 ISBN 978-0-85729-694-8 (eBook)
DOI 10.1007/978-0-85729-694-8
Springer London Dordrecht Heidelberg New York

British Library Cataloguing in Publication Data
A catalogue record for this book is available from the British Library

Cover design: eStudio Calamar, Berlin/Figueres

Printed on acid-free paper

Springer is part of Springer Science+Business Media (www.springer.com)

Preface

This book is concerned with the identification, evaluation, prediction and optimization of operation, reliability, availability and safety of technical systems related to their operation processes. The main emphasis is on multi-state systems composed of ageing components and changing their structures and their component's reliability and safety characteristics during the operation processes. In this book these systems are called the complex technical systems. The practical importance of such an approach in reliability and safety identification, assessment and prediction, and analyzing the effectiveness of operation processes of real technical systems is evident.

Most real technical systems are very complex and it is difficult to analyze their reliability, availability and safety. Large numbers of components and subsystems and their operating complexity cause the identification, evaluation, prediction and optimization of their reliability, availability and safety to be complicated. The complexity of the systems' operation processes and their influence on changing in time the systems' structures and their components' reliability characteristics are very often met in real practice. We meet complex technical systems, for instance, in piping transportation of water, gas, oil and various chemical substances. Complex technical systems are also used in electrical energy distribution, in telecommunication, in rope transportation, in maritime transport and in shipyard and port transport systems using belt conveyers and elevators. Rope transportation systems such as port elevators and ship-rope elevators used in shipyards during ship docking and undocking are model examples of such systems.

Taking into account the importance of the safety and operating process effectiveness of such systems it seems reasonable to expand the two-state approach to multi-state approach in their reliability and safety analysis. The assumption that the systems are composed of multi-state components with reliability states or safety states degrading in time gives the possibility for more precise analysis of their reliability, safety and operation processes' effectiveness. This assumption allows us to distinguish a system reliability or safety critical state to exceed which is either dangerous for the environment or does not assure the necessary level of its operation process effectiveness. Then, an important system reliability or safety characteristic is the time to the moment of exceeding the system reliability or safety critical state and its distribution, which is called the system risk function. This distribution is strictly related to the system multi-state reliability function and the system multi-state safety function and these are the basic characteristics of the multi-state system. In the case of large systems, the determination of the exact reliability functions of the systems and the system risk functions leads us to very complicated formulae that are often useless for reliability practitioners. One of the important techniques in this situation is the asymptotic approach to system reliability evaluation. In this approach, instead of the preliminary complex formula for the system reliability function, after assuming that the number of system components tends to infinity and finding the limit reliability of the system, we obtain its simplified form. This aspect of complex technical systems is also considered in this book.

The convenient tools for analyzing these problems are semi-Markov modeling the systems' operation processes and multi-state approach to the systems' reliability evaluation that is proposed in this book.

The aim of this book is to present the general reliability, availability and safety analytical models of complex non-repairable and repairable multi-state technical systems related to their operation processes that are developed by the author and to apply them practically to real industrial systems and processes. The integrated general models of complex industrial systems, linking their reliability, availability and safety models with their operation process models and considering variable in different operation states their reliability and safety structures and their components reliability and safety parameters are proposed. The common usage of the multi-state systems' reliability and availability models and the semi-Markov model for the systems' operation processes in order to construct the joint general system reliability and availability models related to their operation process is the main idea of the book. The joint models linking the reliability and safety models of the multi-state systems and their varying in time operation processes models are constructed and applied in the reliability, availability and safety analysis of real complex technical systems. These joint reliability and safety models together with the linear programming are proposed for the reliability, availability and safety optimization and operation cost analysis of the complex technical systems.

The models and methods proposed in this book proposed in the book models and methods are applied to reliability, availability and safety analysis, identification, prediction and optimization of the port and maritime transportation systems related to varying in time their operation processes, structures and components reliability and safety parameters.

This book delivers a complete elaboration of the state of art on the method of reliability and safety identification, evaluation, prediction and optimization for as wide as possible a range of complex technical systems. Pointing out the possibility of this method's extensive practical application in the operating processes of these systems is also an important reason for this book. This book contains complete actual solutions of the formulated problems for the considered complex technical systems in the case of the exponential multi-state reliability functions of their components. This assumption is necessary in the analytical approach to the subject considered.

This book consists of a Preface, Chaps. 1–7 presenting the main theoretical results and their practical applications, Summary and Appendices including statistical tables that are used in the book. References suitable to the subjects considered in the particular chapters are provided at the end of each chapter.

In the Preface, the book introduction and its contents are presented.

In Chap. 1, the basic notions of ageing multi-state systems reliability analysis are introduced. The system components and the system multi-state reliability functions are defined. The mean values and variances of the multi-state systems' lifetimes in the reliability state subsets and the mean values of their lifetimes in the particular reliability states are defined. The multi-state system risk function and the moment of exceeding by the system the critical reliability state are introduced. The multi-state series, parallel, "m out of n", consecutive "m out of n: F", series-parallel, parallel-series, series-"m out of k", "m_i out of l_i"-series, series-consecutive "m out of k: F" and consecutive "m_i out of l_i: F"-series reliability structures of the multi-state systems with degrading (ageing) components are defined and their reliability functions determined. As a particular case, the reliability functions of the considered multi-state systems composed of components having exponential reliability functions are determined. Moreover, the multi-state approach to analysis and defining the basic notions of the systems' safety is proposed. The system safety function and the system risk function that allow to define basic safety structures of the multi-state systems composed of components with degrading safety states are introduced. Applications of the proposed multi-state system reliability and safety models to the evaluation and prediction of the reliability characteristics of an exemplary system and a port oil transportation system and the safety characteristics of a maritime ferry technical system are presented as well.

In Chap. 2, the operation process of the complex technical system is considered and its operation states are introduced. The semi-Markov process is used to construct a general probabilistic model of the considered complex technical systems operation processes. To build this model the vector of probabilities of the system operation process straying at the initial operation states, the matrix of probabilities of the system operation process transitions between the operation states, the matrix of conditional distribution functions and the matrix of conditional density functions of the system operation process conditional sojourn times at the operation states are defined. To describe the system operation process the conditional sojourn times at the particular operation states, the uniform distribution, the triangular distribution, the double trapezium distribution, the quasi-trapezium distribution, the exponential distribution, the Weibull distribution and the chimney distribution are suggested and introduced. Based on these assumptions from the constructed model, the main characteristics of the system operation process are found. The mean values of the system operation process conditional sojourn times at the particular operation states having these distributions are given. Moreover, the distribution functions of the system operation process unconditional sojourn times at the particular operation states, the mean values of the system operation process unconditional sojourn times at the particular operation states, the limit values of the transient probabilities of the system operation process at the particular operation states and the approximate mean values of the system operation process total sojourn times at the particular operation states for the fixed sufficiently large system operation time are determined. Applications of the proposed model for the evaluation and prediction of the operation processes' characteristics of the exemplary system, the port oil transportation system and the maritime ferry technical system are presented as well.

In Chap. 3, there are presented the general reliability, availability and safety analytical models of complex non-repairable and repairable multi-state technical systems related to their operation processes. They are the integrated general models of complex technical systems, linking their multi-state reliability, availability and safety models with their operation process models and considering variable at the different operation states their reliability and safety structures and their components reliability and safety parameters. The conditional reliability and safety functions at the system particular operation states and independent of the system particular operation states the unconditional reliability and safety function and the risk function of the complex technical systems are defined. These joint models of the reliability, availability and safety and the variable in time system operation processes are constructed for multi-state series, parallel, "*m* out of *n*", consecutive "*m* out of *n*: F", series-parallel, parallel-series, series-"*m* out of *k*", "m_i out of l_i"-series, series-consecutive "*m* out of *k*: F" and consecutive "m_i out of l_i: F"-series systems. The joint models are applied for determining the reliability, availability, renewal and safety characteristics of these systems related to their varying in time reliability and safety structures and their components' reliability and safety characteristics. Under the assumption that the considered systems are exponential, the unconditional reliability and safety functions of these systems are determined. Moreover, in the case of large-scale systems, the possibility of combining the results coming from these joint models and the results concerning the limit reliability functions of the considered systems is briefly presented. The proposed models and methods are applied for the reliability and availability analysis, evaluation and prediction of the exemplary technical system and the port oil piping transportation system and for the safety analysis, evaluation and prediction of the maritime ferry technical system related to varying in time their operation processes, structures and components' reliability and safety parameters.

In Chap. 4, there are presented the methods of identification of the operation processes of complex technical systems. These are the methods and procedures for estimating the unknown basic parameters of the system operation process semi-Markov model and identifying the distributions of the conditional system operation process sojourn times at the operation states. The formulae estimating the probabilities of the system operation process straying at the operation states at the initial moment, the probabilities of the system operation process transitions between the operation states and the parameters of the distributions suitable and typical for the description of the system operation process conditional sojourn times at the operation states are given. Namely, the parameters of the uniform distribution, the triangular distribution, the double trapezium distribution, the quasi-trapezium distribution, the exponential distribution, the Weibull's distribution and the chimney distribution are estimated using the statistical methods such as the method of moments and the maximum likelihood method. The chi-square goodness-of-fit test is described and proposed to be applied for verifying the hypotheses about these distributions' choice validity. The procedure of statistical data sets' uniformity analysis based on Kolmogorov–Smirnov test is proposed to be applied to the empirical conditional sojourn times at the operation states coming from different realizations of the same complex technical system operation process. The applications of the proposed statistical methods of the unknown parameters identification of the complex technical system operation process model for determining the operation parameters of the exemplary system, the port oil transportation system and the maritime ferry technical system are presented. The procedure of testing the uniformity of statistical data sets is applied for the realizations of the conditional sojourn times at the operation states of the ferry technical system collected at two different operating conditions.

In Chap. 5, there are presented procedures and formulae estimating the unknown parameters of the complex technical system components reliability and safety models on the basis of statistical data coming from the components reliability and safety states changing processes. The maximum likelihood method is applied to estimating the unknown intensities of departures from the reliability and safety state subsets of the multi-state system components having different exponential reliability functions at various system operation states. This method is applied to the statistical data collected in different kinds of empirical experiments, including cases of small number of realizations and non-completed investigations. The goodness-of-fit test applied to verify the hypotheses concerned with the exponential forms of the multi-state reliability functions of the particular components of the complex technical systems at the variable operations conditions is presented. In the case of lack of data coming from the components' reliability and safety states changing processes, the simplified method of estimating the unknown intensities of departures from the reliability and safety state subsets based on the expert opinions is proposed. Applications of the proposed statistical methods of the unknown parameters identification of the complex technical system components reliability and safety models to determine the reliability parameters of the exemplary system and the port oil transportation system components and the safety parameters of the maritime ferry technical system components are presented as well.

In Chap. 6, the methods based on the results of the joint model linking a semi-Markov modeling of the system operation processes with a multi-state approach to system reliability and safety and the linear programming are proposed to the complex technical systems at the variable operation conditions operation, reliability, availability and safety optimization and cost analysis. The method of optimization of the complex technical systems operation processes determining the optimal values of limit transient probabilities at the system operation states that maximize the system lifetimes in the reliability or safety state subsets is proposed.

The way of operation cost analysis of the complex technical system at the variable operation conditions and its application for the evaluation of the cost before and after this system operation process optimization is presented. The methods of corrective and preventive maintenance policy maximizing the availability and minimizing the renovation cost of the complex technical systems at the variable operation conditions are presented as well. The proposed methods are applied to the operation, reliability and availability optimization and operation cost analysis of the exemplary technical system and the port oil piping transportation system and to the operation and safety optimization of the maritime ferry technical system related to varying in time their operation processes, structures and components reliability and safety parameters. The procedures of the corrective and preventive maintenance policy optimization are proposed and applied to the exemplary system, the port oil piping transportation system and the ferry technical system.

In the complementary Chap. 7, the methods based on the results included in the book that are not used in previous chapters, are applied to the reliability, availability and safety identification, prediction and optimization of various kinds of systems. Direct applications of the results stated in the propositions of Chap. 3 to the reliability, renewal and availability characteristics prognosis of the series and parallel systems operating at the variable operation conditions are presented. The results of the asymptotic approach to the reliability analysis of large systems are applied to the reliability prediction of the large parallel-series system operating at the varying conditions. The safety analysis of the steel cover is performed using the results concerned with the consecutive "*m* out of *n*: F" systems. Finally, the comprehensive application of the methods of the operation, reliability, renewal and availability modeling, identification, prediction and optimization to the container gantry crane is performed to practically illustrate the overall approach to reliability and safety analysis of any complex technical systems.

This book concludes with a Summary that contains the evaluation of the presented results, the formulation of open problems concerned with complex technical systems reliability and safety and the perspective of further investigations on the considered problems.

Contents

Chapter 1
Modeling Reliability and Safety of Multistate Systems with Ageing Components

1.1 Introduction

Taking into account the importance of the safety and operating process effectiveness of real technical systems it seems reasonable to expand the two-state approach [14, 16, 21] to multistate approach [1–6, 8–11, 19, 20, 22–30] in their reliability and safety analysis. The assumption that the systems are composed of multistate components with reliability states or safety states degrading in time [7, 12–15, 18, 27–29] gives the possibility for a more precise analysis of their reliability, safety and operational processes' effectiveness. This assumption allows us to distinguish a system reliability or safety critical state to exceed, which is either dangerous for the environment or does not assure the necessary level of its operation process effectiveness. Then, an important system reliability or safety characteristic is the time to the moment of exceeding the system reliability or safety critical state and its distribution, which is called the system risk function. This distribution is strictly related to the system multistate reliability function and safety function that are basic characteristics of the multistate system. The reliability and safety models of the typical multistate system structures considered here can be applied in the reliability, availability and safety analysis of real complex technical systems. They may be successfully applied, for instance, to reliability, availability and safety analysis, identification, prediction and optimization of the port and maritime transportation systems.

1.2 Reliability Analysis of Multistate Systems

In multistate reliability analysis, to define the system with degrading components, we assume that:

K. Kołowrocki and J. Soszyńska-Budny, *Reliability and Safety of Complex Technical Systems and Processes*, Springer Series in Reliability Engineering,
DOI: 10.1007/978-0-85729-694-8_1, © Springer-Verlag London Limited 2011

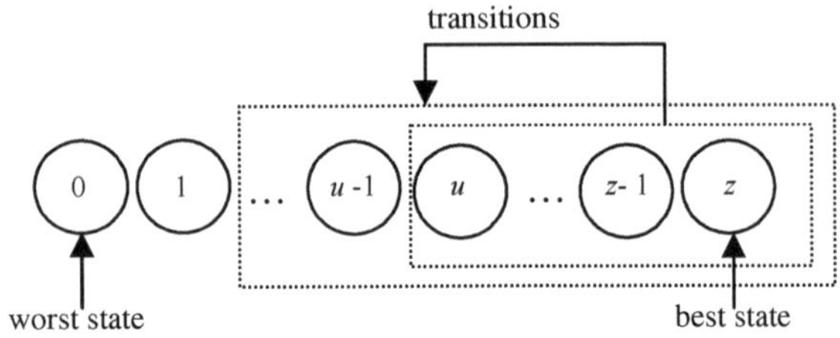

Fig. 1.1 Illustration of a system and components reliability states changing

- n is the number of system components,
- $E_i, i = 1, 2, \ldots, n$, are the components of a system,
- all components and a system under consideration have the reliability state set $\{0, 1, \ldots, z\}, z \geq 1$,
- the reliability states are ordered, the reliability state 0 is the worst and the reliability state z is the best,
- $T_i(u), i = 1, 2, \ldots, n$, are independent random variables representing the lifetimes of components E_i in the reliability state subset $\{u, u+1, \ldots, z\}$, while they were in the reliability state z at the moment $t = 0$,
- $T(u)$ is a random variable representing the lifetime of a system in the reliability state subset $\{u, u+1, \ldots, z\}$ while it was in the reliability state z at the moment $t = 0$,
- the system states degrade with time t,
- $E_i(t)$ is a component E_i reliability state at the moment t, $t \in \langle 0, \infty)$, given that it was in the reliability state z at the moment $t = 0$,
- $S(t)$ is a system S reliability state at the moment t, $t \in \langle 0, \infty)$, given that it was in the reliability state z at the moment $t = 0$.

The above assumptions mean that the reliability states of the system with degrading components may be changed in time only from better to worse [7, 14, 15, 27–29]. The way in which the components and the system reliability states change is illustrated in Fig. 1.1.

Definition 1.1 A vector

$$R_i(t, \cdot) = [R_i(t, 0), R_i(t, 1), \ldots, R_i(t, z)], \quad t \in \langle 0, \infty), \quad i = 1, 2, \ldots, n, \qquad (1.1)$$

where

$$R_i(t, u) = P(E_i(t) \geq u | E_i(0) = z) = P(T_i(u) > t), \quad t \in \langle 0, \infty), \quad u = 0, 1, \ldots, z, \qquad (1.2)$$

is the probability that the component E_i is in the reliability state subset $\{u, u+1, \ldots, z\}$ at the moment t, $t \in \langle 0, \infty)$, while it was in the reliability state z at the moment $t = 0$, is called the multistate reliability function of a component E_i.

The reliability functions $R_i(t, u), t \in \langle 0, \infty), u = 0, 1, \ldots, z$, defined by (1.2) are called the coordinates of the component $E_i, i = 1, 2, \ldots, n$, multistate reliability function $R_i(t, \cdot)$ given by (1.1). Thus, the relationship between the distribution function $F_i(t,u)$ of the component $E_i, i = 1,2,\ldots,n$, lifetime $T_i(u)$ in the reliability state subset $\{u, u+1,\ldots,z\}$ and the coordinate $R_i(t,u)$ of its multistate reliability function is given by

$$F_i(t,u) = P(T_i(u) \le t) = 1 - P(T_i(u) > t) = 1 - R_i(t,u), \quad t \in \langle 0, \infty), \\ u = 0,1,\ldots,z.$$

Under *Definition 1.1* and the agreements, we have the following property of the component multistate reliability function coordinates

$$R_i(t,0) \ge R_i(t,1) \ge \cdots \ge R_i(t,z), \quad t \in \langle 0,\infty), \ i = 1,2,\ldots,n.$$

Further, if we denote by

$$p_i(t,u) = P(E_i(t) = u | E_i(0) = z), \quad t \in \langle 0,\infty), \ u = 0,1,\ldots,z,$$

the probability that the component E_i is in the reliability state u at the moment t, while it was in the reliability state z at the moment $t = 0$, then by (1.1)

$$R_i(t,0) = 1, R_i(t,z) = p_i(t,z), \quad t \in \langle 0,\infty), \ i = 1,2,\ldots,n, \tag{1.3}$$

and

$$p_i(t,u) = R_i(t,u) - R_i(t,u+1), \quad u = 0,1,\ldots,z-1, \quad t \in \langle 0,\infty), \\ i = 1,2,\ldots,n. \tag{1.4}$$

Moreover, if

$$R_i(t,u) = 1 \text{ for } t \le 0, \quad u = 1,2,\ldots,z, \ i = 1,2,\ldots,n,$$

then

$$\mu_i(u) = \int_0^\infty R_i(t,u)\mathrm{d}t, \quad u = 1,2,\ldots,z, \ i = 1,2,\ldots,n, \tag{1.5}$$

is the mean lifetime of the component E_i in the reliability state subset $\{u, u+1,\ldots,z\}$,

$$\sigma_i(u) = \sqrt{n_i(u) - [\mu_i(u)]^2}, \quad u = 1,2,\ldots,z, \ i = 1,2,\ldots,n, \tag{1.6}$$

where

$$n_i(u) = 2\int_0^\infty tR_i(t,u)\mathrm{d}t, \quad u = 1,2,\ldots,z, \ i = 1,2,\ldots,n, \tag{1.7}$$

is the standard deviation of the component E_i lifetime in the reliability state subset $\{u, u+1,\ldots,z\}$ and

$$\overline{\mu}_i(u) = \int_0^\infty p_i(t,u)\mathrm{d}t, \quad u = 1,2,\ldots,z, \quad i = 1,2,\ldots,n, \tag{1.8}$$

is the mean lifetime of the component E_i in the reliability state u, in the case when the integrals defined by (1.5), (1.7) and (1.8) are convergent.

Next, according to (1.3), (1.4), (1.5) and (1.8), we have

$$\overline{\mu}_i(u) = \mu_i(u) - \mu_i(u+1), \quad u = 0,1,\ldots,z-1, \quad \overline{\mu}_i(z) = \mu_i(z), \quad i = 1,2,\ldots,n. \tag{1.9}$$

Definition 1.2 A vector

$$\boldsymbol{R}(t,\ \cdot) = [\boldsymbol{R}(t,0), \boldsymbol{R}(t,1), \ldots, \boldsymbol{R}(t,z)], \quad t \in \langle 0,\infty), \tag{1.10}$$

where

$$\boldsymbol{R}(t,u) = P(S(t) \geq u | S(0) = z) = P(T(u) > t), \quad t \in \langle 0,\infty), \quad u = 0,1,\ldots,z, \tag{1.11}$$

is the probability that the system is in the reliability state subset $\{u, u+1,\ldots,z\}$ at the moment t, $t \in \langle 0,\infty)$, while it was in the reliability state z at the moment $t = 0$, is called the multistate reliability function of this system.

The reliability functions $\boldsymbol{R}(t,u), t \in \langle 0,\infty), u = 0,1,\ldots,z,$ defined by (1.11) are called the coordinates of the system multistate reliability function $\boldsymbol{R}(t,\ \cdot)$ given by (1.10). Consequently, the relationship between the distribution function $\boldsymbol{F}(t,u)$ of the system S lifetime $T(u)$ in the reliability state subset $\{u, u+1,\ldots,z\}$ and the coordinate $\boldsymbol{R}(t,u)$ of its multistate reliability function is given by

$$\boldsymbol{F}(t,u) = P(T(u) \leq t) = 1 - P(T(u) > t) = 1 - \boldsymbol{R}(t,u),$$
$$t \in \langle 0,\infty), \quad u = 0,1,\ldots,z.$$

The exemplary graph of a four-state (z = 3) system reliability function

$$\boldsymbol{R}(t,\ \cdot) = [1, \boldsymbol{R}(t,1), \boldsymbol{R}(t,2), \boldsymbol{R}(t,3)], \quad t \in \langle 0,\infty),$$

is shown in Fig. 1.2.

Under *Definition 1.2*, we have

$$\boldsymbol{R}(t,0) \geq \boldsymbol{R}(t,1) \geq \cdots \geq \boldsymbol{R}(t,z), \quad t \in \langle 0,\infty),$$

and if

$$p(t,u) = P(S(t) = u | S(0) = z), \quad t \in \langle 0,\infty), \quad u = 0,1,\ldots,z, \tag{1.12}$$

is the probability that the system is in the state u at the moment t, $t \in \langle 0,\infty)$, while it was in the state z at the moment $t = 0$, then

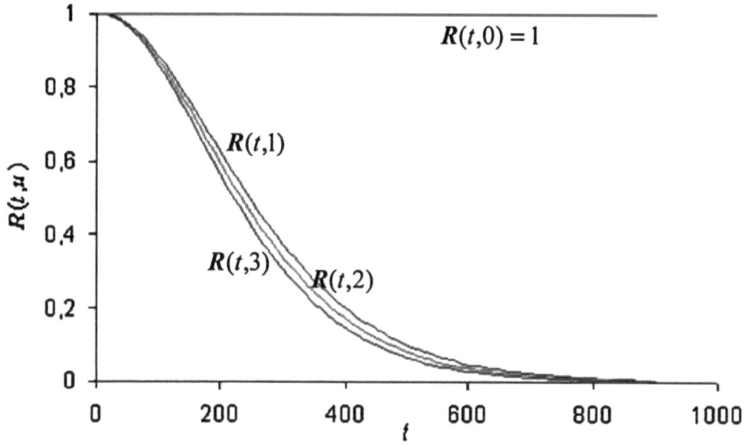

Fig. 1.2 The graph of a four-state system reliability function $\boldsymbol{R}(t,\cdot)$ coordinates

$$\boldsymbol{R}(t,0) = 1, \quad \boldsymbol{R}(t,z) = p(t,z), \qquad t \in \langle 0,\infty), \tag{1.13}$$

and

$$p(t,u) = \boldsymbol{R}(t,u) - \boldsymbol{R}(t,u+1), \quad u = 0,1,\ldots,z-1, \;\; t \in \langle 0,\infty). \tag{1.14}$$

Moreover, if

$$\boldsymbol{R}(t,u) = 1 \text{ for } t \leq 0, \quad u = 1,2,\ldots,z,$$

then

$$\mu(u) = \int_0^\infty \boldsymbol{R}(t,u)\mathrm{d}t, \quad u = 1,2,\ldots,z, \tag{1.15}$$

is the mean lifetime of the system in the state subset $\{u, u+1,\ldots,z\}$,

$$\sigma(u) = \sqrt{n(u) - [\mu(u)]^2}, \quad u = 1,2,\ldots,z, \tag{1.16}$$

where

$$n(u) = 2\int_0^\infty t\boldsymbol{R}(t,u)\mathrm{d}t, \quad u = 1,2,\ldots,z, \tag{1.17}$$

is the standard deviation of the system lifetime in the reliability state subset $\{u, u+1,\ldots,z\}$ and moreover

$$\bar{\mu}(u) = \int_0^\infty p(t,u)\mathrm{d}t, \quad u = 1,2,\ldots,z, \tag{1.18}$$

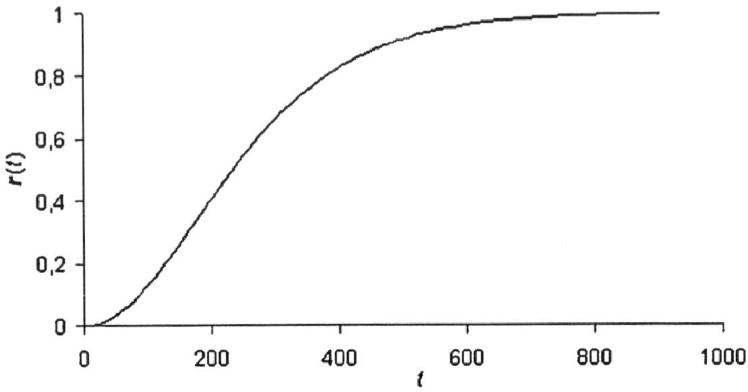

Fig. 1.3 The graph of a four-state system risk function *r(t)*

is the mean lifetime of the system in the state u while the integrals (1.15), (1.17) and (1.18) are convergent.

Additionally, according to (1.13–1.15) and (1.18), we get the following relationship

$$\bar{\mu}(u) = \mu(u) - \mu(u+1), \quad u = 0, 1, \ldots, z-1, \ \bar{\mu}(z) = \mu(z). \tag{1.19}$$

Definition 1.3 A probability

$$\boldsymbol{r}(t) = P(S(t) < r | S(0) = z) = P(T(r) \le t), \quad t \in \langle 0, \infty),$$

that the system is in the subset of reliability states worse than the critical state $r, r \in \{1, \ldots, z\}$ while it was in the state z at the moment $t = 0$ is called a risk function of the multistate system or, in short, a risk [14].

Under this definition, from (1.2), we have

$$\boldsymbol{r}(t) = 1 - P(S(t) \ge r | S(0) = z) = 1 - \boldsymbol{R}(t, r), \quad t \in \langle 0, \infty), \tag{1.20}$$

and if τ is the moment when the system risk exceeds a permitted level δ, then

$$\tau = \boldsymbol{r}^{-1}(\delta), \tag{1.21}$$

where $\boldsymbol{r}^{-1}(t)$, if it exists, is the inverse function of the system risk function $\boldsymbol{r}(t)$.

The exemplary graph of a four-state system risk function for the critical reliability state $r = 2$

$$\boldsymbol{r}(t) = 1 - \boldsymbol{R}(t, 2), \quad t \in \langle 0, \infty),$$

corresponding to the reliability function illustrated in Fig. 1.2 is shown in Fig. 1.3.

Now, after introducing the notion of the multistate reliability analysis, we may define the basic the multistate reliability structures.

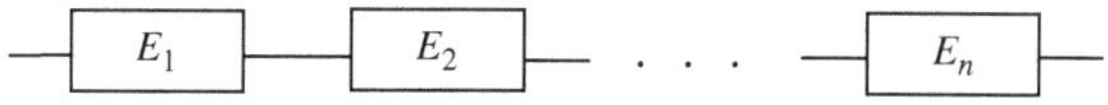

Fig. 1.4 The scheme of a series system

Definition 1.4 A multistate system is called a series if its lifetime $T(u)$ in the reliability state subset $\{u, u+1, \ldots, z\}$ is given by

$$T(u) = \min_{1 \le i \le n} \{T_i(u)\}, \quad u = 1, 2, \ldots, z.$$

The number n is called the system structure shape parameter.

The above definition means that a multistate series system is in the reliability state subset $\{u, u+1, \ldots, z\}$ if and only if all its n components are in this subset of reliability states. This meaning is very close to the definition of a two-state series system that is not failed if all its components are not failed. This fact can justify the same reliability structure scheme presented in Fig. 1.4 for a two-state series system and for a multistate series system.

It is easy to work out that the reliability function of the multistate series system is given by the vector [14]

$$\overline{\boldsymbol{R}}_n(t, \cdot) = [1, \overline{\boldsymbol{R}}_n(t, 1), \ldots, \overline{\boldsymbol{R}}_n(t, z)] \tag{1.22}$$

with the coordinates

$$\overline{\boldsymbol{R}}_n(t, u) = \prod_{i=1}^{n} R_i(t, u), \quad t \in \langle 0, \infty), \quad u = 1, 2, \ldots, z. \tag{1.23}$$

Definition 1.5 A multistate system is called parallel if its lifetime $T(u)$ in the reliability state subset $\{u, u+1, \ldots, z\}$ is given by

$$T(u) = \max_{1 \le i \le n} \{T_i(u)\}, \quad u = 1, 2, \ldots, z.$$

The number n is called the system structure shape parameter.

The above definition means that the multistate parallel system is in the reliability state subset $\{u, u+1, \ldots, z\}$ if and only if at least one of its n components is in this subset of states. This meaning is very close to the definition of a two-state parallel system that is not failed if at least one of its components is not failed, which can justify the same reliability structure scheme presented in Fig. 1.5 for a two-state parallel system and for a multistate parallel system.

The reliability function of the multistate parallel system is given by the vector [14]

$$\boldsymbol{R}_n(t, \cdot) = [1, \boldsymbol{R}_n(t, 1), \ldots, \boldsymbol{R}_n(t, z)], \tag{1.24}$$

with the coordinates

$$\boldsymbol{R}_n(t, u) = 1 - \prod_{i=1}^{n} F_i(t, u), \quad t \in \langle 0, \infty), \quad u = 1, 2, \ldots, z. \tag{1.25}$$

Fig. 1.5 The scheme of a parallel system

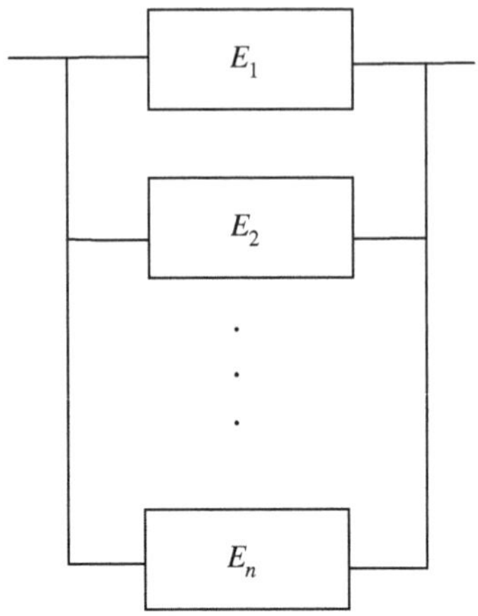

Definition 1.6 A multistate system is called an "*m* out of *n*" system if its lifetime $T(u)$ in the reliability state subset $\{u, u+1, \ldots, z\}$ is given by

$$T(u) = T_{(n-m+1)}(u), \quad m = 1, 2, \ldots, n, \ u = 1, 2, \ldots, z,$$

where $T_{(n-m+1)}(u)$ is the *m*th maximal order statistic in the sequence of the component lifetimes $T_1(u), T_2(u), \ldots, T_n(u), u = 1, 2, \ldots, z$.

The above definition means that the multistate "*m* out of *n*" system is in the reliability state subset $\{u, u+1, \ldots, z\}$ if and only if at least *m* out of its *n* components are in this reliability state subset and it is a multistate parallel system if $m = 1$ and it is a multistate series system if $m = n$. The numbers *m* and *n* are called the system structure shape parameters. The scheme of an "*m* out of *n*" multistate system, justified in an analogous way as in the case of a multistate series system and a multistate parallel system, is given in Fig. 1.6, where $i_1, i_2, \ldots, i_n \in \{1, 2, \ldots, n\}$ and $i_a \neq i_b$ for $a \neq b$.

It can be simply shown that the reliability function of the multistate "*m* out of *n*" system is given either by the vector [14]

$$\boldsymbol{R}_n^m(t, \cdot) = [1, \boldsymbol{R}_n^m(t, 1), \ldots, \boldsymbol{R}_n^m(t, z)], \tag{1.26}$$

with the coordinates

$$\boldsymbol{R}_n^m(t, u) = 1 - \sum_{\substack{r_1, r_2, \ldots, r_n = 0 \\ r_1 + r_2 + \cdots + r_n \leq m-1}}^{1} [R_i(t, u)]^{r_i} [F_i(t, u)]^{1-r_i}, \quad t \in \langle 0, \infty), \quad u = 1, 2, \ldots, z, \tag{1.27}$$

or by the vector

$$\overline{\boldsymbol{R}}_n^{\overline{m}}(t, \cdot) = [1, \overline{\boldsymbol{R}}_n^{\overline{m}}(t, 1), \ldots, \overline{\boldsymbol{R}}_n^{\overline{m}}(t, z)], \tag{1.28}$$

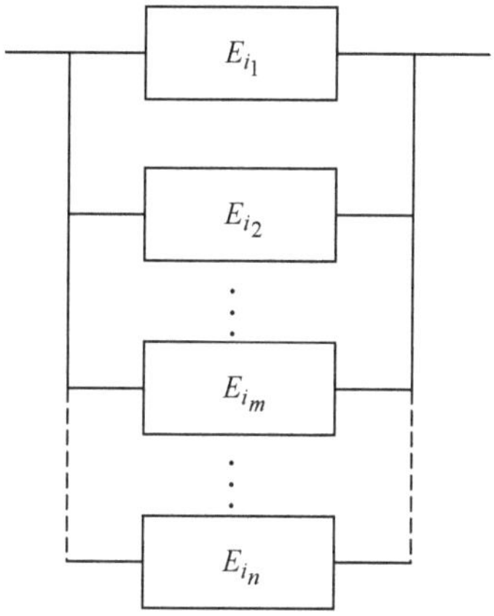

Fig. 1.6 The scheme of an "m out of n" system

with the coordinates

$$\overline{\boldsymbol{R}}_n^{\overline{m}}(t,u) = \sum_{\substack{r_1,r_2,\ldots,r_n=0\\ r_1+r_2+\cdots+r_n\le\overline{m}}}^{1} [F_i(t,u)]^{r_i}[R_i(t,u)]^{1-r_i}, \quad t\in\langle 0,\infty),$$

$$\overline{m} = n - m, \quad u = 1,2,\ldots,z. \tag{1.29}$$

Definition 1.7 A multistate system is called a consecutive "m out of n: F" system if it is out of the reliability state subset $\{u, u+1,\ldots,z\}$ if and only if at least its m neighbouring components out of its n components arranged in a sequence of $E_1, E_2,\ldots, E_n$, are out of this reliability state subset. The numbers m and n are called the system structure shape parameters.

After denoting by

$$\boldsymbol{CR}_n^m(t,u) = P(S(t)\ge u|S(0)=z) = P(T(u)>t), \quad t\in\langle 0,\infty), \; u=0,1,\ldots,z, \tag{1.30}$$

the probability that the consecutive "m out of n: F" system is in the reliability state subset $\{u, u+1,\ldots,z\}$ at the moment t, $t\in\langle 0,\infty)$, while it was in the reliability state z at the moment $t = 0$ and by

$$\boldsymbol{CF}_n^m(t,u) = 1 - \boldsymbol{CR}_n^m(t,u) = P(T(u)\le t), \quad t\in\langle 0,\infty), \; u=0,1,\ldots,z, \tag{1.31}$$

the distribution function of the lifetime $T(u)$ of this system in the reliability state subset $\{u, u+1,\ldots,z\}$, while it was in the state z at the moment $t = 0$, we conclude that the reliability function of the consecutive "m out of n: F" system is the given by the vector

$$\boldsymbol{CR}_n^m(t,\cdot) = [1, \boldsymbol{CR}_n^m(t,1), \boldsymbol{CR}_n^m(t,2),\ldots, \boldsymbol{CR}_n^m(t,z)], \tag{1.32}$$

with the coordinates given by the following recurrent formula [7, 14, 21]

$$\boldsymbol{CR}_{\boldsymbol{n}}^{m}(t,u) = \begin{cases} 1 & \text{for } n<m, \\ 1-\prod_{i=1}^{n} F_i(t,u) & \text{for } n=m, \\ R_n(t,u)\boldsymbol{CR}_{n-1}^{m}(t,u) + \sum_{i=1}^{m-1} R_{n-i}(t,u)CR_{n-i-1}^{m}(t,u) & \\ \times \prod_{j=n-i+1}^{n} F_j(t,u) & \text{for } n>m, \end{cases} \tag{1.33}$$

for $t \geq 0, u = 1, 2, \ldots, z$.

Other basic multistate reliability structures with components degrading in time series-parallel, parallel-series, series-"m out of k", "m_i out of l_i"-series, series-consecutive "m out of k: F" and consecutive "m_i out of l_i: F"-series systems. To define them, we assume that:

- k is the number of the system subsystems,
- $l_i, i = 1, 2, \ldots, k$, are the numbers of the subsystem components,
- $E_{ij}, i = 1, 2, \ldots, k, j = 1, 2, \ldots, l_i, k, l_1, l_2, \ldots, l_k \in N$, are components of a system,
- all components E_{ij} have the same reliability state set as before $\{0, 1, \ldots, z\}$,
- $T_{ij}(u), i = 1, 2, \ldots, k, j = 1, 2, \ldots, l_i, k, l_1, l_2, \ldots, l_k \in N$, are independent random variables representing the lifetimes of components E_{ij} in the reliability state subset $\{u, u+1, \ldots, z\}$, while they were in the state z at the moment $t = 0$,
- $E_{ij}(t)$ is a component E_{ij} reliability state at the moment $t, t \in \langle 0, \infty)$, while they were in the state z at the moment $t = 0$.

Definition 1.8 A vector

$$R_{ij}(t, \cdot) = [R_{ij}(t,0), R_{ij}(t,1), \ldots, R_{ij}(t,z)], \quad t \in \langle 0, \infty), \\ i = 1, 2, \ldots, k, \ \ j = 1, 2, \ldots, l_i, \tag{1.34}$$

where

$$R_{ij}(t,u) = P(E_{ij}(t) \geq u | E_{ij}(0) = z) = P(T_{ij}(u) > t), \quad t \in \langle 0, \infty), \\ u = 0, 1, \ldots, z, \tag{1.35}$$

is the probability that the component E_{ij} is in the reliability state subset $\{u, u+1, \ldots, z\}$ at the moment $t, t \in \langle 0, \infty)$, while it was in the reliability state z at the moment $t = 0$, is called the multistate reliability function of a component E_{ij}.

The reliability functions $R_{ij}(t,u), t \in \langle 0, \infty)$, $u = 0, 1, \ldots, z$, defined by (1.35) are called the coordinates of the component $E_{ij}, i = 1, 2, \ldots, k; j = 1, 2, \ldots, l_i$, multistate reliability function $R_{ij}(t, \cdot)$ given by (1.34). Thus, the relationship between the distribution function $F_{ij}(t,u)$ of the component $E_{ij}, i = 1, 2, \ldots, k$,

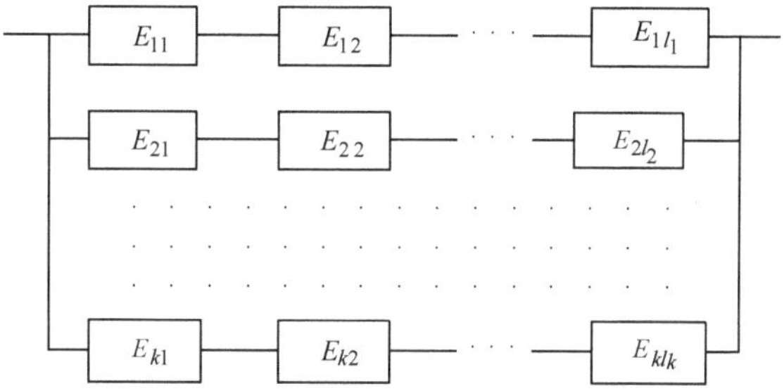

Fig. 1.7 The scheme of a series-parallel system

$j = 1, 2, \ldots, l_i$, lifetime $T_{ij}(u)$ in the reliability state subset $\{u, u+1, \ldots, z\}$ and the coordinate $R_{ij}(t, u)$ of its multistate reliability function is given by

$$F_{ij}(t,u) = P(T_{ij}(u) \le t) = 1 - P(T_{ij}(u) > t) = 1 - R_{ij}(t,u), \quad t \in \langle 0, \infty),$$
$$u = 0, 1, \ldots, z.$$

Definition 1.9 A multistate system is called series-parallel if its lifetime $T(u)$ in the state subset $\{u, u+1, \ldots, z\}$ is given by

$$T(u) = \max_{1 \le i \le k}\{\min_{1 \le j \le l_i}\{T_{ij}(u)\}\}, \quad u = 1, 2, \ldots, z.$$

The above definition means that the multistate series-parallel system is composed of k multistate series subsystems and it is in the reliability state subset $\{u, u+1, \ldots, z\}$ if and only if at least one out of its k series subsystems is in this reliability state subset. In this definition, $l_i, i = 1, 2, \ldots, k$, denote the number of components in the series subsystems. The numbers k and $l_1, l_2, \ldots, l_k$ are called the system structure shape parameters. Joining the justification for the reliability structure schemes of the multistate series system and the multistate parallel system leads to the scheme of a multistate series-parallel system given in Fig. 1.7.

The reliability function of the multistate series-parallel system is given by the vector [14]

$$\boldsymbol{R}_{k;l_1,l_2,\ldots,l_k}(t, \cdot) = [1, \boldsymbol{R}_{k;l_1,l_2,\ldots,l_k}(t,1), \ldots, \boldsymbol{R}_{k;l_1,l_2,\ldots,l_k}(t,z)], \tag{1.36}$$

with the coordinates

$$\boldsymbol{R}_{k;l_1,l_2,\ldots,l_k}(t,u) = 1 - \prod_{i=1}^{k}\left[1 - \prod_{j=1}^{l_i} R_{ij}(t,u)\right], \quad t \in \langle 0, \infty), \ u = 1, 2, \ldots, z, \tag{1.37}$$

where k is the number of series subsystems linked in parallel and l_i are the numbers of components in the series subsystems.

Definition 1.10 A multistate system is called parallel-series if its lifetime $T(u)$ in the reliability state subset $\{u, u+1, \ldots, z\}$ is given by

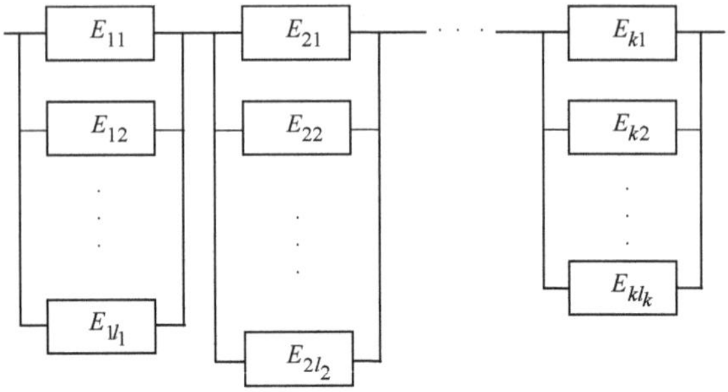

Fig. 1.8 The scheme of a parallel-series system

$$T(u) = \min_{1 \le i \le k} \{ \max_{1 \le j \le l_i} \{T_{ij}(u)\}\}, \quad u = 1, 2, \ldots, z.$$

The above definition means that the multistate parallel-series system is composed of k multistate parallel subsystems and it is in the reliability state subset $\{u, u+1, \ldots, z\}$ if and only if all its k parallel subsystems are in this reliability state subset. In this definition $l_i, i = 1, 2, \ldots, k$, denote the number of components in the parallel subsystems. The numbers k and $l_1, l_2, \ldots, l_k$ are called the system structure shape parameters. Joining the justification for the reliability structure schemes of the multistate parallel system and the multistate series system leads to the scheme of a multistate parallel-series system given in Fig. 1.8.

The reliability function of the multistate parallel-series system is given by the vector [14]

$$\overline{\boldsymbol{R}}_{k;l_1,l_2,\ldots,l_k}(t, \cdot) = [1, \overline{\boldsymbol{R}}_{k;l_1,l_2,\ldots,l_k}(t, 1), \ldots, \overline{\boldsymbol{R}}_{k;l_1,l_2,\ldots,l_k}(t, z)], \tag{1.38}$$

with the coordinates

$$\overline{\boldsymbol{R}}_{k;l_1,l_2,\ldots,l_k}(t, u) = \prod_{i=1}^{k} \left[1 - \prod_{j=1}^{l_i} F_{ij}(t, u) \right], \quad t \in \langle 0, \infty), \quad u = 1, 2, \ldots, z, \tag{1.39}$$

where k is the number of its parallel subsystems linked in series and $l_i, i = 1, 2, \ldots, k$, are the number of components in the parallel subsystems.

Definition 1.11 A multistate system is called a series-"m out of k" system if its lifetime $T(u)$ in the state subset $\{u, u+1, \ldots, z\}$ is given by

$$T(u) = T_{(k-m+1)}(u), \quad m = 1, 2, \ldots, k, \quad u = 1, 2, \ldots, z,$$

where $T_{(k-m+1)}(u)$ is the mth maximal order statistic in the set of random variables

$$T_i(u) = \min_{1 \le j \le l_i} \{T_{ij}(u)\}, \quad i = 1, 2, \ldots, k, \quad u = 1, 2, \ldots, z.$$

The above definition means that the multistate series-"m out of k" system is composed of k multistate series subsystems and it is in the reliability state subset

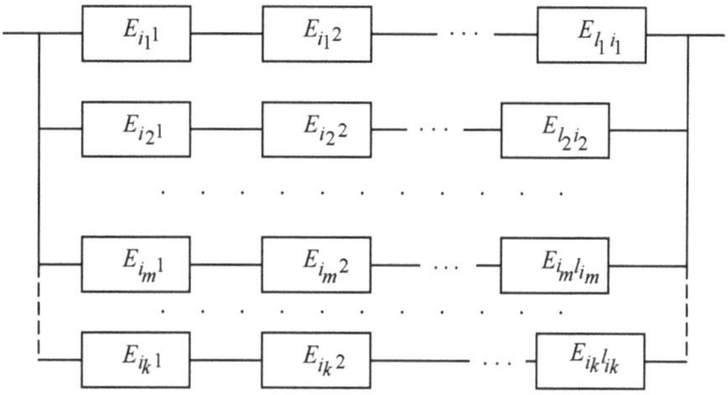

Fig. 1.9 The scheme of a series-"m out of k" system

$\{u, u+1, \ldots, z\}$ if and only if at least m out of its k series subsystems are in this reliability state subset. In this definition $l_i, i = 1, 2, \ldots, k$, denote the number of components in the series subsystems. The numbers m, k and $l_1, l_2, \ldots, l_k$ are called the system structure shape parameters. Joining the justification for the reliability structure schemes of the multistate series system and the multistate "m out of k" system leads to the scheme of a multistate series-"m out of k" system given in Fig. 1.9, where $i_1, i_2, \ldots, i_k \in \{1, 2, \ldots, k\}$ and $i_a \neq i_b$ for $a \neq b$.

The reliability function of the multistate series-"m out of k" system is given either by the vector [14]

$$\boldsymbol{R}^{m}_{k;l_1,l_2,\ldots,l_k}(t,\cdot) = [1,\ \boldsymbol{R}^{m}_{k;l_1,l_2,\ldots,l_k}(t,1), \ldots, \boldsymbol{R}^{m}_{k;,l_1,l_2,\ldots,l_k}(t,z)], \tag{1.40}$$

with the coordinates

$$\boldsymbol{R}^{m}_{k;l_1,l_2,\ldots,l_k}(t,u) = 1 - \sum_{\substack{r_1,r_2,\ldots,r_k=0 \\ r_1+r_2+\cdots+r_k \le m-1}}^{1} \prod_{i=1}^{k} \left[\prod_{j=1}^{l_i} R_{ij}(t,u)\right]^{r_i} \left[1 - \prod_{j=1}^{l_i} R_{ij}(t,u)\right]^{1-r_i},$$

$$t \in \langle 0, \infty), \quad u = 1, 2, \ldots, z, \tag{1.41}$$

or by the vector

$$\overline{\boldsymbol{R}}^{\overline{m}}_{k;l_1,l_2,\ldots,l_k}(t,\cdot) = [1,\ \overline{\boldsymbol{R}}^{\overline{m}}_{k;l_1,l_2,\ldots,l_k}(t,1), \ldots, \overline{\boldsymbol{R}}^{\overline{m}}_{k;l_1,l_2,\ldots,l_k}(t,z)], \tag{1.42}$$

with the coordinates

$$\overline{\boldsymbol{R}}^{\overline{m}}_{k;l_1,l_2,\ldots,l_k}(t,u) = \sum_{\substack{r_1,r_2,\ldots,r_k=0 \\ r_1+r_2+\cdots+r_k \le \overline{m}}}^{1} \prod_{i=1}^{k} \left[1 - \prod_{j=1}^{l_i} R_{ij}(t,u)\right]^{r_i} \left[\prod_{j=1}^{l_i} R_{ij}(t,u)\right]^{1-r_i},$$

$$t \in \langle 0, \infty), \quad \overline{m} = k - m, \quad u = 1, 2, \ldots, z. \tag{1.43}$$

Definition 1.12 A multistate system is called an "m_i out of l_i"-series system if its lifetime $T(u)$ in the state subset $\{u, u+1, \ldots, z\}$ is given by

Fig. 1.10 The scheme of an "m_i out of l_i"-series system

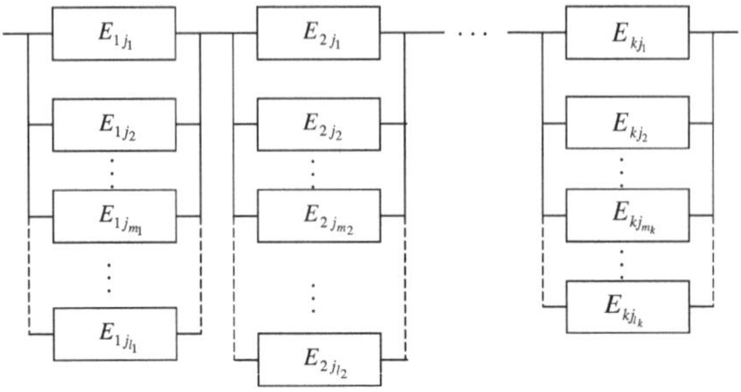

$$T(u) = \min_{1\le i\le k} T_{(l_i-m_i+1)}(u), \quad m_i = 1,2,\ldots,l_i, \quad u = 1,2,\ldots,z,$$

where $T_{(l_i-m_i+1)}(u)$ is the m_ith maximal order statistic in the set of random variables

$$T_{i1}(u), T_{i2}(u), \ldots, T_{il_i}(u), \quad i = 1,2,\ldots,k, \quad u = 1,2,\ldots,z.$$

The above definition means that the multistate "m_i out of l_i"-series system is composed of k subsystems that are multistate "m_i out of l_i" systems and it is in the reliability state subset $\{u, u+1,\ldots,z\}$ if all its "m_i out of l_i" subsystems are in this reliability state subset. In this definition $l_i, i = 1,2,\ldots,k$, denote the number of components in the "m_i out of l_i" subsystems. The numbers k, $m_1, m_2, \ldots, m_k$ and $l_1, l_2, \ldots, l_k$ are called the system structure shape parameters. Joining the justification for the reliability structure schemes of the multistate "m_i out of l_i" system and the multistate series system schemes lead to the scheme of a multistate of an "m_i out of l_i"-series system given in Fig. 1.10, where $j_1, j_2, \ldots, j_{l_i} \in \{1,2,\ldots,l_i\}$, for $i = 1,2,\ldots,k$ and $j_a \ne j_b$ for $a \ne b$.

The reliability function of the multistate "m_i out of l_i"-series system is given either by the vector [14]

$$\overline{\boldsymbol{R}^{m_1,m_2,\ldots,m_k}_{k;l_1,l_2,\ldots,l_k}}(t,\cdot) = [1, \overline{\boldsymbol{R}^{m_1,m_2,\ldots,m_k}_{k;l_1,l_2,\ldots,l_k}}(t,1), \ldots, \overline{\boldsymbol{R}^{m_1,m_2,\ldots,m_k}_{k;l_1,l_2,\ldots,l_k}}(t,z)], \tag{1.44}$$

where

$$\overline{\boldsymbol{R}^{m_1,m_2,\ldots,m_k}_{k;l_1,l_2,\ldots,l_k}}(t,u) = \prod_{i=1}^{k}\left[1 - \sum_{\substack{r_1,r_2,\ldots,r_{l_i}=0\\ r_1+r_2+\cdots+r_{l_i}\le m_i-1}}^{1} \left[\prod_{j=1}^{l_i} R_{ij}(t,u)\right]^{r_i} \left[1 - \prod_{j=1}^{l_i} R_{ij}(t,u)\right]^{1-r_i}\right],$$

$$t \in \langle 0,\infty), \quad u = 1,2,\ldots,z, \tag{1.45}$$

or by the vector

$$\overline{\boldsymbol{R}^{\overline{m}_1,\overline{m}_2,\ldots,\overline{m}_k}_{k;l_1,l_2,\ldots,l_k}}(t,\cdot) = \left[1, \overline{\boldsymbol{R}^{\overline{m}_1,\overline{m}_2,\ldots,\overline{m}_k}_{k;l_1,l_2,\ldots,l_k}}(t,1),\ldots,\overline{\boldsymbol{R}^{\overline{m}_1,\overline{m}_2,\ldots,\overline{m}_k}_{k;l_1,l_2,\ldots,l_k}}(t,z)\right], \tag{1.46}$$

where

$$\overline{\boldsymbol{R}^{\overline{m}_1,\overline{m}_2,\ldots,\overline{m}_k}_{k;l_1,l_2,\ldots,l_k}}(t,u) = \prod_{i=1}^{k}\left[\sum_{\substack{r_1,r_2,\ldots,r_{l_i}=0\\ r_1+r_2+\cdots+r_{l_i}\le \overline{m}_i}}^{1}\left[1-\prod_{j=1}^{l_i}R_{ij}(t,u)\right]^{r_i}\left[\prod_{j=1}^{l_i}R_{ij}(t,u)\right]^{1-r_i}\right],$$

$$t\in\langle 0,\infty),\quad \overline{m}_i = l_i - m_i,\quad i=1,2,\ldots,k,\quad u=1,2,\ldots,z. \tag{1.47}$$

To define a multistate series-consecutive "m out of k: F" system, we assume that components $E_{i1}, E_{i2}, \ldots, E_{il_i}, i = 1, 2, \ldots, k$, create a series subsystems $S_i, i = 1, 2, \ldots, k$, and that these subsystems are arranged in a sequence $S_1, S_2, \ldots, S_k$.

Definition 1.13 A multistate system is called series-consecutive "m out of k: F" system if it is out of the reliability state subset $\{u, u+1, \ldots, z\}$ if and only if at least its m neighboring series subsystems out of k its series subsystems arranged in a sequence $S_1, S_2, \ldots, S_k$, are out of this reliability state subset. The numbers m, k and $l_1, l_2, \ldots, l_k$ are called the system structure shape parameters.

According to the above definition and formulae (1.22–1.23) the coordinates of the multistate reliability function of the subsystem $S_i, i = 1, 2, \ldots, k$, are given by

$$\overline{\boldsymbol{R}}_{il_i}(t,u) = \prod_{j=1}^{l_i} R_{ij}(t,u),\quad t\in\langle 0,\infty),\quad u=1,2,\ldots,z,$$

and its lifetime distribution functions are given by

$$\overline{\boldsymbol{F}}_{il_i}(t,u) = 1-\overline{\boldsymbol{R}}_{il_i}(t,u) = 1-\prod_{j=1}^{l_i} R_{ij}(t,u),$$
$$t\in\langle 0,\infty),\quad u=1,2,\ldots,z,\quad i=1,2,\ldots,k.$$

Hence and considering (1.32–1.33), after denoting by

$$\boldsymbol{CR}^{m}_{k;l_1,l_2,\ldots,l_k}(t,u) = P(T(u) > t),\quad t\in\langle 0,\infty),\quad u=1,2,\ldots,z,$$

the coordinates of the reliability function of the multistate series-consecutive "m out of k: F" system, we conclude that it is the vector

$$\boldsymbol{CR}^{m}_{k;l_1,l_2,\ldots,l_k}(t,\cdot) = [1, \boldsymbol{CR}^{m}_{k;l_1,l_2,\ldots,l_k}(t,1), \boldsymbol{CR}^{m}_{k;l_1,l_2,\ldots,l_k}(t,2),\ldots,\boldsymbol{CR}^{(m)}_{k,l_1,l_2,\ldots,l_k}(t,z)], \tag{1.48}$$

with the coordinates given by the following recurrent formula

$$
\boldsymbol{CR}^{m}_{k;l_1,l_2,\ldots,l_k}(t,u) = \begin{cases} 1 & \text{for } k<m, \\ 1-\prod_{i=1}^{k}\left[1-\prod_{j=1}^{l_i} R_{ij}(t,u)\right] & \text{for } k=m, \\ \left[\prod_{j=1}^{l_k} R_{kj}(t,u)\right]\boldsymbol{CR}^{m}_{k-1;l_1,l_2,\ldots,l_k}(t,u) & \\ +\sum_{j=1}^{m-1}\left[\prod_{v=1}^{l_{k-j}} R_{k-j,v}(t,u)\right]\boldsymbol{CR}^{m}_{k-j-1;l_1,l_2,\ldots,l_k}(t,u) & \\ \times\prod_{i=k-j+1}^{k}\left[1-\prod_{v=1}^{l_i} R_{iv}(t,u)\right] & \text{for } k>m, \end{cases} \tag{1.49}
$$

for $t \geq 0, u = 1, 2, \ldots, z$.

To define a multistate consecutive "*m* out of *l*: *F*"-series system, we assume that components $E_{i1}, E_{i2}, \ldots, E_{il_i}, i = 1, 2, \ldots, k$, create a consecutive "$m_i$ out of l_i: *F*" subsystem $S_i, i = 1, 2, \ldots, k$, and that these subsystems $S_1, S_2, \ldots, S_k$, create a series system.

Definition 1.14 A multistate system is called a consecutive "m_i out of l_i: *F*"-series system if it is out of the reliability state subset $\{u, u+1, \ldots, z\}$ if and only if at least one of its consecutive "m_i out of l_i: *F*" subsystems $S_i, i = 1, 2, \ldots, k$, is out of this reliability state subset. The numbers k and $m_i, l_i, i = 1, 2, \ldots, k$, are called the system structure shape parameters.

According to the above definition and formulae (1.32–1.33), the multistate reliability function of the subsystem $S_i, i = 1, 2, \ldots, k$, is the vector

$$
\boldsymbol{CR}^{m_i}_{i,l_i}(t,\cdot) = [1, \boldsymbol{CR}^{m_i}_{i,l_i}(t,1), \boldsymbol{CR}^{m_i}_{i,l_i}(t,2), \ldots, \boldsymbol{CR}^{m_i}_{i,l_i}(t,z)], \tag{1.50}
$$

with the coordinates given by the following recurrent formula

$$
\boldsymbol{CR}^{m_i}_{i,l_i}(t,u) = \begin{cases} 1 & \text{for } l_i<m_i, \\ 1-\prod_{j=1}^{l_i} F_{ij}(t,u) & \text{for } l_i=m_i, \\ R_{il_i}(t,u)\boldsymbol{CR}^{m_i}_{i,l_i-1}(t,u)+\sum_{j=1}^{m_i-1} R_{i,l_i-j}(t,u) & \\ \boldsymbol{CR}^{m_i}_{i,l_i-j-1}(t,u)\times\prod_{v=l_i-j+1}^{l_i} F_{iv}(t,u) & \text{for } l_i>m_i, \end{cases} \tag{1.51}
$$

for $t \geq 0, u = 1, 2, \ldots, z$. Hence, denoting by

$$
\bar{\boldsymbol{C}}\bar{\boldsymbol{R}}^{m_1,m_2,\ldots,m_k}_{k;l_1,l_2,\ldots,l_k}(t,u) = P(T(u)>t), \quad t \in \langle 0, \infty), \quad u = 1, 2, \ldots, z,
$$

the coordinates of the reliability function of the multistate consecutive "m_i out of l_i: *F*"-series system, we conclude that it is given by the vector

$$
\bar{\boldsymbol{C}}\bar{\boldsymbol{R}}^{m_1,m_2,\ldots,m_k}_{k;l_1,l_2,\ldots,l_k}(t,\cdot) = [1, \bar{\boldsymbol{C}}\bar{\boldsymbol{R}}^{m_1,m_2,\ldots,m_k}_{k;l_1,l_2,\ldots,l_k}(t,1), \bar{\boldsymbol{C}}\bar{\boldsymbol{R}}^{m_1,m_2,\ldots,m_k}_{k;l_1,l_2,\ldots,l_k}(t,2), \ldots, \bar{\boldsymbol{C}}\bar{\boldsymbol{R}}^{m_1,m_2,\ldots,m_k}_{k;l_1,l_2,\ldots,l_k}(t,z)], \tag{1.52}
$$

with the coordinates

$$\bar{CR}^{m_1,m_2,\ldots,m_k}_{k;l_1,l_2,\ldots,l_k}(t,u) = \prod_{i=1}^{k} CR^{m_i}_{i,l_i}(t,u) \quad \text{for } t \geq 0, \quad u = 1,2,\ldots,z, \tag{1.53}$$

where $\boldsymbol{CR}^{(m_i)}_{i,l_i}(t,u), i = 1,2,\ldots,k$, are given by the recurrent formula (1.51).

Further, for the systems composed of components having multistate exponential reliability functions, we formulate the particular cases of the fixed above formulae for the considered system structures in the form of the following proposition.

Proposition 1.1 *If components of the multistate system have the exponential reliability functions*

$$R_i(t,\cdot) = [1, R_i(t,1),\ldots,R_i(t,z)], \quad t \in (-\infty,\infty), \tag{1.54}$$

where

$$R_i(t,u) = 1 \quad \text{for } t < 0,$$
$$R_i(t,u) = \exp[-\lambda_i(u)t] \quad \text{for } t \geq 0, \quad \lambda_i(u) > 0, \quad i = 1,2,\ldots,n, \quad u = 1,2,\ldots,z, \tag{1.55}$$

in the case of the multistate series, parallel, "m out of n", consecutive "m out of n" systems and respectively

$$R_{ij}(t,\cdot) = [1, R_{ij}(t,1),\ldots,R_{ij}(t,z)], \quad t \in (-\infty,\infty), \tag{1.56}$$

where

$$R_{ij}(t,u) = 1 \quad \text{for } t < 0,$$
$$R_{ij}(t,u) = \exp[-\lambda_{ij}(u)t] \quad \text{for } t \geq 0, \quad \lambda_{ij}(u) > 0, \quad i = 1,2,\ldots,n,$$
$$j = 1,2,\ldots,l_i, \quad u = 1,2,\ldots,z, \tag{1.57}$$

in the case of the series-parallel, parallel-series, series-"m out of k", "m_i out of l_i"-series, series-consecutive "m out of k" and consecutive "m_i out of l_i"-series systems, then its multistate reliability function is given by the vector:

(i) *for a series system*

$$\overline{\boldsymbol{R}}_n(t,\cdot) = [1, \overline{\boldsymbol{R}}_n(t,1),\ldots,\overline{\boldsymbol{R}}_n(t,z)], \tag{1.58}$$

where

$$\overline{\boldsymbol{R}}_n(t,u) = 1 \quad \text{for } t < 0, \quad \overline{\boldsymbol{R}}_n(t,u) = \exp\left[-\sum_{i=1}^{n} \lambda_i(u)t\right] \quad \text{for } t \geq 0, \tag{1.59}$$
$$u = 1,2,\ldots,z,$$

(ii) *for a parallel system*

$$\boldsymbol{R}_n(t,\ \cdot) = [1, \boldsymbol{R}_n(t,1),\ldots,\boldsymbol{R}_n(t,z)], \tag{1.60}$$

where

$$\boldsymbol{R}_n(t,u) = 1 \quad \text{for } t<0, \quad \boldsymbol{R}_n(t,u) = 1 - \prod_{i=1}^{n}[1-\exp[-\lambda_i(u)t]] \quad \text{for } t \geq 0,$$

$$u = 1,2,\ldots,z, \tag{1.61}$$

(iii) *for an "m out of n" system*

$$\boldsymbol{R}_n^m(t,\ \cdot) = [1, \boldsymbol{R}_n^m(t,1),\ldots,\boldsymbol{R}_n^m(t,z)], \tag{1.62}$$

where

$$\boldsymbol{R}_n^m(t,u) = 1 \quad \text{for } t<0,$$

$$\boldsymbol{R}_n^m(t,u) = 1 - \sum_{\substack{r_1,r_2,\ldots,r_n=0 \\ r_1+r_2+\cdots+r_n \leq m-1}}^{1} \prod_{i=1}^{n} \exp[-r_i\lambda_i(u)t][1-\exp[-\lambda_i(u)t]]^{1-r_i} \quad \text{for } t \geq 0,$$

$$u = 1,2,\ldots,z, \tag{1.63}$$

or

$$\overline{\boldsymbol{R}}_n^{\overline{m}}(t,\cdot) = [1, \overline{\boldsymbol{R}}_n^{\overline{m}}(t,1),\ldots,\overline{\boldsymbol{R}}_n^{\overline{m}}(t,z)], \tag{1.64}$$

where

$$\overline{\boldsymbol{R}}_n^{\overline{m}}(t,u) = 1 \quad \text{for } t<0,$$

$$\overline{\boldsymbol{R}}_n^{\overline{m}}(t,u) = \sum_{\substack{r_1,r_2,\ldots,r_n=0 \\ r_1+r_2+\cdots+r_n \leq \overline{m}}}^{1} \prod_{i=1}^{n}[1-\exp[-\lambda_i(u)t]]^{r_i}\exp[-(1-r_i)\lambda_i(u)t] \quad \text{for } t \geq 0,$$

$$\overline{m} = n-m, \quad u = 1,2,\ldots,z, \tag{1.65}$$

(iv) *for a consecutive "m out of n: F" system*

$$\boldsymbol{CR}_n^m(t,\cdot) = [1, \boldsymbol{CR}_n^m(t,1),\ldots,\boldsymbol{CR}_n^m(t,z)], \tag{1.66}$$

where

$$\boldsymbol{CR}_n^m(t,u) = 1 \quad \text{for } t<0,$$

$$\boldsymbol{CR}_n^m(t,u) = \begin{cases} 1 & \text{for } n<m, \\ 1-\prod_{i=1}^{n}[1-\exp[-\lambda_i(u)t]] & \text{for } n=m, \\ \exp[-\lambda_n(u)t]\boldsymbol{CR}_{n-1}^m(t,u) & \\ +\sum_{i=1}^{m-1}\exp[-\lambda_{n-i}(u)t]\boldsymbol{CR}_{n-i-1}^m(t,u) & \\ \times \prod_{j=n-i+1}^{n}[1-\exp[-\lambda_j(u)t]] & \text{for } n>m, \end{cases} \tag{1.67}$$

for $t \geq 0, \quad u = 1,2,\ldots,z,$

(v) *for a series-parallel system*

$$\boldsymbol{R}_{k;l_1,l_2,\ldots,l_k}(t,\cdot) = [1, \boldsymbol{R}_{k;l_1,l_2,\ldots,l_k}(t,1), \ldots, \boldsymbol{R}_{k;l_1,l_2,\ldots,l_k}(t,z)], \tag{1.68}$$

where

$$\boldsymbol{R}_{k;l_1,l_2,\ldots,l_k}(t,u) = 1 \quad \text{for} \quad t<0,$$
$$\boldsymbol{R}_{k;l_1,l_2,\ldots,l_k}(t,u) = 1 - \prod_{i=1}^{k}\left[1 - \exp\left[-\sum_{j=1}^{l_i}\lambda_{ij}(u)t\right]\right] \quad \text{for } t \geq 0, \quad u = 1,2,\ldots,z, \tag{1.69}$$

(vi) *for a parallel-series system*

$$\overline{\boldsymbol{R}}_{k;l_1,l_2,\ldots,l_k}(t,\cdot) = [1, \overline{\boldsymbol{R}}_{k;l_1,l_2,\ldots,l_k}(t,1), \ldots, \overline{\boldsymbol{R}}_{k;l_1,l_2,\ldots,l_k}(t,z)] \tag{1.70}$$

where

$$\overline{\boldsymbol{R}}_{k;l_1,l_2,\ldots,l_k}(t,u) = 1 \quad \text{for } t<0,$$
$$\overline{\boldsymbol{R}}_{k;l_1,l_2,\ldots,l_k}(t,u) = \prod_{i=1}^{k}\left[1 - \prod_{j=1}^{l_i}[1-\exp[-\lambda_{ij}(u)t]]\right] \quad \text{for } t \geq 0, \quad u = 1,2,\ldots,z, \tag{1.71}$$

(vii) *for a series-"m out of k" system*

$$\boldsymbol{R}^{m}_{k;l_1,l_2,\ldots,l_k}(t,\cdot) = [1, \boldsymbol{R}^{m}_{k;l_1,l_2,\ldots,l_k}(t,1), \ldots, \boldsymbol{R}^{m}_{k;l_1,l_2,\ldots,l_k}(t,z)], \tag{1.72}$$

where

$$\boldsymbol{R}^{m}_{k;l_1,l_2,\ldots,l_k}(t,u) = 1 \text{ for } t<0,$$
$$\boldsymbol{R}^{m}_{k;l_1,l_2,\ldots,l_k}(t,u) = 1 - \sum_{\substack{r_1,r_2,\ldots,r_k=0 \\ r_1+r_2+\cdots+r_k \leq m-1}}^{1} \prod_{i=1}^{k} \exp\left[-r_i\sum_{j=1}^{l_i}\lambda_{ij}(u)t\right]$$
$$\left[1-\exp\left[-\sum_{j=1}^{l_i}\lambda_{ij}(u)t\right]\right]^{1-r_i} \quad \text{for } t \geq 0, \quad u = 1,2,\ldots,z, \tag{1.73}$$

or

$$\overline{\boldsymbol{R}}^{\overline{m}}_{k;l_1,l_2,\ldots,l_k}(t,\cdot) = [1, \overline{\boldsymbol{R}}^{\overline{m}}_{k;l_1,l_2,\ldots,l_k}(t,1), \ldots, \overline{\boldsymbol{R}}^{\overline{m}}_{k;l_1,l_2,\ldots,l_k}(t,z)], \tag{1.74}$$

where

$$\overline{\boldsymbol{R}}^{\overline{m}}_{k;l_1,l_2,\ldots,l_k}(t,u) = 1 \quad \text{for } t<0,$$

$$\boldsymbol{R}^{\overline{m}}_{k;l_1,l_2,\ldots,l_k}(t,u) = \sum_{\substack{r_1,r_2,\ldots,r_k=0 \\ r_1+r_2+\cdots+r_k \le \overline{m}}}^{1} \prod_{i=1}^{k}\left[1-\exp\left[-\sum_{j=1}^{l_i}\lambda_{ij}(u)t\right]\right]^{r_i}\exp\left[-(1-r_i)\sum_{j=1}^{l_i}\lambda_{ij}(u)t\right]$$

$$\text{for } t \ge 0, \quad \overline{m} = k - m, \quad u = 1,2,\ldots,z, \tag{1.75}$$

(viii) *for an "m_i out of l_i"-series system*

$$\overline{\boldsymbol{R}^{m_1,m_2,\ldots,m_k}_{k;l_1,l_2,\ldots,l_k}}(t,\cdot) = [1, \overline{\boldsymbol{R}^{m_1,m_2,\ldots,m_k}_{k;l_1,l_2,\ldots,l_k}}(t,1),\ldots, \overline{\boldsymbol{R}^{m_1,m_2,\ldots,m_k}_{k;l_1,l_2,\ldots,l_k}}(t,z)], \tag{1.76}$$

where

$$\overline{R^{m_1,m_2,\ldots,m_k}_{k;l_1,l_2,\ldots,l_k}}(t,u) = 1 \text{ for } t<0,$$

$$\overline{R^{m_1,m_2,\ldots,m_k}_{k;l_1,l_2,\ldots,l_k}}(t,u) = \prod_{i=1}^{k}\left[1 - \sum_{\substack{r_1,r_2,\ldots,r_{l_i}=0 \\ r_1+r_2+\cdots+r_{l_i} \le m_i-1}}^{1} \exp\left[-r_i\sum_{j=1}^{l_i}\lambda_{ij}(u)t\right]\right.$$
$$\left.\times \prod_{j=1}^{li}\left[1-\exp\left[-\lambda_{ij}(u)t\right]\right]^{1-r_i}\right] \quad \text{for } t \ge 0, \quad u = 1,2,\ldots,z, \tag{1.77}$$

or

$$\overline{\boldsymbol{R}^{\overline{m}_1,\overline{m}_2,\ldots,\overline{m}_k}_{k;l_1,l_2,\ldots,l_k}}(t,\cdot) = \left[1, \overline{\boldsymbol{R}^{\overline{m}_1,\overline{m}_2,\ldots,\overline{m}_k}_{k;l_1,l_2,\ldots,l_k}}(t,1),\ldots, \overline{\boldsymbol{R}^{\overline{m}_1,\overline{m}_2,\ldots,\overline{m}_k}_{k;l_1,l_2,\ldots,l_k}}(t,z)\right], \tag{1.78}$$

where

$$\overline{\boldsymbol{R}^{\overline{m}_1,\overline{m}_2,\ldots,\overline{m}_k}_{k;l_1,l_2,\ldots,l_k}}(t,u) = 1 \text{ for } t<0,$$

$$\overline{\boldsymbol{R}^{\overline{m}_1,\overline{m}_2,\ldots,\overline{m}_k}_{k;l_1,l_2,\ldots,l_k}}(t,u) = \prod_{i=1}^{k}\left[\sum_{\substack{r_1,r_2,\ldots,r_{l_i}=0 \\ r_1+r_2+\cdots+r_{l_i} \le \overline{m}_i}}^{1} \prod_{j=1}^{li}[1-\exp[-\lambda_{ij}(u)t]]^{r_i}\right.$$
$$\left.\times \exp\left[-(1-r_i)\sum_{j=1}^{l_i}\lambda_{ij}(u)t\right]\right] \quad \text{for } t \ge 0, \quad \overline{m}_i = l_i - m_i,$$
$$i = 1,2,\ldots,k, \quad u = 1,2,\ldots,z, \tag{1.79}$$

(ix) *for a series-consecutive "m out of k: F" system*

$$\boldsymbol{CR}^{m}_{k;l_1,l_2,\ldots,l_k}(t,\cdot) = [1, \boldsymbol{CR}^{m}_{k;l_1,l_2,\ldots,l_k}(t,1),\ldots, \mathrm{CR}^{m}_{k;l_1,l_2,\ldots,l_k}(t,z)], \tag{1.80}$$

where

$$\boldsymbol{CR}^{m}_{k;l_1,l_2,\ldots,l_k}(t,u) = 1 \quad \text{for } t<0,$$

$$\boldsymbol{CR}^{m}_{k;l_1,l_2,\ldots,l_k}(t,u) = \begin{cases} 1 & \text{for } k<m, \\ 1-\prod_{i=1}^{k}\left[1-\exp\left[-\sum_{j=1}^{l_i}\lambda_{ij}(u)t\right]\right] & \text{for } k=m, \\ \exp\left[-\sum_{j=1}^{l_k}\lambda_{kj}(u)t\right]\boldsymbol{CR}^{m}_{k-1;l_1,l_2,\ldots,l_k}(t,u) & \\ +\sum_{j=1}^{m-1}\left[\exp\left[-\sum_{v=1}^{l_{k-j}}\lambda_{k-j,v}(u)t\right]\right]\boldsymbol{CR}^{m}_{k-j-1;l_1,l_2,\ldots,l_k}(t,u) & \\ \times\prod_{i=k-j+1}^{k}\left[1-\exp\left[-\sum_{v=1}^{l_i}\lambda_{iv}(u)t\right]\right] & \text{for } k>m, \end{cases}$$

$$\text{for } t\geq 0, \quad u=1,2,\ldots,z. \tag{1.81}$$

(x) *for a consecutive "m_i out of l_i: F"-series system*

$$\bar{\boldsymbol{C}}\bar{\boldsymbol{R}}^{m_1,m_2,\ldots,m_k}_{k;l_1,l_2,\ldots,l_k}(t,\cdot) = [1, \bar{\boldsymbol{C}}\bar{\boldsymbol{R}}^{m_1,m_2,\ldots,m_k}_{k;l_1,l_2,\ldots,l_k}(t,1),\ldots,\bar{\boldsymbol{C}}\bar{\boldsymbol{R}}^{m_1,m_2,\ldots,m_k}_{k;l_1,l_2,\ldots,l_k}(t,z)], \tag{1.82}$$

where

$$\bar{\boldsymbol{C}}\bar{\boldsymbol{R}}^{m_1,m_2,\ldots,m_k}_{k;l_1,l_2,\ldots,l_k}(t,u) = 1 \quad \text{for } t<0,$$

$$\bar{\boldsymbol{C}}\bar{\boldsymbol{R}}^{m_1,m_2,\ldots,m_k}_{k;l_1,l_2,\ldots,l_k}(t,u) = \prod_{i=1}^{k} CR^{m_i}_{i,l_i}(t,u) \quad \text{for } t\geq 0, \quad u=1,2,\ldots,z, \tag{1.83}$$

where
$\boldsymbol{CR}^{m_i}_{i,l_i}(t,u), i=1,2,\ldots,k$, *are given by*

$$\boldsymbol{CR}^{m_i}_{i,l_i}(t,u) = \begin{cases} 1 & \text{for } l_i<m_i, \\ 1-\prod_{j=1}^{l_i}[1-\exp[-\lambda_{ij}(u)t)] & \text{for } l_i=m_i, \\ \exp[-\lambda_{il_i}(u)t]\boldsymbol{CR}^{m_i}_{i,l_i-1}(t,u) & \\ +\sum_{j=1}^{m_i-1}\exp[-\lambda_{il_i-j}(u)t]\boldsymbol{CR}^{m_i}_{l_i-j-1}(t,u) & \\ \left.\prod_{v=l_i-j+1}^{l_i}[1-\exp[-\lambda_{iv}(u)t]]\right] & \text{for } l_i>m_i, \end{cases} \tag{1.84}$$

$$\text{for } t\geq 0, \quad u=1,2,\ldots,z.$$

1.3 Safety Analysis of Multistate Systems

The multistate approach to the system safety analysis allows us to define the multistate safety functions of the system components, the multistate system safety function and the system risk function. Further, it is possible to define the basic safety structures of the multistate systems of components with degrading safety states and to determine their safety functions [17]. Namely, considering the safety of multistate systems, similarly as in the multistate systems reliability analysis, we assume that all components and a system S have the same set of safety states $\{0, 1, \ldots, z\}, z \geq 1$, that are ordered from the worst safety state 0 up to the best safety state z. In multistate safety analysis, the independent random variables $T_i(u), i = 1, 2, \ldots, n, u = 0, 1, 2, \ldots, z$, representing in Sect. 1.2 the lifetimes of components $E_i, i = 1, 2, \ldots, n$, in the reliability state subset $\{u, u+1, \ldots, z\}$, are representing the lifetimes of components $E_i, i = 1, 2, \ldots, n$, in the safety state subset $\{u, u+1, \ldots, z\}$, while they were in the safety state z at the moment $t = 0$. Moreover, $T(u)$ is a random variable representing the lifetime of a system in the safety states subset $\{u, u+1, \ldots, z\}$, while it was in the state z at the moment $t = 0$. Additionally, under these agreements, the vector

$$\boldsymbol{R}_i(t, \cdot) = [\boldsymbol{R}_i(t, 0), \boldsymbol{R}_i(t, 1), \ldots, \boldsymbol{R}_i(t, z)], \quad t \in \langle 0, \infty), \quad i = 1, 2, \ldots, n,$$

called in the previous discussion the multistate reliability function of a component E_i is called the multistate safety function of a component and is denoted by

$$s_i(t, \cdot) = [s_i(t, 0), s_i(t, 1), \ldots, s_i(t, z)], \quad t \in \langle 0, \infty), \quad i = 1, 2, \ldots, n.$$

Similarly, the vector

$$\boldsymbol{R}(t, \cdot) = [\boldsymbol{R}(t, 0), \boldsymbol{R}(t, 1), \ldots, \boldsymbol{R}(t, z)], \quad t \in \langle 0, \infty),$$

called earlier the multistate reliability function of a system with the notations introduced in this section is called the multistate safety function of a system and is denoted by

$$\boldsymbol{s}(t, \cdot) = [\boldsymbol{s}(t, 0), \boldsymbol{s}(t, 1), \ldots, \boldsymbol{s}(t, z)], \quad t \in \langle 0, \infty).$$

According to the above agreements, denoting by $E_i(t)$ a component E_i safety state at the moment t, $t \in \langle 0, \infty)$, we introduce the following definition.

Definition 1.15 A vector

$$s_i(t, \cdot) = [s_i(t, 0), s_i(t, 1), \ldots, s_i(t, z)], \quad t \in \langle 0, \infty), \quad i = 1, 2, \ldots, n, \tag{1.85}$$

where

$$s_i(t, u) = P(E_i(t) \geq u | E_i(0) = z) = P(T_i(u) > t) \quad \text{for } t \in \langle 0, \infty), \\ u = 0, 1, \ldots, z, \quad i = 1, 2, \ldots, n, \tag{1.86}$$

is the probability that the component E_i is in the safety state subset $\{u, u+1, \ldots, z\}$ at the moment $t, t \in \langle 0, \infty)$, while it was in the safety state z at the moment $t = 0$, is called the multistate safety function of a component E_i.

Similarly, denoting by $S(t)$ a system safety state at the moment $t, t \in \langle 0, \infty)$, we can define a multistate system safety function [17].

Definition 1.16 A vector

$$\boldsymbol{s}(t, \cdot) = [\boldsymbol{s}(t,0), \boldsymbol{s}(t,1), \ldots, \boldsymbol{s}(t,z)], \quad t \in \langle 0, \infty), \tag{1.87}$$

where

$$\boldsymbol{s}(t,u) = P(S(t) \geq u | S(0) = z) = P(T(u) > t) \quad \text{for } t \in \langle 0, \infty), \quad u = 0, 1, \ldots, z, \tag{1.88}$$

is the probability that the system is in the safety state subset $\{u, u+1, \ldots, z\}$ at the moment $t, t \in \langle 0, \infty)$, while it was in the safety state z at the moment $t = 0$, is called the multistate safety function of a system.

The assumption that the system is composed of multistate components with degrading in time safety states allows us to distinguish a system safety critical state to exceed which is either dangerous for the environment or does not assure the necessary level of its operational process effectiveness. Then, an important system safety characteristic is the time to the moment of exceeding the system safety critical state and its distribution, which is called the system risk function defined below.

Definition 1.17 A probability

$$\boldsymbol{r}(t) = P(S(t) < r | S(0) = z) = P(T(r) \leq t), \quad t \in \langle 0, \infty), \tag{1.89}$$

that the system is in the subset of safety states worse than the critical safety state $r, r \in \{1, \ldots, z\}$ while it was in the safety state z at the moment $t = 0$ is called a risk function of the multistate system.

Under this definition, considering (1.88) and (1.89), we have

$$\boldsymbol{r}(t) = 1 - P(S(t) \geq r | S(0) = z) = 1 - \boldsymbol{s}_n(t, r), \quad t \in \langle 0, \infty), \tag{1.90}$$

and, if τ is the moment when the system risk function exceeds a permitted level $\delta, \delta \in \langle 0, 1 \rangle$, then

$$\tau = \boldsymbol{r}^{-1}(\delta), \tag{1.91}$$

where $\boldsymbol{r}^{-1}(t)$, if it exists, is the inverse function of the system risk function $\boldsymbol{r}(t)$ given by (1.90).

The multistate approach to system safety analysis also allows us to define the basic safety structures of the multistate systems of components with degrading safety states and to determine their safety functions. Below, we define two typical system multistate safety structures.

Definition 1.18 A multistate system is called a series system if it is in the safety state subset $\{u, u+1, \ldots, z\}$ if and only if all its components are in this subset of safety states.

Definition 1.19 A multistate system is called a parallel system if it is in the safety state subset $\{u, u+1, \ldots, z\}$ if and only if at least one of its components is in this subset of safety states.

From *Definitions 1.18* and *1.19*, it follows that it is reasonable to present the schemes of the series and parallel multistate systems' safety structures in the same way as the schemes of their reliability structures presented respectively, in Figs. 1.4 and 1.5.

Under these definitions, it is easy to work out the following results [17].

Proposition 1.2 *The safety function of the multistate series system is given by*

$$\bar{s}_n(t,\cdot) = [1, \bar{s}_n(t,1), \ldots, \bar{s}_n(t,z)], \quad t \in \langle 0, \infty), \tag{1.92}$$

where

$$\bar{s}_n(t,u) = \prod_{i=1}^{n} s_i(t,u), \quad t \in \langle 0, \infty), \ u = 1, 2, \ldots, z. \tag{1.93}$$

Proposition 1.3 *The safety function of the multistate parallel system is given by*

$$s_n(t,\cdot) = [1, s_n(t,1), \ldots, s_n(t,z)], \quad t \in \langle 0, \infty), \tag{1.94}$$

where

$$s_n(t,u) = 1 - \prod_{i=1}^{n} [1 - s_i(t,u)], \quad t \in \langle 0, \infty), \ u = 1, 2, \ldots, z. \tag{1.95}$$

Analogically as in Sect. 1.2, we can define other multistate system safety structures composed of components with degrading safety states and determine their safety functions. The appropriate formulae for the safety functions of the system safety structures, defined similarly as the system reliability structures considered in Sect. 1.2, can simply be obtained by replacing, in the formula for these systems reliability functions, the components' reliability functions with their safety functions and the systems' reliability functions with their safety functions.

1.4 Applications

1.4.1 Reliability of Exemplary System

We consider an exemplary system S that consists of two subsystems S_1, S_2 with the reliability structure presented in Fig. 1.11.

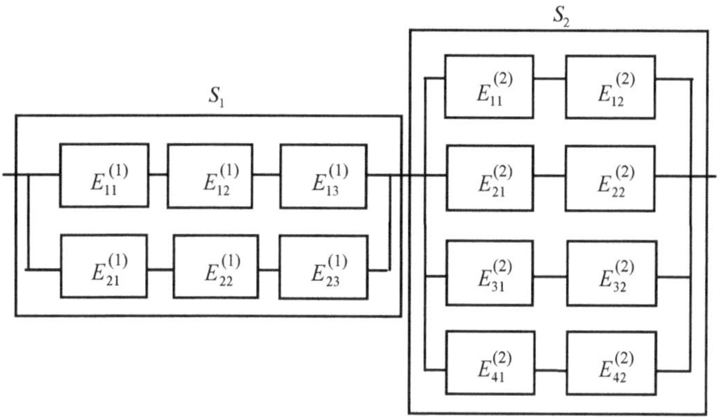

Fig. 1.11 The scheme of the exemplary system S reliability structure

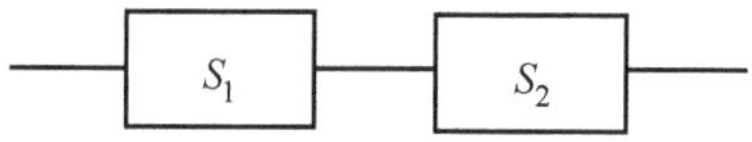

Fig. 1.12 The general scheme of the exemplary system S reliability structure

Subsystem S_1 is composed of two series subsystems, each composed of three components, denoted respectively by

$$E_{ij}^{(1)}, \quad i = 1, 2, \; j = 1, 2, 3.$$

Subsystem S_2 is composed of four series subsystems, each composed of two components, denoted respectively by

$$E_{ij}^{(2)}, \quad i = 1, 2, 3, 4, \; j = 1, 2.$$

Subsystems S_1 and S_2 form a general series reliability structure of the system S presented in Fig. 1.12.

Further, we assume that the system is a series system composed of subsystems S_1 and S_2, with the scheme shown in Fig. 1.12, while subsystem S_1 is a series-parallel system with the scheme given in Fig. 1.11 and subsystem S_2 illustrated in Fig. 1.11 is a series-"2 out of 4" system. Moreover, we arbitrarily distinguish the following four reliability states ($z = 3$) of the system and its components:

- a reliability state 3—the system operation is fully effective,
- a reliability state 2—the system operation is less effective because of ageing,
- a reliability state 1—the system operation is less effective because of ageing and more dangerous for the environment,
- a reliability state 0—the system is destroyed.

To satisfy the assumption on ageing, we assume that transitions are possible between the components' reliability states only from better to worse. Moreover, we assume that the system and its components' critical reliability state is $r = 2$.

From the above, the subsystems $S_\upsilon, \upsilon = 1, 2$, are composed of four-state, i.e. $z = 3$, components $E_{ij}^{(\upsilon)}, \upsilon = 1, 2$, having the four-state reliability functions

$$R_{ij}^{(\upsilon)}(t, \cdot) = [1, R_{ij}^{(\upsilon)}(t, 1), \quad R_{ij}^{(\upsilon)}(t, 2), \quad R_{ij}^{(\upsilon)}(t, 3)],$$

with the coordinates which by assumption are exponentials of the form

$$R_{ij}^{(\upsilon)}(t, 1) = \exp[-\lambda_{ij}^{(\upsilon)}(1)t], \quad R_{ij}^{(\upsilon)}(t, 2) = \exp[-\lambda_{ij}^{(\upsilon)}(2)t], \quad R_{ij}^{(\upsilon)}(t, 3) = \exp[-\lambda_{ij}^{(\upsilon)}(3)t].$$

Thus, the considered exemplary system is a four-state series system composed of the four-state subsystems S_1 and S_2. The subsystem S_1 is a four-state series-parallel system and the subsystem S_2 is a four-state series-"2 out of 4" system.

The subsystem S_1 consists of $k = 2$ identical four-state series subsystems, each composed of three components, i.e. $l_1 = l_2 = 3$, with the exponential reliability functions that are partially identified in Chap. 5 and given below. In both series subsystems of the subsystem S_1 there are respectively:

- the components $E_{i1}^{(1)}, i = 1, 2$, with the conditional reliability function coordinates

$$R_{i1}^{(1)}(t, 1) = \exp[-0.0008t], \quad R_{i1}^{(1)}(t, 2) = \exp[-0.0009t],$$
$$R_{i1}^{(1)}(t, 3) = \exp[-0.0009t], \quad i = 1, 2;$$

- the components $E_{i2}^{(1)}, i = 1, 2$, with the conditional reliability function co-ordinates

$$R_{i2}^{(1)}(t, 1) = \exp[-0.0011t], \quad R_{i2}^{(1)}(t, 2) = \exp[-0.0011t],$$
$$R_{i2}^{(1)}(t, 3) = \exp[-0.0012t], \quad i = 1, 2;$$

- the components $E_{i3}^{(1)}, i = 1, 2$, with the conditional reliability function co-ordinates

$$R_{i3}^{(1)}(t, 1) = \exp[-0.0011t], \quad R_{i3}^{(1)}(t, 2) = \exp[-0.0011t],$$
$$R_{i3}^{(1)}(t, 3) = \exp[-0.0011t], \quad i = 1, 2.$$

The subsystem S_1 is a four-state series-parallel system and according to (1.36–1.37) and (1.68–1.69) its four-state reliability function is given by

$$\boldsymbol{R}^{(1)}(t, \cdot) = [1, \boldsymbol{R}^{(1)}(t, 1), \quad \boldsymbol{R}^{(1)}(t, 2), \quad \boldsymbol{R}^{(1)}(t, 3)], \quad t \geq 0,$$

where

$$\boldsymbol{R}^{(1)}(t,1) = \boldsymbol{R}_{2;3,3}(t,1) = 1 - \prod_{i=1}^{2}\left[1 - \prod_{j=1}^{3} R_{ij}^{(1)}(t,1)\right]$$
$$= 1 - \prod_{i=1}^{2}\left[1 - \exp\left[-\sum_{j=1}^{3}\lambda_{ij}^{(1)}(1)t\right]\right]$$
$$= 1 - [1 - \exp[-[0.0008 + 0.0011 + 0.0011]t]^2$$
$$= 1 - [1 - \exp[-0.003t]]^2$$
$$= 2\exp[-0.003t] - \exp[-0.006t], \tag{1.96}$$

$$\boldsymbol{R}^{(1)}(t,2) = \boldsymbol{R}_{2;3,3}(t,2) = 1 - \prod_{i=1}^{2}\left[1 - \prod_{j=1}^{3} R_{ij}{}^{(1)}(t,2)\right]$$
$$= 1 - \prod_{i=1}^{2}\left[1 - \exp\left[-\sum_{j=1}^{3}\lambda_{ij}^{(1)}(2)t\right]\right]$$
$$= 1 - [1 - \exp[-0.0009 + 0.0011 + 0.0011]t]^2$$
$$= 1 - [1 - \exp[-0.0031t]]^2$$
$$= 2\exp[-0.0031t] - \exp[-0.0062t], \tag{1.97}$$

$$\boldsymbol{R}^{(1)}(t,3) = \boldsymbol{R}_{2;3,3}(t,3) = 1 - \prod_{i=1}^{2}\left[1 - \prod_{j=1}^{3} R_{ij}^{(1)}(t,3)\right]$$
$$= 1 - \prod_{i=1}^{2}\left[1 - \exp\left[-\sum_{j=1}^{3}\lambda_{ij}^{(1)}(3)t\right]\right]$$
$$= 1 - [1 - \exp[-0.0009 + 0.0012 + 0.0011]t]^2$$
$$= 1 - [1 - \exp[-0.0032t]]^2$$
$$= 2\exp[-0.0032t] - \exp[-0.0064t]. \tag{1.98}$$

The subsystem S_2 consists of $k = 4$ identical four-state series subsystems, each composed of two components, i.e. $l_1 = l_2 = l_3 = l_4 = 2$, with the exponential reliability functions given below. In all series subsystems of the subsystem S_2 there are respectively:

- the components $E_{i1}^{(2)}, i = 1,2,3,4,$ with the conditional reliability function co-ordinates

$$R_{i1}^{(2)}(t,1) = \exp[-0.0013t], \quad R_{i1}^{(2)}(t,2) = \exp[-0.0014t],$$
$$R_{i1}^{(2)}(t,3) = \exp[-0.0015t], \quad i = 1,2,3,4;$$

- the components $E_{i2}^{(2)}, i = 1,2,3,4,$ with the conditional reliability function co-ordinates

$$\boldsymbol{R}_{i2}^{(2)}(t,1) = \exp[-0.0015t], \quad R_{i2}^{(2)}(t,2) = \exp[-0.0016t],$$
$$R_{i2}^{(2)}(t,3) = \exp[-0.0018t], \quad i = 1,2,3,4.$$

The subsystem S_2 is a four-state series-"2 out of 4" system and according to (1.40–1.41) and (1.72–1.73) its four-state reliability function is given by

$$\boldsymbol{R}^{(2)}(t,\cdot) = [1, \boldsymbol{R}^{(2)}(t,1),\ \boldsymbol{R}^{(2)}(t,2),\ \boldsymbol{R}^{(2)}(t,3)], \quad t \geq 0, \tag{1.99}$$

where

$$\begin{aligned}
\boldsymbol{R}^{(2)}(t,1) = \boldsymbol{R}^{2}_{4;2,2,2,2}(t,1) &= 1 - \sum_{\substack{r_1,r_2,r_3,r_4=0\\ r_1+r_2+r_3+r_4 \leq 1}}^{1} \prod_{i=1}^{4} \left[\prod_{j=1}^{2} R_{ij}{}^{(2)}(t,1)\right]^{r_i} \left[1 - \prod_{j=1}^{2} R_{ij}{}^{(2)}(t,1)\right]^{1-r_i} \\
&= 1 - \sum_{\substack{r_1,r_2,r_3,r_4=0\\ r_1+r_2+r_3+r_4 \leq 1}}^{1} \prod_{i=1}^{4} \exp\left[-r_i \sum_{j=1}^{2} \lambda_{ij}^{(2)}(1)t\right]\left[1 - \exp\left[-\sum_{j=1}^{2} \lambda_{ij}^{(2)}(1)t\right]\right]^{1-r_i} \\
&= 1 - \left[\prod_{j=1}^{2} R_{1j}^{(2)}(t,1)\right]^{0}\left[1 - \prod_{j=1}^{2} R_{1j}^{(2)}(t,1)\right]^{1}\left[\prod_{j=1}^{2} R_{2j}^{(2)}(t,1)\right]^{0}\left[1 - \prod_{j=1}^{2} R_{2j}^{(2)}(t,1)\right]^{1} \\
&\quad \times \left[\prod_{j=1}^{2} R_{3j}^{(2)}(t,1)\right]^{0}\left[1 - \prod_{j=1}^{2} R_{3j}^{(2)}(t,1)\right]^{1}\left[\prod_{j=1}^{2} R_{4j}^{(2)}(t,1)\right]^{0}\left[1 - \prod_{j=1}^{2} R_{4j}^{(2)}(t,1)\right]^{1} \\
&\quad - \left[\prod_{j=1}^{2} R_{1j}^{(2)}(t,1)\right]^{1}\left[1 - \prod_{j=1}^{2} R_{1j}^{(2)}(t,1)\right]^{0}\left[\prod_{j=1}^{2} R_{2j}^{(2)}(t,1)\right]^{0}\left[1 - \prod_{j=1}^{2} R_{2j}^{(2)}(t,1)\right]^{1} \\
&\quad \times \left[\prod_{j=1}^{2} R_{3j}^{(2)}(t,1)\right]^{0}\left[1 - \prod_{j=1}^{2} R_{3j}^{(2)}(t,1)\right]^{1}\left[\prod_{j=1}^{2} R_{4j}^{(2)}(t,1)\right]^{0}\left[1 - \prod_{j=1}^{2} R_{4j}^{(2)}(t,1)\right]^{1} \\
&\quad - \left[\prod_{j=1}^{2} R_{1j}^{(2)}(t,1)\right]^{0}\left[1 - \prod_{j=1}^{2} R_{1j}^{(2)}(t,1)\right]^{1}\left[\prod_{j=1}^{2} R_{2j}^{(2)}(t,1)\right]^{1}\left[1 - \prod_{j=1}^{2} R_{2j}^{(2)}(t,1)\right]^{0} \\
&\quad \times \left[\prod_{j=1}^{2} R_{3j}^{(2)}(t,1)\right]^{0}\left[1 - \prod_{j=1}^{2} R_{3j}^{(2)}(t,1)\right]^{1}\left[\prod_{j=1}^{2} R_{4j}^{(2)}(t,1)\right]^{0}\left[1 - \prod_{j=1}^{2} R_{4j}^{(2)}(t,1)\right]^{1} \\
&\quad - \left[\prod_{j=1}^{2} R_{1j}^{(2)}(t,1)\right]^{0}\left[1 - \prod_{j=1}^{2} R_{1j}^{(2)}(t,1)\right]^{1}\left[\prod_{j=1}^{2} R_{2j}^{(2)}(t,1)\right]^{0}\left[1 - \prod_{j=1}^{2} R_{2j}^{(2)}(t,1)\right]^{1} \\
&\quad \times \left[\prod_{j=1}^{2} R_{3j}^{(2)}(t,1)\right]^{1}\left[1 - \prod_{j=1}^{2} R_{3j}^{(2)}(t,1)\right]^{0}\left[\prod_{j=1}^{2} R_{4j}^{(2)}(t,1)\right]^{0}\left[1 - \prod_{j=1}^{2} R_{4j}^{(2)}(t,1)\right]^{1}
\end{aligned}$$

$$
\begin{aligned}
&- \left[\prod_{j=1}^{2} R_{1j}^{(2)}(t,1)\right]^{0} \left[1-\prod_{j=1}^{2} R_{1j}^{(2)}(t,1)\right]^{1} \left[\prod_{j=1}^{2} R_{2j}^{(2)}(t,1)\right]^{0} \left[1-\prod_{j=1}^{2} R_{2j}^{(2)}(t,1)\right]^{1} \\
&\times \left[\prod_{j=1}^{2} R_{3j}^{(2)}(t,1)\right]^{0} \left[1-\prod_{j=1}^{2} R_{3j}^{(2)}(t,1)\right]^{1} \left[\prod_{j=1}^{2} R_{4j}^{(2)}(t,1)\right]^{1} \left[1-\prod_{j=1}^{2} R_{4j}^{(2)}(t,1)\right]^{0} \\
={}& 1-\exp[-0[\lambda_{11}^{(2)}(1)+\lambda_{12}^{(2)}(1)]t][1-\exp[-[\lambda_{11}^{(2)}(1)+\lambda_{12}^{(2)}(1)]t]]^{1} \\
&\times \exp[-0[\lambda_{21}^{(2)}(1)+\lambda_{22}^{(2)}(1)]t][1-\exp[-[\lambda_{21}^{(2)}(1)+\lambda_{22}^{(2)}(1)]t]]^{1} \\
&\times \exp[-0[\lambda_{31}^{(2)}(1)+\lambda_{32}^{(2)}(1)]t][1-\exp[-[\lambda_{31}^{(2)}(1)+\lambda_{32}^{(2)}(1)]t]]^{1} \\
&\times \exp[-0[\lambda_{41}^{(2)}(1)+\lambda_{42}^{(2)}(1)]t][1-\exp[-[\lambda_{41}^{(2)}(1)+\lambda_{42}^{(2)}(1)]t]]^{1} \\
&- \exp[-1[\lambda_{11}^{(2)}(1)+\lambda_{12}^{(2)}(1)]t][1-\exp[-[\lambda_{11}^{(2)}(1)+\lambda_{12}^{(2)}(1)]t]]^{0} \\
&\times \exp[-0[\lambda_{21}^{(2)}(1)+\lambda_{22}^{(2)}(1)]t][1-\exp[-[\lambda_{21}^{(2)}(1)+\lambda_{22}^{(2)}(1)]t]]^{1} \\
&\times \exp[-0[\lambda_{31}^{(2)}(1)+\lambda_{32}^{(2)}(1)]t][1-\exp[-[\lambda_{31}^{(2)}(1)+\lambda_{32}^{(2)}(1)]t]]^{1} \\
&\times \exp[-0[\lambda_{41}^{(2)}(1)+\lambda_{42}^{(2)}(1)]t][1-\exp[-[\lambda_{41}^{(2)}(1)+\lambda_{42}^{(2)}(1)]t]]^{1} \\
&- \exp[-0[\lambda_{11}^{(2)}(1)+\lambda_{12}^{(2)}(1)]t][1-\exp[-[\lambda_{11}^{(2)}(1)+\lambda_{12}^{(2)}(1)]t]]^{1} \\
&\times \exp[-1[\lambda_{21}^{(2)}(1)+\lambda_{22}^{(2)}(1)]t][1-\exp[-[\lambda_{21}^{(2)}(1)+\lambda_{22}^{(2)}(1)]t]]^{0} \\
&\times \exp[-0[\lambda_{31}^{(2)}(1)+\lambda_{32}^{(2)}(1)]t][1-\exp[-[\lambda_{31}^{(2)}(1)+\lambda_{32}^{(2)}(1)]t]]^{1} \\
&\times \exp[-0[\lambda_{41}^{(2)}(1)+\lambda_{42}^{(2)}(1)]t][1-\exp[-[\lambda_{41}^{(2)}(1)+\lambda_{42}^{(2)}(1)]t]]^{1} \\
&- \exp[-0[\lambda_{11}^{(2)}(1)+\lambda_{12}^{(2)}(1)]t][1-\exp[-[\lambda_{11}^{(2)}(1)+\lambda_{12}^{(2)}(1)]t]]^{1} \\
&\times \exp[-0[\lambda_{21}^{(2)}(1)+\lambda_{22}^{(2)}(1)]t][1-\exp[-[\lambda_{21}^{(2)}(1)+\lambda_{22}^{(2)}(1)]t]]^{1} \\
&\times \exp[-1[\lambda_{31}^{(2)}(1)+\lambda_{32}^{(2)}(1)]t][1-\exp[-[\lambda_{31}^{(2)}(1)+\lambda_{32}^{(2)}(1)]t]]^{0} \\
&\times \exp[-0[\lambda_{41}^{(2)}(1)+\lambda_{42}^{(2)}(1)]t][1-\exp[-[\lambda_{41}^{(2)}(1)+\lambda_{42}^{(2)}(1)]t]]^{1} \\
&- \exp[-0[\lambda_{11}^{(2)}(1)+\lambda_{12}^{(2)}(1)]t][1-\exp[-[\lambda_{11}^{(2)}(1)+\lambda_{12}^{(2)}(1)]t]]^{1} \\
&\times \exp[-0[\lambda_{21}^{(2)}(1)+\lambda_{22}^{(2)}(1)]t][1-\exp[-[\lambda_{21}^{(2)}(1)+\lambda_{22}^{(2)}(1)]t]]^{1} \\
&\times \exp[-0[\lambda_{31}^{(2)}(1)+\lambda_{32}^{(2)}(1)]t][1-\exp[-[\lambda_{31}^{(2)}(1)+\lambda_{32}^{(2)}(1)]t]]^{1} \\
&\times \exp[-1[\lambda_{41}^{(2)}(1)+\lambda_{42}^{(2)}(1)]t][1-\exp[-[\lambda_{41}^{(2)}(1)+\lambda_{42}^{(2)}(1)]t]]^{0} \\
={}& 1-[1-\exp[-[0.0013+0.0015]t]]^{4} \\
&- 4\exp[-1[0.0013+0.0015]t][1-\exp[-[0.0013+0.0015]t]^{3} \\
={}& 1-[1-\exp[-0.0028t]]^{4} - 4\exp[-0.0028t][1-\exp[-0.0028t]]^{3} \\
={}& 6\exp[-0.0056t] - 8\exp[-0.0084t] + 3\exp[-0.0112t],
\end{aligned}
\tag{1.100}
$$

$$\boldsymbol{R}^{(2)}(t,2)=\boldsymbol{R}^{2}_{4;2,2,2,2}(t,2)=1-\sum_{\substack{r_1,r_2,r_3,r_4=0\\ r_1+r_2+r_3+r_4\le 1}}^{1}\prod_{i=1}^{4}\left[\prod_{j=1}^{2}R_{ij}^{(2)}(t,2)\right]^{r_i}\left[1-\prod_{j=1}^{2}R_{ij}^{(2)}(t,2)\right]^{1-r_i}$$
$$=1-\sum_{\substack{r_1,r_2,r_3,r_4=0\\ r_1+r_2+r_3+r_4\le 1}}^{1}\prod_{i=1}^{4}\exp\left[-r_i\sum_{j=1}^{2}\lambda_{ij}^{(2)}(2)t\right]\left[1-\exp\left[-\sum_{j=1}^{2}\lambda_{ij}^{(2)}(2)t\right]\right]^{1-r_i}$$
$$=1-[1-\exp[-[0.0014+0.0016]t]]^4$$
$$-4\exp[-1[0.0014+0.0016]t][1-\exp[-[0.0014+0.0016]t]^3$$
$$=1-[1-\exp[-0.003t]]^4-4\exp[-0.003t][1-\exp[-0.003t]]^3$$
$$=6\exp[-0.006t]-8\exp[-0.009t]+3\exp[-0.012t], \quad (1.101)$$

$$\boldsymbol{R}^{(2)}(t,3)=\boldsymbol{R}^{2}_{4;2,2,2,2}(t,3)=1-\sum_{\substack{r_1,r_2,r_3,r_4=0\\ r_1+r_2+r_3+r_4\le 1}}^{1}\prod_{i=1}^{4}\left[\prod_{j=1}^{2}R_{ij}^{(2)}(t,3)\right]^{r_i}\left[1-\prod_{j=1}^{2}R_{ij}^{(2)}(t,3)\right]^{1-r_i}$$
$$=1-\sum_{\substack{r_1,r_2,r_3,r_4=0\\ r_1+r_2+r_3+r_4\le 1}}^{1}\prod_{i=1}^{4}\exp\left[-r_i\sum_{j=1}^{2}\lambda_{ij}^{(2)}(3)t\right]\left[1-\exp\left[-\sum_{j=1}^{2}\lambda_{ij}^{(2)}(3)t\right]\right]^{1-r_i}$$
$$=1-[1-\exp[-[0.0015+0.0018]t]]^4$$
$$-4\exp[-1[0.0015+0.0018]t][1-\exp[-[0.0015+0.0018]t]^3$$
$$=1-[1-\exp[-0.0033t]]^4-4\exp[-0.0033t][1-\exp[-0.0033t]]^3$$
$$=6\exp[-0.0066t]-8\exp[-0.0099t]+3\exp[-0.0132t]. \quad (1.102)$$

Considering that the system is a four-state series system, after applying (1.22–1.23), its reliability function is given by

$$\boldsymbol{R}(t,\cdot)=[1,\ \boldsymbol{R}(t,1),\boldsymbol{R}(t,2),\boldsymbol{R}(t,3)],\quad t\ge 0, \quad (1.103)$$

where by (1.96–1.98) and (1.100–1.102) we have

$$\boldsymbol{R}(t,1)=\overline{\boldsymbol{R}}_2(t,1)=\boldsymbol{R}^{(1)}(t,1)\boldsymbol{R}^{(2)}(t,1)$$
$$=12\exp[-0.0086t]-16\exp[-0.0114t]$$
$$+6\exp[-0.0142t]-6\exp[-0.0116t]+8\exp[-0.0144t]-3\exp[-0.0172t], \quad (1.104)$$

$$\boldsymbol{R}(t,2)=\overline{\boldsymbol{R}}_2(t,2)=\boldsymbol{R}^{(1)}(t,2)\boldsymbol{R}^{(2)}(t,2)$$
$$=12\exp[-0.0091t]-16\exp[-0.0121t]$$
$$+6\exp[-0.0151t]-6\exp[-0.0122t]+8\exp[-0.0152t]-3\exp[-0.0182t], \quad (1.105)$$

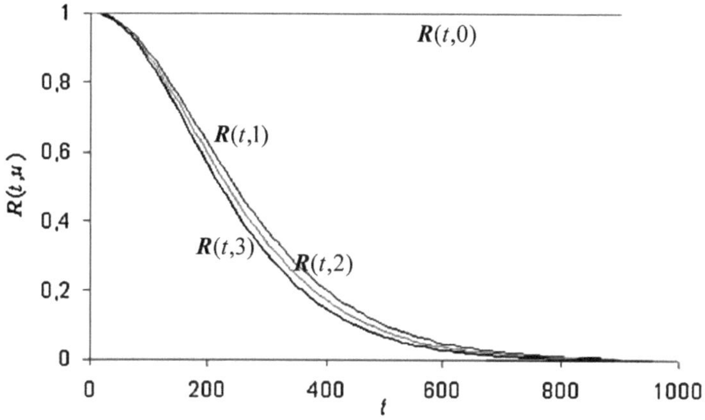

Fig. 1.13 The graph of the exemplary system S reliability function $\boldsymbol{R}(t,\cdot)$ coordinates

$$
\begin{aligned}
\boldsymbol{R}(t,3) = \overline{\boldsymbol{R}}_2(t,3) &= \boldsymbol{R}^{(1)}(t,3)\boldsymbol{R}^{(2)}(t,3)\\
&= 12\exp[-0.0098t] - 16\exp[-0.0131t]\\
&\quad + 6\exp[-0.0164t] - 6\exp[-0.013t] + 8\exp[-0.0163t] - 3\exp[-0.0196t].
\end{aligned}
\tag{1.106}
$$

The coordinates of the exemplary four-state system reliability function are illustrated in Fig. 1.13.

The expected values and standard deviations of the system lifetimes in the reliability state subsets $\{1,2,3\}, \{2,3\}, \{3\}$, calculated from the results given by (1.104–1.106), according to (1.15–1.17), are:

$$\mu(1) \cong 278.27, \quad \mu(2) \cong 263.39, \quad \mu(3) \cong 245.16, \tag{1.107}$$

$$\sigma(1) \cong 167.47, \quad \sigma(2) \cong 158.19, \quad \sigma(3) \cong 146.75. \tag{1.108}$$

The detailed way of finding the above mean value standard deviations is illustrated below for the mean value $\mu(1)$ and the standard deviation $\sigma(1)$ of the exemplary system lifetime in the reliability states subset $\{1,2,3\}$.

According to (1.15), using (1.104) and applying the direct integration, we successively get

$$
\begin{aligned}
\mu(1) = \int_0^\infty R(t,1)\mathrm{d}t = \int_0^\infty &[12\exp[-0.0086t] - 16\exp[-0.0114t]\\
&+6\exp[-0.0142t] - 6\exp[-0.0116t] + 8\exp[-0.0144t] - 3\exp[-0.0172t]]\mathrm{d}t\\
= \Big[&-\frac{12}{0.0086}\exp[-0.0086t] + \frac{16}{0.0114}\exp[-0.0114t]
\end{aligned}
$$

$$
\begin{aligned}
&-\frac{6}{0.0142}\exp[-0.0142t]+\frac{6}{0.0116}\exp[-0.0116t]\\
&\left.-\frac{8}{0.0144}\exp[-0.0144t]+\frac{3}{0.0172}\exp[-0.0172t]\right]_0^{\infty}\\
&=0+\frac{12}{0.0086}-\frac{16}{0.0114}+\frac{6}{0.0142}-\frac{6}{0.0116}\\
&+\frac{8}{0.0144}-\frac{3}{0.0172}\cong 278.27.
\end{aligned}
$$

According to (1.17), using (1.104) and applying the integration by parts, we successively get

$$
\begin{aligned}
n(1)&=2\int_0^{\infty} t\overline{\boldsymbol{R}}(t,1)\mathrm{d}t=2\int_0^{\infty} t[12\exp[-0.0086t]-16\exp[-0.0114t]\\
&+6\exp[-0.0142t]-6\exp[-0.0116t]+8\exp[-0.0144t]-3\exp[-0.0172t]]\mathrm{d}t\\
&=2\Bigg[-\frac{12}{[0.0086]^2}\exp[-0.0086t]+\frac{16}{[0.0114]^2}\exp[-0.0114t]\\
&-\frac{6}{[0.0142]^2}\exp[-0.0142t]+\frac{6}{[0.0116]^2}\exp[-0.0116t]\\
&\left.-\frac{8}{[0.0144]^2}\exp[-0.0144t]+\frac{3}{[0.0172]^2}\exp[-0.0172t]\right]_0^{\infty}\\
&=2\Bigg[0+\frac{12}{[0.0086]^2}-\frac{16}{[0.0114]^2}+\frac{6}{[0.0142]^2}-\frac{6}{[0.0116]^2}\\
&+\frac{8}{[0.0144]^2}-\frac{3}{[0.0172]^2}\Bigg]\cong 105481.83.
\end{aligned}
$$

Hence, after applying (1.16), we get

$$
\sigma(1)=\sqrt{n(1)-[\mu(1)]^2}=\sqrt{105481.83-[278.27]^2}\cong 167.47.
$$

Further, using (1.107) and from (1.19), it follows that the mean values of the system lifetimes in the particular reliability states are:

$$
\begin{aligned}
&\overline{\mu}(1)\cong 278.27-263.39=14.88,\quad \overline{\mu}(2)\cong 263.39-245.16=18.23,\\
&\overline{\mu}(3)\cong 245.16.
\end{aligned}
\tag{1.109}
$$

As the critical reliability state is $r = 2$, then the system risk function, according to (1.20), is given by

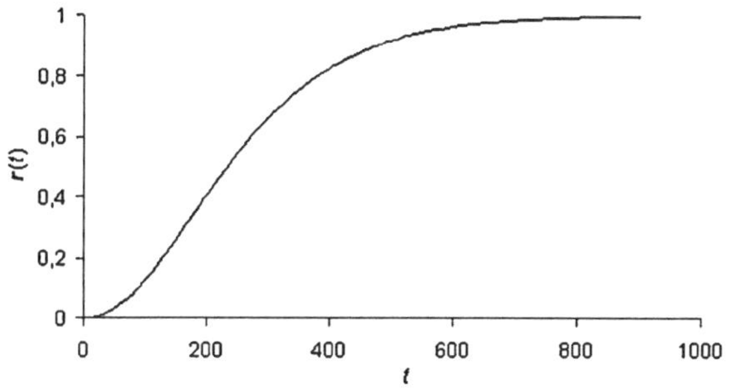

Fig. 1.14 The graph of the risk function $\mathbf{r}(t)$ of the exemplary system S

$$\begin{aligned} \boldsymbol{r}(t) &= 1 - \boldsymbol{R}(t,2) \\ &= 1 - [12\exp[-0.0091t] - 16\exp[-0.0121t] + 6\exp[-0.0151t] \\ &\quad - 6\exp[-0.0122t] + 8\exp[-0.0152t] - 3\exp[-0.0182t]], \text{ for } t \geq 0. \end{aligned} \tag{1.110}$$

Hence, the moment when the system risk function exceeds a permitted level, for instance $\delta = 0.05$, by (1.21), is

$$\tau = \boldsymbol{r}^{-1}(\delta) \cong 63.8. \tag{1.111}$$

The exemplary four-state system risk function is illustrated in Fig. 1.14.

1.4.2 Reliability of Port Oil Piping Transportation System

The considered oil piping transportation system operating at one of the Baltic Oil Terminals is designated for the reception from ships, the storage and sending the oil products by carriages or cars. It is also designated for receiving from carriages or cars, the storage and loading of tankers with oil products such as petrol and oil. The considered terminal is composed of three parts A, B and C, linked by the piping transportation system and the pier. The scheme of this terminal is presented in Fig. 1.15.

The unloading of tankers is performed at the pier placed in the port. The pier is connected with terminal part A through the transportation subsystem S_1 built with two piping lines composed of steel pipe segments of diameter 600 mm. In part A there is a supporting station fortifying tanker pumps and making possible further transport of oil by the subsystem S_2 to the terminal part B. Subsystem S_2 is built with two piping lines composed of steel pipe segments of diameter 600 mm. The terminal part B is connected with the terminal part C by the subsystem S_3. Subsystem S_3 is built with one piping line composed of steel pipe segments of diameter 500 mm and two piping lines composed of steel pipe segments of

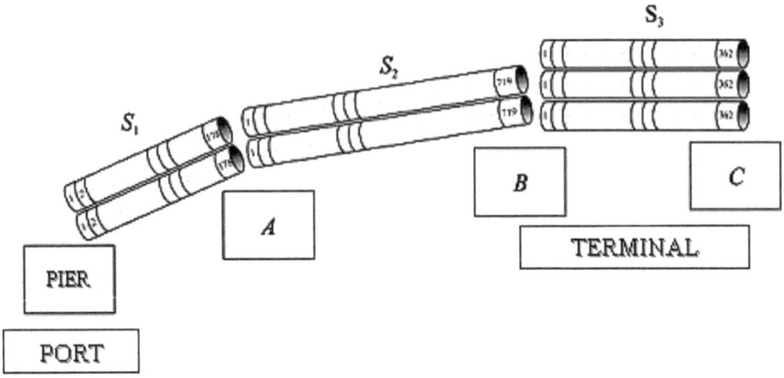

Fig. 1.15 The scheme of the port oil transportation system

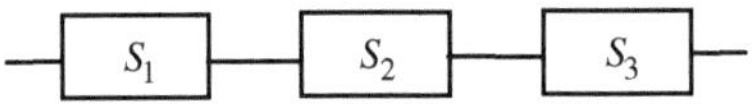

Fig. 1.16 General scheme of the port oil pipeline system reliability structure

diameter 350 mm. Terminal part C is designated for loading rail cisterns with oil products and for wagons sending to the railway station of the port and further to the interior of the country.

Thus, the port oil pipeline transportation system consists of three subsystems:

- subsystem S_1 composed of two identical pipelines, each composed of 178 pipe segments of length 12 m and two valves,
- subsystem S_2 composed of two identical pipelines, each composed of 717 pipe segments of length 12 m and to valves,
- subsystem S_3 composed of two identical and one different pipelines, each composed of 360 pipe segments of either 10 m or 7.5 m length and two valves.

Subsystems S_1, S_2, S_3, indicated in Fig. 1.15 form a general series port oil pipeline system reliability structure as presented in Fig. 1.16.

The system is a series system composed of two series-parallel subsystems S_1, S_2, each containing two pipelines and one series-"2 out of 3" subsystem S_3.

After considering the comments and opinions coming from experts, taking into account the effectiveness and safety aspects of the operation of the oil pipeline transportation system, we distinguish the following three reliability states ($z = 2$) of the system and its components:

- a reliability state 2—piping operation is fully safe,
- a reliability state 1—piping operation is less safe and more dangerous because of the possibility of environment pollution,
- a reliability state 0—piping is destroyed.

Moreover, based on expert opinion, we assume that transitions are possible between the components' reliability states only from better to worse ones and we assume that the system and its components' critical reliability state is $r = 1$.

From the above, the subsystems $S_\upsilon, \upsilon = 1, 2, 3$, are composed of three-state, i.e. $z = 2$, components $E_{ij}^{(\upsilon)}, \upsilon = 1, 2, 3$, having reliability functions

$$R_{ij}^{(\upsilon)}(t, \cdot) = [1, R_{ij}^{(\upsilon)}(t, 1), R_{ij}^{(\upsilon)}(t, 2)],$$

with the coordinates which by assumption are exponentials of the form

$$R_{ij}^{(\upsilon)}(t, 1) = \exp[-\lambda_{ij}^{(\upsilon)}(1)t], \quad R_{ij}^{(\upsilon)}(t, 2) = \exp[-\lambda_{ij}^{(\upsilon)}(2)t].$$

Thus, the system is a three-state series system composed of two three-state series-parallel subsystems S_1, S_2, each containing two pipelines and one three-state series-"2 out of 3" subsystem S_3.

Subsystem S_1 consists of $k = 2$ identical pipelines, each composed of 178 components $E_{ij}^{(1)}, i = 1, 2;\ j = 1, 2, \ldots, 178$, i.e. $l_1 = l_2 = 178$, with the exponential reliability functions partly identified in Chap. 5 on the basis of data coming from experts and are given below. In each pipeline there are:

- 176 pipe segments with the multistate reliability function co-ordinates

$$R_{ij}^{(1)}(t, 1) = \exp[-0.0062t], \quad R_{ij}^{(1)}(t, 2) = \exp[-0.0088t],$$
$$i = 1, 2; \ \ j = 1, 2, \ldots, 176,$$

- 2 valves with the multistate reliability function co-ordinates

$$R_{ij}^{(1)}(t, 1) = \exp[-0.0167t], \quad R_{ij}^{(1)}(t, 2) = \exp[-0.0182t],$$
$$i = 1, 2; \ \ j = 177, 178.$$

Subsystem S_1 is a three-state series-parallel system and according to (1.36–1.37) and (1.68–1.69) its three-state reliability function is given by

$$\boldsymbol{R}^{(1)}(t, \cdot) = [1, \boldsymbol{R}^{(1)}(t, 1), \boldsymbol{R}^{(1)}(t, 2)], \quad t \geq 0, \tag{1.112}$$

where

$$\begin{aligned} \boldsymbol{R}^{(1)}(t, 1) = \boldsymbol{R}_{2;178,178}(t, 1) &= 1 - \prod_{i=1}^{2}\left[1 - \prod_{j=1}^{178} R_{ij}^{(1)}(t, 1)\right] \\ &= 1 - [1 - \exp[-[176 \cdot 0.0062 + 2 \cdot 0.0167]t]]^2 \\ &= 1 - [1 - \exp[-1.1246t]]^2 \\ &= 2\exp[-1.1246t] - \exp[-2.2492t], \end{aligned} \tag{1.113}$$

$$\begin{aligned}\boldsymbol{R}^{(1)}(t,2) &= \boldsymbol{R}_{2;178,178}(t,2) = 1 - \prod_{i=1}^{2}\left[1 - \prod_{j=1}^{178} R_{ij}^{(1)}(t,2)\right]\\ &= 1 - [1 - \exp[-[176 \cdot 0.0088 + 2 \cdot 0.0182]t]]^2\\ &= 1 - [1 - \exp[-1.5852t]]^2\\ &= 2\exp[-1.5852t] - \exp[-3.1704t]. \end{aligned} \tag{1.114}$$

Subsystem S_2 consists of $k = 2$ identical pipelines, each composed of 719 components $E_{ij}^{(2)}, i = 1,2;\ j = 1,2,\ldots,719$, i.e. $l_1 = l_2 = 719$, with the exponential reliability functions identified on the basis of data coming from experts and given below. In each pipeline there are:

- 717 pipe segments with the multistate reliability function co-ordinates

$$R_{ij}^{(2)}(t,1) = \exp[-0.0062t], \quad R_{ij}^{(2)}(t,2) = \exp[-0.0088t],$$
$$i = 1,2;\ \ j = 1,2,\ldots,717,$$

- 2 valves with the multistate reliability function co-ordinates

$$R_{ij}^{(2)}(t,1) = \exp[-0.0166t], \quad R_{ij}^{(2)}(t,2) = \exp[-0.0181t],$$
$$i = 1,2;\ \ j = 718,719.$$

Thus, subsystem S_2 is a three-state series-parallel system and according to (1.36–1.37) and (1.68–1.69) its three-state reliability function is given by

$$\boldsymbol{R}^{(2)}(t,\cdot) = [1, \boldsymbol{R}^{(2)}(t,1), \boldsymbol{R}^{(2)}(t,2)], \quad t \geq 0, \tag{1.115}$$

where

$$\begin{aligned}\boldsymbol{R}^{(2)}(t,1) &= \boldsymbol{R}_{2;719,719}(t,1) = 1 - \prod_{i=1}^{2}\left[1 - \prod_{j=1}^{719} R_{ij}^{(2)}(t,1)\right]\\ &= 1 - [1 - \exp[-[717 \times 0.0062 + 2 \times 0.0166]t]]^2\\ &= 1 - [1 - \exp[-4.4786t]]^2\\ &= 2\exp[-4.4786t] - \exp[-8.9572t], \end{aligned} \tag{1.116}$$

$$\begin{aligned}\boldsymbol{R}^{(2)}(t,2) &= \boldsymbol{R}_{2;719,719}(t,2) = 1 - \prod_{i=1}^{2}\left[1 - \prod_{j=1}^{719} R_{ij}^{(2)}(t,2)\right]\\ &= 1 - [1 - \exp[-[717 \times 0.0088 + 2 \times 0.0181]t]]^2\\ &= 1 - [1 - \exp[-6.3458t]]^2\\ &= 2\exp[-6.3458t] - \exp[-12.6916t]. \end{aligned} \tag{1.117}$$

Subsystem S_3 consists of $k = 3$ pipelines, two pipelines of the first type and one pipeline of the second type, each composed of $l = 362$ components $E_{ij}^{(3)}, i = 1,2,3, j = 1,2,\ldots,362$, i.e. $l_1 = l_2 = l_3 = 362$, with the exponential reliability functions identified on the basis of data coming from experts and are given below. In each pipeline of the first type there are:

- 360 pipe segments with the multistate reliability function co-ordinates

$$R_{ij}^{(3)}(t,1) = \exp[-0.0059t], \quad R_{ij}^{(3)}(t,2) = \exp[-0.0074t],$$
$$i = 1,2; \;\; j = 1,2,\ldots,360,$$

- 2 valves with the multistate reliability function co-ordinates

$$R_{ij}^{(3)}(t,1) = \exp[-0.0166t], \quad R_{ij}^{(3)}(t,2) = \exp[-0.0181t],$$
$$i = 1,2; \;\; j = 361,362.$$

In the pipeline of the second type there are:

- 360 pipe segments with the multistate reliability function co-ordinates

$$R_{ij}^{(3)}(t,1) = \exp[-0.0071t], \quad R_{ij}^{(3)}(t,2) = \exp[-0.0079t],$$
$$i = 3; \;\; j = 1,2,\ldots,360,$$

- 2 valves with the multistate reliability function co-ordinates

$$R_{ij}^{(3)}(t,1) = \exp[-0.0166t], \quad R_{ij}^{(3)}(t,2) = \exp[-0.0181t],$$
$$i = 3; \;\; j = 361,362.$$

Subsystem S_3 is a three-state series-"2 out of 3" system and according to (1.40–1.41) and (1.72–1.73) its multistate reliability function is given by

$$\boldsymbol{R}^{(3)}(t,\cdot) = [1, \boldsymbol{R}^{(3)}(t,1), \boldsymbol{R}^{(3)}(t,2)], \quad t \geq 0, \tag{1.118}$$

where

$$\begin{aligned}
\boldsymbol{R}^{(3)}(t,1)\bar{\boldsymbol{R}}^{2}_{3;362,362,362}(t,1) &= 1 - \sum_{\substack{r_1,r_2,r_3=0\\ r_1+r_2+r_3\leq 1}}^{1} \prod_{i=1}^{3} \left[\prod_{j=1}^{362} R_{ij}^{(3)}(t,1)\right]^{r_i} \left[1 - \prod_{j=1}^{362} R_{ij}^{(3)}(t,1)\right]^{1-r_i} \\
&= 1 - [1 - \exp[-[360\times 0.0059 + 2\times 0.0166]t]]^2 \\
&\quad \times [1 - \exp[-[360\times 0.0071 + 2\times 0.0166]t]]^1 \\
&\quad - 2\exp[-1[360\times 0.0059 + 2\times 0.0166]t] \\
&\quad \times [1 - \exp[-[360\times 0.0059 + 2\times 0.0166]t]]^1 \\
&\quad \times [1 - \exp[-[360\times 0.0071 + 2\times 0.0166]t]]^1 \\
&\quad - \exp[-[360\times 0.0071 + 2\times 0.0166]t] \\
&\quad \times [1 - \exp[-[360\times 0.0059 + 2\times 0.0166]t]]^2 \\
&= \exp[-4.3144t] + 2\exp[-4.7464t] - 2\exp[-6.9036t],
\end{aligned} \tag{1.119}$$

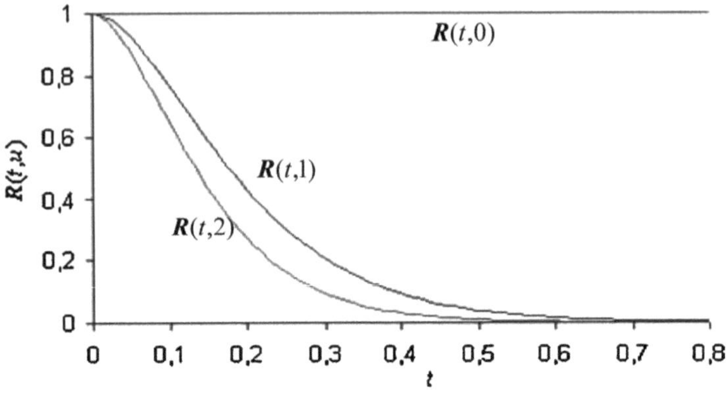

Fig. 1.17 The graph of the port oil piping transportation system reliability function $\boldsymbol{R}(t,\cdot)$ coordinates

$$\begin{aligned}
\boldsymbol{R}^{(3)}(t,2) &= \boldsymbol{R}^{2}_{3;362,362,362}(t,2) = 1 - \sum_{\substack{r_1,r_2,r_3=0\\ r_1+r_2+r_3\le 1}}^{1} \prod_{i=1}^{3}\left[\prod_{j=1}^{362} R^{(3)}_{ij}(t,2)\right]^{r_i}\left[1-\prod_{j=1}^{362} R^{(3)}_{ij}(t,2)\right]^{1-r_i}\\
&= 1-[1-\exp[-[360\times 0.0074+2\times 0.0181]t]]^2\\
&\quad \times [1-\exp[-[360\times 0.0079+2\times 0.0181]t]]^1\\
&\quad -2\exp[-1[360\times 0.0074+2\times 0.0181]t]\\
&\quad \times [1-\exp[-[360\times 0.0074+2\times 0.0181]t]]^1\\
&\quad \times [1-\exp[-[360\times 0.0079+2\times 0.0181]t]]^1\\
&\quad -\exp[-[360\times 0.0079+2\times 0.0181]t]\\
&\quad \times [1-\exp[-[360\times 0.0074+2\times 0.0181]t]]^2\\
&= \exp[-5.4004t]+2\exp[-5.5804t]-2\exp[-8.2806t]. \qquad (1.120)
\end{aligned}$$

Considering that the pipeline system is a three-state series system, after applying (1.22–1.23), its reliability function is given by

$$\boldsymbol{R}(t,\cdot) = [1,\ \boldsymbol{R}(t,1)R(t,2)], \quad t \ge 0, \qquad (1.121)$$

where by (1.112–1.114), (1.115–1.117) and (1.118–1.120) we have

$$\begin{aligned}
\boldsymbol{R}(t,1) &= \overline{\boldsymbol{R}}_3(t,1) = \boldsymbol{R}^{(1)}(t,1)\boldsymbol{R}^{(2)}(t,1)\boldsymbol{R}^{(3)}(t,1)\\
&= 4\exp[-9.9176t] + 8\exp[-10.3496t] - 8\exp[-12.5078t]\\
&\quad - 2\exp[-14.396t] - 4\exp[-14.8282t] + 4\exp[-16.9864t]\\
&\quad - 2\exp[-11.0422t] - 4\exp[-11,4742t] + 4\exp[-13.6324t]\\
&\quad + \exp[-15.5208t] + 2\exp[-15.9528t] - 2\exp[-18.111t], \qquad (1.122)
\end{aligned}$$

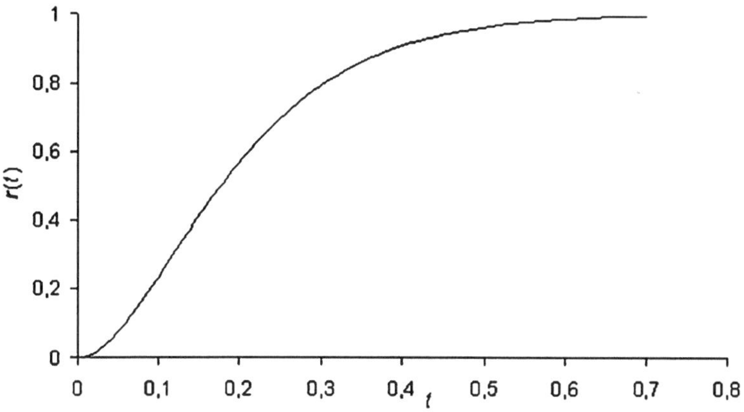

Fig. 1.18 The graph of the risk function **r**(t) of the port oil piping transportation system

$$\begin{aligned}\boldsymbol{R}(t,2) = \overline{\boldsymbol{R}}_3(t,2) = \boldsymbol{R}^{(1)}(t,2)\boldsymbol{R}^{(2)}(t,2)\boldsymbol{R}^{(3)}(t,2) \\ = 4\exp[-13.3314t] + 8\exp[-13.5114t] - 8\exp[-16.2116t] \\ - 2\exp[-19.6772t] - 4\exp[-19.8572t] + 4\exp[-22.5574t] \\ - 2\exp[-14.9166t] - 4\exp[-15,0966t] + 4\exp[-17.7968t] \\ + \exp[-21.2624t] + 2\exp[-21.4424t] - 2\exp[-24.1426t]. \qquad (1.123)\end{aligned}$$

The graph of the three-state piping system reliability function is shown in Fig. 1.17.

The expected values and standard deviations of the pipeline system lifetimes in the reliability state subsets $\{1,2\}, \{2\}$, calculated from the results given by (1.121–1.123), according to (1.15–1.17), respectively are:

$$\mu(1) \cong 0.207, \quad \mu(2) \cong 0.156 \text{ year}, \qquad (1.124)$$

$$\sigma(1) \cong 0.137, \quad \sigma(2) \cong 0.104 \text{ year}, \qquad (1.125)$$

and further, using (1.124), by (1.19), it follows that the mean values of the piping lifetimes in the particular reliability states are:

$$\overline{\mu}(1) \cong 0.051, \quad \overline{\mu}(2) \cong 0.156 \text{ year}. \qquad (1.126)$$

As the critical reliability state is $r = 1$, then the pipeline system risk function, according to (1.20), is given by

$$\begin{aligned}\boldsymbol{r}(t) = 1 - \boldsymbol{R}(t,1) \\ = 1 - [4\exp[-9.9176t] + 8\exp[-10.3496t] - 8\exp[-12.5078t] \\ - 2\exp[-14.396t] - 4\exp[-14.8282t] + 4\exp[-16.9864t] \\ - 2\exp[-11.0422t] - 4\exp[-11,4742t] + 4\exp[-13.6324t] \\ + \exp[-15.5208t] + 2\exp[-15.9528t] - 2\exp[-18.111t]] \quad \text{for } t \geq 0. \\ (1.127)\end{aligned}$$

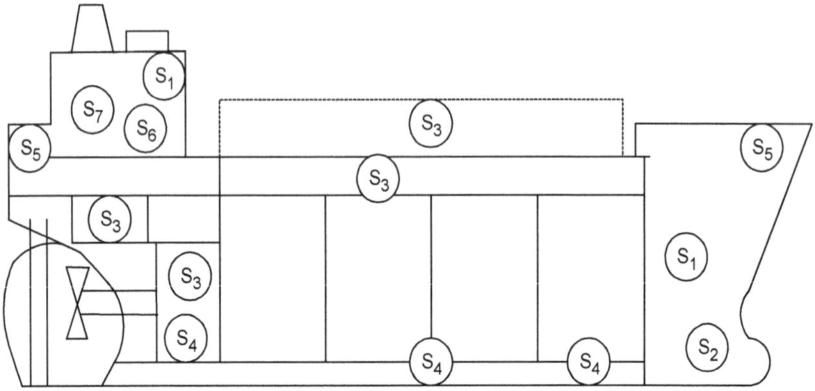

Fig. 1.19 Subsystems having an essential influence on the ferry safety

Hence, and from (1.21), the moment when the piping system risk function exceeds a permitted level, for instance $\delta = 0.05$, is

$$\tau = \boldsymbol{r}^{-1}(\delta) \cong 0.04. \tag{1.128}$$

The graph of the risk function $\mathbf{r}(t)$ of the three-state pipeline system is shown in Fig. 1.18.

1.4.3 Safety of Maritime Ferry Technical System

The considered maritime ferry is a passenger Ro-Ro ship operating in the Baltic Sea between Gdynia and Karlskrona ports on a regular everyday line. We assume that the ferry is composed of a number of main subsystems having an essential influence on its safety. These subsystems are illustrated in Figs. 1.19 and 1.20.

On the scheme of the ferry presented in Fig. 1.19 are distinguished its following subsystems:

S_1—a navigational subsystem,
S_2—a propulsion and controlling subsystem,
S_3—a loading and unloading subsystem,
S_4—a stability control subsystem,
S_5—an anchoring and mooring subsystem,
S_6—a protection and rescue subsystem,
S_7—a social subsystem.

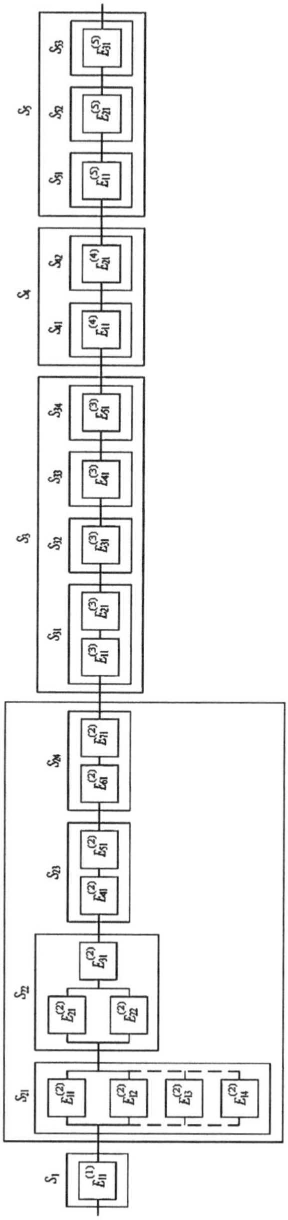

Fig. 1.20 The detailed scheme of the ferry technical system structure

In the safety analysis of the ferry, we omit the protection and rescue subsystem S_6 and the social subsystem S_7 and we consider its strictly technical subsystems S_1, S_2, S_3, S_4 and S_5 only, further called the ferry technical system.

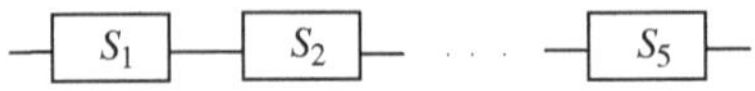

Fig. 1.21 The general scheme of the ferry technical system safety structure

The navigational subsystem S_1 is composed of one general component $E_{11}^{(1)}$, which is equipped with GPS, AIS, speed log, gyrocompass, magnetic compass, echo sounding system, paper and electronic charts, radar, ARPA, communication system and other subsystems.

The propulsion and controlling subsystem S_2 is composed of:

- subsystem S_{21} which consists of four main engines $E_{11}^{(2)}, E_{12}^{(2)}, E_{13}^{(2)}, E_{14}^{(2)}$;
- subsystem S_{22} which consists of three thrusters $E_{21}^{(2)}, E_{22}^{(2)}, E_{31}^{(2)}$;
- subsystem S_{23} which consists of twin pitch propellers $E_{41}^{(2)}, E_{51}^{(2)}$;
- subsystem S_{24} which consists of twin directional rudders $E_{61}^{(2)}, E_{71}^{(2)}$.

The loading and unloading subsystem S_3 is composed of:

- subsystem S_{31} which consists of two remote upper trailer decks to main deck $E_{11}^{(3)}, E_{21}^{(3)}$;
- subsystem S_{32} which consists of one remote fore car deck to main deck $E_{31}^{(3)}$;
- subsystem S_{33} which consists of passenger gangway to Gdynia Terminal $E_{41}^{(3)}$;
- subsystem S_{34} which consists of passenger gangway to Karlskrona Terminal $E_{51.}^{(3)}$.

The stability control subsystem S_4 is composed of:

- subsystem S_{41} which consists of an anti-heeling system $E_{11}^{(4)}$, which is used in port during loading operations;
- subsystem S_{42} which consists of an anti-heeling system $E_{21}^{(4)}$, which is used at sea for stabilizing ships rolling.

The anchoring and mooring subsystem S_5 is composed of:

- subsystem S_{51} which consists of aft mooring winches $E_{11}^{(5)}$;
- subsystem S_{52} which consists of fore mooring and anchor winches $E_{21}^{(5)}$;
- subsystem S_{53} which consists of fore mooring winches $E_{31}^{(5)}$.

The detailed scheme of these subsystems and components is illustrated in Fig. 1.20.

The subsystems S_1, S_2, S_3, S_4, S_5, indicated in Fig. 1.20 form a general series safety structure of the ferry technical system presented in Fig. 1.21.

After discussion with experts, taking into account the safety of the operation of the ferry, we distinguish the following five safety states $(z = 4)$ of the ferry technical system and its components:

- a safety state 4—the ferry operation is fully safe,
- a safety state 3—the ferry operation is less safe and more dangerous because of the possibility of environment pollution,
- a safety state 2—the ferry operation is less safe and more dangerous because of the possibility of environment pollution and causing small accidents,
- a safety state 1—the ferry operation is much less safe and much more dangerous because of the possibility of serious environment pollution and causing extensive accidents,
- a safety state 0—the ferry technical system is destroyed.

Moreover, based on expert opinion, we assume that transitions are possible between the components' safety states only from better to worse ones and we assume that the system and its components critical safety state is $r = 2$.

From the above, subsystems $S_\upsilon, \upsilon = 1, 2, 3, 4, 5$, are composed of five-state, i.e. $z = 4$, components $E_{ij}^{(\upsilon)}, \upsilon = 1, 2, 3, 4, 5$, having the safety functions

$$s_{ij}^{(\upsilon)}(t, \cdot) = [1, s_{ij}^{(\upsilon)}(t, 1), s_{ij}^{(\upsilon)}(t, 2), s_{ij}^{(\upsilon)}(t, 3), s_{ij}^{(\upsilon)}(t, 4)],$$

with coordinates which by assumption are exponentials of the form

$$s_{ij}^{(\upsilon)}(t, 1) = \exp[-\lambda_{ij}^{(\upsilon)}(1)t], \quad s_{ij}^{(\upsilon)}(t, 2) = \exp[-\lambda_{ij}^{(\upsilon)}(2)t],$$
$$s_{ij}^{(\upsilon)}(t, 3) = \exp[-\lambda_{ij}^{(\upsilon)}(3)t], \quad s_{ij}^{(\upsilon)}(t, 4) = \exp[-\lambda_{ij}^{(\upsilon)}(4)t].$$

Thus, the system is a five-state series system composed of five five-state subsystems S_1, S_2, S_3, S_4, S_5 with safety structures presented in Fig. 1.21.

The subsystem S_1 consists of one component $E_{ij}^{(1)}, i = 1; j = 1$, i.e. we may consider it either as a series system composed of $n = 1$ components or for instance as a parallel-series system with parameters $k = 1, l_1 = 1$, with the exponential safety functions partly identified in Chap. 5 on the basis of data coming from experts and are given below. The coordinates of the subsystem S_1 component five-state safety function are:

$$s_{11}^{(1)}(t, 1) = \exp[-0.033t], \quad s_{11}^{(1)}(t, 2) = \exp[-0.04t], \quad s_{11}^{(1)}(t, 3) = \exp[-0.045t],$$
$$s_{11}^{(1)}(t, 4) = \exp[-0.05t].$$

Thus, the subsystem S_1 safety function is identical with the safety function of its component, i.e.

$$\boldsymbol{s}^{(1)}(t, \cdot) = [1, \boldsymbol{s}^{(1)}(t, 1), \boldsymbol{s}^{(1)}(t, 2), \boldsymbol{s}^{(1)}(t, 3), \boldsymbol{s}^{(1)}(t, 4)], \quad t \in \langle 0, \infty), \qquad (1.129)$$

where, also according to the formulae (1.38–1.39) and (1.70–1.71), after replacing in them the components' reliability functions with their safety functions, we have

$$\boldsymbol{s}^{(1)}(t, u) = \boldsymbol{s}_{1;1}(t, u)$$
$$= \prod_{i=1}^{1}\left[1 - \prod_{j=1}^{1}\left[1 - s_{ij}^{(1)}(t, u)\right]\right] = s_{11}^{(1)}(t, u), \quad t \in \langle 0, \infty), \quad u = 1, 2, 3, 4, \qquad (1.130)$$

and particularly

$$\mathbf{s}^{(1)}(t,1) = \mathbf{s}_{1;1}(t,1) = \exp[-0.033t], \tag{1.131}$$

$$\mathbf{s}^{(1)}(t,2) = \mathbf{s}_{1;1}(t,2) = \exp[-0.04t], \tag{1.132}$$

$$\mathbf{s}^{(1)}(t,3) = \mathbf{s}_{1;1}(t,3) = \exp[-0.045t], \tag{1.133}$$

$$\mathbf{s}^{(1)}(t,4) = \mathbf{s}_{1;1}(t,4) = \exp[-0.05t]. \tag{1.134}$$

Subsystem S_2 is a five-state parallel-series system composed of components $E_{ij}^{(2)}, i = 1,2,\ldots,k; j = 1,2,\ldots,l_i; k = 7, l_1 = 4, l_2 = 2, l_3 = 1, l_4 = 1, l_5 = 1, l_6 = 1, l_7 = 1$, with the exponential safety functions identified on the basis of data coming from experts and are given below. The coordinates of the subsystem S_2 components' five-state safety functions are:

$$s_{1j}^{(2)}(t,1) = \exp[-0.033t], \quad s_{1j}^{(2)}(t,2) = \exp[-0.04t], \quad s_{1j}^{(2)}(t,3) = \exp[-0.05t],$$
$$s_{1j}^{(2)}(t,4) = \exp[-0.055t], \quad j = 1,2,3,4,$$
$$s_{2j}^{(2)}(t,1) = \exp[-0.066t], \quad s_{2j}^{(2)}(t,2) = \exp[-0.07t], \quad s_{2j}^{(2)}(t,3) = \exp[-0.075t],$$
$$s_{2j}^{(2)}(t,4) = \exp[-0.08t], \quad j = 1,2,$$
$$s_{31}^{(2)}(t,1) = \exp[-0.066t], \quad s_{31}^{(2)}(t,2) = \exp[-0.07t], \quad s_{31}^{(2)}(t,3) = \exp[-0.075t],$$
$$s_{31}^{(2)}(t,4) = \exp[-0.08t],$$
$$s_{i1}^{(2)}(t,1) = \exp[-0.033t], \quad s_{i1}^{(2)}(t,2) = \exp[-0.04t], \quad s_{i1}^{(2)}(t,3) = \exp[-0.045t],$$
$$s_{i1}^{(2)}(t,4) = \exp[-0.05t], \quad i = 4,5,6,7.$$

Hence, according to the formulae (1.38–1.39) and (1.70–1.71), after replacing in them the components' reliability functions with their safety functions, the subsystem S_2 safety function is given by

$$\mathbf{s}^{(2)}(t,\cdot) = [1, \mathbf{s}^{(2)}(t,1), \mathbf{s}^{(2)}(t,2), \mathbf{s}^{(2)}(t,3), \mathbf{s}^{(2)}(t,4)], \quad t \in \langle 0, \infty), \tag{1.135}$$

where

$$\begin{aligned} \mathbf{s}^{(2)}(t,u) &= \mathbf{s}_{7;4,2,1,1,1,1,1}(t,u) \\ &= \prod_{i=1}^{7}\left[1 - \prod_{j=1}^{l_i}[1 - s_{ij}^{(2)}(t,u)]\right], \quad t \in \langle 0, \infty), \quad u = 1,2,3,4, \end{aligned} \tag{1.136}$$

and particularly

$$
\begin{aligned}
s^{(2)}(t,1) &= s_{7;4,2,1,1,1,1,1}(t,1) \\
&= [6[\exp[-0.033t]]^2[1-\exp[-0.033t]]^2 + 4[\exp[-0.033t]]^3[1-\exp[-0.033t]] \\
&\quad + [\exp[-0.033t]]^4][1-[1-\exp[-0.066t]]^2]\exp[-0.066t]\exp[-0.033t] \\
&\quad \times \exp[-0.033t]\exp[-0.033t]\exp[-0.033t] \\
&= 12\exp[-0.33t] + 8\exp[-0.429t] - 16\exp[-0.363t] - 3\exp[-0.462t]
\end{aligned} \tag{1.137}
$$

$$
\begin{aligned}
s^{(2)}(t,2) &= s_{7;4,2,1,1,1,1,1}(t,2) \\
&= [6[\exp[-0.04t]]^2[1-\exp[-0.04t]]^2 + 4[\exp[-0.04t]]^3[1-\exp[-0.04t]] \\
&\quad + [\exp[-0.04t]]^4][1-[1-\exp[-0.07t]]^2]\exp[-0.07t]\exp[-0.04t] \\
&\quad \times \exp[-0.04t]\exp[-0.04t]\exp[-0.04t] \\
&= 12\exp[-0.38t] + 8\exp[-0.49t] + 6\exp[-0.46t] - 16\exp[-0.42t] \\
&\quad - 6\exp[-0.45t] - 3\exp[-0.53t]
\end{aligned} \tag{1.138}
$$

$$
\begin{aligned}
s^{(2)}(t,3) &= s_{7;4,2,1,1,1,1,1}(t,3) \\
&= [6[\exp[-0.05t]]^2[1-\exp[-0.05t]]^2 + 4[\exp[-0.05t]]^3[1-\exp[-0.05t]] \\
&\quad + [\exp[-0.05t]]^4][1-[1-\exp[-0.075t]]^2]\exp[-0.075t]\exp[-0.045t] \\
&\quad \times \exp[-0.045t]\exp[-0.045t]\exp[-0.045t] \\
&= 12\exp[-0.43t] + 8\exp[-0.555t] + 6\exp[-0.53t] - 16\exp[-0.48t] \\
&\quad - 6\exp[-0.505t] - 3\exp[-0.605t]
\end{aligned} \tag{1.139}
$$

$$
\begin{aligned}
s^{(2)}(t,4) &= s_{7;4,2,1,1,1,1,1}(t,4) \\
&= [6[\exp[-0.055t]]^2[1-\exp[-0.055t]]^2 + 4[\exp[-0.055t]]^3[1-\exp[-0.055t]] \\
&\quad + [\exp[-0.055t]]^4][1-[1-\exp[-0.08t]]^2]\exp[-0.08t]\exp[-0.05t] \\
&\quad \times \exp[-0.05t]\exp[-0.05t]\exp[-0.05t] \\
&= 12\exp[-0.47t] + 8\exp[-0.605t] + 6\exp[-0.58t] - 16\exp[-0.525t] \\
&\quad - 6\exp[-0.55t] - 3\exp[-0.66t].
\end{aligned} \tag{1.140}
$$

Subsystem S_3 is a five-state series system composed of $n = 5$ components which can also be considered as a parallel-series system composed of components $E_{ij}^{(3)}, i = 1, 2, \ldots, k; j = l_i; k = 5, l_1 = 1, l_2 = 1, l_3 = 1, l_4 = 1, l_5 = 1,$ with the exponential safety functions identified on the basis of data coming from experts and are given below. The coordinates of subsystem S_3 components' five-state safety functions are:

$$s_{11}^{(3)}(t,1) = \exp[-0.02t], \quad s_{11}^{(3)}(t,2) = \exp[-0.03t], \quad s_{11}^{(3)}(t,3) = \exp[-0.035t],$$
$$s_{11}^{(3)}(t,4) = \exp[-0.04t],$$
$$s_{21}^{(3)}(t,1) = \exp[-0.02t], \quad s_{21}^{(3)}(t,2) = \exp[-0.025t], \quad s_{21}^{(3)}(t,3) = \exp[-0.03t],$$
$$s_{21}^{(3)}(t,4) = \exp[-0.04t],$$
$$s_{31}^{(3)}(t,1) = \exp[-0.033t], \quad s_{31}^{(3)}(t,2) = \exp[-0.04t], \quad s_{31}^{(3)}(t,3) = \exp[-0.045t],$$
$$s_{31}^{(3)}(t,4) = \exp[-0.05t],$$
$$s_{41}^{(3)}(t,1) = \exp[-0.033t], \quad s_{41}^{(3)}(t,2) = \exp[-0.04t], \quad s_{41}^{(3)}(t,3) = \exp[-0.045t],$$
$$s_{41}^{(3)}(t,4) = \exp[-0.05t],$$
$$s_{51}^{(3)}(t,1) = \exp[-0.033t], \quad s_{51}^{(3)}(t,2) = \exp[-0.04t], \quad s_{51}^{(3)}(t,3) = \exp[-0.045t],$$
$$s_{51}^{(3)}(t,4) = \exp[-0.05t].$$

Hence, according to the formulae (1.38–1.39) and (1.70–1.71), after replacing in them the components' reliability functions with their safety functions, the subsystem S_3 five-state safety function is given by

$$\boldsymbol{s}^{(3)}(t,\cdot) = [1, \boldsymbol{s}^{(3)}(t,1), \boldsymbol{s}^{(3)}(t,2), \boldsymbol{s}^{(3)}(t,3), \boldsymbol{s}^{(3)}(t,4)], \quad t \in \langle 0, \infty), \tag{1.141}$$

where

$$\boldsymbol{s}^{(3)}(t,u) = \boldsymbol{s}_{5;1,1,1,1,1}(t,u)$$
$$= \prod_{i=1}^{5}\left[1 - \prod_{j=1}^{1}[1 - s_{ij}(t,u)]\right] = \prod_{i=1}^{5} s_{i1}(t,u), \quad t \in \langle 0, \infty), \quad u = 1,2,3,4, \tag{1.142}$$

and particularly

$$\begin{aligned}\boldsymbol{s}^{(3)}(t,1) &= \boldsymbol{s}_{5;1,1,1,1,1}(t,1)\\ &= \exp[-0.02t]\exp[-0.02t]\exp[-0.033t]\exp[-0.033]\exp[-0.033]\\ &= \exp[-0.139t],\end{aligned} \tag{1.143}$$

$$\begin{aligned}\boldsymbol{s}^{(3)}(t,2) &= \boldsymbol{s}_{5;1,1,1,1,1}(t,2)\\ &= \exp[-0.03t]\exp[-0.025t]\exp[-0.04t]\exp[-0.04t]\exp[-0.04t]\\ &= \exp[-0.175t],\end{aligned} \tag{1.144}$$

$$\begin{aligned}\boldsymbol{s}^{(3)}(t,3) &= \boldsymbol{s}_{5;1,1,1,1,1}(t,3)\\ &= \exp[-0.035t]\exp[-0.03t]\exp[-0.045t]\exp[-0.045t]\exp[-0.045t]\\ &= \exp[-0.200t], \end{aligned} \tag{1.145}$$

$$\begin{aligned}\boldsymbol{s}^{(3)}(t,4) &= \boldsymbol{s}_{5;1,1,1,1,1}(t,4)\\ &= \exp[-0.04t]\exp[-0.04t]\exp[-0.05t]\exp[-0.05t]\exp[-0.05t]\\ &= \exp[-0.230t], \end{aligned} \tag{1.146}$$

Subsystem S_4 is a five-state series system composed of $n = 2$ components that can also be considered as a parallel-series system composed of components $E_{ij}^{(4)}, i = 1,\ldots,k; j = l_i, k = 2, l_1 = 1, l_2 = 1$, with the exponential safety functions identified on the basis of data coming from experts and are given below. The coordinates of subsystem S_4 components' multistate safety functions are:

$$s_{11}^{(4)}(t,1) = \exp[-0.05t], \quad s_{11}^{(4)}(t,2) = \exp[-0.06t], \quad s_{11}^{(4)}(t,3) = \exp[-0.065t],$$
$$s_{11}^{(4)}(t,4) = \exp[-0.07t],$$
$$s_{21}^{(4)}(t,1) = \exp[-0.033t], \quad s_{21}^{(4)}(t,2) = \exp[-0.04t], \quad s_{21}^{(4)}(t,3) = \exp[-0.045t],$$
$$s_{21}^{(4)}(t,4) = \exp[-0.05t].$$

Hence, according to the formulae (1.38–1.39) and (1.70–1.71), after replacing in them the components' reliability functions with their safety functions, the subsystem S_4 five-state safety function is given by

$$\boldsymbol{s}^{(4)}(t,\cdot) = [1, \boldsymbol{s}^{(4)}(t,1),\ \boldsymbol{s}^{(4)}(t,2),\ \boldsymbol{s}^{(4)}(t,3),\ \boldsymbol{s}^{(4)}(t,4)], \quad t \in \langle 0, \infty), \tag{1.147}$$

where

$$\begin{aligned}\boldsymbol{s}^{(4)}(t,u) &= \boldsymbol{s}_{2;1,1}(t,u)\\ &= \prod_{i=1}^{2}\left[1 - \prod_{j=1}^{1}[1 - s_{ij}(t,u)]\right] = \prod_{i=1}^{2} s_{ij}(t,u), \quad t \in \langle 0,\infty), \quad u = 1,2,3,4,\end{aligned} \tag{1.148}$$

and particularly

$$\boldsymbol{s}^{(4)}(t,1) = \boldsymbol{s}_{2;1,1}(t,1) = \exp[-0.05t]\exp[-0.033t] = \exp[-0.083t], \tag{1.149}$$

$$\boldsymbol{s}^{(4)}(t,2) = \boldsymbol{s}_{2;1,1}(t,2) = \exp[-0.06t]\exp[-0.04t] = \exp[-0.100t], \tag{1.150}$$

$$s^{(4)}(t,3) = s_{2;1,1}(t,3) = \exp[-0.065t]\exp[-0.045t] = \exp[-0.110t], \quad (1.151)$$

$$s^{(4)}(t,4) = s_{2;1,1}(t,4) = \exp[-0.07t]\exp[-0.05t] = \exp[-0.120t]. \quad (1.152)$$

Subsystem S_5 is a five-state series system composed of $n = 3$ components that can also be considered as a parallel-series system composed of components $E_{ij}^{(5)}, i = 1,2,\ldots,k; j = l_i; k = 3, l_1 = 1, l_2 = 1, l_3 = 1$, with the exponential safety functions identified on the basis of data coming from experts and are given below. The coordinates of subsystem S_5 components' five-state safety functions are:

$$s_{11}^{(5)}(t,1) = \exp[-0.033t], \quad s_{11}^{(5)}(t,2) = \exp[-0.04t], \quad s_{11}^{(5)}(t,3) = \exp[-0.045t],$$
$$s_{11}^{(5)}(t,4) = \exp[-0.05t],$$
$$s_{21}^{(5)}(t,1) = \exp[-0.033t], \quad s_{21}^{(5)}(t,2) = \exp[-0.04t], \quad s_{21}^{(5)}(t,3) = \exp[-0.05t],$$
$$s_{21}^{(5)}(t,4) = \exp[-0.055t],$$
$$s_{31}^{(5)}(t,1) = \exp[-0.033t], \quad s_{31}^{(5)}(t,2) = \exp[-0.04t], \quad s_{31}^{(5)}(t,3) = \exp[-0.05t],$$
$$s_{31}^{(5)}(t,4) = \exp[-0.06t].$$

Hence, according to the formulae (1.38–1.39) and (1.70–1.71), after replacing in them the components' reliability functions with their safety functions, the subsystem S_5 five-state safety function is given by

$$s^{(5)}(t,\cdot) = [1, s^{(5)}(t,1), s^{(5)}(t,2), s^{(5)}(t,3), s^{(5)}(t,4)], \quad t \in \langle 0,\infty), \quad (1.153)$$

where

$$s^{(5)}(t,u) = s_{3;1,1,1}(t,u)$$
$$= \prod_{i=1}^{3}\left[1 - \prod_{j=1}^{1}[1 - s_{ij}(t,u)]\right] = \prod_{i=1}^{3} s_{i1}(t,u), t \in \langle 0,\infty), \quad u = 1,2,3,4, \quad (1.154)$$

and particularly

$$s^{(5)}(t,1) = s_{3;1,1,1}(t,1) = \exp[-0.033t]\exp[-0.033t]\exp[-0.033t]$$
$$= \exp[-0.099t], \quad (1.155)$$

$$s^{(5)}(t,2) = s_{3;1,1,1}(t,2) = \exp[-0.04t]\exp[-0.04t]\exp[-0.04t]$$
$$= \exp[-0.12t], \quad (1.156)$$

$$s^{(5)}(t,3) = s_{3;1,1,1}(t,3) = \exp[-0.045t]\exp[-0.05t]\exp[-0.05t]$$
$$= \exp[-0.145t], \quad (1.157)$$

$$s^{(5)}(t,4) = s_{3;1,1,1}(t,4) = \exp[-0.05t]\exp[-0.055t]\exp[-0.06t]$$
$$= \exp[-0.165t]. \tag{1.158}$$

Considering that the ferry technical system is a five-state series system, after applying (1.92–1.93), its safety function is given by

$$s(t,\cdot) = [1, s(t,1), s(t,2), s(t,3), s(t,4)], \quad t \geq 0, \tag{1.159}$$

where by (1.131–1.134), (1.137–1.140), (1.143–1.146), (1.149–1.152) and (1.155–1.158), we have

$$s(t,u) = \bar{s}_5(t,u) = s^{(1)}(t,u)s^{(2)}(t,u)s^{(3)}(t,u)s^{(4)}(t,u)s^{(5)}(t,u) \quad \text{for } u = 1,2,3,4,$$

and particularly

$$s(t,1) = \exp[-0.033t][12\exp[-0.33t] + 8\exp[-0.429t] - 16\exp[-0.363t]$$
$$- 3\exp[-0.462t]]\exp[-0.139t]\exp[-0.083t]\exp[-0.099t]$$
$$= 12\exp[-0.684t] + 8\exp[-0.783t] - 16\exp[-0.717t] - 3\exp[-0.816t], \tag{1.160}$$

$$s(t,2) = \exp[-0.040t][12\exp[-0.38t] + 8\exp[-0.49t] + 6\exp[-0.46t]$$
$$- 16\exp[-0.42t] - 6\exp[-0.45t] - 3\exp[-0.53t]]\exp[-0.175t]$$
$$\times \exp[-0.100t]\exp[-0.12t]$$
$$= 12\exp[-0.815t] + 8\exp[-0.925t] + 6\exp[-0.895t]$$
$$- 16\exp[-0.855t] - 6\exp[-0.885t] - 3\exp[-0.965t], \tag{1.161}$$

$$s(t,3) = \exp[-0.045t][12\exp[-0.43t] + 8\exp[-0.555t] + 6\exp[-0.53t]$$
$$- 16\exp[-0.48t] - 6\exp[-0.505t] - 3\exp[-0.605t]]\exp[-0.200t]$$
$$\times \exp[-0.110t]\exp[-0.145t]$$
$$= 12\exp[-0.930t] + 8\exp[-1.055t] + 6\exp[-1.030t]$$
$$- 16\exp[-0.980t] - 6\exp[-1.005t] - 3\exp[-1.105t], \tag{1.162}$$

$$s(t,4) = \exp[-0.05t][12\exp[-0.47t] + 8\exp[-0.605t] + 6\exp[-0.58t]$$
$$- 16\exp[-0.525t] - 6\exp[-0.55t] - 3\exp[-0.66t]]\exp[-0.230t]$$
$$\times \exp[-0.120t]\exp[-0.165t]$$
$$= 12\exp[-1.035t] + 8\exp[-1.170t] + 6\exp[-1.145t]$$
$$- 16\exp[-1.090t] - 6\exp[-1.115t] - 3\exp[-1.225]. \tag{1.163}$$

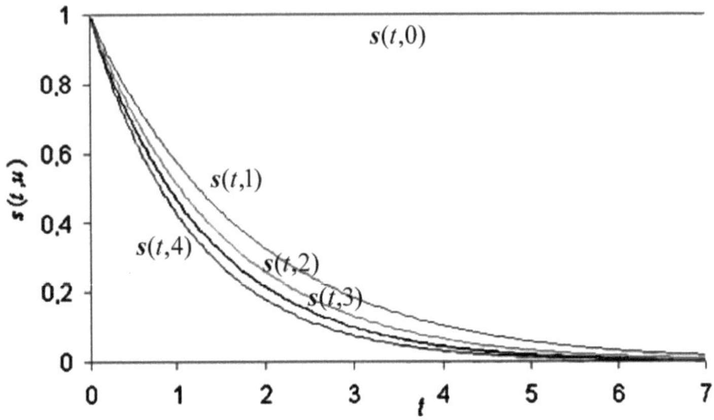

Fig. 1.22 The graph of the ferry technical system safety function $s(t, \cdot)$ coordinates

The safety function of the ferry five-state technical system is presented in Fig. 1.22.

The expected values and standard deviations of the ferry technical system lifetimes in the safety state subsets calculated from the results given by (1.160–1.163), according to the formulae (1.15–1.17), after replacing in them the system reliability function with its safety functions, are:

$$\mu(1) \cong 1.770, \quad \mu(2) \cong 1.476, \quad \mu(3) \cong 1.300, \quad \mu(4) \cong 1.164 \text{ year}, \quad (1.164)$$

$$\sigma(1) \cong 1.733, \quad \sigma(2) \cong 1.447, \quad \sigma(3) \cong 1.277, \quad \sigma(4) \cong 1.144 \text{ year}, \quad (1.165)$$

and further, using (1.164), from (1.19), the mean values of the ferry technical system conditional lifetimes in the particular safety states are:

$$\bar{\mu}(1) \cong 0.294, \quad \bar{\mu}(2) \cong 0.176, \quad \bar{\mu}(3) \cong 0.136, \quad \bar{\mu}(4) \cong 1.164 \text{ year}. \quad (1.166)$$

As the critical safety state is $r = 2$, then the system risk function, according to (1.90), is given by

$$\begin{aligned} \boldsymbol{r}(t) &= 1 - \boldsymbol{s}(t,2) \\ &= 1 - [12\exp[-0.815t] + 8\exp[-0.925t] + 6\exp[-0.895t] \\ &\quad - 16\exp[-0.855t] - 6\exp[-0.885t] - 3\exp[-0.965t]], \quad \text{for } t \geq 0. \end{aligned} \quad (1.167)$$

Hence, the moment when the system risk function exceeds a permitted level, for instance $\delta = 0.05$, by (1.91), is

$$\tau = \boldsymbol{r}^{-1}(\delta) \cong 0.077. \quad (1.168)$$

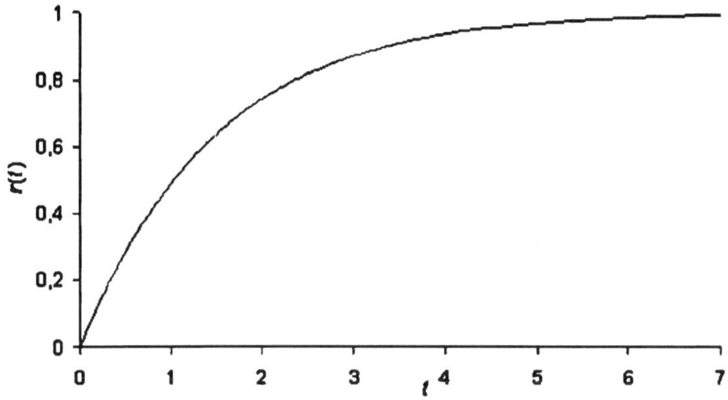

Fig. 1.23 The graph of the risk function $\boldsymbol{r}(t)$ of the ferry technical system

The graph of the risk function $r(t)$ of the ferry five-state technical system is presented in Fig. 1.23.

1.5 Summary

The proposed models in this chapter for reliability and safety evaluation and prediction of the considered systems are the bases for the considerations in further chapters of the book. These system reliability and safety models, together with the models of the system operation process presented in Chap. 2, are used in Chap. 3 for constructing the integrated joint general reliability, availability and safety models of complex technical systems related to their operation processes. The methods of estimation of these models' unknown parameters where evaluations are used in the appliqué part of this chapter are given in Chap. 5. The models applied here, in their particular cases, for the reliability and respectively safety analysis and prediction of the exemplary technical system, the port transportation system and the maritime transportation technical system operating in constant operation conditions will also be applied in Chap. 3 to reliability, availability and safety analysis and prediction of these systems operating under variable operation conditions.

References

1. Amari SV, Misra RB (1997) Comment on: dynamic reliability analysis of coherent multistate systems. IEEE Trans Reliab 46:460–461
2. Aven T (1985) Reliability evaluation of multistate systems with multistate components. IEEE Trans Reliab 34:473–479

3. Aven T (1993) On performance measures for multistate monotone systems. Reliab Eng and Sys Saf 41:259–266
4. Aven T, Jensen U (1999) Stochastic models in reliability. Springer, New York
5. Barlow RE, Wu AS (1978) Coherent systems with multi-state components. Math Oper Res 4:275–281
6. Brunelle RD, Kapur KC (1999) Review and classification of reliability measures for multistate and continuum models. IEEE Trans 31:1117–1180
7. Guze S, Kołowrocki K (2008) Reliability analysis of multi-state ageing consecutive "k out of n: F" systems. Int J Mater Struct Reliab 6(1):47–60
8. Huang J, Zuo MJ, Wu Y (2000) Generalized multi-state k-out-of-n:G systems. IEEE Trans Reliab 49:105–111
9. Hudson JC, Kapur KC (1982) Reliability theory for multistate systems with multistate components. Microelectron Reliab 22:1–7
10. Hudson JC, Kapur KC (1983) Reliability analysis of multistate systems with multistate components. Trans Inst Ind Eng 15:127–135
11. Hudson J, Kapur K (1985) Reliability bounds for multistate systems with multistate components. Oper Res 33:735–744
12. Kołowrocki K (1998) On applications of asymptotic reliability functions to the reliability and risk evaluation of pipelines. Int J Press Vessel Pip 75:545–558
13. Kołowrocki K (2003) An asymptotic approach to reliability evaluation of large multi-state systems with applications to piping transportation systems. Int J Press Vessel Pip 80:59–73
14. Kołowrocki K (2004) Reliability of large systems. Elsevier, Amsterdam
15. Kołowrocki K (2008) Reliability and risk analysis of multi-state systems with degrading components. Summer safety & reliability seminars. J Pol Saf Reliab Assoc 2(2):205–216
16. Kołowrocki K (2008) Reliability of large systems. In: Encyclopedia of quantitative risk analysis and assessment, vol 4. Wiley, New York
17. Kołowrocki K, Soszyńska J (2010) Safety and risk evaluation of a ferry technical system. Summer safety & reliability seminars. J Pol Saf Reliab Assoc 4 (1):159–172
18. Kołowrocki K, Soszyńska J, Judziński M, Dziula P (2007) On multi-state safety analysis in shipping. Int J Reliab Qual Saf Eng Sys Reliab Saf 14(6):547–567
19. Kossow A, Preuss W (1995) Reliability of linear consecutively-connected systems with multistate components. IEEE Trans Reliab 44:518–522
20. Lisnianski A, Levitin G (2003) Multi-State System Reliability. Assessment, Optimisation and Applications. World Scientific Publishing Co. Pte. Ltd., London
21. Malinowski J (2005) Algorithms for reliability evaluation of various kind of network systems. WIT, Warsaw (in Polish)
22. Meng F (1993) Component-relevancy and characterisation in multi-state systems. IEEE Trans Reliab 42:478–483
23. Natvig B (1982) Two suggestions of how to define a multi-state coherent system. Adv Appl Probab 14:434–455
24. Natvig B (1984) Multi-state coherent systems. In: Encyclopaedia of statistical sciences. Wiley, New York
25. Natvig B, Streller A (1984) The steady-state behaviour of multistate monotone systems. J Appl Probab 21:826–835
26. Ohio F, Nishida T (1984) On multistate coherent systems. IEEE Trans Reliab 33: 284–287
27. Xue J (1985) On multi-state system analysis. IEEE Trans Reliab 34:329–337
28. Xue J, Yang K (1995) Dynamic reliability analysis of coherent multi-state systems. IEEE Trans Reliab 4(44):683–688
29. Xue J, Yang K (1995) Symmetric relations in multi-state systems. IEEE Trans Reliab 4(44):689–693
30. Yu K, Koren I, Guo Y (1994) Generalised multistate monotone coherent systems. IEEE Trans Reliab 43:242–250

Chapter 2
Modeling Complex Technical Systems Operation Processes

2.1 Introduction

The operation processes of real technical systems are very complex and it is difficult to analyze these systems' reliability, safety and availability with respect to their operation conditions changing in time that are often essential in this analysis. The complexity of the systems' operation processes and their influence on changing in time the systems' structures and their components' reliability and safety characteristics is often very difficult to fix. Usually, the system environment and infrastructure have either an explicit or an implicit strong influence on the system operation process. As a rule, some of the initiating environment events and infrastructure conditions define a set of different operation states of the technical systems in which the systems change their reliability and safety structure and their components reliability and safety parameters. A convenient tool for analyzing this problem is semi-Markov modeling [1–6, 10, 12] of the systems operation processes proposed in this chapter.

2.2 Semi-Markov Model of System Operation Process

We assume that the system during its operation process takes $v, v \in N$, different operation states $z_1, z_2, \ldots, z_v$. Further, we define the system operation process $Z(t)$, $t \in \langle 0, +\infty)$, with discrete operation states from the set $\{z_1, z_2, \ldots, z_v\}$. Moreover, we assume that the system operation process $Z(t)$ is a semi-Markov process [3, 5, 6, 10, 12] with the conditional sojourn times θ_{bl} at the operation states z_b when its next operation state is z_l, $b, l = 1, 2, \ldots, v$, $b \neq l$. Under these assumptions, the system operation process may be described by:

K. Kołowrocki and J. Soszyńska-Budny, *Reliability and Safety of Complex Technical Systems and Processes*, Springer Series in Reliability Engineering, DOI: 10.1007/978-0-85729-694-8_2, © Springer-Verlag London Limited 2011

- the vector of the initial probabilities $p_b(0) = P(Z(0) = z_b)$, $b = 1, 2, \ldots, v$, of the system operation process $Z(t)$ stay at particular operation states at the moment $t = 0$

$$[p_b(0)]_{1\times v} = [p_1(0), p_2(0), \ldots, p_v(0)]; \tag{2.1}$$

- the matrix of probabilities p_{bl}, b, $l = 1, 2, \ldots, v$, $b \neq l$, of the system operation process $Z(t)$ transitions between the operation states z_b and z_l

$$[p_{bl}]_{v\times v} = \begin{bmatrix} p_{11}\ p_{12} \cdots p_{1v} \\ p_{21}\ p_{22} \cdots p_{2v} \\ \cdots \\ p_{v1}\ p_{v2} \cdots p_{vv} \end{bmatrix}, \tag{2.2}$$

where by formal agreement

$$p_{bb} = 0 \quad \text{for } b = 1, 2, \ldots, v;$$

- the matrix of conditional distribution functions $H_{bl}(t) = P(\theta_{bl} < t)$, $b, l = 1, 2, \ldots, v$, $b \neq l$, of the system operation process $Z(t)$ conditional sojourn times θ_{bl} at the operation states

$$[H_{bl}(t)]_{v\times v} = \begin{bmatrix} H_{11}(t)\ H_{12}(t) \ldots H_{1v}(t) \\ H_{21}(t)\ H_{22}(t) \ldots H_{2v}(t) \\ \ldots \\ H_{v1}(t)\ H_{v2}(t) \ldots H_{vv}(t) \end{bmatrix}, \tag{2.3}$$

where by formal agreement

$$H_{bb}(t) = 0 \quad \text{for } b = 1, 2, \ldots, v.$$

We introduce the matrix of the conditional density functions of the system operation process $Z(t)$ conditional sojourn times θ_{bl} at the operation states corresponding to the conditional distribution functions $H_{bl}(t)$

$$[h_{bl}(t)]_{v\times v} = \begin{bmatrix} h_{11}(t)\ h_{12}(t) \ldots h_{1v}(t) \\ h_{21}(t)\ h_{22}(t) \ldots h_{2v}(t) \\ \ldots \\ h_{v1}(t)\ h_{v2}(t) \ldots h_{vv}(t) \end{bmatrix}, \tag{2.4}$$

where

$$h_{bl}(t) = \frac{\mathrm{d}}{\mathrm{d}t}[H_{bl}(t)] \quad \text{for } b, l = 1, 2, \ldots, v,\ b \neq l,$$

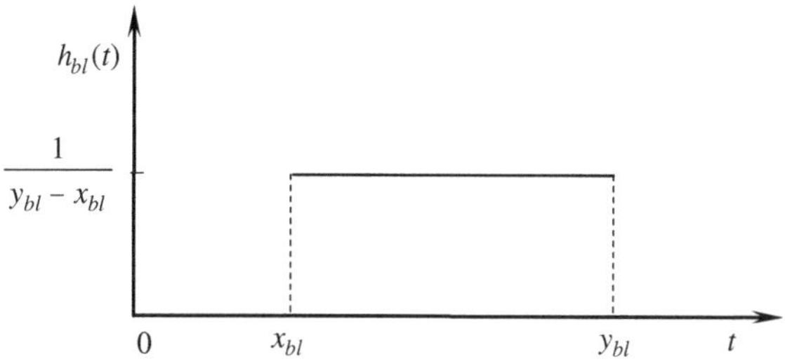

Fig. 2.1 The graph of the uniform distribution's density function

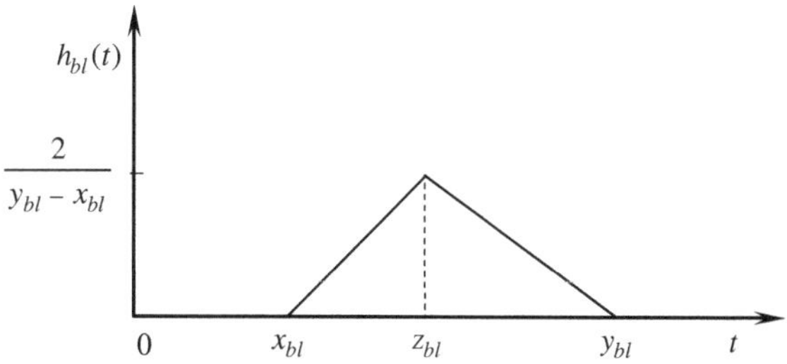

Fig. 2.2 The graph of the triangular distribution's density function

and by formal agreement

$$h_{bb}(t) = 0 \quad \text{for } b = 1, 2, \ldots, v.$$

We assume that the suitable and typical distributions suitable to describe the system operation process $Z(t)$ conditional sojourn times θ_{bl}, $b, l = 1, 2, \ldots, v$, $b \neq l$, in the particular operation states are [9]:

- the uniform distribution with a density function (Fig. 2.1)

$$h_{bl}(t) = \begin{cases} 0, & t < x_{bl} \\ \dfrac{1}{y_{bl} - x_{bl}}, & x_{bl} \leq t \leq y_{bl} \\ 0, & t > y_{bl}, \end{cases} \tag{2.5}$$

where

$$0 \leq x_{bl} < y_{bl} < +\infty;$$

- the triangular distribution with a density function (Fig. 2.2)

$$h_{bl}(t) = \begin{cases} 0, & t < x_{bl} \\ \dfrac{2}{y_{bl} - x_{bl}} \dfrac{t - x_{bl}}{z_{bl} - x_{bl}}, & x_{bl} \leq t \leq z_{bl} \\ \dfrac{2}{y_{bl} - x_{bl}} \dfrac{y_{bl} - t}{y_{bl} - z_{bl}}, & z_{bl} \leq t \leq y_{bl} \\ 0, & t > y_{bl}, \end{cases} \tag{2.6}$$

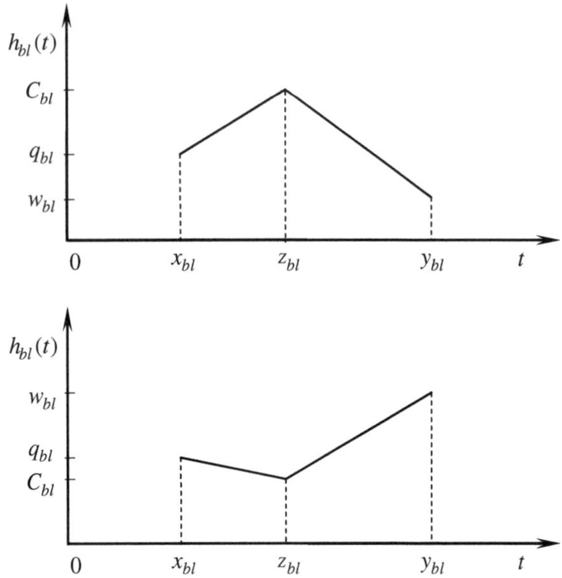

Fig. 2.3 The graphs of the double trapezium distribution's density functions

where

$$0 \le x_{bl} \le z_{bl} \le y_{bl} < +\infty;$$

- the double trapezium distribution with a density function (Fig. 2.3)

$$h_{bl}(t) = \begin{cases} 0, & t < x_{bl} \\ q_{bl} + \left[\dfrac{2 - q_{bl}(z_{bl} - x_{bl}) - w_{bl}(y_{bl} - z_{bl})}{y_{bl} - x_{bl}} - q_{bl}\right]\dfrac{t - x_{bl}}{z_{bl} - x_{bl}}, & x_{bl} \le t \le z_{bl} \\ w_{bl} + \left[\dfrac{2 - q_{bl}(z_{bl} - x_{bl}) - w_{bl}(y_{bl} - z_{bl})}{y_{bl} - x_{bl}} - w_{bl}\right]\dfrac{y_{bl} - t}{y_{bl} - z_{bl}}, & z_{bl} \le t \le y_{bl} \\ 0, & t > y_{bl}, \end{cases} \tag{2.7}$$

where

$$0 \le x_{bl} \le z_{bl} \le y_{bl} < +\infty, \quad 0 \le q_{bl} < +\infty, \quad 0 \le w_{bl} < +\infty,$$

$$0 \le q_{bl}(z_{bl} - x_{bl}) + w_{bl}(y_{bl} - z_{bl}) \le 2;$$

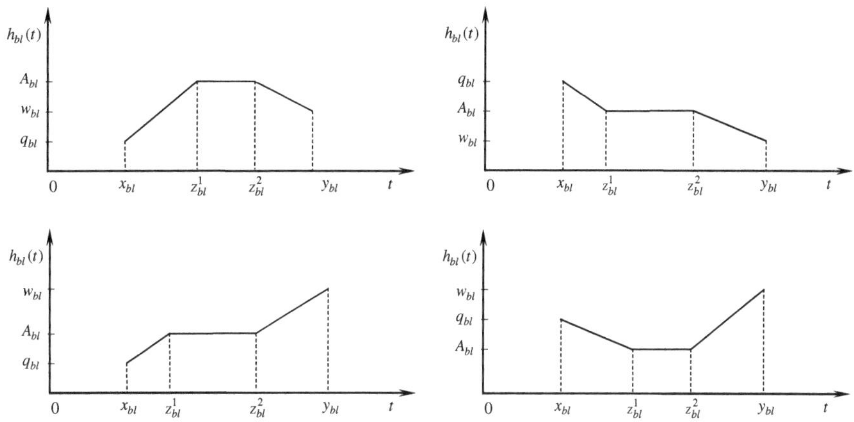

Fig. 2.4 The graphs of the quasi-trapezium distribution's density functions

- the quasi-trapezium distribution with a density function (Fig. 2.4)

$$h_{bl}(t) = \begin{cases} 0, & t < x_{bl} \\ q_{bl} + \dfrac{A_{bl} - q_{bl}}{z_{bl}^1 - x_{bl}}(t - x_{bl}), & x_{bl} \le t \le z_{bl}^1 \\ A_{bl}, & z_{bl}^1 \le t \le z_{bl}^2 \\ w_{bl} + \dfrac{A_{bl} - w_{bl}}{y_{bl} - z_{bl}^2}(y_{bl} - t), & z_{bl}^2 \le t \le y_{bl} \\ 0, & t > y_{bl}, \end{cases} \tag{2.8}$$

where

$$A_{bl} = \frac{2 - q_{bl}(z_{bl}^1 - x_{bl}) - w_{bl}(y_{bl} - z_{bl}^2)}{z_{bl}^2 - z_{bl}^1 + y_{bl} - x_{bl}},$$

Fig. 2.5 The graph of the exponential distribution's density function

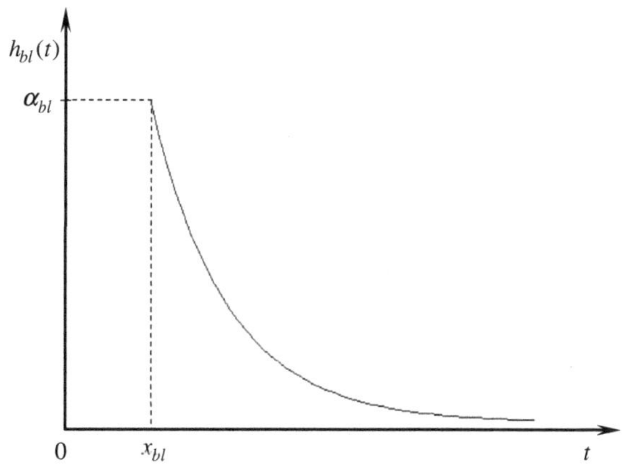

Fig. 2.6 The graphs of the Weibull distribution's density functions

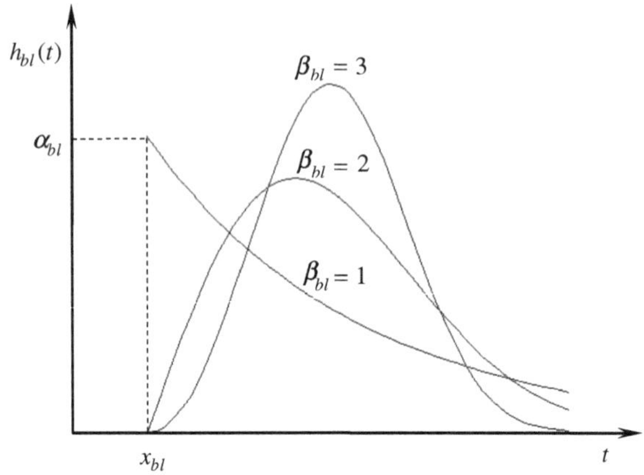

$$0 \le x_{bl} \le z_{bl}^1 \le z_{bl}^2 \le y_{bl} < +\infty, \quad 0 \le q_{bl} < +\infty, \quad 0 \le w_{bl} < +\infty;$$

- the exponential distribution with a density function (Fig. 2.5)

$$h_{bl}(t) = \begin{cases} 0, & t < x_{bl} \\ \alpha_{bl} \exp[-\alpha_{bl}(t - x_{bl})], & t \ge x_{bl}, \end{cases} \tag{2.9}$$

where

$$0 \le \alpha_{bl} < +\infty;$$

- the Weibull distribution with a density function (Fig. 2.6)

$$h_{bl}(t) = \begin{cases} 0, & t < x_{bl} \\ \alpha_{bl}\beta_{bl}(t - x_{bl})^{\beta_{bl}-1} \exp[-\alpha_{bl}(t - x_{bl})^{\beta_{bl}}], & t \ge x_{bl}, \end{cases} \tag{2.10}$$

where

$$0 \le \alpha_{bl} < +\infty, \quad 0 \le \beta_{bl} < +\infty;$$

- the chimney distribution with a density function (Fig. 2.7)

$$h_{bl}(t) = \begin{cases} 0, & t < x_{bl} \\ \dfrac{A_{bl}}{z_{bl}^1 - x_{bl}}, & x_{bl} \le t \le z_{bl}^1 \\ \dfrac{C_{bl}}{z_{bl}^2 - z_{bl}^1}, & z_{bl}^1 \le t \le z_{bl}^2 \\ \dfrac{D_{bl}}{y_{bl} - z_{bl}^2}, & z_{bl}^2 \le t \le y_{bl} \\ 0, & t > y_{bl}, \end{cases} \tag{2.11}$$

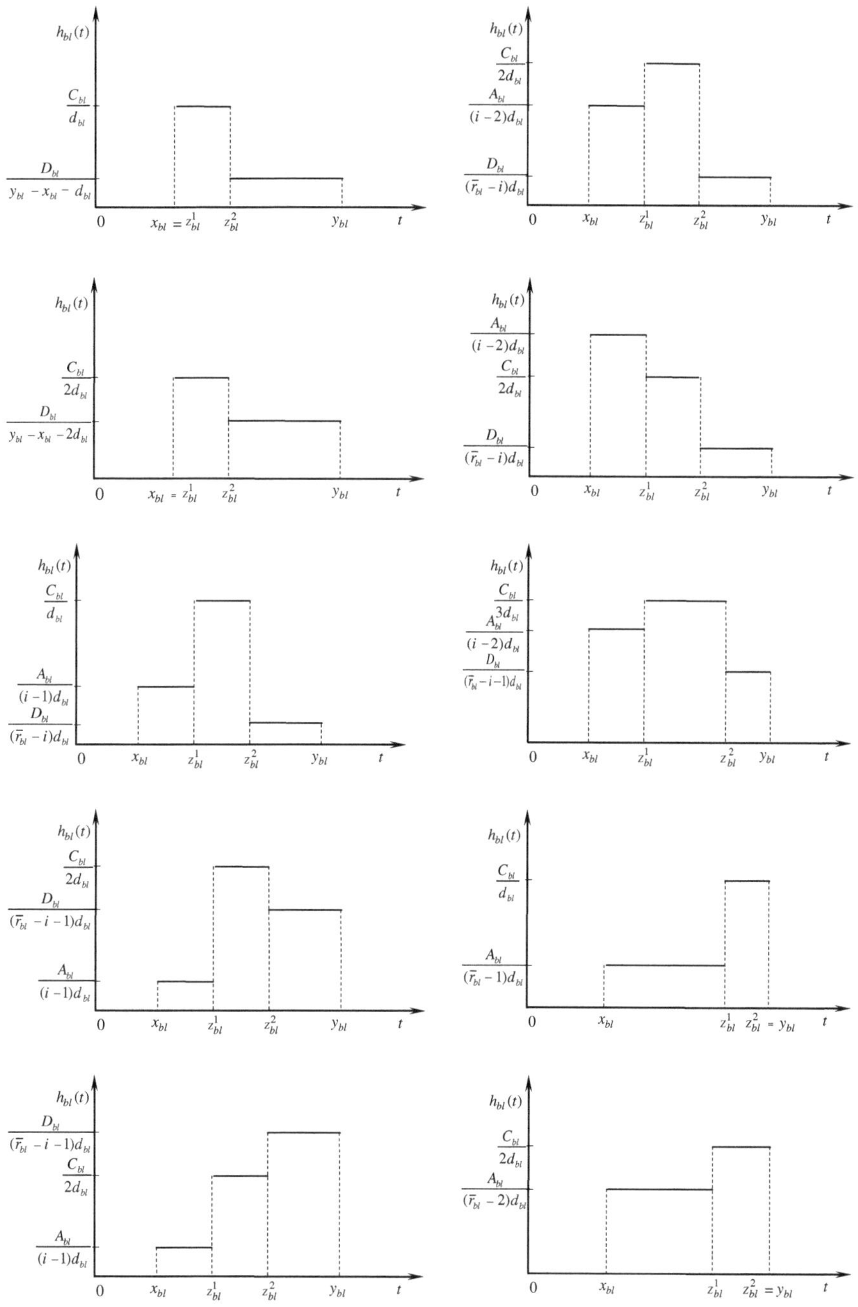

Fig. 2.7 The graphs of the chimney distribution's density functions (the meanings of d_{bl}, i and $\bar{r}_{bl}$ are explained in Chap. 4)

where

$$0 \le x_{bl} \le z_{bl}^{1} \le z_{bl}^{2} \le y_{bl} < +\infty, \quad A_{bl} \ge 0, \quad C_{bl} \ge 0, \quad D_{bl} \ge 0,$$
$$A_{bl} + C_{bl} + D_{bl} = 1.$$

As the mean values of the conditional sojourn times θ_{bl} are given by [10]

$$M_{bl} = E[\theta_{bl}] = \int_{0}^{\infty} t\mathrm{d}H_{bl}(t) = \int_{0}^{\infty} th_{bl}(t)\mathrm{d}t, \quad b,l = 1,2,\ldots,v, \quad b \neq l, \tag{2.12}$$

then for the distinguished distributions (2.5–2.11), the mean values of the system operation process *Z*(*t*) conditional sojourn times θ_{bl}, $b,l = 1,2,\ldots,v$, $b \neq l$, at the particular operation states are respectively given by [7]:

- for the uniform distribution

$$M_{bl} = E[\theta_{bl}] = \frac{x_{bl} + y_{bl}}{2}; \tag{2.13}$$

- for the triangular distribution

$$M_{bl} = E[\theta_{bl}] = \frac{x_{bl} + y_{bl} + z_{bl}}{3}; \tag{2.14}$$

- for the double trapezium distribution

$$\begin{aligned} M_{bl} = E[\theta_{bl}] &= \frac{x_{bl} + y_{bl} + z_{bl}}{3} + \frac{w_{bl}(y_{bl})^2 - q_{bl}(x_{bl})^2}{2} \\ &+ \frac{w_{bl} + q_{bl}}{6}\left[(x_{bl}z_{bl} - y_{bl}z_{bl}) + \frac{x_{bl}y_{bl}(x_{bl} + y_{bl})}{y_{bl} - x_{bl}}\right] \\ &- \frac{(x_{bl})^3 q_{bl} + (y_{bl})^3 w_{bl}}{3(y_{bl} - x_{bl})}; \end{aligned} \tag{2.15}$$

- for the quasi-trapezium distribution

$$\begin{aligned} M_{bl} = E[\theta_{bl}] &= \frac{q_{bl}}{2}\left[(z_{bl}^{1})^2 - (x_{bl})^2\right] - \frac{A_{bl} - q_{bl}}{6}\left[2(z_{bl}^{1})^2 - 5x_{bl}z_{bl}^{1} - (x_{bl})^2\right] \\ &+ \frac{A_{bl}}{2}\left[(z_{bl}^{2})^2 - (z_{bl}^{1})^2\right] + \frac{w_{bl}}{2}\left[(y_{bl})^2 - (z_{bl}^{2})^2\right] \\ &- \frac{w_{bl} - A_{bl}}{6}\left[2(z_{bl}^{2})^2 - 5y_{bl}z_{bl}^{2} - (y_{bl})^2\right]; \end{aligned} \tag{2.16}$$

- for the exponential distribution

$$M_{bl} = E[\theta_{bl}] = x_{bl} + \frac{1}{\alpha_{bl}}; \tag{2.17}$$

- for the Weibull distribution

$$M_{bl} = E[\theta_{bl}] = x_{bl} + \alpha_{bl}^{-\frac{1}{\beta_{bl}}}\Gamma\left(1 + \frac{1}{\beta_{bl}}\right), \tag{2.18}$$

where

$$\Gamma(u) = \int_0^{+\infty} t^{u-1} e^{-t} \mathrm{d}t, \quad u > 0,$$

is the gamma function;

- for the chimney distribution

$$M_{bl} = E[\theta_{bl}] = \frac{1}{2}\left[A_{bl}(x_{bl} + z_{bl}^1) + C_{bl}(z_{bl}^1 + z_{bl}^2) + D_{bl}(z_{bl}^2 + y_{bl})\right]. \tag{2.19}$$

From the formula for total probability, it follows that the unconditional distribution functions of the sojourn times θ_b, $b = 1, 2, \ldots, v$, of the system operation process $Z(t)$ at the operation states z_b, $b = 1, 2, \ldots, v$, are given by [3, 5, 6, 10, 12]

$$H_b(t) = \sum_{l=1}^{v} p_{bl} H_{bl}(t), \quad b = 1, 2, \ldots, v. \tag{2.20}$$

Hence, the mean values $E[\theta_b]$ of the system operation process $Z(t)$ unconditional sojourn times θ_b, $b = 1, 2, \ldots, v$, at the operation states are given by

$$M_b = E[\theta_b] = \sum_{l=1}^{v} p_{bl} M_{bl}, \quad b = 1, 2, \ldots, v, \tag{2.21}$$

where M_{bl} are defined by the formula (2.12) in a case of any distribution of sojourn times θ_{bl} and by the formulae (2.13–2.19) in the cases of particular defined respectively by (2.5–2.11) distributions of these sojourn times.

The limit values of the system operation process $Z(t)$ transient probabilities at the particular operation states

$$p_b(t) = P(Z(t) = z_b), \quad t \in \langle 0, +\infty), \quad b = 1, 2, \ldots, v,$$

are given by [3, 5, 6, 10–12]

$$p_b = \lim_{t \to \infty} p_b(t) = \frac{\pi_b M_b}{\sum_{l=1}^{v} \pi_l M_l}, \quad b = 1, 2, \ldots, v, \tag{2.22}$$

where M_b, $b = 1, 2, \ldots, v$, are given by (2.21), while the steady probabilities π_b of the vector $[\pi_b]_{1xv}$ satisfy the system of equations

$$\begin{cases} [\pi_b] = [\pi_b][p_{bl}] \\ \sum_{l=1}^{v} \pi_l = 1. \end{cases} \tag{2.23}$$

In the case of a periodic system operation process, the limit transient probabilities p_b, $b = 1, 2, \ldots, v$, at the operation states defined by (2.22), are the

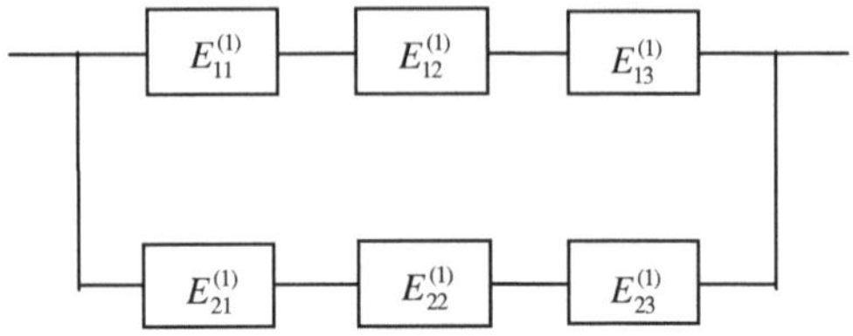

Fig. 2.8 The scheme of the subsystem S_1 reliability structure

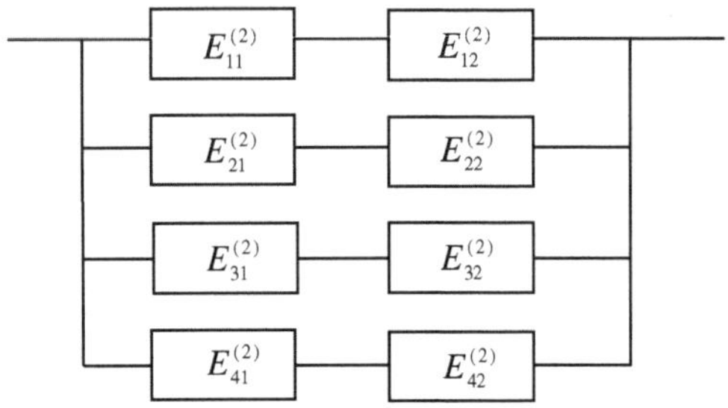

Fig. 2.9 The scheme of the subsystem S_2 reliability structure

long-term proportions of the system operation process $Z(t)$ sojourn times at the particular operation states z_b, $b = 1, 2, \ldots, v$.

Other interesting characteristics of the system operation process $Z(t)$ possible to obtain are its total sojourn times $\hat{\theta}_b$ at the particular operation states z_b, $b = 1, 2, \ldots, v$, during the fixed system operation time. It is well known [3, 5, 6, 10–12] that the system operation process total sojourn times $\hat{\theta}_b$ at the particular operation states z_b, for sufficiently large operation time θ, have approximately normal distributions with the expected value given by

$$E[\hat{\theta}_b] = p_b\theta, \quad b = 1, 2, \ldots, v, \tag{2.24}$$

where p_b are given by (2.22).

2.3 Applications

2.3.1 Operation Process of Exemplary System

The considered exemplary system S is described in Sect. 1.4.1. We assumed there that the system is a series system composed of the subsystems S_1 and S_2, with the general scheme shown in Fig. 1.12. Now, we additionally assume that the subsystem S_1 is a series-parallel system with the scheme given in Fig. 2.8 and the subsystem S_2 illustrated in Fig. 2.9 is either a series-parallel system or a series-"2 out of 4" system.

Subsystems S_1 and S_2 form a general series reliability structure of the exemplary system presented in Fig. 1.12. However, this system reliability structure and its subsystems and components' reliability depend on its changing in time operation states [6, 13].

Under the assumption that the system operation conditions are changing in time, we arbitrarily fix the number of system operation process states $v = 4$ and distinguish the following as its operation states:

- an operation state z_1—the system is composed of subsystem S_1 with the scheme shown in Fig. 2.8 which is a series-parallel system,
- an operation state z_2—the system is composed of subsystem S_2 with the scheme shown in Fig. 2.9 which is a series-parallel system,
- an operation state z_3—the system is a series system with the scheme shown in Fig. 1.12 composed of subsystems S_1 and S_2 which are series-parallel systems with the schemes respectively given in Figs. 2.8 and 2.9,
- an operation state z_4—the system is a series system with the scheme shown in Fig. 1.12 composed of subsystems S_1 and S_2, while subsystem S_1 is a series-parallel system with the scheme given in Fig. 2.8 and subsystem S_2 is a series-"2 out of 4" system with the scheme given in Fig. 2.9.

The influence of the above system operation states changing on the changes of the exemplary system reliability structure is indicated in these operation state definitions.

As we do not have real values of the parameters of the exemplary system operation process, we arbitrarily assume that the probabilities p_{bl} of the exemplary system operation process transitions from operation state z_b into the operation state z_l are given in the matrix below

$$[p_{bl}] = \begin{bmatrix} 0 & 0.22 & 0.32 & 0.46 \\ 0.20 & 0 & 0.30 & 0.50 \\ 0.12 & 0.16 & 0 & 0.72 \\ 0.48 & 0.22 & 0.30 & 0 \end{bmatrix}. \tag{2.25}$$

For the same reason, we arbitrarily fix the conditional mean values $M_{bl} = E[\theta_{bl}]$, $b, l = 1, 2, 3, 4$, of the exemplary system sojourn times at the particular operation states as follows:

$$\begin{aligned} &M_{12} = 192, \quad M_{13} = 480, \quad M_{14} = 200, \\ &M_{21} = 96, \quad M_{23} = 81, \quad M_{24} = 55, \\ &M_{31} = 870, \quad M_{32} = 480, \quad M_{34} = 300, \\ &M_{41} = 325, \quad M_{42} = 510, \quad M_{43} = 438. \end{aligned} \tag{2.26}$$

In this way, the exemplary system operation process is defined and we may find its main characteristics. Namely, applying (2.21), (2.25) and (2.26), the unconditional mean sojourn times at the particular operation states are given by:

$$\begin{aligned} M_1 &= E[\theta_1] = p_{12}M_{12} + p_{13}M_{13} + p_{14}M_{14} \\ &= 0.22 \times 192 + 0.32 \times 480 + 0.46 \times 200 = 287.84, \end{aligned} \tag{2.27}$$

$$M_2 = E[\theta_2] = p_{21}M_{21} + p_{23}M_{23} + p_{24}M_{24}$$
$$= 0.20 \times 96 + 0.30 \times 81 + 0.50 \times 55 = 71.00, \tag{2.28}$$

$$M_3 = E[\theta_3] = p_{31}M_{31} + p_{32}M_{32} + p_{34}M_{34}$$
$$= 0.12 \times 870 + 0.16 \times 480 + 0.72 \times 300 = 397.20, \tag{2.29}$$

$$M_4 = E[\theta_4] = p_{41}M_{41} + p_{42}M_{42} + p_{43}M_{43}$$
$$= 0.48 \times 325 + 0.22 \times 510 + 0.30 \times 438 = 399.60. \tag{2.30}$$

Further, according to (2.23), the system of equations

$$\begin{cases} [\pi_1, \pi_2, \pi_3, \pi_4] = [\pi_1, \pi_2, \pi_3, \pi_4][p_{bl}]_{4\times 4} \\ \pi_1 + \pi_2 + \pi_3 + \pi_4 = 1, \end{cases}$$

after considering (2.25), takes the form

$$\begin{cases} \pi_1 = 0.20\pi_2 + 0.12\pi_3 + 0.48\pi_4 \\ \pi_2 = 0.22\pi_1 + 0.16\pi_3 + 0.22\pi_4 \\ \pi_3 = 0.32\pi_1 + 0.30\pi_2 + 0.30\pi_4 \\ \pi_4 = 0.46\pi_1 + 0.50\pi_2 + 0.72\pi_3 \\ \pi_1 + \pi_2 + \pi_3 + \pi_4 = 1. \end{cases}$$

The approximate solutions of the above system of equations are:

$$\pi_1 \cong 0.236, \quad \pi_2 \cong 0.169, \quad \pi_3 \cong 0.234, \quad \pi_4 \cong 0.361. \tag{2.31}$$

After considering the result (2.31) and (2.27–2.30), we have

$$\sum_{l=1}^{4} \pi_l M_l \cong 0.236 \times 287.84 + 0.169 \times 71.00 + 0.234 \times 397.20 + 0.361 \times 399.60$$
$$= 317.13,$$

and according to (2.22), the limit values of the exemplary system operation process transient probabilities $p_b(t)$ at the operation states z_b are given by

$$p_1 = \frac{0.236 \times 287.84}{317.13} \cong 0.214, \quad p_2 = \frac{0.169 \times 71.00}{317.13} \cong 0.038,$$
$$p_3 = \frac{0.234 \times 397.20}{317.13} \cong 0.293, \quad p_4 = \frac{0.361 \times 399.60}{317.13} \cong 0.455. \tag{2.32}$$

Hence, the expected values of the total sojourn times $\hat{\theta}_b$, $b = 1, 2, 3, 4$, of the exemplary system operation process at the particular operation states z_b, $b = 1, 2, 3, 4$, during the fixed operation time $\theta = 1$ year $= 365$ days, after applying (2.24), amount to:

$$
\begin{aligned}
E[\hat{\theta}_1] &= 0.214 \times 1 = 0.214 \text{ year} = 78.1 \text{ days},\\
E[\hat{\theta}_2] &= 0.038 \times 1 = 0.038 \text{ year} = 13.9 \text{ days},\\
E[\hat{\theta}_3] &= 0.293 \times 1 = 0.293 \text{ year} = 106.9 \text{ days},\\
E[\hat{\theta}_4] &= 0.455 = 0.455 \text{ year} = 166.1 \text{ days}.
\end{aligned} \tag{2.33}
$$

2.3.2 Operation Process of Port Oil Piping Transportation System

The considered port oil piping transportation system is described in Sect. 1.4.2. The system is composed of subsystems S_1, S_2 and S_3 with the scheme shown in Fig. 2.15. The system is a series system composed of two series-parallel subsystems S_1, S_2, each containing two pipelines and one series-"2 out of 3" subsystem S_3. Subsystems S_1, S_2 and S_3 form a general series port oil pipeline system reliability structure presented in Fig. 2.16. However, the pipeline system reliability structure and its subsystems and components reliability depend on its changing in time operation states [7, 9].

Taking into account the expert opinion on the varying in time operation process of the considered piping system, we distinguish the following as its eight operation states [7, 9]:

- operation state z_1— transport of one kind of medium from the terminal part B to part C using two out of three pipelines of subsystem S_3,
- operation state z_2—transport of one kind of medium from the terminal part C to part B using one out of three pipelines of subsystem S_3,
- operation state z_3—transport of one kind of medium from the terminal part B through part A to pier using one out of two pipelines of subsystem S_1 and one out of two pipelines of subsystem S_2,
- operation state z_4—transport of one kind of medium from the pier through parts A and B to part C using one out of two pipelines of subsystem S_1, one out of two pipelines in subsystem S_2 and two out of three pipelines of subsystem S_3,
- operation state z_5—transport of one kind of medium from the pier through part A to B using one out of two pipelines of subsystem S_1.and one out of two pipelines of subsystem S_2,
- operation state z_6—transport of one kind of medium from the terminal part B to C using two out of three pipelines of subsystem S_3, and simultaneously transport one kind of medium from the pier through part A to B using one out of two pipelines of subsystem S_1 and one out of two pipelines of subsystem S_2,
- operation state z_7—transport of one kind of medium from the terminal part B to C using one out of three pipelines of subsystem S_3, and simultaneously transport a second kind of medium from the terminal part C to B using one out of three pipelines of subsystem S_3.

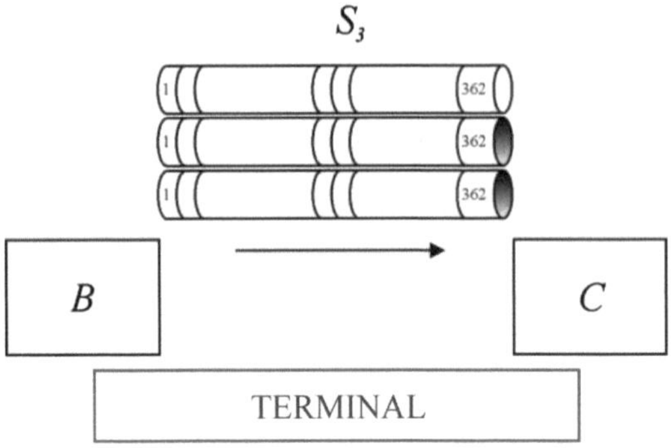

Fig. 2.10 The scheme of the port oil piping transportation system at the operation states z_1 and z_7

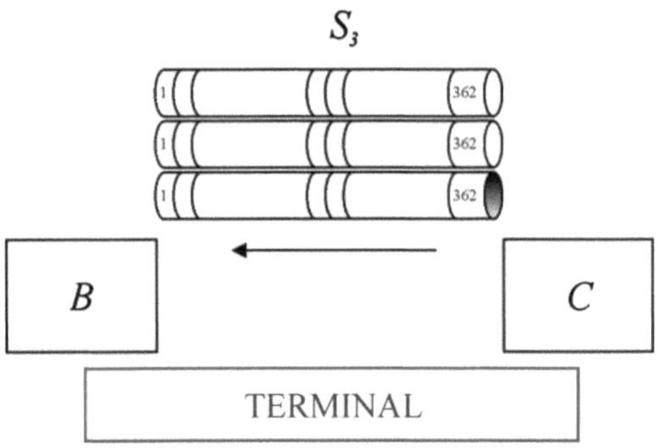

Fig. 2.11 The scheme of the port oil piping transportation system at operation state z_2

The influence of the above system operation states changing on the changes of the pipeline system reliability structure is as follows.

At system operation states z_1 and z_7, the system is composed of subsystem S_3, that is a series-"2 out of 3" system containing three series subsystems with the scheme shown in Fig. 2.10.

At system operation state z_2, the system is composed of a series-parallel subsystem S_3, which contains three pipelines with the scheme shown in Fig. 2.11.

At the system operation states z_3 and z_5, the system is series and composed of two series-parallel subsystems S_1, S_2, each containing two pipelines with the scheme shown in Fig. 2.12.

At system operation states z_4 and z_6, the system is series and composed of two series-parallel subsystems S_1, S_2, each containing two pipelines and one series-"2 out of 3" subsystem S_3 with the scheme shown in Fig. 2.13.

To identify the unknown parameters of the port oil piping transportation system operation process the suitable statistical data coming from its real realizations should be collected. The lack of sufficient statistical data about the port oil piping transportation system operation process causes that it is not possible to estimate exactly its operation parameters. However, even on the basis of the fragmentary statistical data coming from experts, the port oil piping transportation system operation process probabilities p_{bl} of transitions from the operation state z_b into the

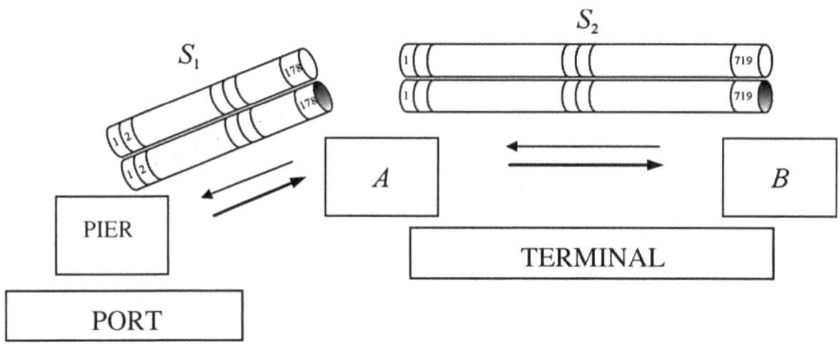

Fig. 2.12 The scheme of port oil piping transportation system at the operation states z_3 and z_5

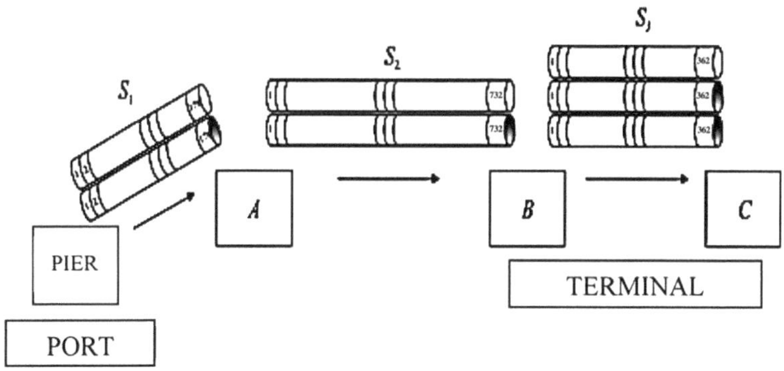

Fig. 2.13 The scheme of the port oil piping transportation system at operation states z_4 and z_6

operation state z_l, $b, l = 1, 2, \ldots, 7$, $b \neq l$, can be evaluated approximately (see Chap. 4 for details). Their approximate evaluations are given in the matrix below

$$[p_{bl}] = \begin{bmatrix} 0 & 0.022 & 0.022 & 0 & 0.534 & 0.111 & 0.311 \\ 0.2 & 0 & 0 & 0 & 0 & 0 & 0.8 \\ 1 & 0 & 0 & 0 & 0 & 0 & 0 \\ 0 & 0 & 0 & 0 & 0 & 0 & 1 \\ 0.488 & 0.023 & 0 & 0.023 & 0 & 0.233 & 0.233 \\ 0.095 & 0 & 0 & 0 & 0.667 & 0 & 0.238 \\ 0.531 & 0.062 & 0 & 0 & 0.219 & 0.188 & 0 \end{bmatrix}. \quad (2.34)$$

Unfortunately, it is not possible to identify completely the matrix of the conditional distribution functions $[H_{bl}(t)]_{7\times7}$ of the sojourn times θ_{bl} for $b, l = 1, 2, \ldots, 7$, $b \neq l$, defined by (2.3), and consequently, it is also not possible to determine the vector $[H_b(t)]_{1\times7}$ of the unconditional distribution functions of the sojourn times θ_b of this system operation process at the operation states z_b, $b = 1, 2, \ldots, 7$, given by (2.20). However, using the methods presented in Chap. 4, on the basis of data coming from practice and collected by experts operating this piping system, some hypotheses on the forms of the distributions describing the system operation process conditional sojourn times θ_{bl}, $b, l = 1, 2, \ldots, 7$, $b \neq l$ at the particular operation states can be formulated and accepted. In this case, having these distributions identified, it is possible to evaluate the mean values $M_{bl} = E[\theta_{bl}]$ of the conditional sojourn times θ_{bl}, at the particular operation states, using either the general formula (2.12) or the particular formulae (2.13–2.19) for the distinguished distributions. Otherwise, if the collected statistical data is not sufficient to test and to accept the forms of the distributions of the piping system operation process conditional sojourn times θ_{bl}, their mean values $M_{bl} = E[\theta_{bl}]$ may be estimated by applying the formula for the empirical man values of the conditional sojourn times at the particular operation states defined by (4.7). As the result of using the last of these two possibilities, the approximate evaluations of these mean values, determined in Chap. 4, are as follows:

$$\begin{aligned}
&M_{12}=1920,\ \ M_{13}=480,\ \ M_{15}=1999.4,\ \ M_{16}=1250,\ \ M_{17}=1129.6,\\
&M_{21}=9960,\ \ M_{27}=810,\ \ M_{31}=575,\ \ M_{47}=380,\ \ M_{51}=874.7,\ \ M_{52}=480,\\
&M_{54}=300,\ \ M_{56}=436.3,\ \ M_{57}=1042.5,\ \ M_{61}=325,\ \ M_{65}=510.7,\ \ M_{67}=438,\\
&M_{71}=850.9,\ \ M_{72}=510,\ \ M_{75}=2585.7,\ \ M_{76}=2380.
\end{aligned} \tag{2.35}$$

In this way, the port oil piping transportation system operation process is approximately defined and we may predict its main characteristics. Namely, applying (2.21), (2.34) and (2.35), the unconditional mean sojourn times of the piping system operation process at the particular operation states are:

$$\begin{aligned}
M_1&=E[\theta_1]=p_{12}M_{12}+p_{13}M_{13}+p_{15}M_{15}+p_{16}M_{16}+p_{17}M_{17}\\
&=0.022\times 1920+0.022\times 480+0.534\times 1999.4+0.111\\
&\quad\times 1250+0.311\times 1129.6\cong 1610.52,\\
M_2&=E[\theta_2]=p_{21}M_{21}+p_{27}M_{27}=0.2\times 9960+0.8\times 810\cong 2640,\\
M_3&=E[\theta_3]=p_{31}M_{31}=1\times 575=575,\ \ M_4=E[\theta_4]=p_{47}M_{47}=1\times 380=380,\\
M_5&=E[\theta_5]=p_{51}M_{51}+p_{52}M_{52}+p_{54}M_{54}+p_{56}M_{56}+p_{57}M_{57}\\
&=0.488\times 874.7+0.023\times 480+0.023\times 300+0.233\\
&\quad\times 436.3+0.233\times 1042.5\cong 789.35,\\
M_6&=E[\theta_6]=p_{61}M_{61}+p_{65}M_{65}+p_{67}M_{67}=0.095\times 325+0.667\times 510.7+0.238\\
&\quad\times 438\cong 475.76,\\
M_7&=E[\theta_7]=p_{71}M_{71}+p_{72}M_{72}+p_{75}M_{75}+p_{76}M_{76}\\
&=0.531\times 850.9+0.062\times 510+0.219\times 2585.7+0.188\times 2380\cong 1497.16.
\end{aligned} \tag{2.36}$$

Considering (2.34) in the system of equations (2.23) that takes the form

$$\begin{cases}
[\pi_1,\pi_2,\pi_3,\pi_4,\pi_5,\pi_6,\pi_7] = [\pi_1,\pi_2,\pi_3,\pi_4,\pi_5,\pi_6,\pi_7][p_{bl}]_{7\times 7}\\
\pi_1+\pi_2+\pi_3+\pi_4+\pi_5+\pi_6+\pi_7 = 1,
\end{cases}$$

we get its following solution

$$\pi_1 \cong 0.291,\ \pi_2 \cong 0.027,\quad \pi_3 \cong 0.006,\quad \pi_4 \cong 0.007,\quad \pi_5 \cong 0.301,$$
$$\pi_6 \cong 0.144,\quad \pi_7 \cong 0.224. \tag{2.37}$$

Hence and from (2.36), after applying (2.22), it follows that the limit values of the piping system operation process transient probabilities $p_b(t)$ at the operation states z_b, $b = 1,2,\ldots,7$, are given by

$$p_1 = 0.395,\quad p_2 = 0.060,\quad p_3 = 0.003,\quad p_4 = 0.002,\quad p_5 = 0.20,$$
$$p_6 = 0.058,\quad p_7 = 0.282. \tag{2.38}$$

Considering the above result, after applying (2.24), the expected values of the total sojourn times $\hat{\theta}_b$, $b = 1,2,\ldots,7$, of the system operation process at the particular operation states z_b, $b = 1,2,\ldots,7$, during the fixed operation time $\theta = 1$ year $=$ 365 days, amount to:

$$\begin{aligned}
&E[\hat{\theta}_1] = 0.395 \text{ year } = 144.2 \text{ days},\quad E[\hat{\theta}_2] = 0.060 \text{ year } = 21.9 \text{ days},\\
&E[\hat{\theta}_3] = 0.003 \text{ year } = 1.1 \text{ day},\quad E[\hat{\theta}_4] = 0.002 \text{ year } = 0.7 \text{ day},\\
&E[\hat{\theta}_5] = 0.20 \text{ year } = 73 \text{ days},\quad E[\hat{\theta}_6] = 0.058 \text{ year } = 21.2 \text{ days},\\
&E[\hat{\theta}_7] = 0.282 \text{ year } = 102.93 \text{ days}.
\end{aligned} \tag{2.39}$$

2.3.3 Operation Process of Maritime Ferry Technical System

The considered technical system of the maritime ferry is described in Sect. 1.4.3. The system is composed of subsystems S_1, S_2, S_3, S_4 and S_5 indicated in Fig. 1.19 with the detailed scheme shown in Fig. 1.20. Subsystems S_1, S_2, S_3, S_4, S_5, form a general series safety structure of the ferry technical system presented in Fig. 1.21. However, the ferry technical system safety structure and the subsystems and components' safety depend on its changing in time operation states [8, 14]. Before indicating this system changing in time safety structure, we define its operation states.

Thus, taking into account the expert opinions concerned with the operation process of the considered ferry technical system, we distinguish the following as its eighteen operation states:

- operation state z_1—loading at Gdynia Port,
- operation state z_2—unmooring operations at Gdynia Port,
- operation state z_3—leaving Gdynia Port and navigation to "GD" buoy,
- operation state z_4—navigation at restricted waters from "GD" buoy to the end of Traffic Separation Scheme,
- operation state z_5—navigation at open waters from the end of Traffic Separation Scheme to "Angoring" buoy,

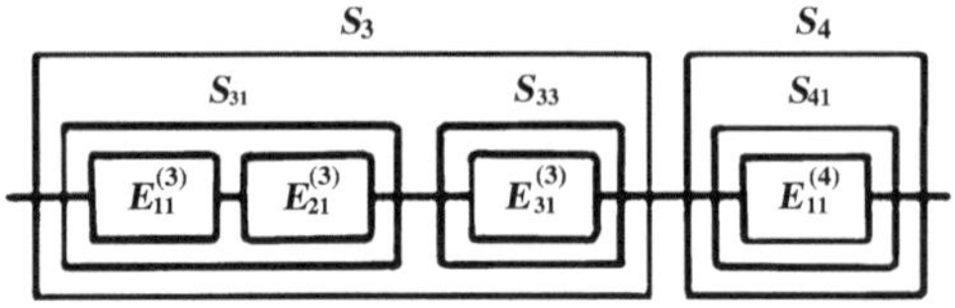

Fig. 2.14 The scheme of the ferry technical system structure at operation states z_1 and z_{18}

- operation state z_6—navigation at restricted waters from "Angoring" buoy to "Verko" Berth at Karlskrona,
- operation state z_7—mooring operations at Karlskrona Port,
- operation state z_8—unloading at Karlskrona Port,
- operation state z_9—loading at Karlskrona Port,
- operation state z_{10}—unmooring operations at Karlskrona Port,
- operation state z_{11}—ferry turning at Karlskrona Port,
- operation state z_{12}—leaving Karlskrona Port and navigation at restricted waters to "Angoring" buoy,
- operation state z_{13}—navigation at open waters from "Angoring" buoy to the entering Traffic Separation Scheme,
- operation state z_{14}—navigation at restricted waters from the entering Traffic Separation Scheme to "GD" buoy,
- operation state z_{15}—navigation from "GD" buoy to turning area,
- operation state z_{16}—ferry turning at Gdynia Port,
- operation state z_{17}—mooring operations at Gdynia Port,
- operation state z_{18}—unloading at Gdynia Port.

The influence of the above defined operation states changing on the changes of the ferry technical system safety structure is as follows.

At operation states z_1 and z_{18}, the ferry technical system is composed of two subsystems S_3 and S_4 forming a series structure shown in Fig. 2.14.

At operation states z_2, z_7, z_{10} and z_{17}, the ferry technical system is composed of three subsystems S_1, S_2 and S_5 forming a series structure shown in Fig. 2.15.

At operation states z_3, z_{11}, z_{15} and z_{16}, the ferry technical system is composed of two subsystems S_1 and S_2 forming a series structure shown in Fig. 2.16.

At operation states z_4, z_5, z_{12}, z_{13} and z_{14}, the ferry technical system is composed of three subsystems S_1, S_2 and S_4 forming a series structure shown in Fig. 2.17.

At operation state z_6, the ferry technical system is composed of three subsystems S_1, S_2 and S_4 forming a series structure shown in Fig. 2.18.

At operation states z_8 and z_9, the ferry technical system is composed of two subsystems S_3 and S_4 forming a series structure shown in Fig. 2.19.

To identify the unknown parameters of the ferry technical system operation process the suitable statistical data coming from its real realizations should be collected. It is possible to collect these data because of the high frequency of the ferry voyages that result in a large number of its technical system operation process realizations. The ferry technical system operation process is very regular in the sense that the operation state changes are from the particular state z_b, $b = 1, 2, \ldots, 17$, to the neighboring state z_{b+1}, $b = 1, 2, \ldots, 17$, and from z_{18} to z_1

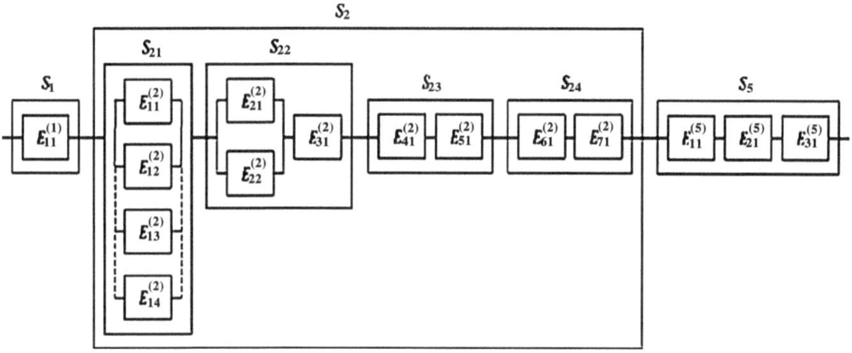

Fig. 2.15 The scheme of the ferry technical system structure at operation states z_2, z_7, z_{10} and z_{17}

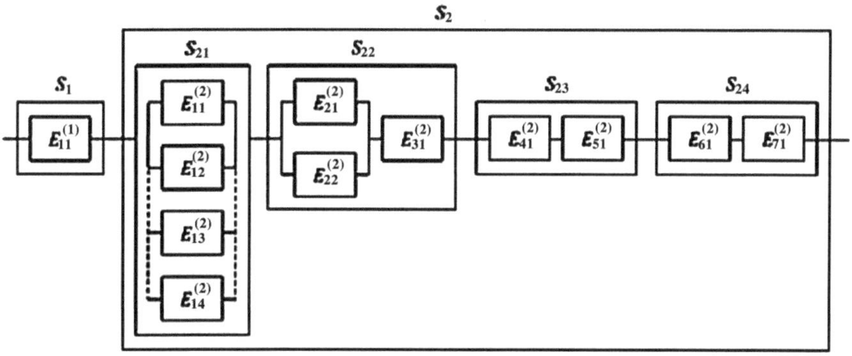

Fig. 2.16 The scheme of the ferry technical system structure at operation states z_3, z_{11}, z_{15} and z_{16}

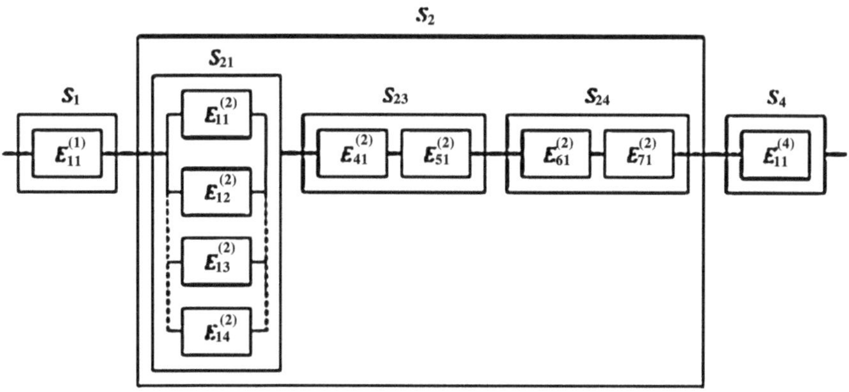

Fig. 2.17 The scheme of the ferry technical system structure at operation states z_4, z_5, z_{12}, z_{13} and z_{14}

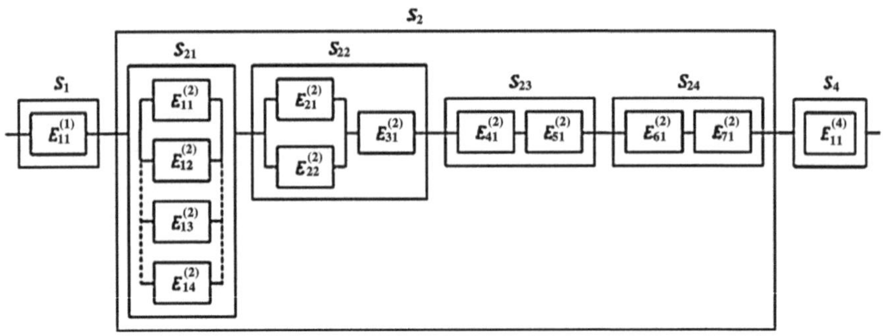

Fig. 2.18 The scheme of the ferry technical system structure at operation state z_6

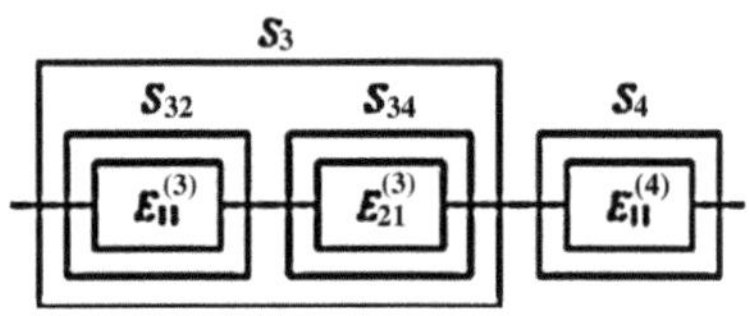

Fig. 2.19 The scheme of the ferry technical system structure at operation states z_8 and z_9

only. Therefore, the probabilities of its transitions between the operation states, according to the methods of their estimation given in Chap. 4, are given by

$$[p_{bl}] = \begin{bmatrix} 0\,1\,0\ldots0\,0 \\ 0\,0\,1\ldots0\,0 \\ \ldots \\ 0\,0\,0\ldots0\,1 \\ 1\,0\,0\ldots0\,0 \end{bmatrix}. \tag{2.40}$$

On the basis of statistical data coming from experts, the matrix $[h_{bl}(t)]_{18\times18}$ of the density functions of the ferry technical system operation process conditional sojourn times θ_{bl}, $b, l = 1, 2, \ldots, 18$, at the particular operation states, defined by (2.4), can be evaluated. In Chap. 4, the forms of the particular density functions $h_{bl}(t)$ of the ferry technical system operation process conditional sojourn times θ_{bl}, $b, l = 1, 2, \ldots, 18$, $b \neq l$, at the particular operation states are identified on the basis of statistical data coming from its operation process realizations and they are as follows:

- the conditional sojourn time θ_{12} have a triangular distribution with the density function

$$h_{12}(t) = \begin{cases} 0, & t<7, \\ 0.00044t - 0.0031, & 7\le t<54, \\ 0.044 - 0.00043t; & 54\le t<103, \\ 0, & t\ge 103; \end{cases} \tag{2.41}$$

- the conditional sojourn time θ_{23} have an exponential distribution with the density function

$$h_{23}(t) = \begin{cases} 0, & t < 1.6, \\ 1.0\exp[-1.0(t-1.6)], & t \geq 1.6; \end{cases} \tag{2.42}$$

- the conditional sojourn time θ_{34} have a chimney distribution with the density function

$$h_{34}(t) = \begin{cases} 0, & t < 26, \\ 0.0119, & 26 \leq t < 34, \\ 0.1667, & 34 \leq t < 38, \\ 0.0198, & 38 \leq t < 50, \\ 0, & t \geq 50; \end{cases}$$

- the conditional sojourn time θ_{45} have a chimney distribution with the density function

$$h_{45}(t) = \begin{cases} 0, & t < 41.7, \\ 0.0259, & 41.7 \leq t < 46.3, \\ 0.0751, & 46.3 \leq t < 55.5, \\ 0.0138, & 55.5 \leq t < 69.3, \\ 0, & t \geq 69.3; \end{cases}$$

- the conditional sojourn time θ_{56} have a chimney distribution with the density function

$$h_{56}(t) = \begin{cases} 0, & t < 467.8, \\ 0.0113, & 467.8 \leq t < 528.6, \\ 0.0025, & 528.6 \leq t < 650.2, \\ 0, & t \geq 650.2; \end{cases}$$

- the conditional sojourn time θ_{67} have a double trapezium distribution with the density function

$$h_{67}(t) = \begin{cases} 0, & t < 31.9, \\ 0.0067t - 0.1747, & 31.9 \leq t < 37.17, \\ 0.0031t - 0.0395, & 37.7 \leq t < 45.1, \\ 0, & t \geq 45.1; \end{cases}$$

- the conditional sojourn time θ_{78} have a double trapezium distribution with the density function

$$h_{78}(t) = \begin{cases} 0, & t < 4.5, \\ -0.0183t + 0.2922, & 4.5 \le t < 7.02, \\ -0.0069t + 0.2122, & 7.02 \le t < 10.5, \\ 0, & t \ge 10.5; \end{cases}$$

- the conditional sojourn time θ_{89} have a chimney distribution with the density function

$$h_{89}(t) = \begin{cases} 0, & t < 0, \\ 0.0096, & 0 \le t < 14.8, \\ 0.0434, & 14.8 \le t < 29.6, \\ 0.0145, & 29.6 \le t < 44.4, \\ 0, & t \ge 44.4; \end{cases}$$

- the conditional sojourn time θ_{910} have a double trapezium distribution with the density function

$$h_{910}(t) = \begin{cases} 0, & t < 14.6, \\ 0.0001t + 0.0109, & 14.6 \le t < 52.26, \\ 0.0062, & 52.2 \le t < 127.4, \\ 0, & t \ge 127.4; \end{cases}$$

- the conditional sojourn time θ_{1011} have a chimney distribution with the density function

$$h_{1011}(t) = \begin{cases} 0, & t < 1.6, \\ 0.5059, & 1.6 \le t < 3.2, \\ 0.0595, & 3.2 \le t < 6.4, \\ 0, & t \ge 6.4; \end{cases}$$

- the conditional sojourn time θ_{1112} have a quasi-trapezium distribution with the density function

$$h_{1112}(t) = \begin{cases} 0, & t < 3.8, \\ -148.899t + 567.4863, & 3.8 \le t < 3.81, \\ 0.181, & 3.81 \le t < 4.48, \\ 0.3773t - 1.5094, & 4.48 \le t < 6.2, \\ 0, & t \ge 6.2; \end{cases}$$

- the conditional sojourn time θ_{1213} have a chimney distribution with the density function

$$h_{1213}(t) = \begin{cases} 0, & t < 18.7, \\ 0.0092, & 18.7 \le t < 21.3, \\ 0.1786, & 21.3 \le t < 26.5, \\ 0.0061, & 26.5 \le t < 34.3, \\ 0, & t \ge 34.3; \end{cases}$$

- the conditional sojourn time θ_{1314} have a chimney distribution with the density function

$$h_{1314}(t) = \begin{cases} 0, & t < 410.2, \\ 0.0014, & 410.2 \le t < 477.4, \\ 0.0184, & 477.4 \le t < 511, \\ 0.0028, & 511 \le t < 611.8, \\ 0, & t \ge 611.8; \end{cases}$$

- the conditional sojourn time θ_{1415} have a double trapezium distribution with the density function

$$h_{1415}(t) = \begin{cases} 0, & t < 36.8, \\ -0.0006t + 0.0518, & 36.8 \le t < 50.14, \\ 0.0003t + 0.0062, & 50.14 \le t < 75.2, \\ 0, & t \ge 75.2; \end{cases}$$

- the conditional sojourn time θ_{1516} have a chimney distribution with the density function

$$h_{1516}(t) = \begin{cases} 0, & t < 29.5, \\ 0.0317, & 29.5 \le t < 32.5, \\ 0.2698, & 32.5 \le t < 35.5, \\ 0.0079, & 35.5 \le t < 47.5, \\ 0, & t \ge 47.5; \end{cases}$$

- the conditional sojourn time θ_{1617} have a chimney distribution with the density function

$$h_{1617}(t) = \begin{cases} 0, & t < 2.7, \\ 0.0397, & 2.7 \le t < 3.9, \\ 0.6944, & 3.9 \le t < 5.1, \\ 0.0992, & 5.1 \le t < 6.3, \\ 0, & t \ge 6.3; \end{cases}$$

- the conditional sojourn time θ_{1718} have an exponential distribution with the density function

$$h_{1718}(t) = \begin{cases} 0, & t<2.3 \\ 0.1779\exp[-0.1779(t-2.3)], & t\geq 2.3; \end{cases}$$

- the conditional sojourn time θ_{181} have a triangular distribution with the density function

$$h_{181}(t) = \begin{cases} 0, & t<0, \\ 0.0023t, & 0\leq t<18.74, \\ -0.0016t+0.0729, & 18.74\leq t<45.59, \\ 0, & t\geq 45.59. \end{cases}$$

Further, for the above accepted distributions, the mean values $M_{bl} = E[\theta_{bl}]$, $b, l = 1, 2, \ldots, 18$, $b \neq l$, defined by (2.12), of the system operation process conditional sojourn times at the particular operation states can be determined by the application of the suitable selected formula from the formulae (2.13–2.19).

For instance, considering (2.6) and (2.41) and applying (2.14), we have

$$M_{12} = \frac{x_{12} + y_{12} + z_{12}}{3} = \frac{7 + 103 + 54}{3} = 54.67. \tag{2.43}$$

Similarly, considering (2.9) and (2.42) and applying (2.17), we have

$$M_{23} = x_{23} + \frac{1}{\alpha_{23}} = 1.6 + \frac{1}{1.0} = 2.60. \tag{2.44}$$

Proceeding in an analogous way, we obtain the remaining mean values of the ferry technical system operation process conditional sojourn times at the operation states that are as follows:

$$\begin{aligned} &M_{34} = 37.33, \quad M_{45} = 52.27, \quad M_{56} = 526.43, \quad M_{67} = 37.16, \\ &M_{78} = 7.02, \quad M_{89} = 23.26, \quad M_{910} = 53.69, \quad M_{1011} = 2.86, \\ &M_{1112} = 4.38, \quad M_{1213} = 24.12, \quad M_{1314} = 508.60, \quad M_{1415} = 50.14, \\ &M_{1516} = 34.43, \quad M_{1617} = 4.59, \quad M_{1718} = 7.92, \quad M_{181} = 18.74. \end{aligned} \tag{2.45}$$

Using the identified parameters of the ferry technical operation process, it is possible to predict its characteristics. Namely, considering (2.40) and (2.43–2.45), by (2.21), the mean values of the ferry technical system operation process unconditional sojourn times at the operation states are:

$$\begin{aligned} &M_1 = 54.67, \quad M_2 = 2.60, \quad M_3 = 37.33, \quad M_4 = 52.27, \quad M_5 = 526.43, \\ &M_6 = 37.16, \quad M_7 = 7.02, \quad M_8 = 23.26, \quad M_9 = 53.69, \quad M_{10} = 2.86, \\ &M_{11} = 4.38, \quad M_{12} = 24.12, \quad M_{13} = 508.60, \quad M_{14} = 50.14, \quad M_{15} = 34.43, \\ &M_{16} = 4.59, \quad M_{17} = 7.92, \quad M_{18} = 18.74. \end{aligned} \tag{2.46}$$

Since from the system of equations (2.23) that takes the form

$$\begin{cases} [\pi_b]_{1\times 18} = [\pi_b]_{1\times 18}[p_{bl}]_{18\times 18} \\ \sum_{l=1}^{18} \pi_l = 1, \end{cases}$$

we get

$$\pi_1 = \pi_2 = \pi_3 = \pi_4 = \pi_5 = \pi_6 = \pi_7 = \pi_8 = \pi_9 = \pi_{10} = \pi_{11} = \pi_{12}$$
$$= \pi_{13} = \pi_{14} = \pi_{15} = \pi_{16} = \pi_{17} = \pi_{18} \cong 0.056, \quad (2.47)$$

then after considering (2.46) and applying (2.22), the long-term proportion of transients p_b of the ferry technical system operation process at the operation states z_b, $b = 1, 2, \ldots, 18$, can be approximated by

$$p_1 = 0.038, \quad p_2 = 0.002, \quad p_3 = 0.026, \quad p_4 = 0.036, \quad p_5 = 0.363,$$
$$p_6 = 0.026, \quad p_7 = 0.005, \quad p_8 = 0.016, \quad p_9 = 0.037, \quad p_{10} = 0.002,$$
$$p_{11} = 0.003, \quad p_{12} = 0.016, \quad p_{13} = 0.351, \quad p_{14} = 0.034, \quad p_{15} = 0.024,$$
$$p_{16} = 0.003, \quad p_{17} = 0.005, \quad p_{18} = 0.013. \quad (2.48)$$

Hence, after applying (2.24), the expected values of the total sojourn times $\hat{\theta}_b$, $b = 1, 2, \ldots, 18$, of the ferry technical system operation process at the particular operation states z_b, $b = 1, 2, \ldots, 18$, during the fixed operation time $\theta = 1$ year = 365 days, approximately amount to:

$$E[\hat{\theta}_1] = 0.038 \text{ year} = 13.9 \text{ days}, \quad E[\hat{\theta}_2] = 0.002 \text{ year} = 0.7 \text{ day},$$
$$E[\hat{\theta}_3] = 0.026 \text{ year} = 9.5 \text{ days}, \quad E[\hat{\theta}_4] = 0.036 \text{ year} = 13.1 \text{ days},$$
$$E[\hat{\theta}_5] = 0.363 \text{ year} = 132.5 \text{ days}, \quad E[\hat{\theta}_6] = 0.026 \text{ year} = 9.5 \text{ days},$$
$$E[\hat{\theta}_7] = 0.005 \text{ year} = 1.8 \text{ days}, \quad E[\hat{\theta}_8] = 0.016 \text{ year} = 5.8 \text{ days},$$
$$E[\hat{\theta}_9] = 0.037 \text{ year} = 13.5 \text{ days}, \quad E[\hat{\theta}_{10}] = 0.002 \text{ year} = 0.7 \text{ day},$$
$$E[\hat{\theta}_{11}] = 0.003 \text{ year} = 1.1 \text{ days}, \quad E[\hat{\theta}_{12}] = 0.016 \text{ year} = 5.8 \text{ days},$$
$$E[\hat{\theta}_{13}] = 0.351 \text{ year} = 128.1 \text{ days}, \quad E[\hat{\theta}_{14}] = 0.034 \text{ year} = 12.4 \text{ days},$$
$$E[\hat{\theta}_{15}] = 0.024 \text{ year} = 8.8 \text{ days}, \quad E[\hat{\theta}_{16}] = 0.003 \text{ year} = 1.1 \text{ days},$$
$$E[\hat{\theta}_{17}] = 0.005 \text{ year} = 1.8 \text{ days}, \quad E[\hat{\theta}_{18}] = 0.013 \text{ year} = 4.7 \text{ days}. \quad (2.49)$$

2.4 Summary

The probabilistic model of the complex system operation processes presented here is the basis for the considerations in the subsequent chapters of the book. The model, together with the reliability and safety models of the systems presented in

Chap. 1, is used in Chap. 3, to construct the integrated general reliability, availability and safety probabilistic models of complex technical systems related to their operation processes. The methods of estimation of this model unknown parameters whose evaluations are used in this chapter are given in Chap. 4. The model of the system operation processes given in this chapter, together with the methods of its unknown parameters estimation given in Chap. 4, deliver the procedures and algorithms that allow us to perform the identification and then to find the basic characteristics of the complex technical systems' operation processes. The final results that can be obtained from this model practically used in the appliqué part of this chapter to the prediction of the main characteristics of the operation processes of the exemplary technical system, the port oil piping transportation system and the maritime ferry technical system justify high reasonability of the proposed approach to the complex technical system operation processes modeling.

References

1. Ferreira F, Pacheco A (2007) Comparison of level-crossing times for Markov and semi-Markov processes. Stat Probab Lett 77(2):151–157
2. Glynn PW, Haas PJ (2006) Laws of large numbers and functional central limit theorems for generalized semi-Markov processes. Stoch Model 22(2):201–231
3. Grabski F (2002) Semi-Markov models of systems reliability and operations analysis. System Research Institute, Polish Academy of Science, Warsaw (in Polish)
4. Guze S, Kołowrocki K, Soszyńska J (2008) Modeling environment and infrastructure influence on reliability and operation processes of port transportation systems. Summer safety & reliability seminars. J Pol Saf Reliab Assoc 2(1):179–188
5. Habibullah MS, Lumanpauw E, Kolowrocki K, Soszynska J, Ming NG (2009) A computational tool for general model of industrial systems. operation processes. Electron J Reliab Risk Anal Theory Appl 2(4):181–191
6. Kołowrocki K (2004) Reliability of large systems. Elsevier, Amsterdam
7. Kolowrocki K, Soszynska J (2009) Modeling environment and infrastructure influence on reliability and operation process of port oil transportation system. Electron J Reliab Risk Anal Theory Appl 2(3):131–142
8. Kolowrocki K, Soszynska J (2009) Safety and risk evaluation of Stena Baltica ferry in variable operation conditions. Electron J Reliab Risk Anal Theory Appl 2(4):168–180
9. Kolowrocki K, Soszynska J (2010) Reliability modeling of a port oil transportation system's operation processes. Int J Perform Eng 6(1):77–87
10. Kolowrocki K, Soszynska J (2010) Reliability, availability and safety of complex technical systems: modelling–identification–prediction–optimization. Summer safety & reliability seminars. J Pol Saf Reliab Assoc 4(1):133–158
11. Limnios N, Oprisan G (2005) Semi-Markov processes and reliability. Birkhauser, Boston
12. Mercier S (2008) Numerical bounds for semi-Markovian quantities and application to reliability. Methodol Comput Appl Probab 10(2):179–198
13. Soszyńska J (2007) Systems reliability analysis in variable operation conditions. PhD Thesis, Gdynia Maritime University-System Research Institute, Warsaw (in Polish)
14. Soszyńska J, Kołowrocki K, Blokus-Roszkowska A, Guze S (2010) Prediction of complex technical systems operation processes. Summer safety & reliability seminars. J Pol Saf Reliab Assoc 4(2):379–510

Chapter 3
Complex Technical Systems, Reliability, Availability and Safety Evaluation and Prediction

3.1 Introduction

Most real technical systems are structurally complex and often have complicated operation processes. Large numbers of components and subsystems and their operating complexity render the evaluation and prediction of their reliability, availability and safety difficult. The time-dependent interactions between systems' operation processes, operation states changing and systems' structures and their components reliability and safety states changing processes are evident features of most real technical systems. The common reliability or safety and operation analysis of these complex technical systems is of great value in industrial practice. The convenient tools for analyzing this problem are presented in Chap. 1, the multistate system's reliability and safety modeling [1–6, 10, 13–16, 23, 37, 39, 40, 42–45, 55–58], commonly used with that presented in Chap. 2, the semi-Markov modeling [7–9, 12, 19, 26, 38, 41, 54] of the systems operation processes, leading to the construction the joint general reliability, availability and safety models of the complex technical systems related to their operation process [20–22, 25, 31, 35, 46–48, 50–52]. The main objective of this chapter is to present the recently developed, mainly by the authors, the general reliability, availability and safety analytical models of complex non-repairable and repairable multistate technical systems related to their operation processes [20–22, 25, 31, 46–48, 50–52] and to apply them practically to real industrial systems and processes [11, 17, 18, 27–30, 32–34, 36, 49, 53]. In the case of large systems, the determination of the exact reliability functions of the systems and the system risk functions leads us to very complicated formulae that are often useless for reliability practitioners. One of the important techniques in this situation is the asymptotic approach [19, 24, 46–48, 51, 52] to system reliability and safety evaluation. This aspect of complex technical systems is also briefly discussed in this chapter.

K. Kołowrocki and J. Soszyńska-Budny, *Reliability and Safety of Complex Technical Systems and Processes*, Springer Series in Reliability Engineering, DOI: 10.1007/978-0-85729-694-8_3,

3.2 Reliability of Multistate Systems at Variable Operation Conditions

We assume that the changes of the operation states of the system operation process $Z(t)$ have an influence on the system multistate components E_i, $i = 1, 2, \ldots, n$, reliability and the system reliability structure as well. Consequently, we denote the system multistate component E_i, $i = 1, 2, \ldots, n$, conditional lifetime in the reliability state subset $\{u, u+1, \ldots, z\}$ while the system is at the operation state z_b, $b = 1, 2, \ldots, v$, by $T_i^{(b)}(u)$ and its conditional reliability function by the vector

$$[R_i(t,\cdot)]^{(b)} = \left[1, [R_i(t,1)]^{(b)}, \ldots, [R_i(t,z)]^{(b)}\right], \tag{3.1}$$

with the coordinates defined by

$$[R_i(t,u)]^{(b)} = P(T_i^{(b)}(u) > t | Z(t) = z_b) \tag{3.2}$$

for $t \in \langle 0, \infty)$, $u = 1, 2, \ldots, z$, $b = 1, 2, \ldots, v$.

The reliability function $[R_i(t,u)]^{(b)}$ is the conditional probability that the component E_i lifetime $T_i^{(b)}(u)$ in the reliability state subset $\{u, u+1, \ldots, z\}$ is greater than t, while the system operation process $Z(t)$ is at the operation state z_b.

Similarly, we denote the system conditional lifetime in the reliability state subset $\{u, u+1, \ldots, z\}$ while the system is at the operation state z_b, $b = 1, 2, \ldots, v$, by $T_i^{(b)}(u)$ and the conditional reliability function of the system by the vector

$$[\boldsymbol{R}(t,\cdot)]^{(b)} = \left[1, [\boldsymbol{R}(t,1)]^{(b)}, \ldots, [\boldsymbol{R}(t,z)]^{(b)}\right], \tag{3.3}$$

with the coordinates defined by

$$[\boldsymbol{R}(t,u)]^{(b)} = P(T^{(b)}(u) > t | Z(t) = z_b) \tag{3.4}$$

for $t \in \langle 0, \infty)$, $u = 1, 2, \ldots, z$, $b = 1, 2, \ldots, v$.

The reliability function $[R(t,u)]^{(b)}$ is the conditional probability that the system lifetime $T^{(b)}(u)$ in the reliability state subset $\{u, u+1, \ldots, z\}$ is greater than t, while the system operation process $Z(t)$ is at the operation state z_b.

Thus, the system conditional lifetimes in the reliability states subset $\{u, u+1, \ldots, z\}$ at the operational state z_b

$$T^{(b)}(u) = T(T_1^{(b)}(u), T_2^{(b)}(u), \ldots, T_n^{(b)}(u))$$

defined for $u = 1, 2, \ldots, z$, $b = 1, 2, \ldots, v$, $n \in N$, are dependent on the components, conditional lifetimes $T_1^{(b)}(u), T_2^{(b)}(u), \ldots, T_n^{(b)}(u)$, in the reliability states subset $\{u, u+1, \ldots, z\}$ at the operation state z_b and the coordinates of the system conditional reliability function at the operation state z_b

$$[\boldsymbol{R}(t,u)]^{(b)} = \boldsymbol{R}([R_1(t,u)]^{(b)}, [R_2(t,u)]^{(b)}, \ldots, [R_n(t,u)]^{(b)})$$

defined for $t \in \langle 0, \infty)$, $u = 1, 2, \ldots, z$, $b = 1, 2, \ldots, v$, $n \in N$, are dependent on the coordinates $[R_1(t,u)]^{(b)}, [R_2(t,u)]^{(b)}, \ldots, [R_n(t,u)]^{(b)}$, of the components conditional reliability functions at the operation state z_b.

Further, we denote the system unconditional lifetime in the reliability state subset $\{u, u+1, \ldots, z\}$ by $T(u)$ and the unconditional reliability function of the system by the vector

$$\boldsymbol{R}(t, \cdot) = [1, \boldsymbol{R}(t,1), \ldots, \boldsymbol{R}(t,z)], \tag{3.5}$$

with the coordinates defined by

$$R(t,u) = P(T(u) > t)$$

for $t \in \langle 0, \infty)$, $u = 1, 2, \ldots, z$.

In the case when the system operation time θ is large enough, the coordinates of the unconditional reliability function of the system defined by (3.5) are given by

$$\boldsymbol{R}(t,u) \cong \sum_{b=1}^{v} p_b [\boldsymbol{R}(t,u)]^{(b)} \quad \text{for } t \geq 0,\ u = 1, 2, \ldots, z, \tag{3.6}$$

where $[\boldsymbol{R}(t,u)]^{(b)}$, $u = 1, 2, \ldots, z$, $b = 1, 2, \ldots, v$, are the coordinates of the system conditional reliability functions defined by (3.3)–(3.4) and p_b, $b = 1, 2, \ldots, v$, are the system operation process limit transient probabilities given by (2.22).

The mean value of the system unconditional lifetime $T(u)$ in the reliability state subset $\{u, u+1, \ldots, z\}$ is given by [25, 31, 50, 51]

$$\mu(u) \cong \sum_{b=1}^{v} p_b \mu_b(u), \quad u = 1, 2, \ldots, z, \tag{3.7}$$

where $\mu_b(u)$ are the mean values of the system conditional lifetimes $T^{(b)}(u)$ in the reliability state subset $\{u, u+1, \ldots, z\}$ at the operation state z_b, $b = 1, 2, \ldots, v$, given by

$$\mu_b(u) = \int_0^\infty [\boldsymbol{R}(t,u)]^{(b)}\, dt, \quad u = 1, 2, \ldots, z, \tag{3.8}$$

$[R(t,u)]^{(b)}$, $u = 1, 2, \ldots, z$, $b = 1, 2, \ldots, v$, are defined by (3.3)–(3.4) and p_b are given by (2.22). Whereas, the variance of the system unconditional lifetime $T(u)$ is given by

$$\sigma^2(u) = 2 \int_0^\infty t\boldsymbol{R}(t,u)\, dt - [\mu(u)]^2, \quad u = 1, 2, \ldots, z, \tag{3.9}$$

where $\boldsymbol{R}(t,u)$, $u = 1,2,\ldots,z$, are given by (3.5)–(3.6) and $\mu(u)$, $u = 0,1,\ldots,z$, are given by (3.7)–(3.8).

Hence, according to (1.19), we get the following formulae for the mean values of the unconditional lifetimes of the system in particular reliability states

$$\bar{\mu}(u) = \mu(u) - \mu(u+1), \quad u = 0,1,\ldots,z-1, \; \bar{\mu}(z) = \mu(z), \tag{3.10}$$

where $\mu(u)$, $u = 0,1,\ldots,z$, are given by (3.7)–(3.8).

Moreover, according (1.20)–(1.21), if r is the system critical reliability state, then the system risk function is given by

$$\boldsymbol{r}(t) = 1 - \boldsymbol{R}(t,r), \quad t \in \langle 0,\infty), \tag{3.11}$$

where $\boldsymbol{R}(t,r)$ is the coordinate of the system unconditional reliability function given by (3.6) for $u = r$ and if τ is the moment when the system risk function exceeds a permitted level δ, then

$$\tau = \boldsymbol{r}^{-1}(\delta), \tag{3.12}$$

where $\boldsymbol{r}^{-1}(t)$, if it exists, is the inverse function of the risk function $\boldsymbol{r}(t)$ given by (3.11).

Further, we assume that the system components E_i, $i = 1,2,\ldots,n$, at the system operation states z_b, $b = 1,2,\ldots,v$, have the exponential reliability functions, i.e. their coordinates are given by

$$[R_i(t,u)]^{(b)} = P(T_i^{(b)}(u) > t | Z(t) = z_b) = \exp\left[-[\lambda_i(u)]^{(b)} t\right]$$

for $t \in \langle 0,\infty)$, $u = 1,2,\ldots,z$, $b = 1,2,\ldots,v$, and we have

$$\begin{aligned}[\boldsymbol{R}(t,u)]^{(b)} &= P(T^{(b)}(u) > t | Z(t) = z_b)\\ &= [\boldsymbol{R}([R_1(t,u)]^{(b)}, [R_2(t,u)]^{(b)}, \ldots, [R_n(t,u)]^{(b)})]^{(b)}\\ &= [\boldsymbol{R}(\exp[-[\lambda_1(u)]^{(b)} t], \exp[-[\lambda_2(u)]^{(b)} t], \ldots, \exp[-[\lambda_n(u)]^{(b)} t])]^{(b)}\end{aligned}$$

for $t \in \langle 0,\infty)$, $u = 1,2,\ldots,z$, $b = 1,2,\ldots,v$, $n \in N$.

The reason for this strong assumption on the system components is that the exponential distribution has "no memory" expressed in the following property:

$$\begin{aligned}[R_i(t_0+t,u)]^{(b)} &= P(T_i^{(b)}(u) > t_0 + t / T_i^{(b)}(u) > t_0 | Z(t) = z_b)\\ &= P(T_i^{(b)}(u) > t_0 + t \cap T_i^{(b)}(u) > t_0 | Z(t) = z_b) / P(T_i^{(b)}(u) > t_0 | Z(t) = z_b)\\ &= P(T_i^{(b)}(u) > t_0 + t | Z(t) = z_b) / P(T_i^{(b)}(u) > t_0 | Z(t) = z_b)\\ &= \exp[-[\lambda_i(u)]^{(b)}(t+t_0) / \exp[-[\lambda_i(u)]^{(b)} t_0] = \exp[-[\lambda_i(u)]^{(b)} t]\\ &= P(T_i^{(b)}(u) > t | Z(t) = z_b) = [R_i(t,u)]^{(b)}\end{aligned}$$

for $t_0 \in \langle 0,\infty)$ and $t \in \langle 0,\infty)$, $u = 1,2,\ldots,z$.

Both, the assumption about the exponential reliability functions of the system components and the above property, justify the following form of the formula (3.6):

$$\begin{aligned}\boldsymbol{R}(t,u) &\cong \sum_{b=1}^{v} p_b[\boldsymbol{R}(t,u)]^{(b)} \\ &= \sum_{b=1}^{v} p_b[\boldsymbol{R}(\exp[-[\lambda_1(u)]^{(b)}t], \exp[-[\lambda_2(u)]^{(b)}t], \ldots, \exp[-[\lambda_n(u)]^{(b)}t])]^{(b)}\end{aligned}$$

for $t \geq 0,\ u = 1, 2, \ldots, z.$

The application of the above formula and the results given in Chap. 1 yield the following results formulated in the form of the following proposition.

Proposition 3.1 *If components of the multistate system at the operation state* $z_b,\ b = 1, 2, \ldots, v$, *have the exponential reliability functions given by*

$$[R_i(t,\cdot)]^{(b)} = \left[1, [R_i(t,1)]^{(b)}, \ldots, [R_i(t,z)]^{(b)}\right], \quad t \in (-\infty, \infty),\ b = 1, 2, \ldots, v, \tag{3.13}$$

where

$$\begin{aligned}&[R_i(t,u)]^{(b)} = 1 \quad \text{for } t<0, \\ &[R_i(t,u)]^{(b)} = \exp[-[\lambda_i(u)]^{(b)}t] \quad \text{for } t \geq 0,\ [\lambda_i(u)]^{(b)} > 0, \\ &\quad i = 1, 2, \ldots, n,\ u = 1, 2, \ldots, z,\ b = 1, 2, \ldots, v,\end{aligned} \tag{3.14}$$

in the case of series, parallel, "m out of n", consecutive "m out of n: F" systems and respectively by

$$\left[R_{ij}(t,\cdot)\right]^{(b)} = \left[1, \left[R_{ij}(t,1)\right]^{(b)}, \ldots, \left[R_{ij}(t,z)\right]^{(b)}\right], \quad t \in (-\infty, \infty),\ b = 1, 2, \ldots, v, \tag{3.15}$$

where

$$\begin{aligned}&[R_{ij}(t,u)]^{(b)} = 1 \quad \text{for } t<0, \\ &[R_{ij}(t,u)]^{(b)} = \exp[-[\lambda_{ij}(u)]^{(b)}t] \quad \text{for } t \geq 0,\ [\lambda_{ij}(u)]^{(b)} > 0, \\ &\quad i = 1, 2, \ldots, k,\ j = 1, 2, \ldots, l_i,\ u = 1, 2, \ldots, z,\ b = 1, 2, \ldots, v,\end{aligned} \tag{3.16}$$

in the case of series–parallel, parallel–series, series-"m out of k", "m_i out of l_i"-series, series-consecutive "m out of k: F*" and consecutive "m_i out of l_i:* F*"-series systems and the system operation time* θ *is large enough, then its multistate unconditional reliability function is given by the vector:*

(i) *for a series system*

$$\overline{\boldsymbol{R}}_n(t,\cdot) = [1, \overline{\boldsymbol{R}}_n(t,1), \ldots, \overline{\boldsymbol{R}}_n(t,z)], \tag{3.17}$$

where

$$\overline{\boldsymbol{R}}_n(t,u) = 1 \quad \textit{for } t < 0,$$
$$\overline{R}_n(t,u) \cong \sum_{b=1}^{v} p_b \exp[-\sum_{i=1}^{n} [\lambda_i(u)]^{(b)} t] \quad \textit{for } t \geq 0,\ u = 1,2,\ldots,z; \tag{3.18}$$

(ii) *for a parallel system*

$$\boldsymbol{R}_n(t,\cdot) = [1, \boldsymbol{R}_n(t,1), \ldots, \boldsymbol{R}_n(t,z)], \tag{3.19}$$

where

$$\boldsymbol{R}_n(t,u) \cong 1 - \sum_{b=1}^{v} p_b \prod_{i=1}^{n} [1 - \exp[-[\lambda_i(u)]^{(b)} t]] \quad \textit{for } t \geq 0, \quad u = 1,2,\ldots,z; \tag{3.20}$$

(iii) *for an "m out of n" system*

$$\boldsymbol{R}_n^m(t,\cdot) = [1, \boldsymbol{R}_n^m(t,1), \ldots, \boldsymbol{R}_n^m(t,z)], \tag{3.21}$$

where

$$R_n^m(t,u) = 1 \quad \textit{for } t < 0,$$
$$R_n^m(t,u) \cong 1 - \sum_{b=1}^{v} p_b \sum_{\substack{r_1,r_2,\ldots,r_n=0 \\ r_1+r_2+\cdots+r_n \leq m-1}}^{1} \prod_{i=1}^{n} \exp[-r_i[\lambda_i(u)]^{(b)} t][1 - \exp[-[\lambda_i(u)]^{(b)} t]]^{1-r_i}$$
$$\textit{for } t \geq 0,\ u = 1,2,\ldots,z, \tag{3.22}$$

or

$$\overline{\boldsymbol{R}}_n^{\overline{m}}(t,\cdot) = [1, \overline{\boldsymbol{R}}_n^{\overline{m}}(t,1), \ldots, \overline{\boldsymbol{R}}_n^{\overline{m}}(t,z)], \tag{3.23}$$

where

$$\overline{\boldsymbol{R}}_n^{\overline{m}}(t,u) = 1 \quad \textit{for } t < 0,$$
$$\overline{\boldsymbol{R}}_n^{\overline{m}}(t,u) \cong \sum_{b=1}^{v} p_b \sum_{\substack{r_1,r_2,\ldots,r_n=0 \\ r_1+r_2+\cdots+r_n \leq \overline{m}}}^{1} \prod_{i=1}^{n} [1 - \exp[-[\lambda_i(u)]^{(b)} t]]^{r_i} \exp[-(1-r_i)[\lambda_i(u)]^{(b)} t]$$
$$\textit{for } t \geq 0,\ \overline{m} = n - m,\ u = 1,2,\ldots,z; \tag{3.24}$$

(iv) *for a consecutive "m out of n: F" system*

$$\boldsymbol{CR}_n^m(t,\cdot) = [1, \boldsymbol{CR}_n^m(t,1), \ldots, \boldsymbol{CR}_n^m(t,z)], \tag{3.25}$$

where

$$\boldsymbol{CR}_n^m(t,u) = 1 \quad \text{for } t < 0,$$

$$\boldsymbol{CR}_n^m(t,u) \cong \begin{cases} 1 & \text{for } n < m, \\ 1 - \sum_{b=1}^{v} p_b \prod_{i=1}^{n} [1 - \exp[-[\lambda_i(u)]^{(b)} t]] & \text{for } n = m, \\ \sum_{b=1}^{v} p_b \exp[-[\lambda_n(u)]^{(b)} t][\boldsymbol{CR}_{n-1}^m(t,u)]^{(b)} & \\ \quad + \sum_{i=1}^{m-1} \exp[-[\lambda_{n-i}(u)]^{(b)} t][\boldsymbol{CR}_{n-i-1}^m(t,u)]^{(b)} & \\ \quad \prod_{j=n-i+1}^{n} [1 - \exp[-[\lambda_j(u)]^{(b)} t]] & \text{for } n > m, \end{cases}$$

for $t \geq 0,\ u = 1, 2, \ldots, z;$

$$\tag{3.26}$$

(v) *for a series–parallel system*

$$\boldsymbol{R}_{k;l_1,l_2,\ldots,l_k}(t,\cdot) = [1, \boldsymbol{R}_{k;l_1,l_2,\ldots,l_k}(t,1), \ldots, \boldsymbol{R}_{k;l_1,l_2,\ldots,l_k}(t,z)], \tag{3.27}$$

where

$$\boldsymbol{R}_{k;l_1,l_2,\ldots,l_k}(t,u) = 1 \quad \text{for } t < 0,$$

$$\boldsymbol{R}_{k;l_1,l_2,\ldots,l_k}(t,u) \cong 1 - \sum_{b=1}^{v} p_b \prod_{i=1}^{k} \left[1 - \exp\left[-\sum_{j=1}^{l_i} [\lambda_{ij}(u)]^{(b)} t\right]\right] \tag{3.28}$$

for $t \geq 0,\ u = 1, 2, \ldots, z;$

(vi) *for a parallel–series system*

$$\overline{\boldsymbol{R}}_{k;l_1,l_2,\ldots,l_k}(t,\cdot) = [1, \overline{\boldsymbol{R}}_{k;l_1,l_2,\ldots,l_k}(t,1), \ldots, \overline{\boldsymbol{R}}_{k;l_1,l_2,\ldots,l_k}(t,z)] \tag{3.29}$$

where

$$\overline{\boldsymbol{R}}_{k;l_1,l_2,\ldots,l_k}(t,u) = 1 \quad \text{for } t < 0,$$

$$\overline{\boldsymbol{R}}_{k;l_1,l_2,\ldots,l_k}(t,u) \cong \sum_{b=1}^{v} p_b \prod_{i=1}^{k} \left[1 - \prod_{j=1}^{l_i} [1 - \exp[-[\lambda_{ij}(u)]^{(b)} t]]\right] \tag{3.30}$$

for $t \geq 0,\ u = 1, 2, \ldots, z;$

(vii) *for a series-"m out of k" system*

$$R_{k;l_1,l_2,\ldots,l_k}^m(t,\cdot) = [1, R_{k;l_1,l_2,\ldots,l_k}^m(t,1), \ldots, R_{k;l_1,l_2,\ldots,l_k}^m(t,z)], \tag{3.31}$$

where

$$\boldsymbol{R}^{m}_{k;l_1,l_2,\ldots,l_k}(t,u) = 1 \quad \textit{for } t<0,$$

$$\boldsymbol{R}^{m}_{k;l_1,l_2,\ldots,l_k}(t,u) \cong 1 - \sum_{b=1}^{v} p_b \sum_{\substack{r_1,r_2,\ldots,r_k=0 \\ r_1+r_2+\cdots+r_k \le m-1}}^{1} \prod_{i=1}^{k} \left[\prod_{j=1}^{l_i} \exp[-[\lambda_{ij}(u)]^{(b)} t]\right]^{r_i} \cdot \left[1 - \prod_{j=1}^{l_i} \exp[-[\lambda_{ij}(u)]^{(b)} t]\right]^{1-r_i} \quad \textit{for } t \ge 0,\ u = 1,2,\ldots,z, \tag{3.32}$$

or

$$\overline{\boldsymbol{R}}^{\overline{m}}_{k;l_1,l_2,\ldots,l_k}(t,\cdot) = [1, \overline{\boldsymbol{R}}^{\overline{m}}_{k;l_1,l_2,\ldots,l_k}(t,1),\ldots,\overline{\boldsymbol{R}}^{\overline{m}}_{k;l_1,l_2,\ldots,l_k}(t,z)], \tag{3.33}$$

where

$$\overline{\boldsymbol{R}}^{\overline{m}}_{k;l_1,l_2,\ldots,l_k}(t,u) = 1 \quad \textit{for } t<0,$$

$$\overline{\boldsymbol{R}}^{\overline{m}}_{k;l_1,l_2,\ldots,l_k}(t,u) \cong \sum_{b=1}^{v} p_b \sum_{\substack{r_1,r_2,\ldots,r_k=0 \\ r_1,r_2+\cdots+r_k \le \overline{m}}}^{1} \prod_{i=1}^{k} \left[1 - \left[\prod_{i=1}^{l_i} \exp[-[\lambda_{ij}(u)]^{(b)} t]\right]\right]^{r_i} \cdot \left[\prod_{j=1}^{l_i} \exp[-[\lambda_{ij}(u)]^{(b)} t]\right]^{1-r_i} \quad \textit{for } t \ge 0,\ \overline{m} = k-m,\ u = 1,2,\ldots,z; \tag{3.34}$$

(viii) *for an "m_i out of l_i"-series system*

$$\overline{\boldsymbol{R}^{m_1,m_2,\ldots,m_k}_{k;l_1,l_2,\ldots,l_k}}(t,\cdot) = \left[1, \overline{\boldsymbol{R}^{m_1,m_2,\ldots,m_k}_{k;l_1,l_2,\ldots,l_k}}(t,1),\ldots,\overline{\boldsymbol{R}^{m_1,m_2,\ldots,m_k}_{k;l_1,l_2,\ldots,l_k}}(t,z)\right], \tag{3.35}$$

where

$$\overline{\boldsymbol{R}^{m_1,m_2,\ldots,m_k}_{k;l_1,l_2,\ldots,l_k}}(t,u) = 1 \quad \textit{for } t$$

$$\overline{\boldsymbol{R}^{m_1,m_2,\ldots,m_k}_{k;l_1,l_2,\ldots,l_k}}(t,u) \cong \sum_{b=1}^{v} p_b \prod_{i=1}^{k} \left[1 - \sum_{\substack{r_1,r_2,\ldots,r_{l_i}=0 \\ r_1+r_2+\ldots+r_{l_i} \le m_i-1}}^{1} \prod_{j=1}^{l_i} \exp[-r_j[\lambda_{ij}(u)]^{(b)} t] \cdot [1 - \exp[-[\lambda_{ij}(u)]^{(b)} t]]^{1-r_j}\right] \quad \textit{for } t \ge 0,\ u = 1,2,\ldots,z, \tag{3.36}$$

or

$$\overline{\boldsymbol{R}^{\overline{m}_1,\overline{m}_2,\ldots,\overline{m}_k}_{k;l_1,l_2,\ldots,l_k}}(t,\cdot) = \left[1, \overline{\boldsymbol{R}^{\overline{m}_1,\overline{m}_2,\ldots,\overline{m}_k}_{k;l_1,l_2,\ldots,l_k}}(t,1),\ldots,\overline{\boldsymbol{R}^{\overline{m}_1,\overline{m}_2,\ldots,\overline{m}_k}_{k;l_1,l_2,\ldots,l_k}}(t,z)\right], \tag{3.37}$$

where

$$\overline{\overline{\boldsymbol{R}}_{k;l_1,l_2,\ldots,l_k}^{\overline{m}_1,\overline{m}_2,\ldots,\overline{m}_k}}(t,u) = 1 \quad \text{for } t < 0,$$

$$\overline{\overline{\boldsymbol{R}}_{k;l_1,l_2,\ldots,l_k}^{\overline{m}_1,\overline{m}_2,\ldots,\overline{m}_k}}(t,u) \cong \sum_{b=1}^{v} p_b \prod_{i=1}^{k} \left[\sum_{\substack{r_1,r_2,\ldots,r_{l_i}=0 \\ r_1+r_2+\cdots+r_{l_i} \geq \overline{m}_i}}^{1} \prod_{j=1}^{l_i} [1-\exp[-[\lambda_{ij}(u)]^{(b)} t]]^{r_j} \cdot \exp[-(1-r_j)[\lambda_{ij}(u)]^{(b)} t] \right] \quad \text{for } t \geq 0,\ \overline{m}_i = l_i - m_i,$$

$$i = 1,2,\ldots,k,\ u = 1,2,\ldots,z; \tag{3.38}$$

(ix) *for a series-consecutive "m out of k: F" system*

$$\boldsymbol{CR}_{k;l_1,l_2,\ldots,l_k}^{m}(t,\cdot) = [1, \boldsymbol{CR}_{k;l_1,l_2,\ldots,l_k}^{m}(t,1),\ldots,\boldsymbol{CR}_{k;l_1,l_2,\ldots,l_k}^{m}(t,z)], \tag{3.39}$$

where

$$\boldsymbol{CR}_{k;l_1,l_2,\ldots,l_k}^{m}(t,u) = 1 \quad \text{for } t < 0,$$

$$\boldsymbol{CR}_{k;l_1,l_2,\ldots,l_k}^{m}(t,u) \cong \sum_{b=1}^{v} p_b [\boldsymbol{CR}_{k;l_1,l_2,\ldots,l_k}^{m}(t,u)]^{(b)} \quad \text{for } t \geq 0,\ u = 1,2,\ldots,z, \tag{3.40}$$

and $[\boldsymbol{CR}_{k;l_1,l_2,\ldots,l_k}^{m}(t,u)]^{(b)}$, $b = 1,2,\ldots,v$, *are given by*

$$\left[\boldsymbol{CR}_{k;l_1,l_2,\ldots,l_k}^{m}(t,u)\right]^{(b)} = \begin{cases} 1 & \text{for } k <; m, \\ 1 - \prod_{i=1}^{k}\left[1-\exp\left[-\sum_{j=1}^{l_i}\lambda_{ij}(u)t\right]\right] & \text{for } k = m, \\ \cdot\exp\left[-\sum_{j=1}^{l_k}[\lambda_{kj}(u)]^{(b)}t\right]\left[\boldsymbol{CR}_{k-1;l_1,l_2,\ldots,l_k}^{m}(t,u)\right]^{(b)} & \\ +\sum_{j=1}^{m-1}\left[\exp\left[-\sum_{v=1}^{l_{k-j}}[\lambda_{k-jv}(u)]^{(b)}t\right]\right]\left[\boldsymbol{CR}_{k-j-1;l_1,l_2,\ldots,l_k}^{m}(t,u)\right]^{(b)} & \\ \cdot\prod_{i=k-j+1}^{k}\left[1-\exp\left[-\sum_{v=1}^{l_i}[\lambda_{iv}(u)]^{(b)}t\right]\right] & \text{for } k > m, \end{cases} \tag{3.41}$$

for $t \geq 0,\ u = 1,2,\ldots,z;$

(x) *for a consecutive "m_i out of l_i: F"-series system*

$$\overline{\boldsymbol{CR}}^{m_1,m_2,\ldots,m_k}_{k;l_1,l_2,\ldots,l_k}(t,\cdot)=\left[1,\overline{\boldsymbol{CR}}^{m_1,m_2,\ldots,m_k}_{k;l_1,l_2,\ldots,l_k}(t,1),\ldots,\overline{\boldsymbol{CR}}^{m_1,m_2,\ldots,m_k}_{k;l_1,l_2,\ldots,l_k}(t,z)\right], \tag{3.42}$$

where

$$\overline{\boldsymbol{CR}}^{m_1,m_2,\ldots,m_k}_{k;l_1,l_2,\ldots,l_k}(t,u)=1 \quad \text{for } t<0,$$
$$\overline{\boldsymbol{CR}}^{m_1,m_2,\ldots,m_k}_{k;l_1,l_2,\ldots,l_k}(t,u)\cong\sum_{b=1}^{v}p_b\prod_{i=1}^{k}[CR^{m_i}_{i,l_i}(t,u)]^{(b)} \quad \text{for } t\geq 0,\ u=1,2,\ldots,z, \tag{3.43}$$

and $[\boldsymbol{CR}^{m_i}_{i,l_i}(t,u)]^{(b)},\ i=1,2,\ldots,k,\ b=1,2,\ldots,v,$ *are given by*

$$\left[\boldsymbol{CR}^{m_i}_{i,l_i}(t,u)\right]^{(b)}$$
$$=\begin{cases}1 & \text{for } l_i<m_i,\\ 1-\prod_{j=1}^{l_i}\left[1-\exp\left[-[\lambda_{ij}(u)]^{(b)}t\right]\right] & \text{for } l_i=m_i,\\ \exp\left[-[\lambda_{il_i}(u)]^{(b)}t\right]\left[\boldsymbol{CR}^{m_i}_{i,l_i-1}(t,u)\right]^{(b)} & \\ +\sum_{j=1}^{m_i-1}\exp\left[-[\lambda_{il_i-j}(u)]^{(b)}t\right]\left[\boldsymbol{CR}^{m_i}_{i,l_i-j-1}(t,u)\right]^{(b)} & \\ \cdot\prod_{v=l_i-j+1}^{l_i}\left[1-\exp\left[-[\lambda_{iv}(u)]^{(b)}t\right]\right] & \text{for } l_i>m_i,\end{cases} \tag{3.44}$$

for $t\geq 0,\ u=1,2,\ldots,z.$

Remark 3.1 The formulae for the reliability functions stated in *Proposition 3.1* are valid for the considered systems under the assumption that they do not change their structures at different operation states $z_b,\ b=1,2,\ldots,v$. This limitation can be simply omitted by the replacement in these formulae the system's structure shape constant parameters n,m,k,m_i,l_i, respectively by their changing at different operation states $z_b,\ b=1,2,\ldots,v$, equivalent structure shape parameters $n^{(b)},m^{(b)},k^{(b)},m_i^{(b)},l_i^{(b)},\ b=1,2,\ldots,v$.

For the exponential complex technical systems, considered in *Proposition 3.1*, we determine the mean values $\mu(u)$ and the standard deviations $\sigma(u)$ of the unconditional lifetimes of the system in the reliability state subsets $\{u,u+1,\ldots,z\}$, $u=1,2,\ldots,z$, the mean values $\overline{\mu}(u)$ of the unconditional lifetimes of the system in the particular reliability states u, $u=1,2,\ldots,z$, the system risk function $\boldsymbol{r}(t)$ and the moment τ when the system risk function exceeds a permitted level δ respectively defined by (3.7)–(3.12), after substituting for $R(t,u)$, $u=1,2,\ldots,z$, the coordinates of the unconditional reliability functions given respectively by (3.17)–(3.44).

3.3 Asymptotic Approach to Reliability of Large Multistate Systems at Variable Operation Conditions

In the case of large systems, the determination of the exact reliability functions of the complex systems and the system risk functions, sometimes, leads us to very complicated formulae that are often useless for reliability practitioners. One of the important techniques that can be useful in this situation is the asymptotic approach [19, 24, 46–48, 52] to system reliability evaluation. In this approach, instead of the preliminary complex formula for the system reliability function, after assuming that the number of system components tends to infinity and finding the limit reliability of the system, we can obtain its simplified form. Moreover, in the case of large systems, the possibility of combining the results of the reliability joint models of complex technical systems and the results concerning the limit reliability functions of the considered systems is possible. In this way, the results concerned with the asymptotic approach to estimation of non-repairable multistate systems at variable operation conditions may be obtained. The main results concerning the asymptotic approach to multistate large systems reliability with ageing components in the constant operation conditions are comprehensively presented in the work [19] and some of these results' extentions to the systems operating at the variable conditions can be found in [46–48, 51–52].

In order to combine the results on the reliability of multistate systems related to their operation processes with the results concerning the limit reliability functions of the multistate systems, and to obtain the results on the asymptotic approach to the evaluation of the large multistate systems reliability at the variable operation conditions, we assume the following definition [51].

Definition 3.1 A reliability function

$$\Re(t,\cdot) = [1, \Re(t,1), \ldots, \Re(t,z)], \quad t \in (-\infty, \infty), \tag{3.45}$$

where

$$\Re(t,u) = \sum_{b=1}^{v} p_b [\Re(t,u)]^{(b)}, \quad u = 1, 2, \ldots, z, \tag{3.46}$$

is called a limit reliability function of a complex multistate system with the reliability function sequence

$$\boldsymbol{R}_n(t,\cdot) = [1, \boldsymbol{R}_n(t,1), \ldots, \boldsymbol{R}_n(t,z)], \quad t \in (-\infty, \infty),\ n \in N, \tag{3.47}$$

where

$$R_n(t,u) \cong \sum_{b=1}^{v} p_b [R_n(t,u)]^{(b)}, \quad u = 1, 2, \ldots, z, \tag{3.48}$$

if there exist normalizing constants

$$a_n^{(b)}(u) > 0, \quad b_n^{(b)}(u) \in (-\infty, \infty), \quad u = 1, 2, \ldots, z, \ b = 1, 2, \ldots, v,$$

such that

$$\lim_{n \to \infty} [\boldsymbol{R}_n(a_n^{(b)}(u)t + b_n^{(b)}(u), u)]^{(b)} = [\Re(t, u)]^{(b)}$$

for all t from the sets of continuity points $C_{[\Re(u)]^{(b)}}$ of the functions $[\Re(t, u)]^{(b)}$, $u = 1, 2, \ldots, z, \ b = 1, 2, \ldots, v$.

Hence, for sufficiently large n, the following approximate formulae are valid

$$\boldsymbol{R}_n(t, \cdot) = [1, \boldsymbol{R}_n(t, 1), \ldots, \boldsymbol{R}_n(t, z)], \quad t \in (-\infty, \infty), \tag{3.49}$$

where

$$\boldsymbol{R}_n(t, u) \cong \sum_{b=1}^{v} p_b \left[\Re\left(\frac{t - b_n^{(b)}(u)}{a_n^{(b)}(u)}, u \right) \right]^{(b)}, \quad t \in (-\infty, \infty), \ u = 1, 2, \ldots, z. \tag{3.50}$$

The following propositions concerned with the large series–parallel and parallel–series exponential systems operating at the variable operation states are exemplary results that can be worked out on the basis of the results included in [19, 24, 48, 52] for the large systems considered in the book.

Proposition 3.2 *If components of the multistate series–parallel regular system at the operation states* z_b, $b = 1, 2, \ldots, v$, *i.e., the system with the structure shape parameters such that*

$$k = k_n^{(b)}, \quad l_1 = l_2 = \cdots = l_k = l_n^{(b)}, \quad b = 1, 2, \ldots, v, \ n \in N,$$

have the exponential reliability functions given by (3.15)–(3.16) *are homogeneous, i.e.,*

$$\left[\lambda_{ij}(u)\right]^{(b)} = [\lambda(u)]^{(b)}, \quad i = 1, 2, \ldots, k_n^{(b)}, \ j = 1, 2, \ldots, l_n^{(b)}, \ b = 1, 2, \ldots, v,$$

then the system unconditional multistate reliability function is given by the approximate formulae, respectively, in the following cases of the system structure shape at the particular operation states:

(i) $k_n^{(b)} = n, \ l_n^{(b)} > 0,$

$$\boldsymbol{R}(t, \cdot) = [1, \boldsymbol{R}(t, 1), \ldots, \boldsymbol{R}(t, z)]$$

where

$$\boldsymbol{R}(t,u) \cong 1 - \sum_{b=1}^{v} p_b \exp[-n\exp[-[\lambda(u)]^{(b)} l_n^{(b)} t]] \quad for\ t \in (-\infty,\infty), \qquad u = 1,2,\ldots,z; \tag{3.51}$$

(ii)

$$k_n^{(b)} \to k^{(b)},\ l_n^{(b)} \to \infty,$$

$$\boldsymbol{R}(t,\cdot) = [1, \boldsymbol{R}(t,1),\ldots,\boldsymbol{R}(t,z)]$$

where

$$\boldsymbol{R}(t,u) \cong \begin{cases} 1 & for\ t<0, \\ 1 - \sum_{b=1}^{v} p_b[1 - \exp[-[\lambda(u)]^{(b)} l_n^{(b)} t]]^{k^{(b)}} & for\ t \geq 0, \end{cases} \qquad u = 1,2,\ldots,z. \tag{3.52}$$

Proposition 3.3 *If components of the multistate parallel–series regular system at the operation states* z_b, $b = 1,2,\ldots,v$, *i.e., the system with the structure shape parameters such that*

$$k = k_n^{(b)}, \quad l_1 = l_2 = \cdots = l_k = l_n^{(b)}, \quad b = 1,2,\ldots,v,\ n \in N,$$

have the exponential reliability functions given by (3.15)–(3.16) *are homogeneous, i.e.,*

$$\left[\lambda_{ij}(u)\right]^{(b)} = [\lambda(u)]^{(b)}, \quad i = 1,2,\ldots,k_n^{(b)},\ j = 1,2,\ldots,l_n^{(b)},\ b = 1,2,\ldots,v,$$

then the system unconditional multistate reliability function is given by the approximate formulae, respectively, in the following cases of the system structure shapes at the particular operation states:

(i) $k_n^{(b)} = n,\ l_n^{(b)} \to l^{(b)},\ l^{(b)} > 0,$

$$\boldsymbol{R}(t,\cdot) = [1, \boldsymbol{R}(t,1),\ldots,\boldsymbol{R}(t,z)]$$

where

$$\boldsymbol{R}(t,u) \cong \begin{cases} 1 & for\ t<0, \\ \sum_{b=1}^{v} p_b \exp[-n([\lambda(u)]^{(b)} t)^{l^{(b)}}] & for\ t \geq 0, \end{cases} \qquad u = 1,2,\ldots,z. \tag{3.53}$$

(ii) $k_n^{(b)} \to k^{(b)},\ l_n^{(b)} \to \infty,$

$$\boldsymbol{R}(t,\cdot) = [1, \boldsymbol{R}(t,1), \ldots, \boldsymbol{R}(t,z)]$$

where

$$\boldsymbol{R}(t,u) \cong \sum_{b=1}^{v} p_b [1 - \exp[-l_n^{(b)} \exp[-[\lambda(u)]^{(b)} t]]]^{k^{(b)}} \quad \textit{for } t \in (-\infty, \infty),$$
$$u = 1, 2, \ldots, z. \tag{3.54}$$

It is possible to obtain similar and more general results for other considered in the book multistate systems after some modification of the results included in [19, 24].

3.4 Renewal and Availability of Multistate Systems at Variable Operation Conditions

The models and methods presented in the previous sections of this chapter can be applied for determining the reliability functions of the considered complex multistate systems at variable operation conditions, the mean values and variances of their lifetimes in the reliability state subsets and in the particular reliability states in the case when they are non-repairable. Combining these results with the results of the classical renewal theory it is possible to obtain the renewal and availability characteristics for repairable complex technical systems with ignored and non-ignored time of renovation. In this section are determined the distributions, the expected values and the variances of the times until the successive exceedings of the reliability critical state and the distributions, the expected values and the variances of the number of exceedings of the reliability critical state at a fixed moment of time for the considered complex technical systems in the case when they are repairable and the time of their renovation is ignored. The distribution functions, the expected values and the variances of the times are also determined until the successive renovations and the distribution functions, the expected values and the variances of the times until the successive exceeding of the reliability critical state and the distributions, the expected values and variances of the number of renovations up to a fixed moment of time and the distributions, the expected values and variances of the number of exceedings of the reliability critical state up to a fixed moment of time, the steady availability coefficients and the availability coefficients in a fixed time interval for the considered systems in the case when they are repairable and the time of their renovation is non-ignored.

3.4.1 Multistate Systems with Ignored Time of Renovation

We assume here that the considered systems after exceeding the critical reliability state are repaired and that the time of their renovation is very small in comparison to their lifetimes in the reliability state subsets, which is not worse than the critical reliability state, and hence we may omit it. Under this assumption, it is possible to obtain the results for the repairable systems with ignored time of renovation formulated in the following proposition [25, 31].

Proposition 3.4 *If components of the multistate repairable system with ignored time of renovation have the exponential reliability functions at the operation states* $z_b,\ b = 1,2,\ldots,v$, *given by* (3.13)–(3.14) *or respectively by* (3.15)–(3.16) *and the system reliability critical state is* $r,\ r \in \{1,2,\ldots,z\}$, *then:*

(a) *the time* $S_N(r)$ *until the Nth exceeding by the system the reliability critical state r, for sufficiently large N, has approximately normal distribution* $N(N\mu(r), \sqrt{N}\sigma(r))$, *i.e.,*

$$F^{(N)}(t,r) = P(S_N < t) \cong F_{N(0,1)}\left(\frac{t - N\mu(r)}{\sqrt{N}\sigma(r)}\right), \quad t \in (-\infty,\infty),\ r \in \{1,2,\ldots,z\}; \tag{3.55}$$

(b) *the expected value and the variance of the time* $S_N(r)$ *until the Nth exceeding by the system the reliability critical state r, for sufficiently large N, are respectively given by*

$$E[S_N(r)] \cong N\mu(r), \quad D[S_N(r)] \cong N\sigma^2(r), \quad r \in \{1,2,\ldots,z\}; \tag{3.56}$$

(c) *the number* $N(t,r)$ *of exceeding by the system the reliability critical state r up to the moment* $t,\ t \geq 0$, *for sufficiently large t, has approximately distribution of the form*

$$P(N(t,r) = N) \cong F_{N(0,1)}\left(\frac{(N+1)\mu(r) - t}{\sigma(r)\sqrt{\frac{t}{\mu(r)}}}\right) - F_{N(0,1)}\left(\frac{N\mu(r) - t}{\sigma(r)\sqrt{\frac{t}{\mu(r)}}}\right),$$

$$N = 0,1,\ldots,\ r \in \{1,2,\ldots,z\}; \tag{3.57}$$

(d) *the expected value and the variance of the number* $N(t,r)$ *of exceeding by the system the reliability critical state r up to the moment* $t,\ t \geq 0$, *for sufficiently large t, are respectively given by*

$$H(t,r) \cong \frac{t}{\mu(r)}, \quad D(t,r) \cong \frac{t}{\mu^3(r)}\sigma^2(r), \quad r \in \{1,2,\ldots,z\}; \tag{3.58}$$

where $\mu(r)$ *and* $\sigma(r)$ *are given by:*

(i) *for a series system*:

$$\mu(r) = \sum_{b=1}^{v} p_b \frac{1}{\sum_{i=1}^{n} \lambda_i^{(b)}(r)}, \tag{3.59}$$

$$\sigma^2(r) = \int_0^{+\infty} 2t\overline{\boldsymbol{R}}_n(t,r)\,\mathrm{d}t - [\mu(r)]^2, \tag{3.60}$$

where $\overline{\boldsymbol{R}}_n(t,r)$ *is given by* (3.18) *for* $u = r,\ r \in \{1,2,\ldots,z\}$;

(ii) *for a parallel system*:

$$\mu(r) = \int_0^{+\infty} \boldsymbol{R}_n(t,r)\,\mathrm{d}t, \tag{3.61}$$

$$[\sigma(r)]^2 = \int_0^{+\infty} 2t\boldsymbol{R}_n(t,r)\,\mathrm{d}t - [\mu(r)]^2, \tag{3.62}$$

where $\boldsymbol{R}_n(t,r)$ *is given by* (3.20) *for* $u = r,\ r \in \{1,2,\ldots,z\}$;

(iii) *for an "m out of n" system*:

$$\mu(r) = \int_0^{+\infty} \boldsymbol{R}_n^m(t,r)\,\mathrm{d}t, \tag{3.63}$$

$$[\sigma(r)]^2 = \int_0^{+\infty} 2t\boldsymbol{R}_n^m(t,r)\,\mathrm{d}t - [\mu(r)]^2, \tag{3.64}$$

where $\boldsymbol{R}_n^m(t,r)$ *is given by* (3.22) *for* $u = r,\ r \in \{1,2,\ldots,z\}$, *or*

$$\mu(r) = \int_0^{+\infty} \overline{\boldsymbol{R}}_n^{\overline{m}}(t,r)\,\mathrm{d}t, \tag{3.65}$$

$$[\sigma(r)]^{(2)} = \int_0^{+\infty} 2t\overline{\boldsymbol{R}}_n^{\overline{m}}(t,r)\,\mathrm{d}t - [\mu(r)]^2, \tag{3.66}$$

where $\overline{\boldsymbol{R}}_n^{\overline{m}}(t,r)$ *is given by* (3.24) *for* $u = r,\ r \in \{1,2,\ldots,z\}$;

(iv) *for a consecutive "m out of n*: F" *system*:

$$\mu(r) = \int_0^{+\infty} \boldsymbol{CR}_n^m(t,r)\,\mathrm{d}t, \tag{3.67}$$

$$[\sigma(r)]^2 = \int_0^{+\infty} 2t\boldsymbol{CR}_n^m(t,r)\,\mathrm{d}t - [\mu(r)]^2, \tag{3.68}$$

where $\boldsymbol{CR}_n^m(t,r)$ *is given by* (3.26) *for* $u = r,\ r \in \{1,2,\ldots,z\}$;

(v) *for a series–parallel system*:

$$\mu(r) = \int_0^{+\infty} \boldsymbol{R}_{k;l_1 l_2,\ldots,l_k}(t,r)\,\mathrm{d}t, \tag{3.69}$$

$$[\sigma(r)]^2 = \int_0^{+\infty} 2t\boldsymbol{R}_{k;l_1,l_2,\ldots,l_k}(t,r)\,\mathrm{d}t - [\mu(r)]^2, \tag{3.70}$$

where $\boldsymbol{R}_{k;l_1,l_2,\ldots,l_k}(t,r)$ *is given by* (3.28) *for* $u = r,\ r \in \{1,2,\ldots,z\}$;

(vi) *for a parallel–series system*:

$$\mu(r) = \int_0^{+\infty} \overline{\boldsymbol{R}}_{k;l_1 l_2,\ldots,l_k}(t,r)\,\mathrm{d}t, \tag{3.71}$$

$$[\sigma(r)]^2 = \int_0^{+\infty} 2t\overline{\boldsymbol{R}}_{k;l_1,l_2,\ldots,l_k}(t,r)\,\mathrm{d}t - [\mu(r)]^2, \tag{3.72}$$

where $\overline{\boldsymbol{R}}_{k;l_1,l_2,\ldots,l_k}(t,r)$ *is given by* (3.30) *for* $u = r,\ r \in \{1,2,\ldots,z\}$;

(vii) *for a series-"m out of k" system*:

$$\mu(r) = \int_0^{+\infty} \boldsymbol{R}^m_{k;l_1,l_2,\ldots,l_k}(t,r)\,\mathrm{d}t, \tag{3.73}$$

$$[\sigma(r)]^2 = \int_0^{+\infty} 2t\boldsymbol{R}^m_{k;l_1,l_2,\ldots,l_k}(t,r)\,\mathrm{d}t - [\mu(r)]^2, \tag{3.74}$$

where $\boldsymbol{R}^m_{k;l_1,l_2,\ldots,l_k}(t,r)$ *is given by* (3.32) *for* $u = r,\ r \in \{1,2,\ldots,z\}$, *or*

$$\mu(r) = \int_0^{+\infty} \overline{\boldsymbol{R}}^{\overline{m}}_{k;l_1,l_2,\ldots,l_k}(t,r)\,\mathrm{d}t, \tag{3.75}$$

$$[\sigma(r)]^2 = \int_0^{+\infty} 2t\overline{\boldsymbol{R}}^{\overline{m}}_{k;l_1,l_2,\ldots,l_k}(t,r)\,\mathrm{d}t - [\mu(r)]^2, \tag{3.76}$$

where $\overline{\boldsymbol{R}}^{\overline{m}}_{k;l_1,l_2,\ldots,l_k}(t,r)$ *is given by* (3.34) *for* $u = r,\ r \in \{1,2,\ldots,z\}$;

(viii) *for an "*m_i *out of* l_i*"-series system*:

$$\mu(r) = \int_0^{+\infty} \overline{\boldsymbol{R}^{m_1,m_2,\ldots,m_k}_{k;l_1,l_2,\ldots,l_k}}(t,r)\,\mathrm{d}t, \tag{3.77}$$

$$[\sigma(r)]^2 = \int_0^{+\infty} 2t\overline{\boldsymbol{R}^{m_1,m_2,\ldots,m_k}_{k;l_1,l_2,\ldots,l_k}}(t,r)\,\mathrm{d}t - [\mu(r)]^2, \tag{3.78}$$

where $\overline{\boldsymbol{R}^{m_1,m_2,\ldots,m_k}_{k;l_1,l_2,\ldots,l_k}}(t,r)$ *is given by* (3.36) *for* $u = r,\ r \in \{1,2,\ldots,z\}$, *or*

$$\mu(r) = \int_0^{+\infty} \overline{\overline{\boldsymbol{R}}^{\overline{m}_1,\overline{m}_2,\ldots,\overline{m}_k}_{k;l_1,l_2,\ldots,l_k}}(t,r)\,\mathrm{d}t, \tag{3.79}$$

$$[\sigma(r)]^2 = \int_0^{+\infty} 2t\overline{\overline{\boldsymbol{R}}^{\overline{m}_1,\overline{m}_2,\ldots,\overline{m}_k}_{k;l_1,l_2,\ldots,l_k}}(t,r)\,\mathrm{d}t - [\mu(r)]^2, \tag{3.80}$$

where $\overline{\overline{\boldsymbol{R}}^{\overline{m}_1,\overline{m}_2,\ldots,\overline{m}_k}_{k;l_1,l_2,\ldots,l_k}}(t,r)$ *is given by* (3.38) *for* $u = r,\ r \in \{1,2,\ldots,z\}$;

(ix) *for a series-consecutive "m out of k: F" system*:

$$\mu(r) = \int_0^{+\infty} \boldsymbol{CR}^{m}_{k;l_1,l_2,\ldots,l_k}(t,r)\,\mathrm{d}t, \tag{3.81}$$

$$[\sigma(r)]^2 = \int_0^{+\infty} 2t\boldsymbol{CR}^{m}_{k;l_1,l_2,\ldots,l_k}(t,r)\,\mathrm{d}t - [\mu(r)]^2, \tag{3.82}$$

where $\boldsymbol{CR}^{m}_{k;l_1,l_2,\ldots,l_k}(t,r)$ *is given by* (3.40)–(3.41) *for* $u = r,\ r \in \{1,2,\ldots,z\}$;

(x) *for a consecutive "m out of l: F"-series system*:

$$\mu(r) = \int_0^{+\infty} \overline{\boldsymbol{CR}}_{k;l_1,l_2,\ldots,l_k}^{m_1,m_2,\ldots,m_k}(t,r)\,\mathrm{d}t, \tag{3.83}$$

$$[\sigma(r)]^2 = \int_0^{+\infty} 2t\overline{\boldsymbol{CR}}_{k;l_1,l_2,\ldots,l_k}^{m_1,m_2,\ldots,m_k}(t,r)\,\mathrm{d}t - [\mu(r)]^2, \tag{3.84}$$

where $\overline{\boldsymbol{CR}}_{k;l_1,l_2,\ldots,l_k}^{m_1,m_2,\ldots,m_k}(t,r)$ *is given by* (3.43)–(3.44) *for* $u = r,\ r \in \{1,2,\ldots,z\}$.

3.4.2 Multistate Systems with Non-Ignored Time of Renovation

We assume here that the considered systems after exceeding the critical reliability state are repaired and that the time of their renovation is not very small in comparison to their lifetimes in the reliability state subsets, which is not worse than the critical reliability state, and hence we may not omit it. Under this assumption, it is possible to obtain the results formulated in the following proposition [25, 31].

Proposition 3.5 *If components of the multistate repairable system with non-ignored time of renovation have the exponential reliability functions at the operation states* z_b, $b = 1,2,\ldots,v$, *given by* (3.13)–(3.14) *or respectively by* (3.15)–(3.16), *the system reliability critical state is* r, $r \in \{1,2,\ldots,z\}$, *and the successive times of the system's renovations are independent and have an identical distribution function with the expected value* $\mu_0(r)$ and the variance $\sigma_0^2(r)$, then:

(a) *the time* $\overline{S}_N(r)$ *until the Nth exceeding by the system the reliability critical state* r, *for sufficiently large* N, *has approximately normal distribution* $N(N\mu(r) + (N-1)\mu_0(r)), \sqrt{N\sigma^2(r) + (N-1)\sigma_0^2(r)})$, *i.e.*,

$$\overline{F}^{(N)}(t,r) = P(\overline{S}_N(r) < t) \cong F_{N(0,1)}\left(\frac{t - N(\mu(r) + \mu_0(r)) + \mu_0(r)}{\sqrt{N(\sigma^2(r) + \sigma_0^2(r)) - \sigma_0^2(r)}}\right),$$
$$t \in (-\infty,\infty),\ r \in \{1,2,\ldots,z\}; \tag{3.85}$$

(b) *the expected value and the variance of the time* $\overline{S}_N(r)$ *until the Nth exceeding by the system the reliability critical state r, for sufficiently large N, are respectively given by*

$$E[\overline{S}_N(r)] \cong N\mu(r) + (N-1)\mu_0(r), \quad D[\overline{S}_N(r)] \cong N\sigma^2(r) + (N-1)\sigma_0^2(r),$$
$$r \in \{1,2,\ldots,z\}; \tag{3.86}$$

(c) *the number* $\overline{N}(t,r)$ *of exceeding by the system the reliability critical state r up to the moment t,* $t \geq 0$, *for sufficiently large t, has approximately distribution of the form*

$$P(\overline{N}(t,r) = N) \cong F_{N(0,1)}\left(\frac{(N+1)(\mu(r)+\mu_0(r)) - t - \mu_0(r)}{\sqrt{\frac{t+\mu_0(r)}{\mu(r)+\mu_0(r)}(\sigma^2(r)+\sigma_0^2(r))}}\right)$$
$$- F_{N(0,1)}\left(\frac{N(\mu(r)+\mu_0(r)) - t - \mu_0(r)}{\sqrt{\frac{t+\mu_0(r)}{\mu(r)+\mu_0(r)}(\sigma^2(r)+\sigma_0^2(r))}}\right), \tag{3.87}$$
$$N = 0,1,\ldots, \; r \in \{1,2,\ldots,z\};$$

(d) *the expected value and the variance of the number* $\overline{N}(t,r)$ *of exceeding by the system the reliability critical state r up to the moment t,* $t \geq 0$, *for sufficiently large t, are respectively given by*

$$\overline{H}(t,r) \cong \frac{t+\mu_0(r)}{\mu(r)+\mu_0(r)}, \quad \overline{D}(t,r) \cong \frac{t+\mu_0(r)}{(\mu(r)+\mu_0(r))^3}(\sigma^2(r)+\sigma_0^2(r)),$$
$$r \in \{1,2,\ldots,z\}; \tag{3.88}$$

(e) *the time* $\overline{\overline{S}}_N(r)$ *until the Nth system's renovation, for sufficiently large N, has approximately normal distribution* $N(N(\mu(r)+\mu_0(r)), \sqrt{N(\sigma^2(r)+\sigma_0^2(r))})$, *i.e.,*

$$\overline{\overline{F}}^{(N)}(t,r) = P(\overline{\overline{S}}_N(r) < t) \cong F_{N(0,1)}\left(\frac{t - N(\mu(r)+\mu_0(r))}{\sqrt{N(\sigma^2(r)+\sigma_0^2(r))}}\right), \quad t \in (-\infty,\infty),$$
$$r \in \{1,2,\ldots,z\}; \tag{3.89}$$

(f) *the expected value and the variance of the time* $\overline{\overline{S}}_N(r)$ *until the Nth system's renovation, for sufficiently large N, are respectively given by*

$$E[\bar{\bar{S}}_N(r)] \cong N(\mu(r) + \mu_0(r)), \quad D[\bar{\bar{S}}_N(r)] \cong N(\sigma^2(r) + \sigma_0^2(r)), \quad r \in \{1, 2, \ldots, z\}; \tag{3.90}$$

(g) *the number* $\bar{\bar{N}}(t, r)$ *of the system's renovations up to the moment* t, $t \geq 0$, *for sufficiently large* t, *has approximately distribution of the form*

$$P(\bar{\bar{N}}(t, r) = N) \cong F_{N(0,1)}\left(\frac{(N+1)(\mu(r) + \mu_0(r)) - t}{\sqrt{\frac{t}{\mu(r)+\mu_0(r)}(\sigma^2(r) + \sigma_0^2(r))}}\right) - F_{N(0,1)}\left(\frac{N(\mu(r) + \mu_0(r)) - t}{\sqrt{\frac{t}{\mu(r)+\mu_0(r)}(\sigma^2(r) + \sigma_0^2(r))}}\right), \quad N = 0, 1, \ldots, \ r \in \{1, 2, \ldots, z\}; \tag{3.91}$$

(h) *the expected value and the variance of the number* $\bar{\bar{N}}(t, r)$ *of system's reno-va-tions up to the moment* t, $t \geq 0$, *for sufficiently large* t, *are respectively given by*

$$\bar{\bar{H}}(t, r) \cong \frac{t}{\mu(r) + \mu_0(r)}, \quad \bar{\bar{D}}(t, r) \cong \frac{t}{(\mu(r) + \mu_0(r))^3}(\sigma^2(r) + \sigma_0^2(r)), \quad r \in \{1, 2, \ldots, z\}; \tag{3.92}$$

(i) *the steady availability coefficient of the system at the moment* t, $t \geq 0$, *for suffi-ciently large* t, *is given by*

$$A(t, r) \cong \frac{\mu(r)}{\mu(r) + \mu_0(r)}, \quad t \geq 0, \ r \in \{1, 2, \ldots, z\}; \tag{3.93}$$

(j) *the steady availability coefficient of the system in the time interval* $\langle t, t+\tau)$, $\tau > 0$, *for sufficiently large* t, *is given by*

$$A(t, \tau, r) \cong \frac{1}{\mu(r) + \mu_0(r)} \int_{\tau}^{\infty} R(t, r)\, dt, \quad t \geq 0, \ \tau > 0, \ r \in \{1, 2, \ldots, z\}; \tag{3.94}$$

where $\mu(r)$ *and* $\sigma(r)$ *are given respectively for the considered system by corresponding formulae selected from* (3.59)–(3.84) *and* $\boldsymbol{R}(t, r)$ *should be replaced by that corresponding to this system formula selected from* (3.17)–(3.44) *for* $u = r$.

3.5 Safety of Multistate Systems at Variable Operation Conditions

We assume that the changes of the system operation process $Z(t)$ states have an influence on the system multistate components E_i, $i = 1, 2, \ldots, n$, safety and the system safety structure as well. We mark by $T_1^{(b)}(u), T_2^{(b)}(u), \ldots, T_n^{(b)}(u)$ the system components $E_1, E_2, \ldots, E_n$ conditional lifetimes in the safety states subset $\{u, u+1, z\}$, $u = 1, 2, \ldots, z$, and by $T^{(b)}(u)$ the system conditional lifetimes in the safety states subset $\{u, u+1, z\}$, $u = 1, 2, \ldots, z$, while the system is at the operation state z_b, $b = 1, 2, \ldots, v$. Further, we define the conditional safety function of the system multistate component E_i, $i = 1, 2, \ldots, n$, while the system is at the operation state z_b, $b = 1, 2, \ldots, v$, by the vector [33, 50]

$$[s_i(t, \cdot)]^{(b)} = \left[1, [s_i(t,1)]^{(b)}, \ldots, [s_i(t,z)]^{(b)}\right], \tag{3.95}$$

where

$$[s_i(t,u)]^{(b)} = P(T_i^{(b)}(u) > t | Z(t) = z_b) \tag{3.96}$$

for $t \in \langle 0, \infty)$, $u = 1, 2, \ldots, z$, $b = 1, 2, \ldots, v$, and the conditional safety function of the multistate system while the system is at the operation state z_b, $b = 1, 2, \ldots, v$, by the vector [33, 50]

$$[\boldsymbol{s}(t, \cdot)]^{(b)} = [1, [\boldsymbol{s}(t,1)]^{(b)}, \ldots, [\boldsymbol{s}(t,z)]^{(b)}, \tag{3.97}$$

where

$$[\boldsymbol{s}(t,u)]^{(b)} = P(T^{(b)}(u) > t | Z(t) = z_b) \tag{3.98}$$

for $t \in \langle 0, \infty)$, $u = 1, 2, \ldots, z$, $b = 1, 2, \ldots, v$.

The system conditional lifetimes

$$T^{(b)}(u) = T(T_1^{(b)}(u), T_2^{(b)}(u), \ldots, T_n^{(b)}(u))$$

at the operation states z_b, defined for $u = 1, 2, \ldots, z$, $b = 1, 2, \ldots, v$, $n \in N$, are dependent on the system components conditional lifetimes $T_1^{(b)}(u), T_2^{(b)}(u), \ldots$, $T_n^{(b)}(u)$, at the operation state z_b and the coordinates of the system conditional multistate safety functions

$$[\boldsymbol{s}(t,u)]^{(b)} = \left[\boldsymbol{s}([s_1(t,u)]^{(b)}, [s_2(t,u)]^{(b)}, \ldots, [s_n(t,u)]^{(b)})\right]^{(b)}$$

at the operational state z_b, defined for $t \in \langle 0, \infty)$, $u = 1, 2, \ldots, z$, $b = 1, 2, \ldots, v$, $n \in N$, are dependent on the components, conditional safety function $[s_1(t,u)]^{(b)}, [s_2(t,u)]^{(b)}, \ldots, [s_n(t,u)]^{(b)}$ at the operation state z_b.

The safety function $[s_i(t,u)]^{(b)}$ is the conditional probability that the component E_i lifetime $T_i^{(b)}(u)$ in the safety state subset $\{u, u+1, \ldots, z\}$ is greater than t, while the process $Z(t)$ is at the operation state z_b. Similarly, the safety function $[\boldsymbol{s}(t,u)]^{(b)}$ is the conditional probability that the system lifetime $T^{(b)}(u)$ in the safety state subset $\{u, u+1, \ldots, z\}$ is greater than t, while the process $Z(t)$ is at the operation state z_b.

Consequently, we mark by $T(u)$ the system unconditional lifetime in the safety states subset $\{u, u+1, z\}$, $u = 1, 2, \ldots, z$, and we define the system unconditional safety function by the vector

$$\boldsymbol{s}(t,\cdot) = [1, \boldsymbol{s}(t,1), \ldots, \boldsymbol{s}(t,z)], \tag{3.99}$$

where

$$\boldsymbol{s}(t,u) = P(T(u) > t) \quad \text{for } t \in \langle 0, \infty), \; u = 1, 2, \ldots, z, \tag{3.100}$$

In the case when the system operation time θ is large enough, the system unconditional safety function is given by [33, 50]

$$\boldsymbol{s}(t,u) \cong \sum_{b=1}^{v} p_b [\boldsymbol{s}(t,u)]^{(b)} \quad \text{for } t \geq 0, \; u = 1, 2, \ldots, z, \tag{3.101}$$

where $[\boldsymbol{s}(t,u)]^{(b)}$, $u = 1, 2, \ldots, z$, $b = 1, 2, \ldots, v$, are the coordinates of the system conditional safety functions defined by (3.97)–(3.98) and p_b, $b = 1, 2, \ldots, v$, are the system operation process limit transient probabilities given by (2.22).

In safety analysis of large multistate systems at the variable operation conditions, we assume the following definitions.

Definition 3.2 A safety function

$$\mathcal{S}(t,\cdot) = [1, \mathcal{S}(t,1), \ldots, \mathcal{S}(t,z)], \quad t \in (-\infty, \infty), \tag{3.102}$$

where

$$\mathcal{S}(t,u) = \sum_{b=1}^{v} p_b [\mathcal{S}(t,u)]^{(b)}, \quad u = 1, 2, \ldots, z, \tag{3.103}$$

is called a limit safety function of a complex multistate system with the safety function sequence

$$\boldsymbol{s}_n(t,\cdot) = [1, \boldsymbol{s}_n(t,1), \ldots, \boldsymbol{s}_n(t,z)], \quad t \in (-\infty, \infty), \; n \in N,$$

where

$$\boldsymbol{s}_n(t,u) \cong \sum_{b=1}^{v} p_b [\boldsymbol{s}_n(t,u)]^{(b)}, \quad u = 1, 2, \ldots, z,$$

if there exist normalizing constants

$$a_n^{(b)}(u) > 0, \quad b_n^{(b)}(u) \in (-\infty, \infty), \quad u = 1, 2, \ldots, z, \; b = 1, 2, \ldots, v,$$

such that

$$\lim_{n \to \infty} [s_n(a_n^{(b)}(u)t + b_n^{(b)}(u), u)]^{(b)} = [\mathcal{S}(t, u)]^{(b)}$$

for all continuity points t of the safety functions $[\mathcal{S}(t, u)]^{(b)}$, $u = 1, 2, \ldots, z$, $b = 1, 2, \ldots, v$.

Hence, for sufficiently large n, the following approximate formulae are valid:

$$s_n(t, \cdot) = [s_n(t, 0), s_n(t, 1), \ldots, s_n(t, z)], \quad t \in (-\infty, \infty), \tag{3.104}$$

where

$$s_n(t, u) \cong \sum_{b=1}^{v} p_b \left[\mathcal{S}\left(\frac{t - b_n^{(b)}(u)}{a_n^{(b)}(u)}, u \right) \right]^{(b)}, \quad t \in (-\infty, \infty), \; u = 1, 2, \ldots, z. \tag{3.105}$$

From the definitions and agreements introduced in this chapter, it follows that all results concerning reliability analysis of multistate systems at the variable operation conditions determined in Sects. 3.2–3.4 can be directly used for safety analysis of the multistate systems at the variable operation conditions after replacing the appropriate notions of the multistate reliability analysis by the corresponding notions of the multistate safety analysis.

3.6 Applications

3.6.1 *Reliability, Renewal and Availability of Exemplary System at Variable Operation Conditions*

The exemplary technical system reliability analysis considered here at the constant operation conditions is performed in Sect. 1.4.1, whereas its varying in time operation process is analyzed in Sect. 2.3.1. In Sect. 2.3.1, it is fixed that the exemplary system reliability structure and its subsystems and components reliability depend on its changing in time operation states. Considering the assumptions and agreements of these sections, we assume that its subsystems S_υ, $\upsilon = 1, 2$, are composed of four-state, i.e. $z = 3$, components $E_{ij}^{(\upsilon)}$, $\upsilon = 1, 2$, having the conditional reliability functions given by the vector

$$\left[R_{ij}^{(\upsilon)}(t,\cdot)\right]^{(b)} = \left[1, \left[R_{ij}^{(\upsilon)}(t,1)\right]^{(b)}, \left[R_{ij}^{(\upsilon)}(t,2)\right]^{(b)}, \left[R_{ij}^{(\upsilon)}(t,3)\right]^{(b)}\right], \quad b = 1,2,3,4,$$

with the exponential co-ordinates

$$[R_{ij}^{(\upsilon)}(t,1)]^{(b)} = \exp[-[\lambda_{ij}^{(\upsilon)}(1)]^{(b)}], \quad [R_{ij}^{(\upsilon)}(t,2)]^{(b)} = \exp[-[\lambda_{ij}^{(\upsilon)}(2)]^{(b)}],$$
$$[R_{ij}^{(\upsilon)}(t,3)]^{(b)} = \exp[-[\lambda_{ij}^{(\upsilon)}(3)]^{(b)}],$$

different at various operation states z_b, $b = 1,2,3,4$, and with the intensities of departure from the reliability state subsets $\{1,2,3\}, \{2,3\}, \{3\}$, respectively

$$\left[\lambda_{ij}^{(\upsilon)}(1)\right]^{(b)}, \quad \left[\lambda_{ij}^{(\upsilon)}(2)\right]^{(b)}, \quad \left[\lambda_{ij}^{(\upsilon)}(3)\right]^{(b)}, \quad b = 1,2,3,4.$$

The influence of the system operation states changing on the changes in the system reliability structure and its components reliability functions is as follows.

At the system operation state z_1, the system is composed of the series–parallel subsystem S_1 with the structure shown in Fig. 2.8, containing two identical series subsystems ($k = 2$), each composed of three components ($l_1 = 3$, $l_2 = 3$) with the exponential reliability functions partly identified in Chap. 5 and given below. In both series subsystems of subsystem S_1 there are respectively:

- the components $E_{i1}^{(1)}$, $i = 1,2$, with the conditional reliability function coordinates

$$[R_{i1}^{(1)}(t,1)]^{(1)} = \exp[-0.0008t], \quad [R_{i1}^{(1)}(t,2)]^{(1)} = \exp[-0.0009t],$$
$$[R_{i1}^{(1)}(t,3)]^{(1)} = \exp[-0.0009t], \quad i = 1,2;$$

- the components $E_{i2}^{(1)}$, $i = 1,2$, with the conditional reliability function co-ordinates

$$[R_{i2}^{(1)}(t,1)]^{(1)} = \exp[-0.0011t], \quad [R_{i2}^{(1)}(t,2)]^{(1)} = \exp[-0.0011t],$$
$$[R_{i2}^{(1)}(t,3)]^{(1)} = \exp[-0.0012t], \quad i = 1,2;$$

- the components $E_{i3}^{(1)}$, $i = 1,2$, with the conditional reliability function co-ordinates

$$[R_{i3}^{(1)}(t,1)]^{(1)} = \exp[-0.0011t], \quad [R_{i3}^{(1)}(t,2)]^{(1)} = \exp[-0.0011t],$$
$$[R_{i3}^{(1)}(t,3)]^{(1)} = \exp[-0.0011t] \quad i = 1,2.$$

Thus, at the operational state z_1, the system is identical with subsystem S_1 that is a four-state series–parallel system with its structure shape parameters $k = 2$, $l_1 = 3$, $l_2 = 3$, and according to (1.36)–(1.37) and (1.68)–(1.69), its conditional reliability function is given by

$$[\boldsymbol{R}(t,\cdot)]^{(1)} = \left[1, [\boldsymbol{R}(t,1)]^{(1)}, [\boldsymbol{R}(t,2)]^{(1)}, [\boldsymbol{R}(t,3)]^{(1)}\right], \quad t \geq 0, \tag{3.106}$$

where

$$\begin{aligned}
[\boldsymbol{R}(t,1)]^{(1)} &= \boldsymbol{R}_{2;3,3}(t,1) = 1 - \prod_{i=1}^{2}\left[1 - \prod_{j=1}^{3}[R_{ij}^{(1)}(t,1)]^{(1)}\right] \\
&= 1 - \prod_{i=1}^{2}\left[1 - \exp\left[-\sum_{j=1}^{3}[\lambda_{ij}^{(1)}(1)]^{(1)}t\right]\right] \\
&= 1 - [1 - \exp[-[0.0008 + 0.0011 + 0.0011]t]]^2 \\
&= 1 - [1 - \exp[-0.003t]]^2 \\
&= 2\exp[-0.003t] - \exp[-0.006t],
\end{aligned} \tag{3.107}$$

$$\begin{aligned}
[\boldsymbol{R}(t,2)]^{(1)} &= \boldsymbol{R}_{2;3,3}(t,2) = 1 - \prod_{i=1}^{2}\left[1 - \prod_{j=1}^{3}[R_{ij}^{(1)}(t,2)]^{(1)}\right] \\
&= 1 - \prod_{i=1}^{2}\left[1 - \exp\left[-\sum_{j=1}^{3}[\lambda_{ij}^{(1)}(2)]^{(1)}t\right]\right] \\
&= 1 - [1 - \exp[-[0.0009 + 0.0011 + 0.0011]t]]^2 \\
&= 1 - [1 - \exp[-0.0031t]]^2 \\
&= 2\exp[-0.0031t] - \exp[-0.0062t],
\end{aligned} \tag{3.108}$$

$$\begin{aligned}
[\boldsymbol{R}(t,3)]^{(1)} &= \boldsymbol{R}_{2;3,3}(t,3) = 1 - \prod_{i=1}^{2}\left[1 - \prod_{j=1}^{3}[R_{ij}^{(1)}(t,3)]^{(1)}\right] \\
&= 1 - \prod_{i=1}^{2}\left[1 - \exp\left[-\sum_{j=1}^{3}[\lambda_{ij}^{(1)}(3)]^{(1)}t\right]\right] \\
&= 1 - [1 - \exp[-[0.0009 + 0.0012 + 0.0011]t]]^2 \\
&= 1 - [1 - \exp[-0.0032t]]^2 \\
&= 2\exp[-0.0032t] - \exp[-0.0064t].
\end{aligned} \tag{3.109}$$

The expected values and standard deviations of the system conditional lifetimes in the reliability state subsets $\{1,2,3\}$, $\{2,3\}$, $\{3\}$ at the operation state z_1, calculated from the results given by (3.107)–(3.109), according to (3.8)–(3.9), respectively are:

$$\mu_1(1) \cong 505, \quad \mu_1(2) \cong 483.87, \quad \mu_1(3) \cong 468.73, \tag{3.110}$$

$$\sigma_1(1) \cong 365.87, \quad \sigma_1(2) \cong 360.66, \quad \sigma_1(3) \cong 349.41, \tag{3.111}$$

and further, using (3.10) and (3.110), the mean values of the conditional lifetimes in the particular reliability states 1, 2, 3 at the operation state z_1, respectively are:

$$\bar{\mu}_1(1)21.13, \quad \bar{\mu}_1(2) \cong 15.14, \quad \bar{\mu}_1(3) \cong 468.7. \tag{3.112}$$

At the system operation state z_2, the system is composed of the series–parallel subsystem S_2 with the structure shown in Fig. 2.9, containing four identical series subsystems ($k = 4$), each composed of two components ($l_1 = 2$, $l_2 = 2$, $l_3 = 2$, $l_4 = 2$) with the exponential reliability functions given below. In all series subsystems of subsystem S_2 there are respectively:

- the components $E_{i1}^{(2)}$, $i = 1, 2, 3, 4$, with the conditional reliability function co-ordinates

$$[R_{i1}^{(2)}(t,1)]^{(2)} = \exp[-0.0013t], \quad [R_{i1}^{(2)}(t,2)]^{(2)} = \exp[-0.0014t],$$
$$[R_{i1}^{(2)}(t,3)]^{(2)} = \exp[-0.0015t], \quad i = 1, 2, 3, 4;$$

- the components $E_{i2}^{(2)}$, $i = 1, 2, 3, 4$, with the conditional reliability function co-ordinates

$$[R_{i2}^{(2)}(t,1)]^{(2)} = \exp[-0.0015t], \quad [R_{i2}^{(2)}(t,2)]^{(2)} = \exp[-0.0016t],$$
$$[R_{i2}^{(2)}(t,3)]^{(2)} = \exp[-0.0017t], \quad i = 1, 2, 3, 4.$$

Thus, at the operation state z_2, the system is identical with subsystem S_2 that is a four-state series–parallel system with its structure shape parameters $k = 4$, $l_1 = 2$, $l_2 = 2$, $l_3 = 2$, $l_4 = 2$ and according to (1.36)–(1.37) and (1.68)–(1.69), its conditional reliability function is given by

$$[\boldsymbol{R}(t,\cdot)]^{(2)} = \left[1, [\boldsymbol{R}(t,1)]^{(2)}, [\boldsymbol{R}(t,2)]^{(2)}, [\boldsymbol{R}(t,3)]^{(2)}\right], \quad t \geq 0, \tag{3.113}$$

where

$$\begin{aligned}
[\boldsymbol{R}(t,1)]^{(2)} &= \boldsymbol{R}_{4;2,2,2,2}(t,1) = 1 - \prod_{i=1}^{4}\left[1 - \prod_{j=1}^{2}[R_{ij}^{(2)}(t,1)]^{(2)}\right] \\
&= 1 - \prod_{i=1}^{4}\left[1 - \exp\left[-\sum_{j=1}^{2}[\lambda_{ij}^{(2)}(1)]^{(2)}t\right]\right] \\
&= 1 - [1 - \exp[-[0.0013 + 0.0015]t]]^4 \\
&= 1 - [1 - \exp[-0.0028t]]^4 \\
&= 4\exp[-0.0028t] - 6\exp[-0.0056t] + 4\exp[-0.0084t] - \exp[-0.0112t],
\end{aligned} \tag{3.114}$$

$$[\boldsymbol{R}(t,2)]^{(2)} = \boldsymbol{R}_{4;2,2,2,2}(t,2) = 1 - \prod_{i=1}^{4}\left[1 - \prod_{j=1}^{2}[R_{ij}^{(2)}(t,2)]^{(2)}\right]$$
$$= 1 - \prod_{i=1}^{4}\left[1 - \exp\left[-\sum_{j=1}^{2}[\lambda_{ij}^{(2)}(2)]^{(2)}t\right]\right]$$
$$= 1 - [1 - \exp[-[0.0014 + 0.0016]t]]^4$$
$$= 1 - [1 - \exp[-0.003t]]^4$$
$$= 4\exp[-0.003t] - 6\exp[-0.006t] + 4\exp[-0.009t] - \exp[-0.012t], \tag{3.115}$$

$$[\boldsymbol{R}(t,3)]^{(2)} = \boldsymbol{R}_{4;2,2,2,2}(t,3) = 1 - \prod_{i=1}^{4}\left[1 - \prod_{j=1}^{2}[R_{ij}^{(2)}(t,3)]^{(2)}\right]$$
$$= 1 - \prod_{i=1}^{4}\left[1 - \exp\left[-\sum_{j=1}^{2}[\lambda_{ij}^{(2)}(3)]^{(2)}t\right]\right]$$
$$= 1 - [1 - \exp[-[0.0015 + 0.0017]t]]^4$$
$$= 1 - [1 - \exp[-0.0032t]]^4$$
$$= 4\exp[-0.0032t] - 6\exp[-0.0064t] + 4\exp[-0.0096t] - \exp[-0.0128t]. \tag{3.116}$$

The expected values and standard deviations of the system conditional lifetimes in the reliability state subsets $\{1,2,3\}$, $\{2,3\}$, $\{3\}$ at the operation state z_2, calculated from the results given by (3.114)–(3.116), according to (3.8)–(3.9), respectively are:

$$\mu_2(1) \cong 744.05, \quad \mu_2(2) \cong 694.44, \quad \mu_2(3) \cong 651.04, \tag{3.117}$$

$$\sigma_2(1) \cong 426.12, \quad \sigma_2(2) \cong 397.76, \quad \sigma_2(3) \cong 372.86, \tag{3.118}$$

and further, using (3.10) and (3.117), the mean values of the conditional lifetimes in the particular reliability states 1, 2, 3at the operation state z_2, respectively are:

$$\bar{\mu}_2(1) \cong 49.61, \quad \bar{\mu}_2(2) \cong 43.4, \quad \bar{\mu}_2(3) \cong 651.04. \tag{3.119}$$

At the system operational state z_3, the system is a series system with the structure shown in Fig. 1.12, composed of two series–parallel subsystems S_1 and S_2 illustrated respectively in Figs. 2.8 and 2.9.

Subsystem S_1 with the structure shown in Fig. 2.8, consists of two identical series subsystems ($k = 2$) each composed of three components ($l_1 = 3,\ l_2 = 3$) with the exponential reliability functions given below. In both series subsystems of subsystem S_1 there are respectively:

- the components $E_{i1}^{(1)}$, $i = 1,2$, with the conditional reliability function co-ordinates

$$[R_{i1}^{(1)}(t,1)]^{(3)} = \exp[-0.0009t], \quad [R_{i1}^{(1)}(t,2)]^{(3)} = \exp[-0.001t],$$
$$[R_{i1}^{(1)}(t,3)]^{(3)} = \exp[-0.001t], \quad i = 1,2;$$

- the components $E_{i2}^{(1)}$, $i = 1,2$, with the conditional reliability function co-ordinates

$$[R_{i2}^{(1)}(t,1)]^{(3)} = \exp[-0.0012t], \quad [R_{i2}^{(1)}(t,2)]^{(3)} = \exp[-0.0012t],$$
$$[R_{i2}^{(1)}(t,3)]^{(3)} = \exp[-0.0013t], \quad i = 1,2;$$

- the components $E_{i3}^{(1)}$, $i = 1,2$, with the conditional reliability function co-ordinates

$$[R_{i3}^{(1)}(t,1)]^{(3)} = \exp[-0.0011t], \quad [R_{i3}^{(1)}(t,2)]^{(3)} = \exp[-0.0012t],$$
$$[R_{i3}^{(1)}(t,3)]^{(3)} = \exp[-0.0012t], \quad i = 1,2.$$

Thus, at the operation state z_3, subsystem S_1 is a four-state series–parallel system with its structure shape parameters $k = 2,\ l_1 = 3,\ l_2 = 3$, and according to (1.36)–(1.37) and (1.68)–(1.69), its conditional reliability function is given by

$$\left[\boldsymbol{R}^{(1)}(t,\cdot)\right]^{(3)} = \left[1, \left[\boldsymbol{R}^{(1)}(t,1)\right]^{(3)}, \left[\boldsymbol{R}^{(1)}(t,2)\right]^{(3)}, \left[\boldsymbol{R}^{(1)}(t,3)\right]^{(3)}\right], \quad t \geq 0, \tag{3.120}$$

where

$$\begin{aligned}[\boldsymbol{R}^{(1)}(t,1)]^{(3)} &= \boldsymbol{R}_{2;3,3}(t,1) = 1 - \prod_{i=1}^{2}\left[1 - \prod_{j=1}^{3}[R_{ij}^{(1)}(t,1)]^{(3)}\right] \\ &= 1 - \prod_{i=1}^{2}\left[1 - \exp\left[-\sum_{j=1}^{3}[\lambda_{ij}^{(1)}(1)]^{(3)}t\right]\right] \\ &= 1 - [1 - \exp[-[0.0009 + 0.0012 + 0.0011]t]]^2 \\ &= 1 - [1 - \exp[-0.0032t]]^2 \\ &= 2\exp[-0.0032t] - \exp[-0.0064t], \end{aligned} \tag{3.121}$$

$$\begin{aligned}[\boldsymbol{R}^{(1)}(t,2)]^{(3)} &= \boldsymbol{R}_{2;3,3}(t,2) = 1 - \prod_{i=1}^{2}\left[1 - \prod_{j=1}^{3}[R_{ij}^{(1)}(t,2)]^{(3)}\right] \\ &= 1 - \prod_{i=1}^{2}\left[1 - \exp\left[-\sum_{j=1}^{3}[\lambda_{ij}^{(1)}(2)]^{(3)}t\right]\right] \\ &= 1 - [1 - \exp[-[0.001 + 0.0012 + 0.0012]t]]^2 \\ &= 1 - [1 - \exp[-0.0034t]]^2 \\ &= 2\exp[-0.0034t] - \exp[-0.0068t], \end{aligned} \tag{3.122}$$

$$[\boldsymbol{R}(t,3)]^{(1)} = \boldsymbol{R}_{2;3,3}(t,3) = 1 - \prod_{i=1}^{2}\left[1 - \prod_{j=1}^{3}[R_{ij}^{(1)}(t,3)]^{(1)}\right]$$
$$= 1 - \prod_{i=1}^{2}\left[1 - \exp\left[-\sum_{j=1}^{3}[\lambda_{ij}^{(1)}(3)]^{(1)}t\right]\right]$$
$$= 1 - [1 - \exp[-[0.001 + 0.0013 + 0.0012]t]]^2$$
$$= 1 - [1 - \exp[-0.0035t]]^2$$
$$= 2\exp[-0.0035t] - \exp[-0.007t]. \tag{3.123}$$

Subsystem S_2 with the structure shown in Fig. 2.9, consists of four identical series subsystems ($k = 4$), each composed of two components ($l_1 = 2,\ l_2 = 2,\ l_3 = 2,\ l_4 = 2$) with the exponential reliability functions given below. In all series subsystems of subsystem S_2 there are respectively:

- the components $E_{i1}^{(2)}$, $i = 1,2,3,4$, with the conditional reliability function co-ordinates
  $$[R_{i1}^{(2)}(t,1)]^{(3)} = \exp[-0.0009t],\quad [R_{i1}^{(2)}(t,2)]^{(3)} = \exp[-0.001t],\quad [R_{i1}^{(2)}(t,3)]^{(3)} = \exp[-0.001t],\quad i = 1,2,3,4;$$
- the components $E_{i2}^{(2)}$, $i = 1,2,3,4$, with the conditional reliability function co-ordinates
  $$[R_{i2}^{(2)}(t,1)]^{(3)} = \exp[-0.0012t],\quad [R_{i2}^{(2)}(t,2)]^{(3)} = \exp[-0.0012t],\quad [R_{i2}^{(2)}(t,3)]^{(3)} = \exp[-0.0013t],\quad i = 1,2,3,4.$$

Thus, at the operation state z_3, subsystem S_2 is a four-state series–parallel system with its structure shape parameters $k = 4,\ l_1 = 2,\ l_2 = 2,\ l_3 = 2,\ l_4 = 2$, and according to (1.36)–(1.37) and (1.68)–(1.69), its conditional reliability function is given by

$$[\boldsymbol{R}^{(2)}(t,\cdot)]^{(3)} = [1, [\boldsymbol{R}^{(2)}(t,1)]^{(3)}, [\boldsymbol{R}^{(2)}(t,2)]^{(3)}, [\boldsymbol{R}^{(2)}(t,3)]^{(3)}],\quad t \geq 0, \tag{3.124}$$

where

$$[\boldsymbol{R}^{(2)}(t,1)]^{(3)} = \boldsymbol{R}_{4;2,2,2,2}(t,1) = 1 - \prod_{i=1}^{4}\left[1 - \prod_{j=1}^{2}[R_{ij}^{(2)}(t,1)]^{(3)}\right]$$
$$= 1 - \prod_{i=1}^{4}\left[1 - \exp\left[-\sum_{j=1}^{2}[\lambda_{ij}^{(2)}(1)]^{(3)}t\right]\right]$$
$$= 1 - [1 - \exp[-[0.0009 + 0.0012]t]]^4$$
$$= 1 - [1 - \exp[-0.0021t]]^4$$
$$= 4\exp[-0.0021t] - 6\exp[-0.0042t] + 4\exp[-0.0063t] - \exp[-0.0084t], \tag{3.125}$$

$$[\boldsymbol{R}^{(2)}(t,2)]^{(3)} = \boldsymbol{R}_{4;2,2,2,2}(t,2) = 1 - \prod_{i=1}^{4}\left[1 - \prod_{j=1}^{2}[R_{ij}^{(2)}(t,2)]^{(3)}\right]$$
$$= 1 - \prod_{i=1}^{4}\left[1 - \exp\left[-\sum_{j=1}^{2}[\lambda_{ij}^{(2)}(2)]^{(3)}t\right]\right]$$
$$= 1 - [1 - \exp[-[0.001 + 0.0012]t]]^4$$
$$= 1 - [1 - \exp[-0.0022t]]^4$$
$$= 4\exp[-0.0022t] - 6\exp[-0.0044t] + 4\exp[-0.0066t] - \exp[-0.0088t], \quad (3.126)$$

$$[\boldsymbol{R}^{(2)}(t,3)]^{(3)} = \boldsymbol{R}_{4;2,2,2,2}(t,3) = 1 - \prod_{i=1}^{4}\left[1 - \prod_{j=1}^{2}[R_{ij}^{(2)}(t,3)]^{(3)}\right]$$
$$= 1 - \prod_{i=1}^{4}\left[1 - \exp\left[-\sum_{j=1}^{2}[\lambda_{ij}^{(2)}(3)]^{(3)}t\right]\right]$$
$$= 1 - [1 - \exp[-[0.001 + 0.0013]t]]^4$$
$$= 1 - [1 - \exp[-0.0023t]]^4$$
$$= 4\exp[-0.0023t] - 6\exp[-0.0046t] + 4\exp[-0.0069t] - \exp[-0.0092t]. \quad (3.127)$$

Considering that the system at the operation state z_3 is a four-state series system composed of subsystems S_1 and S_2, after applying (1.22)–(1.23) and (3.120)–(3.123) and (3.124)–(3.127), its conditional reliability function is given by

$$[\boldsymbol{R}(t,\cdot)]^{(3)} = [1, [\boldsymbol{R}(t,1)]^{(3)}, [\boldsymbol{R}(t,2)]^{(3)}, [\boldsymbol{R}(t,3)]^{(3)}], \quad t \geq 0, \quad (3.128)$$

where

$$[\boldsymbol{R}(t,1)]^{(3)} = \overline{\boldsymbol{R}}_2(t,1) = [\boldsymbol{R}^{(1)}(t,1)]^{(3)}[\boldsymbol{R}^{(2)}(t,1)]^{(3)}$$
$$= 8\exp[-0.0053t] - 12\exp[-0.0074t] + 8\exp[-0.0095t] - 2\exp[-0.0116t]$$
$$- 4\exp[-0.0085t] + 6\exp[-0.0106t] - 4\exp[-0.0127t] + \exp[-0.0148t], \quad (3.129)$$

$$[\boldsymbol{R}(t,2)]^{(3)} = \overline{\boldsymbol{R}}_2(t,2) = [\boldsymbol{R}^{(1)}(t,2)]^{(3)}[\boldsymbol{R}^{(2)}(t,2)]^{(3)}$$
$$= 8\exp[-0.0056t] - 12\exp[-0.0078t] + 8\exp[-0.01t] - 2\exp[-0.0122t],$$
$$- 4\exp[-0.009t] + 6\exp[-0.0112t] - 4\exp[-0.0134t] + \exp[-0.0156t], \quad (3.130)$$

$$[\boldsymbol{R}(t,3)]^{(3)} = \overline{\boldsymbol{R}}_2(t,3) = [\boldsymbol{R}^{(1)}(t,3)]^{(3)}[\boldsymbol{R}^{(2)}(t,3)]^{(3)}$$
$$= 8\exp[-0.0058t] - 12\exp[-0.0081t] + 8\exp[-0.0104t] - 2\exp[-0.0127t],$$
$$- 4\exp[-0.0093t] + 6\exp[-0.00116t] - 4\exp[-0.0139t] + \exp[-0.0162t]. \tag{3.131}$$

The expected values and standard deviations of the system conditional lifetimes in the reliability state subsets $\{1,2,3\}$, $\{2,3\}$, $\{3\}$ at the operation state z_3, calculated from the results given by (3.128)–(3.131), according to (3.8)–(3.9), respectively are:

$$\mu_3(1) \cong 405.56, \quad \mu_3(2) \cong 383.04, \quad \mu_3(3) \cong 370.67, \tag{3.132}$$

$$\sigma_3(1) \cong 264.58, \quad \sigma_3(2) \cong 250.39, \quad \sigma_3(3) \cong 241.78, \tag{3.133}$$

and further, using (3.10) and (3.132), the mean values of the conditional lifetimes in the particular reliability states 1, 2, 3at the operation state z_3, respectively are:

$$\overline{\mu}_3(1)22.52, \quad \overline{\mu}_3(2) \cong 12.37, \quad \overline{\mu}_3(3) \cong 370.67. \tag{3.134}$$

At the system operation state z_4, the system is a series system with the scheme shown in Fig. 1.12, composed of subsystems S_1 and S_2 illustrated respectively in Figs. 2.8 and 2.9, whereas subsystem S_1 is a series–parallel system and subsystem S_2 is a series-"2 out of 4" system.

Subsystem S_1 consists of two identical series subsystems ($k = 2$), each composed of three components ($l_1 = 3$, $l_2 = 3$) with the exponential reliability functions the same as at the operation state z_3. Thus, according to (3.120)–(3.123), subsystem S_1 unconditional reliability function at the operation state z_4, is given by

$$[\boldsymbol{R}^{(1)}(t,\cdot)]^{(4)} = [1, [\boldsymbol{R}^{(1)}(t,1)]^{(4)}, [\boldsymbol{R}^{(1)}(t,2)]^{(4)}, [\boldsymbol{R}^{(1)}(t,3)]^{(4)}], \quad t \geq 0, \tag{3.135}$$

where

$$[\boldsymbol{R}^{(1)}(t,1)]^{(4)} = 2\exp[-0.0032t] - \exp[-0.0064t], \tag{3.136}$$

$$[\boldsymbol{R}^{(1)}(t,2)]^{(4)} = 2\exp[-0.0034t] - \exp[-0.0068t], \tag{3.137}$$

$$[\boldsymbol{R}(t,3)]^{(4)} = 2\exp[-0.0035t] - \exp[-0.007t]. \tag{3.138}$$

Subsystem S_2 consists of four identical series subsystems ($k = 4$), each composed of two components ($l_1 = 2$, $l_2 = 2$, $l_3 = 2$, $l_4 = 2$) with the exponential reliability functions given below and is a series-"2 out of 4" system ($m = 2$). In all series subsystems of subsystem S_2 there are respectively:

- the components $E_{i1}^{(2)}$, $i = 1, 2, 3, 4$, with the conditional reliability function co-ordinates

$$[R_{i1}^{(2)}(t,1)]^{(4)} = \exp[-0.0013t], \quad [R_{i1}^{(2)}(t,2)]^{(4)} = \exp[-0.0014t],$$
$$[R_{i1}^{(2)}(t,3)]^{(4)} = \exp[-0.0015t], \quad i = 1, 2, 3, 4;$$

- the components $E_{i2}^{(2)}$, $i = 1, 2, 3, 4$, with the conditional reliability function co-ordinates

$$[R_{i2}^{(2)}(t,1)]^{(4)} = \exp[-0.0015t], \quad [R_{i2}^{(2)}(t,2)]^{(4)} = \exp[-0.0016t],$$
$$[R_{i2}^{(2)}(t,3)]^{(4)} = \exp[-0.0018t], \quad i = 1, 2, 3, 4.$$

Thus, at the operation state z_4, subsystem S_2 is a four-state series-"2 out of 4" system, with its structure shape parameters $k = 4$, $m = 2$, $l_1 = 2$, $l_2 = 2$, $l_3 = 2$, $l_4 = 2$, and according to (1.40)–(1.41) and (1.72)–(1.73), its conditional reliability function is given by

$$[\boldsymbol{R}^{(2)}(t,\cdot)]^{(4)} = [1, [\boldsymbol{R}^{(2)}(t,1)]^{(4)}, [\boldsymbol{R}^{(2)}(t,2)]^{(4)}, [\boldsymbol{R}^{(2)}(t,3)]^{(4)}], \quad t \geq 0, \tag{3.139}$$

where

$$\begin{aligned}
[\boldsymbol{R}^{(2)}(t,1)]^{(4)} &= \boldsymbol{R}^{2}_{4;2,2,2,2}(t,1) = 1 - \sum_{\substack{r_1,r_2,r_3,r_4=0 \\ r_1+r_2+r_3+r_4 \leq 1}}^{1} \prod_{i=1}^{4} \left[\prod_{j=1}^{2} [R_{ij}^{(2)}(t,1)]^{(4)} \right]^{r_i} \\
&\quad \times \left[1 - \prod_{j=1}^{2} [R_{ij}^{(2)}(t,1)]^{(4)} \right]^{1-r_i} \\
&= 1 - \sum_{\substack{r_1,r_2,r_3,r_4=0 \\ r_1+r_2+r_3+r_4 \leq 1}}^{1} \prod_{i=1}^{4} \exp\left[-r_i \sum_{j=1}^{2} [\lambda_{ij}^{(2)}(1)]^{(4)} t \right] \\
&\quad \times \left[1 - \exp\left[-\sum_{j=1}^{2} [\lambda_{ij}^{(2)}(1)]^{(4)} t \right] \right]^{1-r_i} \\
&= 1 - [1 - \exp[-[0.0013 + 0.0015]t]]^{4} \\
&\quad - 4\exp[-1[0.0013 + 0.0015]t][1 - \exp[-[0.0013 + 0.0015]t]]^{3} \\
&= 1 - [1 - \exp[-0.0028t]]^{4} - 4\exp[-0.0028t][1 - \exp[-0.0028t]]^{3} \\
&= 6\exp[-0.0056t] - 8\exp[-0.0084t] + 3\exp[-0.0112t],
\end{aligned} \tag{3.140}$$

$$
\begin{aligned}
[\boldsymbol{R}^{(2)}(t,2)]^{(4)} &= \boldsymbol{R}^{2}_{4;2,2,2,2}(t,2) = 1 - \sum_{\substack{r_1,r_2,r_3,r_4=0\\ r_1+r_2+r_3+r_4\le 1}}^{1} \prod_{i=1}^{4}\left[\prod_{j=1}^{2}[R_{ij}^{(2)}(t,2)]^{(4)}\right]^{r_i} \\
&\quad \times \left[1-\prod_{j=1}^{2}[R_{ij}^{(2)}(t,2)]^{(4)}\right]^{1-r_i} \\
&= 1 - \sum_{\substack{r_1,r_2,r_3,r_4=0\\ r_1+r_2+r_3+r_4\le 1}}^{1} \prod_{i=1}^{4} \exp\left[-r_i\sum_{j=1}^{2}[\lambda_{ij}^{(2)}(2)]^{(4)}t\right] \\
&\quad \times \left[1-\exp\left[-\sum_{j=1}^{2}[\lambda_{ij}^{(2)}(2)]^{(4)}t\right]\right]^{1-r_i} \\
&= 1-[1-\exp[-[0.0014+0.0016]t]]^4 \\
&\quad -4\exp[-1[0.0014+0.0016]t][1-\exp[-[0.0014+0.0016]t]]^3 \\
&= 1-[1-\exp[-0.003t]]^4-4\exp[-0.003t][1-\exp[-0.003t]]^3 \\
&= 6\exp[-0.006t]-8\exp[-0.009t]+3\exp[-0.012t],
\end{aligned} \tag{3.141}
$$

$$
\begin{aligned}
[\boldsymbol{R}^{(2)}(t,3)]^{(4)} &= \boldsymbol{R}^{2}_{4;2,2,2,2}(t,3) = 1 - \sum_{\substack{r_1,r_2,r_3,r_4=0\\ r_1+r_2+r_3+r_4\le 1}}^{1} \prod_{i=1}^{4}\left[\prod_{j=1}^{2}[R_{ij}^{(2)}(t,3)]^{(4)}\right]^{r_i} \\
&\quad \times \left[1-\prod_{j=1}^{2}[R_{ij}^{(2)}(t,3)]^{(4)}\right]^{1-r_i} \\
&= 1 - \sum_{\substack{r_1,r_2,r_3,r_4=0\\ r_1+r_2+r_3+r_4\le 1}}^{1} \prod_{i=1}^{4} \exp\left[-r_i\sum_{j=1}^{2}[\lambda_{ij}^{(2)}(3)]^{(4)}t\right] \\
&\quad \times \left[1-\exp\left[-\sum_{j=1}^{2}[\lambda_{ij}^{(2)}(3)]^{(4)}t\right]\right]^{1-r_i} \\
&= 1-[1-\exp[-[0.0015+0.0018]t]]^4 \\
&\quad -4\exp[-1[0.0015+0.0018]t][1-\exp[-[0.0015+0.0018]t]]^3 \\
&= 1-[1-\exp[-0.0033t]]^4-4\exp[-0.0033t][1-\exp[-0.0033t]]^3 \\
&= 6\exp[-0.0066t]-8\exp[-0.0099t]+3\exp[-0.0132t].
\end{aligned} \tag{3.142}
$$

Considering that the system at the operation state z_4 is a four-state series system composed of subsystems S_1 and S_2, after applying (1.22)–(1.23) and (3.135)–(3.138) and (3.139)–(3.142), its conditional reliability function is given by

$$[\boldsymbol{R}(t,\cdot)]^{(4)} = [1, [\boldsymbol{R}(t,1)]^{(4)}, [\boldsymbol{R}(t,2)]^{(4)}, [\boldsymbol{R}(t,3)]^{(4)}], \quad t \geq 0, \tag{3.143}$$

where

$$\begin{aligned}[\boldsymbol{R}(t,1)]^{(4)} &= \overline{\boldsymbol{R}}_2(t,1) = [\boldsymbol{R}^{(1)}(t,1)]^{(4)}[\boldsymbol{R}^{(2)}(t,1)]^{(4)} \\ &= 12\exp[-0.0088t] - 16\exp[-0.0116t] + 6\exp[-0.0144t] \\ &\quad - 6\exp[-0.012t] + 8\exp[-0.0148t] - 3\exp[-0.0176t],\end{aligned} \tag{3.144}$$

$$\begin{aligned}[\boldsymbol{R}(t,2)]^{(4)} &= \overline{\boldsymbol{R}}_2(t,2) = [\boldsymbol{R}^{(1)}(t,2)]^{(4)}[\boldsymbol{R}^{(2)}(t,2)]^{(4)} \\ &= 12\exp[-0.0094t] - 16\exp[-0.0124t] + 6\exp[-0.0154t] \\ &\quad - 6\exp[-0.0128t] + 8\exp[-0.0158t] - 3\exp[-0.0188t],\end{aligned} \tag{3.145}$$

$$\begin{aligned}[\boldsymbol{R}(t,3)]^{(4)} &= \overline{\boldsymbol{R}}_2(t,3) = [\boldsymbol{R}^{(1)}(t,3)]^{(4)}[\boldsymbol{R}^{(2)}(t,3)]^{(4)} \\ &= 12\exp[-0.0101t] - 16\exp[-0.0134t] + 6\exp[-0.0167t] \\ &\quad - 6\exp[-0.0136t] + 8\exp[-0.0169t] - 3\exp[-0.0202t].\end{aligned} \tag{3.146}$$

The expected values and standard deviations of the system conditional lifetimes in the reliability state subsets $\{1,2,3\}$, $\{2,3\}$, $\{3\}$ at the operation state z_4, calculated from the results given by (3.143)–(3.146), according to (3.8)–(3.9), respectively are:

$$\mu_4(1) \cong 271.08, \quad \mu_4(2) \cong 253.88, \quad \mu_4(3) \cong 237.05, \tag{3.147}$$

$$\sigma_4(1) \cong 163.8, \quad \sigma_4(2) \cong 153.35, \quad \sigma_4(3) \cong 142.58, \tag{3.148}$$

and further, using (3.10) and (3.147), the mean values of the conditional lifetimes in the particular reliability states 1, 2, 3at the operation state z_4, respectively are:

$$\overline{\mu}_4(1) \cong 17.20, \quad \overline{\mu}_4(2) \cong 16.83, \quad \overline{\mu}_4(3) \cong 237.05. \tag{3.149}$$

In the case when the system operation time is large enough its unconditional four-state reliability function is given by the vector

$$\boldsymbol{R}(t,\cdot) = [1, \boldsymbol{R}(t,1), \boldsymbol{R}(t,2), \boldsymbol{R}(t,3)], \quad t \geq 0, \tag{3.150}$$

where according to (3.6) and considering the exemplary system operation process transient probabilities at the operation states determined by (2.32), the vector co-ordinates are given respectively by

$$\begin{aligned}\boldsymbol{R}(t,1) &= p_1[\boldsymbol{R}(t,1)]^{(1)} + p_2[\boldsymbol{R}(t,1)]^{(2)} + p_3[\boldsymbol{R}(t,1)]^{(3)} + p_4[\boldsymbol{R}(t,1)]^{(4)} \\ &= 0.214 \times [\boldsymbol{R}(t,1)]^{(1)} + 0.038 \times [\boldsymbol{R}(t,1)]^{(2)} + 0.293 \times [\boldsymbol{R}(t,1)]^{(3)} \\ &\quad + 0.455 \times [\boldsymbol{R}(t,1)]^{(4)} \quad \text{for } t \geq 0,\end{aligned} \tag{3.151}$$

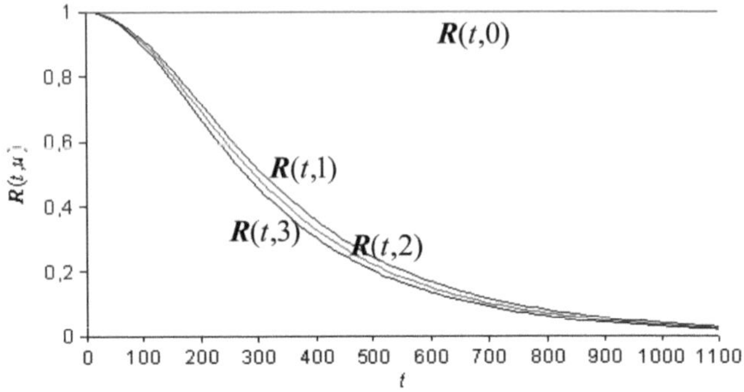

Fig. 3.1 The graph of the exemplary system reliability function $\boldsymbol{R}(t,\cdot)$ coordinates

$$\begin{aligned}\boldsymbol{R}(t,2) &= p_1[\boldsymbol{R}(t,2)]^{(1)} + p_2[\boldsymbol{R}(t,2)]^{(2)} + p_3[\boldsymbol{R}(t,2)]^{(3)} + p_4[\boldsymbol{R}(t,2)]^{(4)} \\ &= 0.214 \times [\boldsymbol{R}(t,2)]^{(1)} + 0.038 \times [\boldsymbol{R}(t,2)]^{(2)} + 0.293 \times [\boldsymbol{R}(t,2)]^{(3)} \\ &\quad + 0.455 \times [\boldsymbol{R}(t,2)]^{(4)} \quad \text{for } t \geq 0, \end{aligned} \tag{3.152}$$

$$\begin{aligned}\boldsymbol{R}(t,3) &= p_1[\boldsymbol{R}(t,3)]^{(1)} + p_2[\boldsymbol{R}(t,3)]^{(2)} + p_3[\boldsymbol{R}(t,3)]^{(3)} + p_4[\boldsymbol{R}(t,3)]^{(4)} \\ &= 0.214 \times [\boldsymbol{R}(t,3)]^{(1)} + 0.038 \times [\boldsymbol{R}(t,3)]^{(2)} + 0.293 \times [\boldsymbol{R}(t,3)]^{(3)} \\ &\quad + 0.455 \times [\boldsymbol{R}(t,3)]^{(4)} \quad \text{for } t \geq 0, \end{aligned} \tag{3.153}$$

where coordinates $[\boldsymbol{R}(t,1)]^{(1)}$, $[\boldsymbol{R}(t,1)]^{(2)}$, $[\boldsymbol{R}(t,1)]^{(3)}$, $[\boldsymbol{R}(t,1)]^{(4)}$ are given by (3.107), (3.114), (3.129), (3.144), $[\boldsymbol{R}(t,2)]^{(1)}$, $[\boldsymbol{R}(t,2)]^{(2)}$, $[\boldsymbol{R}(t,2)]^{(3)}$, $[\boldsymbol{R}(t,2)]^{(4)}$ are given by (3.108), (3.115), (3.130), (3.145) and $[\boldsymbol{R}(t,3)]^{(1)}$, $[\boldsymbol{R}(t,3)]^{(2)}$, $[\boldsymbol{R}(t,3)]^{(3)}$, $[\boldsymbol{R}(t,3)]^{(4)}$ are given by (3.109), (3.116), (3.131), (3.146).

The graph of the four-state exemplary system reliability function is illustrated in Fig. 3.1.

The expected values and standard deviations of the system unconditional lifetimes in the reliability state subsets $\{1,2,3\}$, $\{2,3\}$, $\{3\}$, calculated from the results given by (3.150)–(3.153), according to (3.7)–(3.9) and considering (2.32), (3.110), (3.117), (3.132), (3.147), respectively are:

$$\begin{aligned}\mu(1) &= p_1\mu_1(1) + p_2\mu_2(1) + p_3\mu_3(1) + p_4\mu_4(1) \\ &= 0.214 \times 505 + 0.038 \times 744.05 + 0.293 \times 405.56 + 0.455 \times 271.08 \cong 378.51,\end{aligned} \tag{3.154}$$

$$\sigma(1) \cong 286.77, \tag{3.155}$$

$$\begin{aligned}\mu(2) &= p_1\mu_1(2) + p_2\mu_2(2) + p_3\mu_3(2) + p_4\mu_4(2) \\ &= 0.214 \times 483.87 + 0.038 \times 694.44 + 0.293 \times 383.04 + 0.455 \times 253.88 \cong 357.68,\end{aligned} \tag{3.156}$$

$$\sigma(2) \cong 275.18, \tag{3.157}$$

$$\mu(3) = p_1\mu_1(3) + p_2\mu_2(3) + p_3\mu_3(3) + p_4\mu_4(3)$$
$$= 0.214 \times 468.73 + 0.038 \times 651.04 + 0.293 \times 370.67 + 0.455 \times 237.05 \cong 341.51, \tag{3.158}$$

$$\sigma(3) \cong 264.78. \tag{3.159}$$

Further, considering (3.10) and (3.154), (3.156) and (3.158), the mean values of the system unconditional lifetimes in the particular reliability states 1, 2, 3, respectively are:

$$\bar{\mu}(1) = \mu(1) - \mu(2) = 20.83, \quad \bar{\mu}(2) = \mu(2) - \mu(3) = 16.17,$$
$$\bar{\mu}(3) = \mu(3) = 341.51. \tag{3.160}$$

Since the critical reliability state is $r = 2$, then the system risk function, according to (3.11), is given by

$$\boldsymbol{r}(t) = 1 - \boldsymbol{R}(t, 2) \quad \text{for } t \geq 0, \tag{3.161}$$

where $\boldsymbol{R}(t, 2)$ is given by (3.152).

Hence, by (3.12), the moment when the system risk function exceeds a permitted level, for instance $\delta = 0.05$, is

$$\tau = \boldsymbol{r}^{-1}(\delta) \cong 70.08. \tag{3.162}$$

The graph of the risk function $\boldsymbol{r}(t)$ of the exemplary four-state system operating at the variable conditions is given in Fig. 3.2.

The reliability characteristics of the exemplary system operating at the variable conditions predicted in this section are different from those determined in Sect. 1.4.1 for this system operating at constant conditions. This fact justifies the sensibility of considering the real systems at the variable operation conditions that is appearing out in a natural way from practice. This approach, upon the good accuracy of the systems' operation processes identification, makes their reliability prediction more precise.

Further, assuming that the system is repaired after the exceeding of its reliability critical state $r = 2$ and that the time of the system renovation is ignored and applying *Proposition 3.4*, we obtain the following results:

(a) the time $S_N(2)$ until the Nth exceeding by the system the reliability critical state $r = 2$, for sufficiently large N, has approximately normal distribution $N(357.68N, 275.18\sqrt{N})$, i.e.,

$$F^{(N)}(t, 2) = P(S_N(2) < t) \cong F_{N(0,1)}\left(\frac{t - 357.68N}{275.18\sqrt{N}}\right), \quad t \in (-\infty, \infty);$$

(b) the expected value and the variance of the time $S_N(2)$ until the Nth exceeding by the system the reliability critical state $r = 2$ are respectively given by

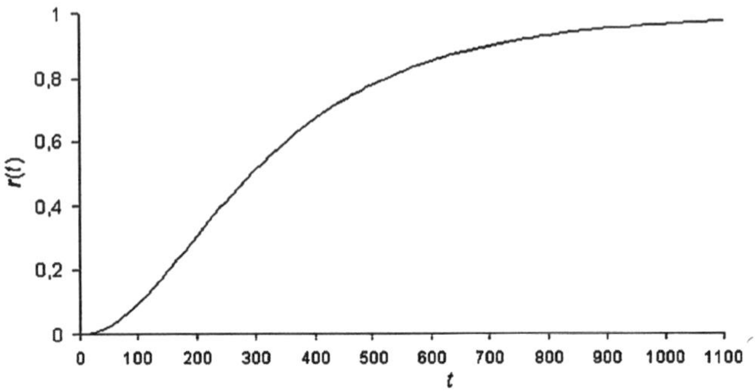

Fig. 3.2 The graph of the exemplary system risk function $\boldsymbol{r}(t)$

$$E[S_N(2)] \cong 357.68N, \quad D[S_N(2)] \cong 75724.03N;$$

(c) the number $N(t,2)$ of exceeding by the system the reliability critical state $r = 2$ up to the moment t, $t \geq 0$, for sufficiently large t, approximately has the distribution of the form

$$P(N(t,2) = N) \cong F_{N(0,1)}\left(\frac{357.68(N+1) - t}{14.55\sqrt{t}}\right) - F_{N(0,1)}\left(\frac{357.68N - t}{14.55\sqrt{t}}\right),$$
$$N = 0, 1, \ldots;$$

(d) the expected value and the variance of the number $N(t,2)$ of exceeding by the system the reliability critical state $r = 2$ up to the moment t, $t \geq 0$, for sufficiently large t, approximately are respectively given by

$$H(t,2) \cong 0.0028t, \quad D(t,2) \cong 0.0016t. \tag{3.163}$$

Assuming that the system is repaired after the exceeding of its reliability critical state $r = 2$ and that the time of the system renovation is non-ignored and it has the mean value $\mu_0(2) = 10$ and the standard deviation $\sigma_0(2) = 5$ and applying *Proposition 3.5*, we obtain the following results:

(a) the time $\overline{S}_N(2)$ until the Nth exceeding by the system the reliability critical state $r = 2$, for sufficiently large N, has approximately normal distribution $N(357.68N + 10(N-1), \sqrt{75724.03N + 25(N-1)})$, i.e.,

$$\overline{F}^{(N)}(t,2) = P(\overline{S}_N(2) < t) \cong F_{N(0,1)}\left(\frac{t - 367.68N + 10}{\sqrt{75749.03N - 25}}\right), \quad t \in (-\infty, \infty);$$

(b) the expected value and the variance of the time $\overline{S}_N(2)$ until the Nth exceeding by the system the reliability critical state $r = 2$, for sufficiently large N, are respectively given by

$$E[\bar{S}_N(2)] \cong 357.68N + 10(N-1), \quad D[\bar{S}_N(2)] \cong 75{,}724.03N + 25(N-1);$$

(c) the number $\overline{N}(t,2)$ of exceeding by the system the reliability critical state $r = 2$ up to the moment t, $t \geq 0$, for sufficiently large t, has approximately distribution of the form

$$P(\overline{N}(t,2) = N) \cong F_{N(0,1)}\left(\frac{367.68(N+1) - t - 10}{14.35\sqrt{t+10}}\right) - F_{N(0,1)}\left(\frac{367.68N - t - 10}{14.35\sqrt{t+10}}\right), \quad N = 0, 1, \ldots;$$

(d) the expected value and the variance of the number $\overline{N}(t,2)$ of exceeding by the system the reliability critical state $r = 2$ up to the moment t, $t \geq 0$, for sufficiently large t, are respectively given by

$$\overline{H}(t,2) \cong \frac{t+10}{367.68}, \quad \overline{D}(t,2) \cong 0.0015(t+10); \tag{3.164}$$

(e) the time $\bar{\bar{S}}_N(2)$ until the Nth system's renovation, for sufficiently large N, has approximately normal distribution $N(367.68N, 275.23\sqrt{N})$, i.e.,

$$\bar{\bar{F}}^{(N)}(t,2) = P(\bar{\bar{S}}_N(2) < t) \cong F_{N(0,1)}\left(\frac{t - 367.68N}{275.23\sqrt{N}}\right), \quad t \in (-\infty, \infty);$$

(f) the expected value and the variance of the time $\bar{\bar{S}}_N(2)$ until the Nth system's renovation, for sufficiently large N, are respectively given by

$$E[\bar{\bar{S}}_N(2)] \cong 367.68N, \quad D[\bar{\bar{S}}_N(2)] \cong 75749.03N;$$

(g) the number $\overline{\overline{N}}(t,2)$ of the system's renovations up to the moment t, $t \geq 0$, for sufficiently large t, has approximately distribution of the form

$$P(\overline{\overline{N}}(t,2) = N) \cong F_{N(0,1)}\left(\frac{367.68(N+1) - t}{14.35\sqrt{t}}\right) - F_{N(0,1)}\left(\frac{367.68N - t}{14.35\sqrt{t}}\right), \quad N = 0, 1, \ldots;$$

(h) the expected value and the variance of the number $\overline{\overline{N}}(t,2)$ of system's renovations up to the moment t, $t \geq 0$, for sufficiently large t, are respectively given by

$$\overline{\overline{H}}(t,2) \cong 0.0027t, \quad \overline{\overline{D}}(t,2) \cong 0.0015t; \tag{3.165}$$

(i) the steady availability coefficient of the system at the moment t, $t \geq 0$, for suffi-ciently large t, is given by

$$A(t,2) \cong 0.97, \quad t \geq 0;$$

(j) the steady availability coefficient of the system in the time interval $\langle t, t+\tau)$, $\tau > 0$, for sufficiently large t, is given by

$$A(t,\tau,2) \cong 0.0027 \int_{\tau}^{\infty} \boldsymbol{R}(t,2)\,\mathrm{d}t, \quad t \ge 0, \ \tau > 0,$$

where $\boldsymbol{R}(t,2)$ is given by (3.152).

The exemplary system reliability, renewal and availability characteristics predicted in this section are used for this system operation and reliability optimization and cost analysis performed in the appliqué part of Chap. 6.

3.6.2 Reliability, Renewal and Availability of Port Oil Piping Transportation System at Variable Operation Conditions

The port oil piping transportation system reliability analysis at the constant operation conditions considered here is performed in Sect. 1.4.2, whereas its varying in time operation process analysis is performed in Sect. 2.3.2. Considering the assumptions and agreements of these two sections, we assume that its subsystems S_υ, $\upsilon = 1,2,3$, are composed of three-state, i.e. $z = 2$, components $E_{ij}^{(\upsilon)}$, $\upsilon = 1,2,3$, with the conditional three-state reliability functions

$$[R_{ij}^{(\upsilon)}(t,\cdot)]^{(b)} = [1, [R_{ij}^{(\upsilon)}(t,1)]^{(b)}, [R_{ij}^{(\upsilon)}(t,2)]^{(b)}], \quad b = 1,2,\ldots,7,$$

with the exponential coordinates

$$[R_{ij}^{(\upsilon)}(t,1)]^{(b)} = \exp[-[\lambda_{ij}^{(\upsilon)}(1)]^{(b)}], \quad [R_{ij}^{(\upsilon)}(t,2)]^{(b)} = \exp[-[\lambda_{ij}^{(\upsilon)}(2)]^{(b)}],$$

different at various operation states z_b, $b = 1,2,\ldots,7$, and with the intensities of departure from the safety state subsets $\{1,2\}$, $\{2\}$, respectively

$$[\lambda_{ij}^{(\upsilon)}(1)]^{(b)}, \quad [\lambda_{ij}^{(\upsilon)}(2)]^{(b)}, \quad b = 1,2,\ldots,7.$$

In Sect. 2.3.2, it is fixed that the port oil piping transportation system reliability structure and its subsystems and components reliability depend on its changing in time operation states. The influence of the system operation states changing on the changes of the system reliability structure and its components reliability functions is as follows.

At the system operation state z_1, the system is composed of subsystem S_3 illustrated in Fig. 2.10, which contains three series subsystems ($k = 3$), each composed of 362 components ($l_1 = 362$, $l_2 = 362$, $l_3 = 362$) with the exponential reliability functions given below and is a series-"2 out of 3" system ($m = 2$).

Subsystem S_3 consists of two pipelines of the first type and one pipeline of the second type. In each pipeline of the first type there are:

- 360 pipe segments with conditional three-state reliability function co-ordinates

$$[R_{ij}^{(3)}(t,1)]^{(1)} = \exp[-0.0059t], \quad [R_{ij}^{(3)}(t,2)]^{(1)} = \exp[-0.0074t],$$
$$i = 1,2,\ j = 1,2,\ldots,360,$$

- two valves with conditional three-state reliability function co-ordinates

$$[R_{ij}^{(3)}(t,1)]^{(1)} = \exp[-0.0166t], \quad [R_{ij}^{(3)}(t,2)]^{(1)} = \exp[-0.0181t],$$
$$i = 1,2,\ j = 361,362.$$

In the pipeline of the second type there are:

- 360 pipe segments with conditional three-state reliability function co-ordinates

$$[R_{ij}^{(3)}(t,1)]^{(1)} = \exp[-0.0071t], \quad [R_{ij}^{(3)}(t,2)]^{(1)} = \exp[-0.0079t],$$
$$i = 3,\ j = 1,2,\ldots,360,$$

- two valves with conditional three-state reliability function co-ordinates

$$[R_{ij}^{(3)}(t,1)]^{(1)} = \exp[-0.0166t], \quad [R_{ij}^{(3)}(t,2)]^{(1)} = \exp[-0.0181t],$$
$$i = 3,\ j = 361,362.$$

Thus, at the operation state z_1, the pipeline system is identical with subsystem S_3 that is a three-state series-"2 out of 3" system with its structure shape parameters $k = 3$, $m = 2$, $l_1 = 362$, $l_2 = 362$, $l_3 = 362$, and according to (1.40)–(1.41) and (1.72)–(1.73), its conditional reliability function is given by

$$[\boldsymbol{R}(t,\cdot)]^{(1)} = [1, [\boldsymbol{R}(t,1)]^{(1)}, [\boldsymbol{R}(t,2)]^{(1)}], \quad t \geq 0, \tag{3.166}$$

where

$$\begin{aligned}
[\boldsymbol{R}(t,1)]^{(1)} &= \boldsymbol{R}^2_{3;362,362,362}(t,1) \\
&= 1 - \sum_{\substack{r_1,r_2,r_3=0 \\ r_1+r_2+r_3 \leq 1}}^{1} \prod_{i=1}^{3} \left[\prod_{j=1}^{362} [R_{ij}^{(3)}(t,1)]^{(1)}\right]^{r_i} \left[1 - \prod_{j=1}^{362} [R_{ij}^{(3)}(t,1)]^{(1)}\right]^{1-r_i} \\
&= 1 - [1 - \exp[-[360 \times 0.0059 + 2 \times 0.0166]t]]^2 \\
&\quad \times [1 - \exp[-[360 \times 0.0071 + 2 \times 0.0166]t]]^1 \\
&\quad - 2\exp[-1[360 \times 0.0059 + 2 \times 0.0166]t]
\end{aligned}$$

$$\times [1 - \exp[-[360 \times 0.0059 + 2 \times 0.0166]t]]^1$$
$$\times [1 - \exp[-[360 \times 0.0071 + 2 \times 0.0166]t]]^1$$
$$- \exp[-[360 \times 0.0071 + 2 \times 0.0166]t] \tag{3.167}$$
$$\times [1 - \exp[-[360 \times 0.0059 + 2 \times 0.0166]t]]^2$$
$$= \exp[-4.3144t] + 2\exp[-4.7464t] - 2\exp[-6.9036t],$$

$$[\boldsymbol{R}(t,2)]^{(1)} = \boldsymbol{R}^2_{3;362,362,362}(t,2)$$
$$= 1 - \sum_{\substack{r_1,r_2,r_3=0 \\ r_1+r_2+r_3 \le 1}}^{1} \prod_{i=1}^{3} \left[\prod_{j=1}^{362} [R^{(3)}_{ij}(t,2)]^{(1)}\right]^{r_i} \left[1 - \prod_{j=1}^{362} [R^{(3)}_{ij}(t,2)]^{(1)}\right]^{1-r_i}$$
$$= 1 - [1 - \exp[-[360 \times 0.0074 + 2 \times 0.0181]t]]^2$$
$$\times [1 - \exp[-[360 \times 0.0079 + 2 \times 0.0181]t]]^1$$
$$- 2\exp[-1[360 \times 0.0074 + 2 \times 0.0181]t]$$
$$\times [1 - \exp[-[360 \times 0.0074 + 2 \times 0.0181]t]]^1$$
$$\times [1 - \exp[-[360 \times 0.0079 + 2 \times 0.0181]t]]^1$$
$$- \exp[-[360 \times 0.0079 + 2 \times 0.0181]t]$$
$$\times [1 - \exp[-[360 \times 0.0074 + 2 \times 0.0181]t]]^2$$
$$= \exp[-5.4004t] + 2\exp[-5.5804t] - 2\exp[-8.2806t]. \tag{3.168}$$

The expected values and standard deviations of the pipeline system conditional lifetimes in the reliability state subsets $\{1,2\}$, $\{2\}$ at the operation state z_1, calculated from the results given by (3.166)–(3.168), according to (3.8)–(3.9), respectively are:

$$\mu_1(1) \cong 0.364, \quad \mu_1(2) \cong 0.302 \text{ year}, \tag{3.169}$$

$$\sigma_1(1) \cong 0.263, \quad \sigma_1(2) \cong 0.219 \text{ year}, \tag{3.170}$$

and further, using (3.10) and (3.169), the mean values of the conditional lifetimes in the particular reliability states 1, 2 at the operation state z_1, respectively are:

$$\bar{\mu}_1(1) \cong 0.062, \quad \bar{\mu}_1(2) \cong 0.302 \text{ year}. \tag{3.171}$$

At the system operation state z_2, the system is composed of subsystem S_3 illustrated in Fig. 2.11, which contains three series subsystems ($k = 3$), each composed of 362 components ($l_1 = 362,\ l_2 = 362,\ l_3 = 362$) with the exponential reliability functions given below and is a series–parallel system.

Subsystem S_3 consists of two pipelines of the first type and one pipeline of the second type with the same reliability functions as at the operation state z_1, having the conditional reliability function co-ordinates

$$[R^{(3)}_{ij}(t,u)]^{(2)} = [R^{(3)}_{ij}(t,u)]^{(1)}, \quad u = 1,2,\ i = 1,2,3,\ j = 1,2,\ldots,362.$$

Thus, at the operation state z_2, the pipeline system is identical with subsystem S_3 that is a three-state series–parallel system with its structure shape parameters $k = 3$, $l_1 = 362$, $l_2 = 362$, $l_3 = 362$, and according to (1.36)–(1.37) and (1.68)–(1.69), its conditional reliability function is given by

$$[\boldsymbol{R}(t,\cdot)]^{(2)} = [1, [\boldsymbol{R}(t,1)]^{(2)}, [\boldsymbol{R}(t,2)]^{(2)}], \quad t \geq 0, \tag{3.172}$$

where

$$\begin{aligned}[\boldsymbol{R}(t,1)]^{(2)} &= \boldsymbol{R}_{3;362,362,362}(t,1) = 1 - \prod_{i=1}^{3}\left[1 - \prod_{j=1}^{362}[R_{ij}^{(2)}(t,1)]^{(2)}\right] \\ &= 1 - [1 - \exp[-[360 \times 0.0059 + 2 \times 0.0166]t]]^2[1 - \exp[-[360 \times 0.0071 \\ &\quad + 2 \times 0.0166]t]]^1 \\ &= 1 - [1 - \exp[-2.1572t]]^2[1 - \exp[-2.5892t]] \\ &= \exp[-2.5892t] + 2\exp[-2.1572t] - 2\exp[-4.7464t] \\ &\quad - \exp[-4.3144t] + \exp[-6.9036t],\end{aligned} \tag{3.173}$$

$$\begin{aligned}[\boldsymbol{R}(t,2)]^{(2)} &= \boldsymbol{R}_{3;362,362,362}(t,2) = 1 - \prod_{i=1}^{3}[1 - \prod_{j=1}^{362}[R_{ij}^{(2)}(t,2)]^{(2)}] \\ &= 1 - [1 - \exp[-[360 \times 0.0074 + 2 \times 0.0181]t]]^2[1 - \exp[-[360 \times 0.0079 \\ &\quad + 2 \times 0.0181]t]]^1 \\ &= 1 - [1 - \exp[-2.7002t]]^2[1 - \exp[-2.8802t]] \\ &= \exp[-2.8802t] + 2\exp[-2.7002t] - 2\exp[-5.5804t] \\ &\quad - \exp[-5.4004t] + \exp[-8.2806t].\end{aligned} \tag{3.174}$$

The expected values and standard deviations of the pipeline system conditional lifetimes in the reliability state subsets $\{1,2\}$, $\{2\}$ at the operation state z_2, calculated from the results given by (3.172)–(3.174), according to (3.8)–(3.9), respectively are:

$$\mu_2(1) \cong 0.805, \quad \mu_2(2) \cong 0.665 \text{ year}, \tag{3.175}$$

$$\sigma_2(1) \cong 0.516, \quad \sigma_2(2) \cong 0.424 \text{ year}, \tag{3.176}$$

and further, using (3.10) and (3.175), the mean values of the conditional lifetimes in the particular reliability states 1, 2 at the operation state z_2, respectively are:

$$\bar{\mu}_2(1) \cong 0.140, \quad \bar{\mu}_2(2) \cong 0.665 \text{ year}. \tag{3.177}$$

At the system operation state z_3, the pipeline system is a series composed of two series–parallel subsystems S_1 and S_2 illustrated in Fig. 2.12.

Subsystem S_1 illustrated in Fig. 2.12, contains two series pipeline subsystems ($k = 2$), each composed of 178 components ($l_1 = 178$, $l_2 = 178$) with the exponential reliability functions identified in Chap. 5 and given below and is a series–parallel system. In each pipeline there are:

- 176 pipe segments with conditional three-state reliability function co-ordinates

$$[R_{ij}^{(1)}(t,1)]^{(3)} = \exp[-0.0062t], \quad [R_{ij}^{(1)}(t,2)]^{(3)} = \exp[-0.0088t],$$
$$i = 1,2, \; j = 1,2,\ldots,176,$$

- two valves with conditional three-state reliability function co-ordinates

$$[R_{ij}^{(1)}(t,1)]^{(3)} = \exp[-0.0167t], \quad [R_{ij}^{(1)}(t,2)]^{(3)} = \exp[-0.0182t],$$
$$i = 1,2, \; j = 177, 178.$$

Thus, at the operation state z_3, subsystem S_1 is a three-state series–parallel system with its structure shape parameters $k = 2$, $l_1 = 178$, $l_2 = 178$, and according to (1.36)–(1.37) and (1.68)–(1.69), its conditional reliability function is given by

$$[\boldsymbol{R}^{(1)}(t,\cdot)]^{(3)} = [1, [\boldsymbol{R}^{(1)}(t,1)]^{(3)}, [\boldsymbol{R}^{(1)}(t,2)]^{(3)}], \quad t \geq 0, \tag{3.178}$$

where

$$\begin{aligned}[\boldsymbol{R}^{(1)}(t,1)]^{(3)} &= \boldsymbol{R}_{2;178,178}(t,1) = 1 - \prod_{i=1}^{2}\left[1 - \prod_{j=1}^{178}[R_{ij}^{(1)}(t,1)]^{(3)}\right] \\ &= 1 - [1 - \exp[-[176 \times 0.0062 + 2 \times 0.0167]t]]^2 \\ &= 1 - [1 - \exp[-1.1246t]]^2 \\ &= 2\exp[-1.1246t] - \exp[-2.2492t],\end{aligned} \tag{3.179}$$

$$\begin{aligned}[\boldsymbol{R}^{(1)}(t,2)]^{(3)} &= \boldsymbol{R}_{2;178,178}(t,2) = 1 - \prod_{i=1}^{2}\left[1 - \prod_{j=1}^{178}[R_{ij}^{(1)}(t,2)]^{(3)}\right] \\ &= 1 - [1 - \exp[-[176 \times 0.0088 + 2 \times 0.0182]t]]^2 \\ &= 1 - [1 - \exp[-1.5852t]]^2 \\ &= 2\exp[-1.5852t] - \exp[-3.1704t].\end{aligned} \tag{3.180}$$

Subsystem S_2 contains two series pipeline subsystems ($k = 2$), each composed of 719 components ($l_1 = 719$, $l_2 = 719$) with the exponential reliability functions given below and is a series–parallel system. In each pipeline there are:

- 717 pipe segments with conditional three-state reliability function co-ordinates

$$[R_{ij}^{(2)}(t,1)]^{(3)} = \exp[-0.0062t], \quad [R_{ij}^{(2)}(t,2)]^{(3)} = \exp[-0.0088t],$$
$$i = 1,2, \; j = 1,2,\ldots,717,$$

- two valves with conditional three-state reliability function co-ordinates

$$[R_{ij}^{(2)}(t,1)]^{(3)} = \exp[-0.0166t], \quad [R_{ij}^{(2)}(t,2)]^{(3)} = \exp[-0.0181t],$$
$$i = 1,2, \ j = 718,719.$$

Thus, at the operation state z_3, subsystem S_2 is a three-state series–parallel system with its structure shape parameters $k = 2, \ l_1 = 719, \ l_2 = 719,$ and according to (1.36)–(1.37) and (1.68)–(1.69), its conditional reliability function is given by

$$[\boldsymbol{R}^{(2)}(t,\cdot)]^{(3)} = [1, [\boldsymbol{R}^{(2)}(t,1)]^{(3)}, [\boldsymbol{R}^{(2)}(t,2)]^{(3)}], \quad t \geq 0, \tag{3.181}$$

where

$$\begin{aligned}[\boldsymbol{R}^{(2)}(t,1)]^{(3)} &= \boldsymbol{R}_{2;719,719}(t,1) = 1 - \prod_{i=1}^{2}\left[1 - \prod_{j=1}^{719}[R_{ij}^{(2)}(t,1)]^{(3)}\right] \\ &= 1 - [1 - \exp[-[717 \times 0.0062 + 2 \times 0.0166]t]]^2 \\ &= 1 - [1 - \exp[-4.4786t]]^2 \\ &= 2\exp[-4.4786t] - \exp[-8.9572t],\end{aligned} \tag{3.182}$$

$$\begin{aligned}[\boldsymbol{R}^{(2)}(t,2)]^{(3)} &= \boldsymbol{R}_{2;719,719}(t,2) = 1 - \prod_{i=1}^{2}\left[1 - \prod_{j=1}^{719}[R_{ij}^{(2)}(t,2)]^{(3)}\right] \\ &= 1 - [1 - \exp[-[717 \times 0.0088 + 2 \times 0.0181]t]]^2 \\ &= 1 - [1 - \exp[-6.3458t]]^2 \\ &= 2\exp[-6.3458t] - \exp[-12.6916t].\end{aligned} \tag{3.183}$$

Considering that the pipeline system at the operation state z_3 is a three-state series system composed of subsystems S_1 and S_2, after applying (1.22)–(1.23) and (3.178)-(3.180) and (3.181)-(3.183), its conditional reliability function is given by

$$[\boldsymbol{R}(t,\cdot)]^{(3)} = [1, [\boldsymbol{R}(t,1)]^{(3)}, [\boldsymbol{R}(t,2)]^{(3)}], \quad t \geq 0, \tag{3.184}$$

where

$$\begin{aligned}[\boldsymbol{R}(t,1)]^{(3)} &= \overline{\boldsymbol{R}}_2(t,1) = [\boldsymbol{R}^{(1)}(t,1)]^{(3)}[\boldsymbol{R}^{(2)}(t,1)]^{(3)} \\ &= 4\exp[-5.6032t] - 2\exp[-6.7278t] - 2\exp[-10.0818t] \\ &\quad + \exp[-11.20604t],\end{aligned} \tag{3.185}$$

$$\begin{aligned}[\boldsymbol{R}(t,2)]^{(3)} &= \overline{\boldsymbol{R}}_2(t,2) = [\boldsymbol{R}^{(1)}(t,2)]^{(3)}[\boldsymbol{R}^{(2)}(t,2)]^{(3)} \\ &= 4\exp[-7.931t] - 2\exp[-14.2768t] - 2\exp[-9.5162t] \\ &\quad + \exp[-15.862t].\end{aligned} \tag{3.186}$$

The expected values and standard deviations of the pipeline system conditional lifetimes in the reliability state subsets $\{1,2\},\{2\}$ at the operation state z_3, calculated from the results given by (3.184)–(3.186), according to (3.8)–(3.9), respectively are:

$$\mu_3(1) \cong 0.307, \quad \mu_3(2) \cong 0.217 \text{ year}, \tag{3.187}$$

$$\sigma_3(1) \cong 0.222, \quad \sigma_3(2) \cong 0.157 \text{ year}, \tag{3.188}$$

and further, using (3.10) and (3.187), the mean values of the conditional lifetimes in the particular reliability states 1, 2 at the operation state z_3, respectively are:

$$\bar{\mu}_3(1) \cong 0.09, \quad \bar{\mu}_3(2) \cong 0.217 \text{ year}. \tag{3.189}$$

At the system operation state z_4, the system is series and composed of two series–parallel subsystems S_1 and S_2, each containing two pipelines, and one series-"2 out of 3" subsystem S_3 containing three pipelines, with the scheme shown in Fig. 2.13.

Subsystem S_1 illustrated in Fig. 2.13, contains two series pipeline subsystems ($k=2$), each composed of 178 components ($l_1 = 178,\ l_2 = 178$), with the same reliability functions as at the operation state z_3. Thus, according to (3.178)–(3.180), subsystem S_1 conditional reliability function at the operation state z_4, is given by

$$[\boldsymbol{R}^{(1)}(t,\cdot)]^{(4)} = [1, [\boldsymbol{R}^{(1)}(t,1)]^{(4)}, [\boldsymbol{R}^{(1)}(t,2)]^{(4)}], \quad t \geq 0, \tag{3.190}$$

where

$$[\boldsymbol{R}^{(1)}(t,1)]^{(4)} = 2\exp[-1.1246t] - \exp[-2.2492t], \tag{3.191}$$

$$[\boldsymbol{R}^{(1)}(t,2)]^{(4)} = 2\exp[-1.5852t] - \exp[-3.1704t]. \tag{3.192}$$

Subsystem S_2 illustrated in Fig. 2.14, contains two series pipeline subsystems ($k=2$), each composed of 719 components ($l_1 = 719,\ l_2 = 719$) with the same reliability functions as at the operation state z_3. Thus, according to (3.181)–(3.183), subsystem S_2 conditional reliability function at the operation state z_4, is given by

$$[\boldsymbol{R}^{(2)}(t,\cdot)]^{(4)} = [1, [\boldsymbol{R}^{(2)}(t,1)]^{(4)}, [\boldsymbol{R}^{(2)}(t,2)]^{(4)}], \quad t \geq 0, \tag{3.193}$$

where

$$[\boldsymbol{R}^{(2)}(t,1)]^{(4)} = 2\exp[-4.4786t] - \exp[-8.9572t], \tag{3.194}$$

$$[\boldsymbol{R}^{(2)}(t,2)]^{(4)} = 2\exp[-6.3458t] - \exp[-12.6916t]. \tag{3.195}$$

At the operation state z_4, subsystem S_3 illustrated in Fig. 2.13, contains three series subsystems ($k=3$), each composed of 362 components ($l_1 = 362$,

$l_2 = 362,\ l_3 = 362$) and is a series-"2 out of 3" system ($m = 2$) which means that it has the same as that at the operation state z_1 reliability structure.

Subsystem S_3 consists of two pipelines of the first type and 1 pipeline of the second type, each composed of 362 components with the same reliability functions as at the operation state z_1. Thus, according to (3.163)–(3.165), subsystem S_3 conditional reliability function at the operation state z_4, is given by

$$[\boldsymbol{R}^{(3)}(t,\cdot)]^{(4)} = [1, [\boldsymbol{R}^{(3)}(t,1)]^{(4)}, [\boldsymbol{R}^{(3)}(t,2)]^{(4)}], \quad t \geq 0, \tag{3.196}$$

where

$$[\boldsymbol{R}^{(3)}(t,1)]^{(4)} = \exp[-4.3144t] + 2\exp[-4.7464t] - 2\exp[-6.9036t], \tag{3.197}$$

$$[\boldsymbol{R}^{(3)}(t,2)]^{(4)} = \exp[-5.4004t] + 2\exp[-5.5804t] - 2\exp[-8.2806t]. \tag{3.198}$$

Considering that the pipeline system at the operation state z_4 is a three-state series system composed of subsystems S_1, S_2 and S_3, after applying (1.22)–(1.23) and (3.190)–(3.192), (3.193)–(3.195) and (3.196)–(3.198), its conditional reliability function is given by

$$[\boldsymbol{R}(t,\cdot)]^{(4)} = [1, [\boldsymbol{R}(t,1)]^{(4)}, [\boldsymbol{R}(t,2)]^{(4)}], \quad t \geq 0, \tag{3.199}$$

where

$$\begin{aligned}
[\boldsymbol{R}(t,1)]^{(4)} &= \overline{\boldsymbol{R}}_3(t,1) = [\boldsymbol{R}^{(1)}(t,1)]^{(4)}[\boldsymbol{R}^{(2)}(t,1)]^{(4)}[\boldsymbol{R}^{(3)}(t,1)]^{(4)} \\
&= 4\exp[-9.9176t] + 8\exp[-10.3496t] - 8\exp[-12.5078t] \\
&\quad - 2\exp[-14.396t] - 4\exp[-14.8282t] + 4\exp[-16.9864t] \\
&\quad - 2\exp[-11.0422t] - 4\exp[-11,4742t] + 4\exp[-13.6324t] \\
&\quad + \exp[-15.5208t] + 2\exp[-15.9528t] - 2\exp[-18.111t],
\end{aligned} \tag{3.200}$$

$$\begin{aligned}
[\boldsymbol{R}(t,2)]^{(4)} &= \overline{\boldsymbol{R}}_3(t,2) = [\boldsymbol{R}^{(1)}(t,2)]^{(4)}[\boldsymbol{R}^{(2)}(t,2)]^{(4)}[\boldsymbol{R}^{(3)}(t,2)]^{(4)} \\
&= 4\exp[-13.3314t] + 8\exp[-13.5114t] - 8\exp[-16.2116t] \\
&\quad - 2\exp[-19.6772t] - 4\exp[-19.8572t] + 4\exp[-22.5574t] \\
&\quad - 2\exp[-14.9166t] - 4\exp[-15,0966t] + 4\exp[-17.7968t] \\
&\quad + \exp[-21.2624t] + 2\exp[-21.4424t] - 2\exp[-24.1426t].
\end{aligned} \tag{3.201}$$

The expected values and standard deviations of the pipeline system conditional lifetimes in the reliability state subsets $\{1,2\}$, $\{2\}$ at the operation state z_4, calculated from the results given by (3.199)–(3.201), according to (3.8)–(3.9), respectively are:

$$\mu_4(1) \cong 0.207, \quad \mu_4(2) \cong 0.156 \text{ year}, \tag{3.202}$$

$$\sigma_4(1) \cong 0.137, \quad \sigma_4(2) \cong 0.104 \text{ year}, \tag{3.203}$$

and further, using (3.10) and (3.202), the mean values of the conditional lifetimes in the particular reliability states 1, 2 at the operation state z_4, respectively are:

$$\bar{\mu}_4(1) \cong 0.051, \quad \bar{\mu}_4(2) \cong 0.156 \text{ year}. \tag{3.204}$$

At the system operation state z_5, the pipeline system is series and composed of two series–parallel subsystems S_1 and S_2 illustrated in Fig. 2.12. The system reliability structure and its components reliability functions are the same as at the operation state z_3. Thus, considering (3.184)–(3.186), its conditional reliability function is given by

$$[\boldsymbol{R}(t,\cdot)]^{(5)} = [1, [\boldsymbol{R}(t,1)]^{(5)}, [\boldsymbol{R}(t,2)]^{(5)}], \quad t \geq 0, \tag{3.205}$$

where

$$[\boldsymbol{R}(t,1)]^{(5)} = 4\exp[-5.6032t] - 2\exp[-6.7278t] - 2\exp[-10.0818t] + \exp[-11.2064t], \tag{3.206}$$

$$[\boldsymbol{R}(t,2)]^{(5)} = 4\exp[-7.931t] - 2\exp[-14.2768t] - 2\exp[-9.5162t] + \exp[-15.8620t]. \tag{3.207}$$

The expected values and standard deviations of the pipeline system conditional lifetimes in the reliability state subsets $\{1,2\}$, $\{2\}$ at the operation state z_5, according to (3.187)–(3.188), respectively are:

$$\mu_5(1) \cong 0.307, \quad \mu_5(2) \cong 0.217 \text{ year}, \tag{3.208}$$

$$\sigma_5(1) \cong 0.222, \quad \sigma_5(2) \cong 0.157 \text{ year}, \tag{3.209}$$

and further, from (3.189), the mean values of the conditional lifetimes in the particular reliability states 1, 2 at the operation state z_5, respectively are:

$$\bar{\mu}_5(1) \cong 0.09, \quad \bar{\mu}_5(2) \cong 0.217 \text{ year}. \tag{3.210}$$

At the system operation state z_6, the system is series and composed of two series–parallel subsystems S_1 and S_2, each containing two pipelines and one series-"2 out of 3" subsystem S_3 containing three pipelines with the scheme shown in Fig. 2.13. The system reliability structure and its components reliability functions are the same as at the operation state z_4. Thus, considering (3.199)–(3.201), its conditional three-state reliability function is given by

$$[\boldsymbol{R}(t,\cdot)]^{(6)} = [1, [\boldsymbol{R}(t,1)]^{(6)}, [\boldsymbol{R}(t,2)]^{(6)}], \quad t \geq 0, \tag{3.211}$$

where

$$[\boldsymbol{R}(t,1)]^{(6)} = 4\exp[-9.9176t] + 8\exp[-10.3496t] - 8\exp[-12.5078t]$$
$$- 2\exp[-14.396t] - 4\exp[-14.8282t] + 4\exp[-16.9864t]$$
$$- 2\exp[-11.0422t] - 4\exp[-11,4742t] + 4\exp[-13.6324t]$$
$$+ \exp[-15.5208t] + 2\exp[-15.9528t] - 2\exp[-18.111t], \quad (3.212)$$

$$[\boldsymbol{R}(t,2)]^{(6)} = 4\exp[-13.3314t] + 8\exp[-13.5114t] - 8\exp[-16.2116t]$$
$$- 2\exp[-19.6772t] - 4\exp[-19.8572t] + 4\exp[-22.5574t]$$
$$- 2\exp[-14.9166t] - 4\exp[-15,0966t] + 4\exp[-17.7968t]$$
$$+ \exp[-21.2624t] + 2\exp[-21.4424t] - 2\exp[-24.1426t]. \quad (3.213)$$

The expected values and standard deviations of the pipeline system conditional lifetimes in the reliability state subsets $\{1,2\}$, $\{2\}$ at the operation state z_6, according to (3.202)–(3.203), respectively are:

$$\mu_6(1) \cong 0.207, \quad \mu_6(2) \cong 0.156 \text{ year}, \quad (3.214)$$

$$\sigma_6(1) \cong 0.137, \quad \sigma_6(2) \cong 0.104 \text{ year}, \quad (3.215)$$

and further, from (3.204), the mean values of the conditional lifetimes in the particular reliability states 1, 2 at the operation state z_6, respectively are:

$$\overline{\mu}_6(1) \cong 0.051, \quad \overline{\mu}_6(2) \cong 0.156 \text{ year}. \quad (3.216)$$

At the system operation state z_7, the system is composed of the subsystem S_3 illustrated in Fig. 2.10. The system reliability structure and its components, reliability functions are the same as at the operation state z_1.

Thus, considering (3.166)–(3.168), its conditional three-state reliability function is given by

$$[\boldsymbol{R}(t,\cdot)]^{(7)} = [1, [\boldsymbol{R}(t,1)]^{(7)}, [\boldsymbol{R}(t,2)]^{(7)}], \quad t \geq 0, \quad (3.217)$$

where

$$[\boldsymbol{R}(t,1)]^{(7)} = \exp[-4.3144t] + 2\exp[-4.7464t] - 2\exp[-6.9036t], \quad (3.218)$$

$$[\boldsymbol{R}(t,2)]^{(7)} = \exp[-5.4004t] + 2\exp[-5.5804t] - 2\exp[-8.2806t]. \quad (3.219)$$

The expected values and standard deviations of the pipeline system conditional lifetimes in the reliability state subsets $\{1,2\}$, $\{2\}$ at the operation state z_7, according to (3.169)–(3.170), respectively are:

$$\mu_7(1) \cong 0.364, \quad \mu_7(2) \cong 0.302 \text{ year}, \quad (3.220)$$

$$\sigma_7(1) \cong 0.263, \quad \sigma_7(2) \cong 0.219 \text{ year}, \quad (3.221)$$

and further, from (3.171), the conditional lifetimes in the particular reliability states 1, 2 at the operation state z_7, respectively are:

$$\bar{\mu}_7(1) \cong 0.062, \quad \bar{\mu}_7(2) \cong 0.302 \text{ year.} \tag{3.222}$$

In the case when the operation time is large enough, the port oil transportation system unconditional reliability function is given by the vector

$$\boldsymbol{R}(t,\cdot) = [1, \boldsymbol{R}(t,1), \boldsymbol{R}(t,2)], \quad t \geq 0, \tag{3.223}$$

where according to (3.6) and considering the pipeline system operation process transient probabilities at the operation states determined by (2.38), the vector co-ordinates are given respectively by

$$\begin{aligned} \boldsymbol{R}(t,1) = {} & 0.395 \times [\boldsymbol{R}(t,1)]^{(1)} + 0.060 \times [\boldsymbol{R}(t,1)]^{(2)} + 0.003 \times [\boldsymbol{R}(t,1)]^{(3)} \\ & + 0.002 \times [\boldsymbol{R}(t,1)]^{(4)} + 0.2 \times [\boldsymbol{R}(t,1)]^{(5)} + 0.058 \times [\boldsymbol{R}(t,1)]^{(6)} \\ & + 0.282 \times [\boldsymbol{R}(t,1)]^{(7)} \quad \text{for } t \geq 0, \end{aligned} \tag{3.224}$$

$$\begin{aligned} \boldsymbol{R}(t,2) = {} & 0.395 \times [\boldsymbol{R}(t,2)]^{(1)} + 0.060 \times [\boldsymbol{R}(t,2)]^{(2)} + 0.003 \times [\boldsymbol{R}(t,2)]^{(3)} \\ & + 0.002 \times [\boldsymbol{R}(t,2)]^{(4)} + 0.2 \times [\boldsymbol{R}(t,2)]^{(5)} + 0.058 \times [\boldsymbol{R}(t,2)]^{(6)} \\ & + 0.282 \times [\boldsymbol{R}(t,2)]^{(7)} \quad \text{for } t \geq 0, \end{aligned} \tag{3.225}$$

where $[R(t,1)]^{(1)}$, $[R(t,1)]^{(2)}$, $[R(t,1)]^{(3)}$, $[R(t,1)]^{(4)}$, $[R(t,1)]^{(5)}$, $[R(t,1)]^{(6)}$, $[R(t,1)]^{(7)}$ are given by (3.167), (3.173), (3.185), (3.200), (3.206), (3.212), (3.218) and $[R(t,2)]^{(1)}$, $[R(t,2)]^{(2)}$, $[R(t,2)]^{(3)}$, $[R(t,2)]^{(4)}$, $[R(t,2)]^{(5)}$, $[R(t,2)]^{(6)}$, $[R(t,2)]^{(7)}$ are given by (3.168), (3.174), (3.186), (3.201), (3.207), (3.213), (3.219).

The graph of the three-state port oil piping transportation system reliability function is presented in Fig. 3.3.

The expected values and standard deviations of the system unconditional lifetimes in the reliability state subsets $\{1,2\}$, $\{2\}$, calculated from the above results given by (3.224)–(3.225), according to (3.7)–(3.9) and considering (2.38), (3.169), (3.175), (3.187), (3.202), (3.208), (3.214), (3.220), respectively are:

$$\begin{aligned} \mu(1) = {} & 0.395 \times 0.364 + 0.060 \times 0.805 + 0.003 \times 0.307 + 0.002 \times 0.207 \\ & + 0.2 \times 0.307 + 0.058 \times 0.207 + 0.282 \times 0.364 \cong 0.370 \text{ year,} \end{aligned} \tag{3.226}$$

$$\sigma(1) \cong 0.308 \text{ year,} \tag{3.227}$$

$$\begin{aligned} \mu(2) = {} & 0.395 \times 0.302 + 0.060 \times 0.665 + 0.003 \times 0.217 + 0.002 \times 0.156 \\ & + 0.2 \times 0.217 + 0.058 \times 0.156 + 0.282 \times 0.302 \cong 0.298 \text{ year,} \end{aligned} \tag{3.228}$$

$$\sigma(2) \cong 0.252 \text{ year,} \tag{3.229}$$

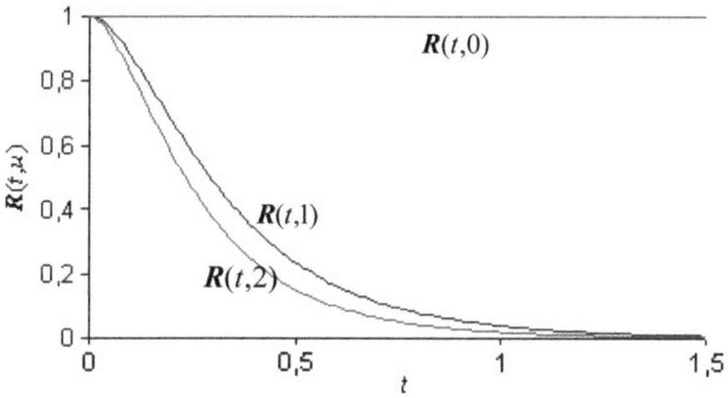

Fig. 3.3 The graph of the port oil piping transportation system reliability function $\boldsymbol{R}(t,\cdot)$ coordinates

and further, considering (3.10) and (3.226) and (3.228), the mean values of the unconditional lifetimes in the particular reliability states 1, 2, respectively are:

$$\bar{\mu}(1) = \mu(1) - \mu(2) = 0.072, \quad \bar{\mu}(2) = \mu(2) = 0.298 \text{ year}. \tag{3.230}$$

Since the critical reliability state is $r = 1$, then according to (3.11), the system risk function is given by

$$\boldsymbol{r}(t) = 1 - \boldsymbol{R}(t,1) \quad \text{for } t \geq 0, \tag{3.231}$$

where $\boldsymbol{R}(t,1)$ is given by (3.224).

Hence, according to (3.12), the moment when the system risk function exceeds a permitted level, for instance $\delta = 0.05$, is

$$\tau = \boldsymbol{r}^{-1}(\delta) \cong 0.066 \text{ year}. \tag{3.232}$$

The graph of the port oil piping transportation system risk function $\boldsymbol{r}(t)$ is presented in Fig. 3.4.

Further, assuming that the oil pipeline system is repaired after the exceeding of its critical reliability state $r = 1$ and that the time of the system renovation is ignored and applying *Proposition 3.4*, we obtain the following results:

(a) the time $S_N(1)$ until the Nth exceeding by the system the reliability critical state $r = 1$, for sufficiently large N, has approximately normal distribution $N(0.370N, 0.308\sqrt{N})$, i.e.,

$$F^{(N)}(t,1) = P(S_N(1) < t) \cong F_{N(0,1)}\left(\frac{t - 0.370N}{0.308\sqrt{N}}\right), \quad t \in (-\infty, \infty);$$

(b) the expected value and the variance of the time $S_N(1)$ until the Nth exceeding by the system the reliability critical state $r = 1$ are respectively given by

$$E[S_N(1)] \cong 0.370N, \quad D[S_N(1)] \cong 0.095N;$$

(c) the number $N(t,1)$ of exceeding by the system the reliability critical state $r = 1$ up to the moment t, $t \geq 0$, for sufficiently large t, approximately has the distribution of the form

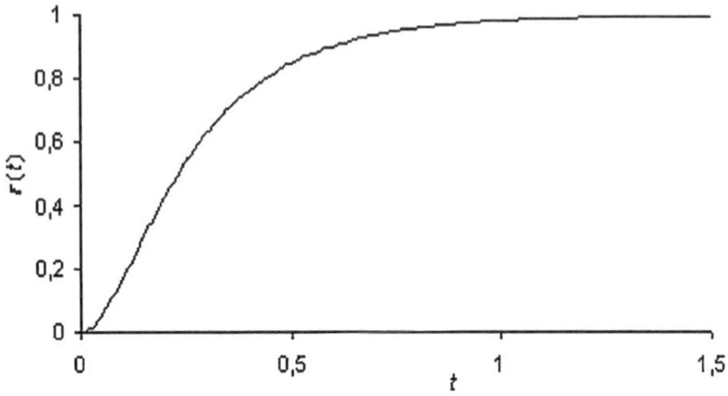

Fig. 3.4 The graph of the port oil piping transportation system risk function $\boldsymbol{r}(t)$

$$P(N(t,1)=N)\cong F_{N(0,1)}\left(\frac{0.370(N+1)-t}{0.506\sqrt{t}}\right)-F_{N(0,1)}\left(\frac{0.370N-t}{0.506\sqrt{t}}\right),$$
$$N=0,1,\ldots;$$

(d) the expected value and the variance of the number $N(t,1)$ of exceeding by the system the reliability critical state $r=1$ up to the moment t, $t\geq 0$, for sufficiently large t, approximately are respectively given by

$$H(t,1)=2.702t,\quad D(t,1)=1.872t. \tag{3.233}$$

Assuming that the oil pipeline system is repaired after the exceeding of its reliability critical state $r=1$ and that the time of the system renovation is not ignored and has the mean value $\mu_0(1)=0.005$ year and the standard deviation $\sigma_0(1)=0.005$ year and applying *Proposition 3.5*, we obtain the following results:

(a) the time $\overline{S}_N(1)$ until the Nth exceeding by the system the reliability critical state $r=1$, for sufficiently large N, has approximately normal distribution $N(0.370N+0.005(N-1),\sqrt{0.095N-0.000025(N-1)})$, i.e.,

$$\overline{F}^{(N)}(t,1)=P(\overline{S}_N(1)<t)\cong F_{N(0,1)}\left(\frac{t-0.375N+0.005}{\sqrt{0.095N-0.000025}}\right),\quad t\in(-\infty,\infty);$$

(b) the expected value and the variance of the time $\overline{S}_N(1)$ until the Nth exceeding by the system the reliability critical state $r=1$, for sufficiently large N, are respectively given by

$$E[\overline{S}_N(1)]\cong 0.370N+0.005(N-1),\quad D[\overline{S}_N(1)]\cong 0.095N+0.000025(N-1);$$

(c) the number $\overline{N}(t,1)$ of exceeding by the system the reliability critical state $r=1$ up to the moment t, $t\geq 0$, for sufficiently large t, has approximately distribution of the form

$$P(\overline{N}(t,1)=N)\cong F_{N(0,1)}\left(\frac{0.375(N+1)-t-0005}{0.503\sqrt{t-0.005}}\right)$$
$$-F_{N(0,1)}\left(\frac{0.375N-t-0.005}{0.503\sqrt{t+0.005}}\right),\quad N=0,1,\ldots;$$

(d) the expected value and the variance of the number $\overline{N}(t,1)$ of exceeding by the system the reliability critical state $r = 1$ up to the moment t, $t \geq 0$, for sufficiently large t, are respectively given by

$$\overline{H}(t,1)\cong 2.667(t+0.005),\quad \overline{D}(t,1)\cong 1.801(t+0.005); \tag{3.234}$$

(e) the time $\overline{\overline{S}}_N(1)$ until the Nth system's renovation, for sufficiently large N, has approximately normal distribution $N(0.375N, 0.308\sqrt{N})$, i.e.,

$$\overline{\overline{F}}^{(N)}(t,1)=P(\overline{\overline{S}}_N(1)<t)\cong F_{N(0,1)}\left(\frac{t-0.375N)}{0.308\sqrt{N}}\right),\quad t\in(-\infty,\infty);$$

(f) the expected value and the variance of the time $\overline{\overline{S}}_N(1)$ until the Nth system's renovation, for sufficiently large N, are respectively given by

$$E[\overline{\overline{S}}_N(1)]\cong 0.375N,\quad D[\overline{\overline{S}}_N(1)]\cong 0.095N;$$

(g) the number $\overline{\overline{N}}(t,1)$ of the system's renovations up to the moment t, $t \geq 0$, for sufficiently large t, has approximately distribution of the form

$$P(\overline{\overline{N}}(t,1)=N)\cong F_{N(0,1)}\left(\frac{0.368(N+1)-t}{0.503\sqrt{t}}\right)$$
$$-F_{N(0,1)}\left(\frac{0.368N-t}{0.503\sqrt{t}}\right),\quad N=0,1,\ldots;$$

(h) the expected value and the variance of the number $\overline{\overline{N}}(t,1)$ of system's renovations up to the moment t, $t \geq 0$, for sufficiently large t, are respectively given by

$$\overline{\overline{H}}(t,1)\cong 2.667t,\quad \overline{\overline{D}}(t,1)\cong 1.801t; \tag{3.235}$$

(i) the steady availability coefficient of the system at the moment t, $t \geq 0$, for sufficiently large t, is given by

$$A(t,1)\cong 0.987,\quad t\geq 0;$$

(j) the steady availability coefficient of the system in the time interval $\langle t, t+\tau)$, $\tau > 0$, for sufficiently large t, is given by

$$A(t,\tau,1)\cong 2.667\int_{\tau}^{\infty}\boldsymbol{R}(t,1)\,dt,\quad t\geq 0,\ \tau>0,$$

where $\boldsymbol{R}(t, 1)$ is given by (3.224).

The port oil piping transportation system operating at varying in time operation conditions reliability and availability evaluation and prediction are approximate because of non-sufficiently exact input data. Although the piping system reliability structures are fixed with high accuracy, the reliability parameters of its components and its operation process parameters are not sufficiently exact because of the lack of statistical data necessary for their estimation. Therefore, the achieved results may only be considered as an illustration of the proposed methods' possibilities of applications in the complex piping systems' reliability and availability analysis and prediction. However, the obtained evaluation may be a very useful example in piping systems' reliability and availability characteristics prediction, especially during the design and when planning and improving their operation processes' safety and effectiveness. To improve the accuracy of the achieved results the statistical data should be collected for an appropriately long period of time and after that the full identification of at least the pipeline oil transportation operation process could be performed and this process, unknown parameters and main characteristics would be determined and used in its reliability, risk and availability for more precise analysis and prediction. The reliability characteristics of the port oil piping transportation system operating at the variable conditions predicted in this section are different from those determined in Sect. 1.4.2 at the system constant operation conditions. This fact suggests the necessity of considering the system at the variable operation conditions that upon the improved accuracy of the system operation process identification, makes its system reliability prediction more precise.

The presented in this section reliability, renewal and availability characteristics of the port oil piping transportation system are used in its operation, reliability, availability and maintenance optimization in the appliqué part of Chap. 6. This example of the considered models practical application justifies their very good placement in the reliability practice of real complex technical systems.

3.6.3 Safety, Renewal and Availability of Maritime Ferry Technical System at Variable Operation Conditions

The ferry technical system safety analysis considered here at the constant operation conditions is performed in Sect. 1.4.3, whereas its changing in time operation process analysis is performed in Sect. 2.3.3. Considering the assumptions and agreements of these sections, we assume that its subsystems S_υ, $\upsilon = 1, 2, \ldots, 5$, are composed of five-state i.e. $z = 4$, components, $E_{ij}^{(\upsilon)}$, $\upsilon = 1, 2, \ldots, 5$, having the conditional safety functions given by the vector

$$[s_{ij}^{(\upsilon)}(t,\cdot)]^{(b)} = [1, [s_{ij}^{(\upsilon)}(t,1)]^{(b)}, [s_{ij}^{(\upsilon)}(t,2)]^{(b)}, [s_{ij}^{(\upsilon)}(t,3)]^{(b)}, [s_{ij}^{(\upsilon)}(t,4)]^{(b)}],$$
$$b = 1,2,\ldots,18,$$

with the exponential coordinates

$$[s_{ij}^{(\upsilon)}(t,1)]^{(b)} = \exp[-[\lambda_{ij}^{(\upsilon)}(1)]^{(b)}t], \quad [s_{ij}^{(\upsilon)}(t,2)]^{(b)} = \exp[-[\lambda_{ij}^{(\upsilon)}(2)]^{(b)}t],$$
$$[s_{ij}^{(\upsilon)}(t,3)]^{(b)} = \exp[-[\lambda_{ij}^{(\upsilon)}(3)]^{(b)}t], \quad [s_{ij}^{(\upsilon)}(t,4)]^{(b)} = \exp[-[\lambda_{ij}^{(\vartheta)}(4)]^{(b)}t],$$
$$b = 1,2,\ldots,18,$$

different in various operation states z_b, $b = 1,2,\ldots,18$, and with the intensities of departure from the safety states subsets $\{1,2,3,4\}$, $\{2,3,4\}$, $\{3,4\}$, $\{4\}$, respectively

$$[\lambda_{ij}^{(\upsilon)}(1)]^{(b)}, \ [\lambda_{ij}^{(\upsilon)}(2)]^{(b)}, \ [\lambda_{ij}^{(\upsilon)}(3)]^{(b)}, \ [\lambda_{ij}^{(\upsilon)}(4)]^{(b)}, \quad b = 1,2,\ldots,18.$$

In Sect. 2.3.3, it is fixed that the ferry technical system safety structure and its subsystems and components safety depend on its changing in time operation states. The influence of the system operation states changing on the changes of the system safety structure fixed in Chap. 2 and its influence on the changes of the system components exponential safety functions partly identified in Chap. 5 are given below.

At the system operation state z_1, the ferry technical system is composed of two subsystems S_3 and S_4 forming a series structure shown in Fig. 2.14.

Subsystem S_3 is a five-state series system composed of $n = 3$ components $E_{11}^{(3)}$, $E_{21}^{(3)}$, $E_{31}^{(3)}$ having the exponential safety functions with coordinates:

$$[s_{11}^{(3)}(t,1)]^{(1)} = \exp[-0.02t], [s_{11}^{(3)}(t,2)]^{(1)} = \exp[-0.03t],$$
$$[s_{11}^{(3)}(t,3)]^{(1)} = \exp[-0.035t], [s_{11}^{(3)}(t,4)]^{(1)} = \exp[-0.04t],$$
$$[s_{21}^{(3)}(t,1)]^{(1)} = \exp[-0.02t], [s_{21}^{(3)}(t,2)]^{(1)} = \exp[-0.025t],$$
$$[s_{21}^{(3)}(t,3)]^{(1)} = \exp[-0.03t], [s_{21}^{(3)}(t,4)]^{(1)} = \exp[-0.04t],$$
$$[s_{31}^{(3)}(t,1)]^{(1)} = \exp[-0.033t], [s_{31}^{(3)}(t,2)]^{(1)} = \exp[-0.04t],$$
$$[s_{31}^{(3)}(t,3)]^{(1)} = \exp[-0.045t], [s_{31}^{(3)}(t,4)]^{(1)} = \exp[-0.05t].$$

Subsystem S_3 is a five-state series system that can also be considered as a five-state parallel–series system with its structure shape parameters $k = 3$, $l_1 = 1$, $l_2 = 1$, $l_3 = 1$. Thus, according either to the formulae (1.38)–(1.39) and (1.70)–(1.71) or to the formulae (1.22)–(1.23) and (1.58)–(1.59), after replacing in them the components' reliability functions by their safety functions, subsystem S_3 conditional safety function at the operation state z_1 is given by

$$[\boldsymbol{s}^{(3)}(t,\cdot)]^{(1)} = [1, [\boldsymbol{s}^{(3)}(t,1)]^{(1)}, [\boldsymbol{s}^{(3)}(t,2)]^{(1)}, [\boldsymbol{s}^{(3)}(t,3)]^{(1)}, [\boldsymbol{s}^{(3)}(t,4)]^{(1)}],$$
$$t \in \langle 0, \infty), \tag{3.236}$$

where

$$[\boldsymbol{s}^{(3)}(t,u)]^{(1)} = \boldsymbol{s}_{3;1,1,1}(t,u)$$
$$= \prod_{i=1}^{3}\left[1 - \prod_{j=1}^{1}[1 - [s_{ij}^{(3)}(t,u)]^{(1)}]\right] = \prod_{i=1}^{3}[s_{i1}^{(3)}(t,u)]^{(1)}, \quad t \in \langle 0, \infty),$$
$$u = 1, 2, 3, 4,$$

and particularly

$$[\boldsymbol{s}^{(3)}(t,1)]^{(1)} = \exp[-0.02t]\exp[-0.02t]\exp[-0.033t] = \exp[-0.073t], \tag{3.237}$$

$$[\boldsymbol{s}^{(3)}(t,2)]^{(1)} = \exp[-0.03t]\exp[-0.025t]\exp[-0.04t] = \exp[-0.095t], \tag{3.238}$$

$$[\boldsymbol{s}^{(3)}(t,3)]^{(1)} = \exp[-0.035t]\exp[-0.03t]\exp[-0.045t] = \exp[-0.110t], \tag{3.239}$$

$$[\boldsymbol{s}^{(3)}(t,4)]^{(1)} = \exp[-0.04t]\exp[-0.04t]\exp[-0.05t] = \exp[-0.130t]. \tag{3.240}$$

Subsystem S_4 consists of one five-state component $E_{11}^{(4)}$ having the exponential safety functions with the coordinates:

$$[s_{11}^{(4)}(t,1)]^{(1)} = \exp[-0.05t], \quad [s_{11}^{(4)}(t,2)]^{(1)} = \exp[-0.06t],$$
$$[s_{11}^{(4)}(t,3)]^{(1)} = \exp[-0.065t], \quad [s_{11}^{(4)}(t,4)]^{(1)} = \exp[-0.07t].$$

Thus, at the operation state z_1, subsystem S_4 conditional safety function is identical with the component $E_{11}^{(4)}$ safety function and is given by

$$[\boldsymbol{s}^{(4)}(t,\cdot)]^{(1)} = [1, [\boldsymbol{s}^{(4)}(t,1)]^{(1)}, [\boldsymbol{s}^{(4)}(t,2)]^{(1)}, [\boldsymbol{s}^{(4)}(t,3)]^{(1)}, [\boldsymbol{s}_{1,1}^{(4)}(t,4)]^{(1)}],$$
$$t \in \langle 0, \infty), \tag{3.241}$$

where

$$[\boldsymbol{s}^{(4)}(t,1)]^{(1)} = \exp[-0.05t], \tag{3.242}$$

$$[\boldsymbol{s}^{(4)}(t,2)]^{(1)} = \exp[-0.06t], \tag{3.243}$$

$$[\boldsymbol{s}^{(4)}(t,3)]^{(1)} = \exp[-0.065t], \tag{3.244}$$

$$[\boldsymbol{s}^{(4)}(t,4)]^{(1)} = \exp[-0.07t]. \tag{3.245}$$

Considering that the ferry technical system at the operation state z_1 is a five-state series system of subsystems S_3 and S_4, after applying (1.22)–(1.23) and (3.236)–(3.240) and (3.241)–(3.245), its conditional safety function is given by

$$[\boldsymbol{s}(t,\cdot)]^{(1)} = [1, [\boldsymbol{s}(t,1)]^{(1)}, [\boldsymbol{s}(t,2)]^{(1)}, [\boldsymbol{s}(t,3)]^{(1)}, [\boldsymbol{s}(t,4)]^{(1)}], \quad t \in \langle 0,\infty), \tag{3.246}$$

where

$$[\boldsymbol{s}(t,u)]^{(1)} = \overline{s}_2(t,u) = [\boldsymbol{s}^{(3)}(t,u)]^{(1)}[\boldsymbol{s}^{(4)}(t,u)]^{(1)}, \quad t \in \langle 0,\infty),\ u = 1,2,3,4,$$

and particularly

$$[\boldsymbol{s}(t,1)]^{(1)} = \exp[-0.073t]\exp[-0.05t] = \exp[-0.123t] \tag{3.247}$$

$$[\boldsymbol{s}(t,2)]^{(1)} = \exp[-0.095t]\exp[-0.06t] = \exp[-0.155t] \tag{3.248}$$

$$[\boldsymbol{s}(t,3)]^{(1)} = \exp[-0.110t]\exp[-0.065t] = \exp[-0.175t], \tag{3.249}$$

$$[\boldsymbol{s}(t,4)]^{(1)} = \exp[-0.130t]\exp[-0.07t] = \exp[-0.20t]. \tag{3.250}$$

The expected values and standard deviations of the ferry technical system conditional lifetimes in the safety state subsets $\{1,2,3,4\}$, $\{2,3,4\}$, $\{3,4\}$, $\{4\}$ at the operation state z_1, calculated from the results given by (3.247)–(3.250), according to (3.8)–(3.9), respectively are:

$$\mu_1(1) \cong 8.13, \quad \mu_1(2) \cong 6.45, \quad \mu_1(3) \cong 5.71, \quad \mu_1(4) \cong 5.00 \text{ years}, \tag{3.251}$$

$$\sigma_1(1) \cong 8.13, \quad \sigma_1(2) \cong 6.45, \quad \sigma_1(3) \cong 5.71, \quad \sigma_1(4) \cong 5.00 \text{ years}, \tag{3.252}$$

and further, using (3.10) and (3.251), the mean values of the ferry technical system conditional lifetimes in the particular safety states 1, 2, 3, 4 at the operation state z_1, respectively are:

$$\overline{\mu}_1(1) \cong 1.68, \quad \overline{\mu}_1(2) \cong 0.74, \quad \overline{\mu}_1(3) \cong 0.71, \quad \overline{\mu}_4(4) \cong 5.00 \text{ years}. \tag{3.253}$$

At the system operation state z_2, the ferry technical system is composed of three subsystems S_1, S_2 and S_5 forming a series structure shown in Fig. 2.15. Subsystem S_1 consists of one five-state component $E_{11}^{(1)}$ having the exponential safety functions partly identified in Chap. 5 with coordinates:

$$[s_{11}^{(1)}(t,1)]^{(2)} = \exp[-0.033t], \quad [s_{11}^{(1)}(t,2)]^{(2)} = \exp[-0.04t],$$
$$[s_{11}^{(1)}(t,3)]^{(2)} = \exp[-0.045t], \quad [s_{11}^{(1)}(t,4)]^{(2)} = \exp[-0.05t].$$

Thus, at the operation state z_2, subsystem S_1 conditional safety function is identical with the component $E_{11}^{(1)}$ safety function and is given by

$$[\boldsymbol{s}^{(1)}(t,\cdot)]^{(2)} = [1, [\boldsymbol{s}^{(1)}(t,1)]^{(2)}, [\boldsymbol{s}^{(1)}(t,2)]^{(2)}, [\boldsymbol{s}^{(1)}(t,3)]^{(2)}, [\boldsymbol{s}^{(1)}(t,4)]^{(2)}],$$
$$t \in \langle 0, \infty), \tag{3.254}$$

where

$$[\boldsymbol{s}^{(1)}(t,1)]^{(2)} = \exp[-0.033t], \tag{3.255}$$

$$[\boldsymbol{s}^{(1)}(t,2)]^{(2)} = \exp[-0.04t], \tag{3.256}$$

$$[\boldsymbol{s}^{(1)}(t,3)]^{(2)} = \exp[-0.045t], \tag{3.257}$$

$$[\boldsymbol{s}^{(1)}(t,4)]^{(2)} = \exp[-0.05t]. \tag{3.258}$$

Subsystem S_2 is a five-state parallel–series system composed of components $E_{ij}^{(2)}$, $i = 1,2,\ldots,k$, $j = 1,2,\ldots,l_i$, $k = 7$, $l_1 = 4$, $l_2 = 2$, $l_3 = 1$, $l_4 = 1$, $l_5 = 1$, $l_6 = 1$, $l_7 = 1$, having the exponential safety functions with coordinates:

$$[s_{1j}^{(2)}(t,1)]^{(2)} = \exp[-0.033t], \quad [s_{1j}^{(2)}(t,2)]^{(2)} = \exp[-0.04t],$$

$$[s_{1j}^{(2)}(t,3)]^{(2)} = \exp[-0.05t], \quad [s_{1j}^{(2)}(t,4)]^{(2)} = \exp[-0.055t], \quad j = 1,2,3,4;$$

$$[s_{2j}^{(2)}(t,1)]^{(2)} = \exp[-0.066t], \quad [s_{2j}^{(2)}(t,2)]^{(2)} = \exp[-0.07t],$$

$$[s_{2j}^{(2)}(t,3)]^{(2)} = \exp[-0.075t], \quad [s_{2j}^{(2)}(t,4)]^{(2)} = \exp[-0.08t], \quad j = 1,2;$$

$$[s_{31}^{(2)}(t,1)]^{(2)} = \exp[-0.066t], \quad [s_{31}^{(2)}(t,2)]^{(2)} = \exp[-0.07t],$$

$$[s_{31}^{(2)}(t,3)]^{(2)} = \exp[-0.075t], \quad [s_{31}^{(2)}(t,4)]^{(2)} = \exp[-0.08t];$$

$$[s_{i1}^{(2)}(t,1)]^{(2)} = \exp[-0.033t], \quad [s_{i1}^{(2)}(t,2)]^{(2)} = \exp[-0.04t],$$

$$[s_{i1}^{(2)}(t,3)]^{(2)} = \exp[-0.045t], \quad [s_{i1}^{(2)}(t,4)]^{(2)} = \exp[-0.05t], \quad i = 4,5,6,7.$$

Subsystem S_2 is a five-state parallel–series system with its structure shape parameters $k = 7$, $l_1 = 4$, $l_2 = 2$, $l_3 = 1$, $l_4 = 1$, $l_5 = 1$, $l_6 = 1$, $l_7 = 1$. Thus, according to the formulae (1.38)–(1.39) and (1.70)–(1.71), after replacing in them the components' reliability functions by their safety functions, subsystem S_2 conditional safety function at the operational state z_2 is given by

$$[\boldsymbol{s}^{(2)}(t,\cdot)]^{(2)} = [1, [\boldsymbol{s}^{(2)}(t,1)]^{(2)}, [\boldsymbol{s}^{(2)}(t,2)]^{(2)}, [\boldsymbol{s}^{(2)}(t,3)]^{(2)}, [\boldsymbol{s}^{(2)}(t,4)]^{(2)}],$$
$$t \in \langle 0, \infty), \tag{3.259}$$

where

$$[\boldsymbol{s}^{(2)}(t,u)]^{(2)} = \boldsymbol{s}_{7;4,2,1,1,1,1,1}(t,u) = \prod_{i=1}^{7}\left[1-\prod_{j=1}^{l_i}[1-[s_{ij}^{(2)}(t,u)]^{(2)}]\right], \quad t\in\langle 0,\infty), \ u=1,2,3,4,$$

and particularly

$$\begin{aligned}[\boldsymbol{s}^{(2)}(t,1)]^{(2)} &= [6[\exp[-0.033t]]^2[1-\exp[-0.033t]]^2+4[\exp[-0.033t]]^3\\ &\quad\times[1-\exp[-0.033t]]+[\exp[-0.033t]]^4][1-[1-\exp[-0.066t]]^2]\\ &\quad\times\exp[-0.066t]\exp[-0.033t]\exp[-0.033t]\exp[-0.033t]\exp[-0.033t]\\ &= 12\exp[-0.33t]+8\exp[-0.429t]-16\exp[-0.363t]-3\exp[-0.462t],\end{aligned} \tag{3.260}$$

$$\begin{aligned}[\boldsymbol{s}^{(2)}(t,2)]^{(2)} &= [6[\exp[-0.04t]]^2[1-\exp[-0.04t]]^2+4[\exp[-0.04t]]^3\\ &\quad\times[1-\exp[-0.04t]]+[\exp[-0.04t]]^4][1-[1-\exp[-0.07t]]^2\\ &\quad\times\exp[-0.07t]\exp[-0.04t]\exp[-0.04t]\exp[-0.04t]\exp[-0.04t]\\ &= 12\exp[-0.38t]+8\exp[-0.49t]+6\exp[-0.46t]\\ &\quad-16\exp[-0.42t]-6\exp[-0.45t]-3\exp[-0.53t],\end{aligned} \tag{3.261}$$

$$\begin{aligned}[\boldsymbol{s}^{(2)}(t,3)]^{(2)} &= [6[\exp[-0.05t]]^2[1-\exp[-0.05t]]^2+4[\exp[-0.05t]]^3\\ &\quad\times[1-\exp[-0.05t]]+[\exp[-0.05t]]^4][1-[1-\exp[-0.075t]]^2]\\ &\quad\times\exp[-0.075t]\exp[-0.045t]\exp[-0.045t]\exp[-0.045t]\exp[-0.045t]\\ &= 12\exp[-0.43t]+8\exp[-0.555t]+6\exp[-0.53t]\\ &\quad-16\exp[-0.48t]-6\exp[-0.505t]-3\exp[-0.605t],\end{aligned} \tag{3.262}$$

$$\begin{aligned}[\boldsymbol{s}^{(2)}(t,4)]^{(2)} &= [6[\exp[-0.055t]]^2[1-\exp[-0.055t]]^2+4[\exp[-0.055t]]^3\\ &\quad\times[1-\exp[-0.055t]]+[\exp]-0.055t]]^4][1-[1-\exp[-0.08t]]^2]\\ &\quad\times\exp[-0.08t]\exp[-0.05t]\exp[-0.05t]\exp[-0.05t]\exp[-0.05t]\\ &= 12\exp[-0.47t]+8\exp[-0.605t]+6\exp[-0.58t]-16\exp[-0.525t]\\ &\quad-6\exp[-0.55t]-3\exp[-0.66t].\end{aligned} \tag{3.263}$$

Subsystem S_5 a five-state series system composed of $n = 3$ components $E_{11}^{(5)}$, $E_{21}^{(5)}$, $E_{31}^{(5)}$ having the exponential safety functions with coordinates:

$$[s_{11}^{(5)}(t,1)]^{(2)} = \exp[-0.033t], \quad [s_{11}^{(5)}(t,2)]^{(2)} = \exp[-0.04t],$$
$$[s_{11}^{(5)}(t,3)]^{(2)} = \exp[-0.045t], \quad [s_{11}^{(5)}(t,4)]^{(2)} = \exp[-0.05t];$$
$$[s_{i1}^{(5)}(t,1)]^{(2)} = \exp[-0.033t], \quad [s_{i1}^{(5)}(t,2)]^{(2)} = \exp[-0.04t],$$
$$[s_{i1}^{(5)}(t,3)]^{(2)} = \exp[-0.05t], \quad [s_{i1}^{(5)}(t,4)]^{(2)} = \exp[-0.055t], \quad i = 2,3.$$

Subsystem S_5 is a five-state series system that can also be considered as a five-state parallel–series system with its structure shape parameters $k = 3,\ l_1 = 1,\ l_2 = 1,\ l_3 = 1$. Thus, according to either the formulae (1.38)–(1.39) and (1.70)–(1.71) or to the formulae (1.22)–(1.23) and (1.58)–(1.59), after replacing in them the components' reliability functions by their safety functions, subsystem S_5 five-state safety function at the operational state z_2 is given by

$$[\boldsymbol{s}^{(5)}(t,\cdot)]^{(2)} = [1, [\boldsymbol{s}^{(5)}(t,1)]^{(2)},\ [\boldsymbol{s}^{(5)}(t,2)]^{(2)},\ [\boldsymbol{s}^{(5)}(t,3)]^{(2)},\ [\boldsymbol{s}^{(5)}(t,4)]^{(2)}],$$
$$t \in \langle 0, \infty), \tag{3.264}$$

where

$$[\boldsymbol{s}^{(5)}(t,u)]^{(2)} = \boldsymbol{s}_{3;1,1,1}(t,u)$$
$$= \prod_{i=1}^{3}\left[1 - \prod_{j=1}^{1}[1 - [s_{ij}^{(5)}(t,u)]^{(2)}]\right] = \prod_{i=1}^{3}[s_{i1}^{(5)}(t,u)]^{(2)}, \quad t \in \langle 0, \infty),$$
$$u = 1,2,3,4,$$

and particularly

$$[\boldsymbol{s}^{(5)}(t,1)]^{(2)} = \exp[-0.033t]\exp[-0.033t]\exp[-0.033t] = \exp[-0.099t], \tag{3.265}$$

$$[\boldsymbol{s}^{(5)}(t,2)]^{(2)} = \exp[-0.04t]\exp[-0.04t]\exp[-0.04t] = \exp[-0.12t], \tag{3.266}$$

$$[\boldsymbol{s}^{(5)}(t,3)]^{(2)} = \exp[-0.045t]\exp[-0.05t]\exp[-0.05t] = \exp[-0.145t], \tag{3.267}$$

$$[\boldsymbol{s}^{(5)}(t,4)]^{(2)} = \exp[-0.05t]\exp[-0.055t]\exp[-0.06t] = \exp[-0.165t]. \tag{3.268}$$

Considering that the ferry technical system at the operation state z_2 is a five-state series system of subsystems S_1, S_2 and S_5, after applying (1.22)–(1.23) and (3.254)–(3.258), (3.259)–(3.263) and (3.264)–(3.268), its conditional safety function is given by

$$[\boldsymbol{s}(t,\cdot)]^{(2)} = [1, [\boldsymbol{s}(t,1)]^{(2)},\ [\boldsymbol{s}(t,2)]^{(2)},\ [\boldsymbol{s}(t,3)]^{(2)},\ [\boldsymbol{s}(t,4)]^{(2)}], \quad t \in \langle 0, \infty), \tag{3.269}$$

where

$$[\mathbf{s}(t,u)]^{(2)} = \bar{\mathbf{s}}_3(t,u) = [\mathbf{s}^{(1)}(t,u)]^{(2)}[\mathbf{s}^{(2)}(t,u)]^{(2)}[\mathbf{s}^{(5)}(t,u)]^{(2)},$$
$$t \in \langle 0,\infty), \quad u = 1,2,3,4,$$

and particularly

$$\begin{aligned}[\mathbf{s}(t,1)]^{(2)} &= \exp[-0.033t][12\exp[-0.33t] + 8\exp[-0.429t] \\ &\quad - 16\exp[-0.363t] - 3\exp[-0.462t]]\exp[-0.099t] \\ &= 12\exp[-0.462t] + 8\exp[-0.561t] - 16\exp[-0.495t] - 3\exp[-0.594t]\end{aligned} \tag{3.270}$$

$$\begin{aligned}[\mathbf{s}(t,2)]^{(2)} &= \exp[-0.04t][12\exp[-0.38t] + 8\exp[-0.49t] + 6\exp[-0.46t] \\ &\quad - 16\exp[-0.42t] - 6\exp[-0.45t] - 3\exp[-0.53t]]\exp[-0.12t] \\ &= 12\exp[-0.54t] + 8\exp[-0.65t] + 6\exp[-0.62t] - 16\exp[-0.58t] \\ &\quad - 6\exp[-0.61t] - 3\exp[-0.69t],\end{aligned} \tag{3.271}$$

$$\begin{aligned}[\mathbf{s}(t,3)]^{(2)} &= \exp[-0.045t][12\exp[-0.43t] + 8\exp[-0.555t] + 6\exp[-0.53t] \\ &\quad - 16\exp[-0.48t] - 6\exp[-0.505t] - 3\exp[-0.605t]]\exp[-0.145t] \\ &= 12\exp[-0.62t] + 8\exp[-0.745t] + 6\exp[-0.72t] - 16\exp[-0.67t] \\ &\quad - 6\exp[-0.695t] - 3\exp[-0.795t],\end{aligned} \tag{3.272}$$

$$\begin{aligned}[\mathbf{s}(t,4)]^{(2)} &= \exp[-0.05t][12\exp[-0.47t] + 8\exp[-0.605t] + 6\exp[-0.58t] \\ &\quad - 16\exp[-0.525t] - 6\exp[-0.55t] - 3\exp[-0.66t]]\exp[-0.165t] \\ &= 12\exp[-0.685t] + 8\exp[-0.82t] + 6\exp[-0.795t] - 16\exp[-0.74t] \\ &\quad - 6\exp[-0.765t] - 3\exp[-0.875t].\end{aligned} \tag{3.273}$$

The expected values and standard deviations of the ferry technical system conditional lifetimes in the safety state subsets $\{1,2,3,4\}$, $\{2,3,4\}$, $\{3,4\}$, $\{4\}$ at the operation state z_2, calculated from the results given by (3.270)–(3.273), according to (3.8)–(3.9), respectively are:

$$\mu_2(1) \cong 2.86, \quad \mu_2(2) \cong 2.43, \quad \mu_2(3) \cong 2.14, \quad \mu_2(4) \cong 1.93 \text{ years}, \tag{3.274}$$

$$\sigma_2(1) \cong 2.74, \quad \sigma_2(2) \cong 2.35, \quad \sigma_2(3) \cong 2.05, \quad \sigma_2(4) \cong 1.85 \text{ years}, \tag{3.275}$$

and further, using (3.10) and (3.274), the mean values of the ferry technical system conditional lifetimes in the particular safety states 1, 2, 3, 4 at the operation state z_2, respectively are:

$$\overline{\mu}_2(1) \cong 0.43, \quad \overline{\mu}_2(2) \cong 0.29, \quad \overline{\mu}_2(3) \cong 0.21, \quad \overline{\mu}_2(4) \cong 1.93 \text{ years.} \tag{3.276}$$

At the system operation state z_3, the ferry technical system is composed of two subsystems S_1 and S_2 forming a series structure shown in Fig. 2.16.

Subsystem S_1 consists of a component $E_{11}^{(1)}$, with the same safety function as at the operation state z_2. Thus, according to (3.254)–(3.258), subsystem S_1 conditional function at the operation state z_3, is given by

$$[\boldsymbol{s}^{(1)}(t,\cdot)]^{(3)} = [1, [\boldsymbol{s}^{(1)}(t,1)]^{(3)}, [\boldsymbol{s}^{(1)}(t,2)]^{(3)}, [\boldsymbol{s}^{(1)}(t,3)]^{(3)}, [\boldsymbol{s}^{(1)}(t,4)]^{(3)}],$$
$$t \in \langle 0, \infty), \tag{3.277}$$

where

$$[\boldsymbol{s}^{(1)}(t,1)]^{(3)} = \exp[-0.033t], \tag{3.278}$$

$$[\boldsymbol{s}^{(1)}(t,2)]^{(3)} = \exp[-0.04t], \tag{3.279}$$

$$[\boldsymbol{s}^{(1)}(t,3)]^{(3)} = \exp[-0.045t], \tag{3.280}$$

$$[\boldsymbol{s}^{(1)}(t,4)]^{(3)} = \exp[-0.05t]. \tag{3.281}$$

Subsystem S_2 is a five-state parallel–series system composed of components $E_{ij}^{(2)}$, $i = 1,2,\ldots,k$, $j = 1,2,\ldots,l_i$, $k = 7$, $l_1 = 4$, $l_2 = 2$, $l_3 = 1$, $l_4 = 1$, $l_5 = 1$, $l_6 = 1$, $l_7 = 1$, having the exponential safety functions with coordinates:

$$[s_{1j}^{(2)}(t,1)]^{(3)} = \exp[-0.033t], \quad [s_{1j}^{(2)}(t,2)]^{(3)} = \exp[-0.04t],$$
$$[s_{1j}^{(2)}(t,3)]^{(3)} = \exp[-0.05t], \quad [s_{1j}^{(2)}(t,4)]^{(3)} = \exp[-0.055t], \quad j = 1,2,3,4;$$
$$[s_{2j}^{(2)}(t,1)]^{(3)} = \exp[-0.033t], \quad [s_{2j}^{(2)}(t,2)]^{(3)} = \exp[-0.04t],$$
$$[s_{2j}^{(2)}(t,3)]^{(3)} = \exp[-0.045t], \quad [s_{2j}^{(2)}(t,4)]^{(3)} = \exp[-0.05t], \quad j = 1,2;$$
$$[s_{i1}^{(2)}(t,1)]^{(3)} = \exp[-0.033t], \quad [s_{i1}^{(2)}(t,2)]^{(3)} = \exp[-0.04t],$$
$$[s_{i1}^{(2)}(t,3)]^{(3)} = \exp[-0.045t], \quad [s_{i1}^{(2)}(t,4)]^{(3)} = \exp[-0.05t], \quad i = 3,4,5,6,7.$$

Subsystem S_2 is a five-state parallel–series system with its structure shape parameters $k = 7$, $l_1 = 4$, $l_2 = 2$, $l_3 = 1$, $l_4 = 1$, $l_5 = 1$, $l_6 = 1$, $l_7 = 1$.

Thus, according to the formulae (1.38)–(1.39) and (1.70)–(1.71), after replacing in them the components' reliability functions by their safety functions, subsystem S_2 conditional safety function at the operation state z_3 is given by

$$[\boldsymbol{s}^{(2)}(t,\cdot)]^{(3)} = [1, [\boldsymbol{s}^{(2)}(t,1)]^{(3)}, [\boldsymbol{s}^{(2)}(t,2)]^{(3)}, [\boldsymbol{s}^{(2)}(t,3)]^{(3)}, [\boldsymbol{s}^{(2)}(t,4)]^{(3)}],$$
$$t \in \langle 0, \infty), \tag{3.282}$$

where

$$[\boldsymbol{s}^{(2)}(t,u)]^{(3)} = \boldsymbol{s}_{7;4,2,1,1,1,1,1}(t,u)$$
$$= \prod_{i=1}^{7}\left[1 - \prod_{j=1}^{l_i}[1 - [s_{ij}^{(2)}(t,u)]^{(3)}]\right], \quad t \in \langle 0, \infty), \ u = 1,2,3,4,$$

and particularly

$$[\boldsymbol{s}^{(2)}(t,1)]^{(3)} = [[6\exp[-0.033t]]^2[1-\exp[-0.033t]]^2 + 4[\exp[-0.033t]]^3$$
$$[1-\exp[-0.033t]] + [\exp[-0.033t]]^4][1-[1-\exp[-0.033t]]^2]$$
$$\times \exp[-0.033t]\exp[-0.033t]\exp[-0.033t]\exp[-0.033t]\exp[-0.033t]$$
$$= 12\exp[-0.258t] + 8\exp[-0.33t] + 6\exp[-0.324t] - 16\exp[-0.291t]$$
$$- 6\exp[-0.297t] - 3\exp[-0.363t], \tag{3.283}$$

$$[\boldsymbol{s}^{(2)}(t,2)]^{(3)} = [[6\exp[-0.04t]]^2[1-\exp[-0.04t]]^2 + 4[\exp[-0.04t]]^3$$
$$[1-\exp[-0.04t]] + [\exp[-0.04t]]^4][1-[1-\exp[-0.04t]]^2]$$
$$\times \exp[-0.04t]\exp[-0.04t]\exp[-0.04t]\exp[-0.04t]\exp[-0.04t]$$
$$= 12\exp[-0.32t] + 14\exp[-0.4t] - 22\exp[-0.36t] - 3\exp[-0.44t], \tag{3.284}$$

$$[\boldsymbol{s}^{(2)}(t,3)]^{(3)} = [[6\exp[-0.05t]]^2[1-\exp[-0.05t]]^2 + 4[\exp[-0.05t]]^3[1-\exp[-0.05t]]$$
$$+ [\exp[-0.05t]]^4][1-[1-\exp[-0.045t]]^2]\exp[-0.045t]\exp[-0.045t]$$
$$\times \exp[-0.045t]\exp[-0.045t]\exp[-0.045t]$$
$$= 12\exp[-0.37t] + 8\exp[-0.465t] + 6\exp[-0.47t] - 16\exp[-0.42t]$$
$$- 6\exp[-0.415t] - 3\exp[-0.515t], \tag{3.285}$$

$$[\boldsymbol{s}^{(2)}(t,4)]^{(3)} = [[6\exp[-0.055t]]^2[1-\exp[-0.055t]]^2 + 4[\exp[-0.055t]]^3$$
$$[1-\exp[-0.055t]][\exp[-0.055t]]^4][1-[1-\exp[-0.05t]]^2]$$
$$\times \exp[-0.05t]\exp[-0.05t] \times \exp[-0.05t]\exp[-0.05t]\exp[-0.05t],$$
$$= 12\exp[-0.41t] + 8\exp[-0.515t] + 6\exp[-0.52t] - 16\exp[-0.465t]$$
$$- 6\exp[-0.46t] - 3\exp[-0.57t]. \tag{3.286}$$

Considering that the ferry technical system at the operation state z_3 is a five-state series system of subsystems S_1 and S_2, after applying (1.22)–(1.23) and (3.277)–(3.281) and (3.282)–(3.286), its conditional five-state safety function is given by

$$[\boldsymbol{s}(t,\cdot)]^{(3)} = [1, [\boldsymbol{s}(t,1)]^{(3)}, [\boldsymbol{s}(t,2)]^{(3)}, [\boldsymbol{s}(t,3)]^{(3)}, [\boldsymbol{s}(t,4)]^{(3)}], \quad t \in \langle 0, \infty), \tag{3.287}$$

where

$$[\boldsymbol{s}(t,u)]^{(3)} = \bar{s}_2(t,u) = [\boldsymbol{s}^{(1)}(t,u)]^{(3)} [\boldsymbol{s}^{(2)}(t,u)]^{(3)}, \quad t \in \langle 0, \infty), \ u = 1,2,3,4,$$

and particularly

$$\begin{aligned}[\boldsymbol{s}(t,1)]^{(3)} &= \exp[-0.033t][12\exp[-0.258t] + 8\exp[-0.33t] + 6\exp[-0.324t] \\ &\quad - 16\exp[-0.291t] - 6\exp[-0.297t] - 3\exp[-0.363t]] \\ &= 12\exp[-0.291t] + 8\exp[-0.363t] + 6\exp[-0.357t] \\ &\quad - 16\exp[-0.324t] - 6\exp[-0.33t] - 3\exp[-0.396t],\end{aligned} \tag{3.288}$$

$$\begin{aligned}[\boldsymbol{s}(t,2)]^{(3)} &= \exp[-0.04t][12\exp[-0.32t] + 14\exp[-0.4t] - 22\exp[-0.36t] \\ &\quad - 3\exp[-0.44t]] = 12\exp[-0.36t] + 14\exp[-0.44t] - 22\exp[-0.4t] \\ &\quad - 3\exp[-0.48t],\end{aligned} \tag{3.289}$$

$$\begin{aligned}[\boldsymbol{s}(t,3)]^{(3)} &= \exp[-0.045t][12\exp[-0.37t] + 8\exp[-0.465t] + 6\exp[-0.47t] \\ &\quad - 16\exp[-0.42t] - 6\exp[-0.415t] - 3\exp[-0.515t]] \\ &= 12\exp[-0.415t] + 8\exp[-0.51t] + 6\exp[-0.515t] - 16\exp[-0.465t] \\ &\quad - 6\exp[-0.46t] - 3\exp[-0.56t],\end{aligned} \tag{3.290}$$

$$\begin{aligned}[\boldsymbol{s}(t,4)]^{(3)} &= \exp[-0.05t][12\exp[-0.41t] + 8\exp[-0.515t] + 6\exp[-0.52t] \\ &\quad - 16\exp[-0.465t] - 6\exp[-0.46t] - 3\exp[-0.57t] \\ &= 12\exp[-0.46t] + 8\exp[-0.565t] + 6\exp[-0.57t] - 16\exp[-0.515t] \\ &\quad - 6\exp[-0.51t] - 3\exp[-0.62t].\end{aligned} \tag{3.291}$$

The expected values and standard deviations of the ferry technical system conditional lifetimes in the safety state subsets $\{1,2,3,4\}$, $\{2,3,4\}$, $\{3,4\}$, $\{4\}$ at the operation state z_3, calculated from the results given by (3.288)–(3.291), according to (3.8)–(3.9), respectively are:

$$\mu_3(1) \cong 4.94, \quad \mu_3(2) \cong 3.9, \quad \mu_3(3) \cong 3.44, \quad \mu_3(4) \cong 3.1 \text{ years}, \tag{3.292}$$

$$\sigma_3(1) \cong 4.6, \quad \sigma_3(2) \cong 3.6, \quad \sigma_3(3) \cong 3.2, \quad \sigma_3(4) \cong 2.9 \text{ years}, \tag{3.293}$$

and further, using (3.10) and (3.292), the mean values of the ferry technical system conditional lifetimes in the particular safety states 1, 2, 3, 4 at the operation state z_3, respectively are:

$$\bar{\mu}_3(1) \cong 1.04, \quad \bar{\mu}_3(2) \cong 0.46, \quad \bar{\mu}_3(3) \cong 0.34, \quad \bar{\mu}_3(4) \cong 3.1 \text{ years}. \tag{3.294}$$

At the system operation state z_4, the ferry technical system is composed of three subsystems S_1, S_2 and S_4 forming a series structure shown in Fig. 2.17.

Subsystem S_1 consists of one component $E_{11}^{(1)}$, with the same safety function as at the operation state z_2. Thus, according to (3.254)–(3.258), subsystem S_1 conditional function at the operation state z_4, is given by

$$[\boldsymbol{s}^{(1)}(t,\cdot)]^{(4)} = [1, [\boldsymbol{s}^{(1)}(t,1)]^{(4)}, [\boldsymbol{s}^{(1)}(t,2)]^{(4)}, [\boldsymbol{s}^{(1)}(t,3)]^{(4)}, [\boldsymbol{s}^{(1)}(t,4)]^{(4)}], \quad t \in \langle 0, \infty), \tag{3.295}$$

where

$$[\boldsymbol{s}^{(1)}(t,1)]^{(4)} = \exp[-0.033t], \tag{3.296}$$

$$[\boldsymbol{s}^{(1)}(t,2)]^{(4)} = \exp[-0.04t], \tag{3.297}$$

$$[\boldsymbol{s}^{(1)}(t,3)]^{(4)} = \exp[-0.045t], \tag{3.298}$$

$$[\boldsymbol{s}^{(1)}(t,4)]^{(4)} = \exp[-0.05t]. \tag{3.299}$$

Subsystem S_2 is a five-state parallel–series system composed of components $E_{ij}^{(2)}$, $i = 1, 2, \ldots, k$, $j = 1, 2, \ldots, l_i$, $k = 5$, $l_1 = 4$, $l_2 = 1$, $l_3 = 1$, $l_4 = 1$, $l_5 = 1$, having the exponential safety functions with coordinates:

$$[s_{1j}^{(2)}(t,1)]^{(4)} = \exp[-0.033t], \quad [s_{1j}^{(2)}(t,2)]^{(4)} = \exp[-0.04t],$$
$$[s_{1j}^{(2)}(t,3)]^{(4)} = \exp[-0.05t], \quad [s_{1j}^{(2)}(t,4)]^{(4)} = \exp[-0.055t], \quad j = 1, 2, 3, 4;$$

$$[s_{i1}^{(2)}(t,1)]^{(4)} = \exp[-0.033t], \quad [s_{i1}^{(2)}(t,2)]^{(4)} = \exp[-0.04t],$$
$$[s_{i1}^{(2)}(t,3)]^{(4)} = \exp[-0.045t], \quad [s_{i1}^{(2)}(t,4)]^{(4)}$$

Subsystem S_2 is a five-state parallel–series system with its structure shape parameters $k = 5$, $l_1 = 4$, $l_2 = 1$, $l_3 = 1$, $l_4 = 1$, $l_5 = 1$. Thus, according to the formulae (1.38)–(1.39) and (1.70)–(1.71), after replacing in them the components' reliability functions by their safety functions, the subsystem S_2 conditional safety function at the operation state z_4 is given by

$$[\boldsymbol{s}^{(2)}(t,\cdot)]^{(4)} = [1, [\boldsymbol{s}^{(2)}(t,1)]^{(4)}, [\boldsymbol{s}^{(2)}(t,2)]^{(4)}, [\boldsymbol{s}^{(2)}(t,3)]^{(4)}, [\boldsymbol{s}^{(2)}(t,4)]^{(4)}], \quad t \in \langle 0, \infty), \tag{3.300}$$

where

$$[\boldsymbol{s}^{(2)}(t,u)]^{(4)} = \boldsymbol{s}_{5;4,1,1,1,1}(t,u) = \prod_{i=1}^{5}\left[1 - \prod_{j=1}^{l_i}[1 - [s_{ij}^{(2)}(t,u)]^{(4)}]\right], \quad t \in \langle 0, \infty), \quad u = 1, 2, 3, 4,$$

and particularly

$$[s^{(2)}(t,1)]^{(4)} = [6[\exp[-0.033t]]^2[1-\exp[-0.033t]]^2 + 4[\exp[-0.033t]]^3$$
$$[1-\exp[-0.033t]] + [\exp[-0.033t]]^4]\exp[-0.033t]\exp[-0.033t]$$
$$\times \exp[-0.033t]\exp[-0.033t]$$
$$= 6\exp[-0.198t] - 8\exp[-0.231t] + 3\exp[-0.264t], \quad (3.301)$$

$$[s^{(2)}(t,2)]^{(4)} = [6[\exp[-0.04t]]^2[1-\exp[-0.04t]]^2 + 4[\exp[-0.04t]]^3[1-\exp[-0.04t]]$$
$$+ [\exp[-0.04t]]^4]\exp[-0.04t]\exp[-0.04t]\exp[-0.04t]\exp[-0.04t]$$
$$= 6\exp[-0.24t] - 8\exp[-0.28t] + 3\exp[-0.32t], \quad (3.302)$$

$$[s^{(2)}(t,3)]^{(4)} = [6[\exp[-0.05t]]^2[1-\exp[-0.05t]]^2 + 4[\exp[-0.05t]]^3$$
$$[1-\exp[-0.05t]] + [\exp[-0.05t]]^4]\exp[-0.045t]$$
$$\times \exp[-0.045t]\exp[-0.045t]\exp[-0.045t]$$
$$= 6\exp[-0.28t] - 8\exp[-0.33t] + 3\exp[-0.38t], \quad (3.303)$$

$$[s^{(2)}(t,4)]^{(4)} = [6[\exp[-0.055t]]^2[1-\exp[-0.055t]]^2 + 4[\exp[-0.055t]]^3$$
$$[1-\exp[-0.055t]] + [\exp[-0.055t]]^4]\exp[-0.05t]\exp[-0.05t]$$
$$\times \exp[-0.05t]\exp[-0.05t]$$
$$= \exp[-0.31t] - 8\exp[-0.365t] + 3\exp[-0.42t]. \quad (3.304)$$

Subsystem S_4 consists of a component $E_{11}^{(4)}$ having the exponential safety function with coordinates:

$$[s_{11}^{(4)}(t,1)]^{(4)} = \exp[-0.05t], \quad [s_{11}^{(4)}(t,2)]^{(4)} = \exp[-0.055t],$$
$$[s_{11}^{(4)}(t,3)]^{(4)} = \exp[-0.06t], \quad [s_{11}^{(4)}(t,4)]^{(4)} = \exp[-0.065t].$$

Thus, at the operation state z_4, subsystem S_4 conditional safety function is identical with the component $E_{11}^{(4)}$ safety function and is given by

$$[s^{(4)}(t,\cdot)]^{(4)} = [1, [s^{(4)}(t,1)]^{(4)}, [s^{(4)}(t,2)]^{(4)}, [s^{(4)}(t,3)]^{(4)}, [s^{(4)}(t,4)]^{(4)}],$$
$$t \in \langle 0,\infty), \quad (3.305)$$

where

$$[s^{(4)}(t,1)]^{(4)} = \exp[-0.05t], \quad (3.306)$$

$$[s^{(4)}(t,2)]^{(4)} = \exp[-0.055t], \quad (3.307)$$

$$[s^{(4)}(t,3)]^{(4)} = \exp[-0.06t], \quad (3.308)$$

$$[s^{(4)}(t,4)]^{(4)} = \exp[-0.065t]. \tag{3.309}$$

Considering that the ferry technical system at the operation state z_4 is a five-state series system of subsystems S_1, S_2 and S_4, after applying (1.22)–(1.23) and (3.295)–(3.299), (3.300)–(3.304) and (3.305)–(3.309), its conditional safety function is given by

$$[s(t,\cdot)]^{(4)} = [1, [s(t,1)]^{(4)}, [s(t,2)]^{(4)}, [s(t,3)]^{(4)}, [s(t,4)]^{(4)}], \quad t \in \langle 0,\infty), \tag{3.310}$$

where

$$[s(t,u)]^{(4)} = \bar{s}_3(t,u) = [s^{(1)}(t,u)]^{(4)} [s^{(2)}(t,u)]^{(4)} [s^{(4)}(t,u)]^{(4)}, \quad t \in \langle 0,\infty),$$
$$u = 1,2,3,4,$$

and particularly

$$\begin{aligned}[s(t,1)]^{(4)} &= \exp[-0.033t][6\exp[-0.198t] - 8\exp[-0.231t] \\ &\quad + 3\exp[-0.264t]]\exp[-0.05t] \\ &= 6\exp[-0.281t] - 8\exp[-0.314t] + 3\exp[-0.347t],\end{aligned} \tag{3.311}$$

$$\begin{aligned}[s(t,2)]^{(4)} &= \exp[-0.04t][6\exp[-0.24t] - 8\exp[-0.28t] \\ &\quad + 3\exp[-0.32t]]\exp[-0.055t] \\ &= 6\exp[-0.335t] - 8\exp[-0.375t] + 3\exp[-0.415t],\end{aligned} \tag{3.312}$$

$$\begin{aligned}[s(t,3)]^{(4)} &= \exp[-0.045t][6\exp[-0.28t] - 8\exp[-0.33t] \\ &\quad + 3\exp[-0.38t]]\exp[-0.06t] \\ &= 6\exp[-0.385t] - 8\exp[-0.435t] + 3\exp[-0.485t],\end{aligned} \tag{3.313}$$

$$\begin{aligned}[s(t,4)]^{(4)} &= \exp[-0.05t][\exp[-0.31t] - 8\exp[-0.365t] \\ &\quad + 3\exp[-0.42t]\exp[-0.065t] \\ &= 6\exp[-0.425t] - 8\exp[-0.48t] + 3\exp[-0.535t].\end{aligned} \tag{3.314}$$

The expected values and standard deviations of the ferry technical system conditional lifetimes in the safety state subsets $\{1,2,3,4\}$, $\{2,3,4\}$, $\{3,4\}$, $\{4\}$ at the operation state z_4, calculated from the results given by (3.311)–(3.314), according to (3.8)–(3.9), respectively are:

$$\mu_4(1) \cong 4.2, \quad \mu_4(2) \cong 3.80, \quad \mu_4(3) \cong 3.38, \quad \mu_4(4) \cong 3.05 \text{ years}, \tag{3.315}$$

$$\sigma_4(1) \cong 4.37, \quad \sigma_4(2) \cong 3.68, \quad \sigma_4(3) \cong 3.12, \quad \sigma_4(4) \cong 2.93 \text{ years}, \tag{3.316}$$

and further, using (3.10) and (3.315), the mean values of the ferry technical system conditional lifetimes in the particular safety states 1, 2, 3, 4 at the operation state z_4, respectively are:

$$\overline{\mu}_4(1) \cong 0.72, \quad \overline{\mu}_4(2) \cong 0.42, \quad \overline{\mu}_4(3) \cong 0.33, \quad \overline{\mu}_4(4) \cong 3.05 \text{ years.} \quad (3.317)$$

At the system operation state z_5, the ferry technical system is composed of three subsystems S_1, S_2 and S_4 forming a series structure shown in Fig. 2.12. The system safety structure and its components safety functions are the same as at the operation state z_4. Thus, considering (3.310)–(3.314), its conditional safety function is given by

$$[\boldsymbol{s}(t,\cdot)]^{(5)} = \left[1, [\boldsymbol{s}(t,1)]^{(5)}, [\boldsymbol{s}(t,2)]^{(5)}, [\boldsymbol{s}(t,3)]^{(5)}, [\boldsymbol{s}(t,4)]^{(5)}\right], \quad t \in \langle 0, \infty), \quad (3.318)$$

where

$$[\boldsymbol{s}(t,1)]^{(5)} = 6\exp[-0.281t] - 8\exp[-0.314t] + 3\exp[-0.347t], \quad (3.319)$$

$$\begin{aligned}[\boldsymbol{s}(t,2)]^{(5)} &= \exp[-0.04t][6\exp[-0.24t] - 8\exp[-0.28t] \\ &\quad + 3\exp[-0.32t]]\exp[-0.055t] \\ &= 6\exp[-0.335t] - 8\exp[-0.375t] + 3\exp[-0.415t], \end{aligned} \quad (3.320)$$

$$\begin{aligned}[\boldsymbol{s}(t,3)]^{(5)} &= \exp[-0.045t][6\exp[-0.28t] - 8\exp[-0.33t] \\ &\quad + 3\exp[-0.38t]]\exp[-0.06t] \\ &= 6\exp[-0.385t] - 8\exp[-0.435t] + 3\exp[-0.485t], \end{aligned} \quad (3.321)$$

$$\begin{aligned}[\boldsymbol{s}(t,4)]^{(5)} &= \exp[-0.05t][\exp[-0.31t] - 8\exp[-0.365t] \\ &\quad + 3\exp[-0.42t]\exp[-0.065t] \\ &= 6\exp[-0.425t] - 8\exp[-0.48t] + 3\exp[-0.535t]. \end{aligned} \quad (3.322)$$

The expected values and standard deviations of the ferry technical system conditional lifetimes in the safety state subsets $\{1,2,3,4\}$, $\{2,3,4\}$, $\{3,4\}$, $\{4\}$ at the operation state z_5, calculated from the results given by (3.319)–(3.322), according to (3.8)–(3.9), respectively are:

$$\mu_5(1) \cong 4.2, \quad \mu_5(2) \cong 3.80, \quad \mu_5(3) \cong 3.38, \quad \mu_5(4) \cong 3.05 \text{ years,} \quad (3.323)$$

$$\sigma_5(1) \cong 4.37, \quad \sigma_5(2) \cong 3.68, \quad \sigma_5(3) \cong 3.12, \quad \sigma_5(4) \cong 2.93 \text{ years,} \quad (3.324)$$

and further, using (3.10) and (3.323), the mean values of the ferry technical system conditional lifetimes in the particular safety states 1, 2, 3, 4 at the operation state z_5, respectively are:

$$\overline{\mu}_5(1) \cong 0.72, \quad \overline{\mu}_5(2) \cong 0.42, \quad \overline{\mu}_5(3) \cong 0.33, \quad \overline{\mu}_5(4) \cong 3.05 \text{ years.} \quad (3.325)$$

At the system operation state z_6, the ferry technical system is composed of three subsystems S_1, S_2 and S_4 forming a series structure shown in Fig. 2.18. Subsystem S_1 consists of a component $E_{11}^{(1)}$, with the same safety function as at the operation state z_2. Thus, according to (3.254)–(3.258), subsystem S_1 conditional safety function at the operation state z_6 is given by

$$[\boldsymbol{s}^{(1)}(t,\cdot)]^{(6)} = [1, [\boldsymbol{s}^{(1)}(t,1)]^{(6)}, [\boldsymbol{s}^{(1)}(t,2)]^{(6)}, [\boldsymbol{s}^{(1)}(t,3)]^{(6)}, [\boldsymbol{s}^{(1)}(t,4)]^{(6)}], \quad t \in \langle 0, \infty), \tag{3.326}$$

where

$$[\boldsymbol{s}^{(1)}(t,1)]^{(6)} = \exp[-0.033t], \tag{3.327}$$

$$[\boldsymbol{s}^{(1)}(t,2)]^{(6)} = \exp[-0.04t], \tag{3.328}$$

$$[\boldsymbol{s}^{(1)}(t,3)]^{(6)} = \exp[-0.045t], \tag{3.329}$$

$$[\boldsymbol{s}^{(1)}(t,4)]^{(6)} = \exp[-0.05t]. \tag{3.330}$$

Subsystem S_2 illustrated in Fig. 2.18, is a five-state parallel–series system composed of components $E_{ij}^{(2)}$, $i = 1,2,\ldots,k$, $j = 1,2,\ldots,l_i$, $k = 7$, $l_1 = 4$, $l_2 = 2$, $l_3 = 1$, $l_4 = 1$, $l_5 = 1$, $l_6 = 1$, $l_7 = 1$, with the same safety functions as at the operation state z_3. Thus, according to (3.282)–(3.286), subsystem S_2 conditional safety function at the operation state z_6 is given by

$$[\boldsymbol{s}^{(2)}(t,\cdot)]^{(6)} = \left[1, [\boldsymbol{s}2)(t,1)]^{(6)}, [\boldsymbol{s}^{(2)}(t,2)]^{(6)}, [\boldsymbol{s}^{(2)}(t,3)]^{(6)}, [\boldsymbol{s}^{(2)}(t,4)]^{(6)}\right], \quad t \in \langle 0, \infty), \tag{3.331}$$

where

$$[\boldsymbol{s}^{(2)}(t,1)]^{(6)} = 12\exp[-0.258t] + 8\exp[-0.33t] + 6\exp[-0.324t] - 16\exp[-0.291t] - 6\exp[-0.297t] - 3\exp[-0.363t], \tag{3.332}$$

$$[\boldsymbol{s}^{(2)}(t,2)]^{(6)} = 12\exp[-0.32t] + 14\exp[-0.4t] - 22\exp[-0.36t] - 3\exp[-0.44t], \tag{3.333}$$

$$[\boldsymbol{s}^{(2)}(t,3)]^{(6)} = 12\exp[-0.37t] + 8\exp[-0.465t] + 6\exp[-0.47t] - 16\exp[-0.42t] - 6\exp[-0.415t] - 3\exp[-0.515t], \tag{3.334}$$

$$[\boldsymbol{s}^{(2)}(t,4)]^{(6)} = 12\exp[-0.41t] + 8\exp[-0.515t] + 6\exp[-0.52t] - 16\exp[-0.465t] - 6\exp[-0.46t] - 3\exp[-0.57t]. \tag{3.335}$$

Subsystem S_4 consists of a component $E_{11}^{(4)}$, with the same safety functions as at the operation state z_4. Thus, according to (3.305)–(3.309), subsystem S_4 conditional safety function at the operation state z_6 is given by

$$[\boldsymbol{s}^{(4)}(t,\cdot)]^{(6)} = [1, [\boldsymbol{s}^{(4)}(t,1)]^{(6)}, [\boldsymbol{s}^{(4)}(t,2)]^{(6)}, [\boldsymbol{s}^{(4)}(t,3)]^{(6)}, [\boldsymbol{s}^{(4)}(t,4)]^{(6)}], \quad t \in \langle 0,\infty), \tag{3.336}$$

where

$$[\boldsymbol{s}^{(4)}(t,1)]^{(6)} = \exp[-0.05t], \tag{3.337}$$

$$[\boldsymbol{s}^{(4)}(t,2)]^{(6)} = \exp[-0.055t], \tag{3.338}$$

$$[\boldsymbol{s}^{(4)}(t,3)]^{(6)} = \exp[-0.06t], \tag{3.339}$$

$$[\boldsymbol{s}^{(4)}(t,4)]^{(6)} = \exp[-0.065t]. \tag{3.340}$$

Considering that the ferry technical system at the operation state z_6 is a five-state series system of subsystems S_1, S_2 and S_4, after applying (1.22)–(1.23) and (3.326)–(3.330), (3.331)–(3.335) and (3.336)–(3.340), its conditional safety function is given by

$$[\boldsymbol{s}(t,\cdot)]^{(6)} = [1, [\boldsymbol{s}(t,1)]^{(6)}, [\boldsymbol{s}(t,2)]^{(6)}, [\boldsymbol{s}(t,3)]^{(6)}, [\boldsymbol{s}(t,4)]^{(6)}], \quad t \in \langle 0,\infty), \tag{3.341}$$

where

$$[\boldsymbol{s}(t,u)]^{(6)} = \bar{s}_3(t,u) = [\boldsymbol{s}^{(1)}(t,u)]^{(6)} [\boldsymbol{s}^{(2)}(t,u)]^{(6)} [\boldsymbol{s}^{(4)}(t,u)]^{(6)}, \quad t \in \langle 0,\infty), \quad u = 1,2,3,4,$$

and particularly

$$\begin{aligned}[\boldsymbol{s}(t,1)]^{(6)} &= \exp[-0.033t][12\exp[-0.258t] + 8\exp[-0.33t] + 6\exp[-0.324t] \\ &\quad - 16\exp[-0.291t] - 6\exp[-0.297t] - 3\exp[-0.363t]][\exp[-0.05t] \\ &= 12\exp[-0.341t] + 8\exp[-0.413t] + 6\exp[-0.407t] - 16\exp[-0.374t] \\ &\quad - 6\exp[-0.38t] - 3\exp[-0.446t],\end{aligned} \tag{3.342}$$

$$\begin{aligned}[\boldsymbol{s}(t,2)]^{(6)} &= \exp[-0.04t][12\exp[-0.32t] + 14\exp[-0.4t] - 22\exp[-0.36t] \\ &\quad - 3\exp[-0.44t]][\exp[-0.055t] \\ &= 12\exp[-0.415t] + 14\exp[-0.495t] - 22\exp[-0.455t] \\ &\quad - 3\exp[-0.535t],\end{aligned} \tag{3.343}$$

$$[s(t,3)]^{(6)} = \exp[-0.045t][12\exp[-0.37t] + 8\exp[-0.465t] + 6\exp[-0.47t] - 16\exp[-0.42t] - 6\exp[-0.415t] - 3\exp[-0.515t]][\exp[-0.06t]] = 12\exp[-0.475t] + 8\exp[-0.57t] + 6\exp[-0.575t] - 16\exp[-0.525t] - 6\exp[-0.52t] - 3\exp[-0.62t], \quad (3.344)$$

$$[s(t,4)]^{(6)} = \exp[-0.05t][12\exp[-0.41t] + 8\exp[-0.515t] + 6\exp[-0.52t] - 16\exp[-0.465t] - 6\exp[-0.46t] - 3\exp[-0.57t]][\exp[-0.065t]] = 12\exp[-0.525t] + 8\exp[-0.63t] + 6\exp[-0.635t] - 16\exp[-0.58t] - 6\exp[-0.575t] - 3\exp[-0.685t]. \quad (3.345)$$

The expected values and standard deviations of the ferry technical system conditional lifetimes in the safety state subsets $\{1,2,3,4\}$, $\{2,3,4\}$, $\{3,4\}$, $\{4\}$ at the operation state z_6, calculated from the results given by (3.342)–(3.345), according to (3.8)–(3.9), respectively are:

$$\mu_6(1) \cong 4.01, \quad \mu_6(2) \cong 3.24, \quad \mu_6(3) \cong 2.88, \quad \mu_6(4) \cong 2.61 \text{ years}, \quad (3.346)$$

$$\sigma_6(1) \cong 3.80, \quad \sigma_6(2) \cong 3.10, \quad \sigma_6(3) \cong 2.70, \quad \sigma_6(4) \cong 2.47 \text{ years}, \quad (3.347)$$

and further, using (3.10) and (3.346) the mean values of the ferry technical system conditional lifetimes in the particular safety states 1, 2, 3, 4 at the operation state z_6, respectively are:

$$\bar{\mu}_6(1) \cong 0.77, \quad \bar{\mu}_6(2) \cong 0.36, \quad \bar{\mu}_6(3) \cong 0.36, \quad \bar{\mu}_6(4) \cong 2.61 \text{ years} \quad (3.348)$$

At the system operation state z_7, the ferry technical system is composed of three subsystems S_1, S_2 and S_5 forming a series structure shown in Fig. 2.15. The system safety structure and its components safety functions are the same as at the operation state z_2.

Thus, considering (3.269)–(3.273), its conditional safety function is given by

$$[s(t,\cdot)]^{(7)} = [1, [s(t,1)]^{(7)}, [s(t,2)]^{(7)}, [s(t,3)]^{(7)}, [s(t,4)]^{(7)}], \quad t \in \langle 0, \infty), \quad (3.349)$$

where

$$[s(t,1)]^{(7)} = 12\exp[-0.462t] + 8\exp[-0.561t] - 16\exp[-0.495t] - 3\exp[-0.594t], \quad (3.350)$$

$$[s(t,2)]^{(7)} = 12\exp[-0.54t] + 8\exp[-0.65t] + 6\exp[-0.62t] - 16\exp[-0.58t] - 6\exp[-0.61t] - 3\exp[-0.69t], \quad (3.351)$$

$$[s(t,3)]^{(7)} = 12\exp[-0.62t] + 8\exp[-0.745t] + 6\exp[-0.72t] - 16\exp[-0.67t] - 6\exp[-0.695t] - 3\exp[-0.795t], \quad (3.352)$$

$$[\boldsymbol{s}(t,4)]^{(7)} = 12\exp[-0.685t] + 8\exp[-0.82t] + 6\exp[-0.795t] - 16\exp[-0.74t] - 6\exp[-0.765t] - 3\exp[-0.875t]. \quad (3.353)$$

The expected values and standard deviations of the ferry technical system conditional lifetimes in the safety state subsets $\{1,2,3,4\}$, $\{2,3,4\}$, $\{3,4\}$, $\{4\}$ at the operation state z_7, calculated from the results given by (3.350)–(3.353), according to (3.8)–(3.9), respectively are:

$$\mu_7(1) \cong 2.86, \quad \mu_7(2) \cong 2.43, \quad \mu_7(3) \cong 2.14, \quad \mu_7(4) \cong 1.93 \text{ years}, \quad (3.354)$$

$$\sigma_7(1) \cong 2.74, \quad \sigma_7(2) \cong 2.35, \quad \sigma_7(3) \cong 2.05, \quad \sigma_7(4) \cong 1.85 \text{ years}, \quad (3.355)$$

and further, using (3.10) and (3.354), the mean values of the ferry technical system conditional lifetimes in the particular safety states 1, 2, 3, 4 at the operation state z_7, respectively are:

$$\bar{\mu}_7(1) \cong 0.43, \quad \bar{\mu}_7(2) \cong 0.29, \quad \bar{\mu}_7(3) \cong 0.21, \quad \bar{\mu}_7(4) \cong 1.93 \text{ years}. \quad (3.356)$$

At the system operation state z_8, the ferry technical system is composed of two subsystems S_3 and S_4 forming a series structure shown in Fig. 2.19.

Subsystem S_3 is a five-state series system composed of $n = 2$ components $E_{11}^{(3)}$, $E_{21}^{(3)}$ having the exponential safety functions identified in Chap. 5 with coordinates:

$$[\boldsymbol{s}_{11}^{(3)}(t,1)]^{(8)} = \exp[-0.02t], \quad [\boldsymbol{s}_{11}^{(3)}(t,2)]^{(8)} = \exp[-0.03t],$$
$$[\boldsymbol{s}_{11}^{(3)}(t,3)]^{(8)} = \exp[-0.035t], \quad [\boldsymbol{s}_{11}^{(3)}(t,4)]^{(8)} = \exp[-0.04t],$$
$$[\boldsymbol{s}_{21}^{(3)}(t,1)]^{(8)} = \exp[-0.033t], \quad [\boldsymbol{s}_{21}^{(3)}(t,2)]^{(8)} = \exp[-0.04t],$$
$$[\boldsymbol{s}_{21}^{(3)}(t,3)]^{(8)} = \exp[-0.045t], \quad [\boldsymbol{s}_{21}^{(3)}(t,4)]^{(8)} = \exp[-0.05t].$$

Subsystem S_3 is a five-state series system that can also be considered as a five-state parallel–series system with its structure shape parameters $k = 2$, $l_1 = 1$, $l_2 = 1$. Thus, according to either the formulae (1.38)–(1.39) and (1.70)–(1.71) or to the formulae (1.22)–(1.23) and (1.58)–(1.59), after replacing in them the components' reliability functions by their safety functions, the subsystem S_3 conditional safety function at the operation state z_8 is given by

$$[\boldsymbol{s}^{(3)}(t,\cdot)]^{(8)} = [1, [\boldsymbol{s}^{(3)}(t,1)]^{(8)}, [\boldsymbol{s}^{(3)}(t,2)]^{(8)}, [\boldsymbol{s}^{(3)}(t,3)]^{(8)}, [\boldsymbol{s}^{(3)}(t,4)]^{(8)}], \quad t \in \langle 0,\infty), \quad (3.357)$$

where

$$[\boldsymbol{s}^{(3)}(t,u)]^{(8)} = \boldsymbol{s}_{3;1,1,1}(t,u) = \prod_{i=1}^{2}\left[1 - \prod_{j=1}^{1}[1 - [\boldsymbol{s}_{ij}^{(3)}(t,u)]^{(8)}]\right] = \prod_{i=1}^{2}[\boldsymbol{s}_{i1}^{(3)}(t,u)]^{(8)},$$
$$t \in \langle 0,\infty), \quad u = 1,2,3,4,$$

and particularly

$$[s^{(3)}(t,1)]^{(8)} = \exp[-0.2t]\exp[-0.033t] = \exp[-0.053t], \tag{3.358}$$

$$[s^{(3)}(t,2)]^{(8)} = \exp[-0.3t]\exp[-0.04t] = \exp[-0.070t], \tag{3.359}$$

$$[s^{(3)}(t,3)]^{(8)} = \exp[-0.35t]\exp[-0.045t] = \exp[-0.080t], \tag{3.360}$$

$$[s^{(3)}(t,4)]^{(8)} = \exp[-0.4t]\exp[-0.05t] = \exp[-0.090t]. \tag{3.361}$$

Subsystem S_4 consists of a component $E_{11}^{(4)}$, with the same safety function as at the operation state z_1. Thus, according to (3.241)–(3.245), subsystem S_4 conditional safety function at the operation state z_6 is given by

$$[s^{(4)}(t,\cdot)]^{(8)} = [1, [s^{(4)}(t,1)]^{(8)},\ [s^{(4)}(t,2)]^{(8)},\ [s^{(4)}(t,3)]^{(8)},\ [s^{(4)}(t,4)]^{(8)}],$$
$$t \in \langle 0,\infty), \tag{3.362}$$

where

$$[s^{(4)}(t,1)]^{(8)} = \exp[-0.05t], \tag{3.363}$$

$$[s^{(4)}(t,2)]^{(8)} = \exp[-0.06t], \tag{3.364}$$

$$[s^{(4)}(t,3)]^{(8)} = \exp[-0.065t], \tag{3.365}$$

$$[s^{(4)}(t,4)]^{(8)} = \exp[-0.07t]. \tag{3.366}$$

Considering that the ferry technical system at the operation state z_8 is a five-state series system of subsystems S_3 and S_4, after applying (1.22)–(1.23) and (3.357)–(3.361) and (3.362)–(3.366), its conditional five-state safety function is given by

$$[s(t,\cdot)]^{(8)} = [1, [s(t,1)]^{(8)},\ [s(t,2)]^{(8)},\ [s(t,3)]^{(8)},\ [s(t,4)]^{(8)}], \quad t \in \langle 0,\infty), \tag{3.367}$$

where

$$[s(t,u)]^{(8)} = \bar{s}_2(t,u) = [s^{(3)}(t,u)]^{(8)}\ [s^{(4)}(t,u)]^{(8)} \quad t \in \langle 0,\infty),\ u = 1,2,3,4,$$

and particularly

$$[s(t,1)]^{(8)} = \exp[-0.053t]\exp[-0.05t] = \exp[-0.103t], \tag{3.368}$$

$$[s(t,2)]^{(8)} = \exp[-0.070t]\exp[-0.06t] = \exp[-0.130t], \tag{3.369}$$

$$[s(t,3)]^{(8)} = \exp[-0.080t]\exp[-0.065t] = \exp[-0.145t], \tag{3.370}$$

$$[\mathbf{s}(t,4)]^{(8)} = \exp[-0.090t]\exp[-0.07t] = \exp[-0.160t]. \tag{3.371}$$

The expected values and standard deviations of the ferry technical system conditional lifetimes in the safety state subsets $\{1,2,3,4\}$, $\{2,3,4\}$, $\{3,4\}$, $\{4\}$ at the operation state z_8, calculated from the results given by (3.68)–(3.371), according to (3.8)–(3.9), respectively are:

$$\mu_8(1) \cong 9.71, \quad \mu_8(2) \cong 7.69, \quad \mu_8(3) \cong 6.88, \quad \mu_8(4) \cong 6.25 \text{ years}, \tag{3.372}$$

$$\sigma_8(1) \cong 9.71, \quad \sigma_8(2) \cong 7.69, \quad \sigma_8(3) \cong 6.88, \quad \sigma_8(4) \cong 6.25 \text{ years}, \tag{3.373}$$

and further, using (3.10) and (3.372), the mean values of the ferry technical system conditional lifetimes in the particular safety states 1, 2, 3, 4 at the operation state z_8, respectively are:

$$\bar{\mu}_8(1) \cong 2.02, \quad \bar{\mu}_8(2) \cong 0.80, \quad \bar{\mu}_8(3) \cong 0.83, \quad \bar{\mu}_8(4) \cong 6.25 \text{ years}. \tag{3.374}$$

At the system operation state z_9, the ferry technical system is composed of two subsystems S_3 and S_4 forming a series structure shown in Fig. 2.19. The system safety structure and its components safety functions are the same as at the operation state z_8. Thus, considering (3.367)–(3.371), its conditional five-state safety function is given by

$$[\mathbf{s}(t,\cdot)]^{(9)} = [1, [\mathbf{s}(t,1)]^{(9)}, [\mathbf{s}(t,2)]^{(9)}, [\mathbf{s}(t,3)]^{(9)}, [\mathbf{s}(t,4)]^{(9)}], \quad t \in \langle 0,\infty), \tag{3.375}$$

where

$$[\mathbf{s}(t,1)]^{(9)} = \exp[-0.103t], \tag{3.376}$$

$$[\mathbf{s}(t,2)]^{(9)} = \exp[-0.130t], \tag{3.377}$$

$$[\mathbf{s}(t,3)]^{(9)} = \exp[-0.145t], \tag{3.378}$$

$$[\mathbf{s}(t,4)]^{(9)} = \exp[-0.160t]. \tag{3.379}$$

The expected values and standard deviations of the ferry technical system conditional lifetimes in the safety state subsets $\{1,2,3,4\}$, $\{2,3,4\}$, $\{3,4\}$, $\{4\}$ at the operation state z_9, calculated from the results given by (3.376)–(3.379), according to (3.8)–(3.9), respectively are:

$$\mu_9(1) \cong 9.71, \quad \mu_9(2) \cong 7.69, \quad \mu_9(3) \cong 6.89, \quad \mu_9(4) \cong 6.25 \text{ years}, \tag{3.380}$$

$$\sigma_9(1) \cong 9.71, \quad \sigma_9(2) \cong 7.69, \quad \sigma_9(3) \cong 6.89, \quad \sigma_9(4) \cong 6.25 \text{ years}, \tag{3.381}$$

and further, using (3.10) and (3.380), the mean values of the ferry technical system conditional lifetimes in the particular safety states 1, 2, 3, 4 at the operation state z_9, respectively are:

$$\overline{\mu}_9(1) \cong 2.02, \quad \overline{\mu}_9(2) \cong 0.80, \quad \overline{\mu}_9(3) \cong 0.64, \quad \overline{\mu}_9(4) \cong 6.25 \text{ years.} \qquad (3.382)$$

At the system operation state z_{10}, the ferry technical system is composed of three subsystems S_1, S_2 and S_5 forming a series structure shown in Fig. 2.15. The system safety structure and its components safety functions are the same as at the operation state z_2. Thus, considering (3.269)–(3.273), its conditional safety function is given by

$$[\boldsymbol{s}(t,\cdot)]^{(10)} = [1, [\boldsymbol{s}(t,1)]^{(10)}, [\boldsymbol{s}(t,2)]^{(10)}, [\boldsymbol{s}(t,3)]^{(10)}, [\boldsymbol{s}(t,4)]^{(10)}], \quad t \in \langle 0, \infty), \qquad (3.383)$$

where

$$[\boldsymbol{s}(t,1)]^{(10)} = 12\exp[-0.462t] + 8\exp[-0.561t] - 16\exp[-0.495t] - 3\exp[-0.594t]. \qquad (3.384)$$

$$[\boldsymbol{s}(t,2)]^{(10)} = 12\exp[-0.54t] + 8\exp[-0.65t] + 6\exp[-0.62t] - 16\exp[-0.58t] - 6\exp[-0.61t] - 3\exp[-0.69t], \qquad (3.385)$$

$$[\boldsymbol{s}(t,3)]^{(10)} = 12\exp[-0.62t] + 8\exp[-0.745t] + 6\exp[-0.72t] - 16\exp[-0.67t] - 6\exp[-0.695t] - 3\exp[-0.795t], \qquad (3.386)$$

$$[\boldsymbol{s}(t,4)]^{(10)} = 12\exp[-0.685t] + 8\exp[-0.82t] + 6\exp[-0.795t] - 16\exp[-0.74t] - 6\exp[-0.765t] - 3\exp[-0.875t]. \qquad (3.387)$$

The expected values and standard deviations of the ferry technical system conditional lifetimes in the safety state subsets $\{1,2,3,4\}$, $\{2,3,4\}$, $\{3,4\}$, $\{4\}$ at the operation state z_{10}, calculated from the results given by (3.384)–(3.387), according to (3.8)–(3.9), respectively are:

$$\mu_{10}(1) \cong 2.86, \quad \mu_{10}(2) \cong 2.43, \quad \mu_{10}(3) \cong 2.14, \quad \mu_{10}(4) \cong 1.93 \text{ years,} \qquad (3.388)$$

$$\sigma_{10}(1) \cong 2.74, \quad \sigma_{10}(2) \cong 2.35, \quad \sigma_{10}(3) \cong 2.05, \quad \sigma_{10}(4) \cong 1.85 \text{ years,} \qquad (3.389)$$

and further, using (3.10) and (3.388), the mean values of the ferry technical system conditional lifetimes in the particular safety states 1, 2, 3, 4 at the operation state z_{10}, respectively are:

$$\overline{\mu}_{10}(1) \cong 0.43, \quad \overline{\mu}_{10}(2) \cong 0.29, \quad \overline{\mu}_{10}(3) \cong 0.21, \quad \overline{\mu}_{10}(4) \cong 1.93 \text{ years.} \qquad (3.390)$$

At the system operation state z_{11}, the ferry technical system is composed of two subsystems S_1 and S_2 forming a series structure shown in Fig. 2.15.

Subsystem S_1 consists of a component $E_{11}^{(1)}$, with the same safety function as at the operation state z_2. Thus, according to (3.254)–(3.258), subsystem S_1 conditional safety function at the operation state z_{11} is given by

$$[s^{(1)}(t,\cdot)]^{(11)} = [1, [s^{(1)}(t,1)]^{(11)}, [s^{(1)}(t,2)]^{(11)}, [s^{(1)}(t,3)]^{(11)}, [s^{(1)}(t,4)]^{(11)}],$$
$$t \in \langle 0,\infty), \tag{3.391}$$

where

$$[s^{(1)}(t,1)]^{(11)} = \exp[-0.033t], \tag{3.392}$$

$$[s^{(1)}(t,2)]^{(11)} = \exp[-0.04t], \tag{3.393}$$

$$[s^{(1)}(t,3)]^{(11)} = \exp[-0.045t], \tag{3.394}$$

$$[s^{(1)}(t,4)]^{(11)} = \exp[-0.05t]. \tag{3.395}$$

Subsystem S_2 is a five-state parallel–series system composed of components $E_{ij}^{(2)}$, $i = 1,2,\ldots,k$, $j = 1,2,\ldots,l_i$, $k = 7$, $l_1 = 4$, $l_2 = 2$, $l_3 = 1$, $l_4 = 1$, $l_5 = 1$, $l_6 = 1$, $l_7 = 1$, with the same safety functions as at the operation state z_2. Thus, according to (3.259)–(3.263), subsystem S_2 conditional safety function at the operation state z_{11} is given by

$$[s^{(2)}(t,\cdot)]^{(11)} = [1, [s^{(2)}(t,1)]^{(11)}, [s^{(2)}(t,2)]^{(11)}, [s^{(2)}(t,3)]^{(11)}, [s^{(2)}(t,4)]^{(11)}],$$
$$t \in \langle 0,\infty), \tag{3.396}$$

where

$$[s^{(2)}(t,1)]^{(11)} = 12\exp[-0.33t] + 8\exp[-0.429t] - 16\exp[-0.363t] - 3\exp[-0.462t], \tag{3.397}$$

$$[s^{(2)}(t,2)]^{(11)} = 12\exp[-0.38t] + 8\exp[-0.49t] + 6\exp[-0.46t] - 16\exp[-0.42t] - 6\exp[-0.45t] - 3\exp[-0.53t], \tag{3.398}$$

$$[s^{(2)}(t,3)]^{(11)} = 12\exp[-0.43t] + 8\exp[-0.555t] + 6\exp[-0.53t] - 16\exp[-0.48t] - 6\exp[-0.505t] - 3\exp[-0.605t], \tag{3.399}$$

$$[s^{(2)}(t,4)]^{(11)} = 12\exp[-0.47t] + 8\exp[-0.605t] + 6\exp[-0.58t] - 16\exp[-0.525t] - 6\exp[-0.55t] - 3\exp[-0.66t]. \tag{3.400}$$

Considering that the ferry technical system at the operation state z_{11} is a five-state series system of subsystems S_1 and S_2, after applying (1.22)–(1.23) and (3.391)–(3.395) and (3.396)–(3.400) its conditional safety function is given by

$$[\mathbf{s}(t,\cdot)]^{(11)} = [1, [\mathbf{s}(t,1)]^{(11)}, \ [\mathbf{s}(t,2)]^{(11)}, [\mathbf{s}(t,3)]^{(11)}, \ [\mathbf{s}(t,4)]^{(11)}], \quad t \in \langle 0, \infty), \tag{3.401}$$

where

$$[\mathbf{s}(t,u)]^{(11)} = \bar{\mathbf{s}}_2(t,u) = [\mathbf{s}^{(1)}(t,u)]^{(11)} \ [\mathbf{s}^{(2)}(t,u)]^{(11)}, \quad t \in \langle 0, \infty), \ u = 1,2,3,4,$$

and particularly

$$\begin{aligned}[\mathbf{s}(t,1)]^{(11)} &= \exp[-0.033t][12\exp[-0.33t] + 8\exp[-0.429t] \\ &\quad - 16\exp[-0.363t] - 3\exp[-0.462t]] \\ &= 12\exp[-0.363t] + 8\exp[-0.462t] - 16\exp[-0.396t] - 3\exp[-0.495t],\end{aligned} \tag{3.402}$$

$$\begin{aligned}[\mathbf{s}(t,2)]^{(11)} &= \exp[-0.04t][12\exp[-0.38t] + 8\exp[-0.49t] + 6\exp[-0.46t] \\ &\quad - 16\exp[-0.42t] - 6\exp[-0.45t] - 3\exp[-0.53t]] \\ &= 12\exp[-0.42t] + 8\exp[-0.53t] + 6\exp[-0.5t] - 16\exp[-0.46t] \\ &\quad - 6\exp[-0.49t] - 3\exp[-0.57t],\end{aligned} \tag{3.403}$$

$$\begin{aligned}[\mathbf{s}(t,3)]^{(11)} &= \exp[-0.045t][12\exp[-0.43t] + 8\exp[-0.555t] + 6\exp[-0.53t] \\ &\quad - 16\exp[-0.48t] - 6\exp[-0.505t] - 3\exp[-0.605t]] \\ &= 12\exp[-0.475t] + 8\exp[-0.6t] + 6\exp[-0.575t] - 16\exp[-0.525t] \\ &\quad - 6\exp[-0.55t] - 3\exp[-0.65t],\end{aligned} \tag{3.404}$$

$$\begin{aligned}[\mathbf{s}(t,4)]^{(11)} &= \exp[-0.05t][12\exp[-0.47t] + 8\exp[-0.605t] + 6\exp[-0.58t] \\ &\quad - 16\exp[-0.525t] - 6\exp[-0.55t] - 3\exp[-0.66t]] \\ &= 12\exp[-0.52t] + 8\exp[-0.655t] + 6\exp[-0.63t] - 16\exp[-0.575t] \\ &\quad - 6\exp[-0.6t] - 3\exp[-0.71t].\end{aligned} \tag{3.405}$$

The expected values and standard deviations of the ferry technical system conditional lifetimes in the safety state subsets $\{1,2,3,4\}$, $\{2,3,4\}$, $\{3,4\}$, $\{4\}$ at the operation state z_{11}, calculated from the results given by (3.402)–(3.405), according to (3.8)–(3.9), respectively are:

$$\mu_{11}(1) \cong 3.91, \quad \mu_{11}(2) \cong 3.37, \quad \mu_{11}(3) \cong 3.07, \quad \mu_{11}(4) \cong 2.76 \text{ years}, \tag{3.406}$$

$$\sigma_{11}(1) \cong 3.64, \quad \sigma_{11}(2) \cong 3.16, \quad \sigma_{11}(3) \cong 2.82, \quad \sigma_{11}(4) \cong 2.57 \text{ years}, \tag{3.407}$$

and further, using (3.10) and (3.406), the mean values of the ferry technical system conditional lifetimes in the particular safety states 1, 2, 3, 4 at the operation state z_{11}, respectively are:

$$\overline{\mu}_{11}(1) \cong 0.54, \quad \overline{\mu}_{11}(2) \cong 0.34, \quad \overline{\mu}_{11}(3) \cong 0.27, \quad \overline{\mu}_{11}(4) \cong 2.76 \text{ years.} \tag{3.408}$$

At the system operation state z_{12}, i.e. at the leaving state, the ferry technical system is composed of three subsystems S_1, S_2 and S_4 forming a series structure shown in Fig. 2.17. The system safety structure and its components safety functions are the same as at the operation state z_4. Thus, considering (3.310)–(3.314), its conditional safety function is given by

$$[\boldsymbol{s}(t,\cdot)]^{(12)} = [1, [\boldsymbol{s}(t,1)]^{(12)}, [\boldsymbol{s}(t,2)]^{(12)}, [\boldsymbol{s}(t,3)]^{(12)}, [\boldsymbol{s}(t,4)]^{(12)}], \quad t \in \langle 0, \infty), \tag{3.409}$$

where

$$[\boldsymbol{s}(t,1)]^{(12)} = 6\exp[-0.281t] - 8\exp[-0.314t] + 3\exp[-0.347t], \tag{3.410}$$

$$[\boldsymbol{s}(t,2)]^{(12)} = 6\exp[-0.335t] - 8\exp[-0.375t] + 3\exp[-0.415t], \tag{3.411}$$

$$[\boldsymbol{s}(t,3)]^{(12)} = 6\exp[-0.385t] - 8\exp[-0.435t] + 3\exp[-0.485t], \tag{3.412}$$

$$[\boldsymbol{s}(t,4)]^{(12)} = 6\exp[-0.425t] - 8\exp[-0.48t] + 3\exp[-0.535t]. \tag{3.413}$$

The expected values and standard deviations of the ferry technical system conditional lifetimes in the safety state subsets $\{1,2,3,4\}$, $\{2,3,4\}$, $\{3,4\}$, $\{4\}$ at the operation state z_{12}, calculated from the results given by (3.410)–(3.413), according to (3.8)–(3.9), respectively are:

$$\mu_{12}(1) \cong 4.2, \quad \mu_{12}(2) \cong 3.80, \quad \mu_{12}(3) \cong 3.38, \quad \mu_{12}(4) \cong 3.05 \text{ years,} \tag{3.414}$$

$$\sigma_{12}(1) \cong 4.37, \quad \sigma_{12}(2) \cong 3.68, \quad \sigma_{12}(3) \cong 3.12, \quad \sigma_{12}(4) \cong 2.93 \text{ years,} \tag{3.415}$$

and further, using (3.10) and (3.414), the mean values of the ferry technical system conditional lifetimes in the particular safety states 1, 2, 3, 4 at the operation state z_{12}, respectively are:

$$\overline{\mu}_{12}(1) \cong 0.72, \quad \overline{\mu}_{12}(2) \cong 0.42, \quad \overline{\mu}_{12}(3) \cong 0.33, \quad \overline{\mu}_{12}(4) \cong 3.05 \text{ years.} \tag{3.416}$$

At the system operation state z_{13}, the ferry technical system is composed of three subsystems S_1, S_2 and S_4 forming a series structure shown in Fig. 2.17. The system safety structure and its components safety functions are the same as at the operation state z_4. Thus, considering (3.310)–(3.314), its conditional safety function is given by

$$[\boldsymbol{s}(t,\cdot)]^{(13)} = \left[1, [\boldsymbol{s}(t,1)]^{(13)}, [\boldsymbol{s}(t,2)]^{(13)}, [\boldsymbol{s}(t,3)]^{(13)}, [\boldsymbol{s}(t,4)]^{(13)}\right], \quad t \in \langle 0, \infty), \tag{3.417}$$

where

$$[s(t,1)]^{(13)} = 6\exp[-0.281t] - 8\exp[-0.314t] + 3\exp[-0.347t], \quad (3.418)$$

$$[s(t,2)]^{(13)} = 6\exp[-0.335t] - 8\exp[-0.375t] + 3\exp[-0.415t], \quad (3.419)$$

$$[s(t,3)]^{(13)} = 6\exp[-0.385t] - 8\exp[-0.435t] + 3\exp[-0.485t], \quad (3.420)$$

$$[s(t,4)]^{(13)} = 6\exp[-0.425t] - 8\exp[-0.48t] + 3\exp[-0.535t]. \quad (3.421)$$

The expected values and standard deviations of the ferry technical system conditional lifetimes in the safety state subsets $\{1,2,3,4\}$, $\{2,3,4\}$, $\{3,4\}$, $\{4\}$ at the operation state z_{13}, calculated from the results given by (3.418)–(3.421), according to (3.8)–(3.9), respectively are:

$$\mu_{13}(1) \cong 4.2, \quad \mu_{13}(2) \cong 3.80, \quad \mu_{13}(3) \cong 3.38, \quad \mu_{13}(4) \cong 3.05 \text{ years}, \quad (3.422)$$

$$\sigma_{13}(1) \cong 4.37, \quad \sigma_{13}(2) \cong 3.68, \quad \sigma_{13}(3) \cong 3.12, \quad \sigma_{13}(4) \cong 2.93 \text{ years}, \quad (3.423)$$

and further, using (3.10) and (3.422), the mean values of the ferry technical system conditional lifetimes in the particular safety states 1, 2, 3, 4 at the operation state z_{13}, respectively are:

$$\bar{\mu}_{13}(1) \cong 0.72, \quad \bar{\mu}_{13}(2) \cong 0.42, \quad \bar{\mu}_{13}(3) \cong 0.33, \quad \bar{\mu}_{13}(4) \cong 3.05 \text{ years}. \quad (3.424)$$

At the system operation state z_{14}, the ferry technical system is composed of three subsystems S_1, S_2 and S_4 forming a series structure shown in Fig. 2.17. The system safety structure and its components safety functions are the same as at the operation state z_4. Thus, considering (3.310)–(3.314), its conditional safety function is given by

$$[s(t,\cdot)]^{(14)} = \left[1, [s(t,1)]^{(14)}, [s(t,2)]^{(14)}, [s(t,3)]^{(14)}, [s(t,4)]^{(14)}\right], \quad t \in \langle 0, \infty), \quad (3.425)$$

where

$$[s(t,1)]^{(14)} = 6\exp[-0.281t] - 8\exp[-0.314t] + 3\exp[-0.347t], \quad (3.426)$$

$$[s(t,2)]^{(14)} = 6\exp[-0.335t] - 8\exp[-0.375t] + 3\exp[-0.415t], \quad (3.427)$$

$$[s(t,3)]^{(14)} = 6\exp[-0.385t] - 8\exp[-0.435t] + 3\exp[-0.485t], \quad (3.428)$$

$$[s(t,4)]^{(14)} = 6\exp[-0.425t] - 8\exp[-0.48t] + 3\exp[-0.535t]. \quad (3.429)$$

The expected values and standard deviations of the ferry technical system conditional lifetimes in the safety state subsets $\{1,2,3,4\}$, $\{2,3,4\}$, $\{3,4\}$, $\{4\}$ at the operation state z_{14}, calculated from the results given by (3.426)–(3.429), according to (3.8)–(3.9), respectively are:

$$\mu_{14}(1) \cong 4.2, \quad \mu_{14}(2) \cong 3.80, \quad \mu_{14}(3) \cong 3.38, \quad \mu_{14}(4) \cong 3.05 \text{ years}, \quad (3.430)$$

$$\sigma_{14}(1) \cong 4.37, \quad \sigma_{14}(2) \cong 3.68, \quad \sigma_{14}(3) \cong 3.12, \quad \sigma_{14}(4) \cong 2.93 \text{ years}, \quad (3.431)$$

and further, using (3.10) and (3.430), the mean values of the ferry technical system conditional lifetimes in the particular safety states 1, 2, 3, 4 at the operation state z_{14}, respectively are:

$$\bar{\mu}_{14}(1) \cong 0.72, \quad \bar{\mu}_{14}(2) \cong 0.42, \quad \bar{\mu}_{14}(3) \cong 0.33, \quad \bar{\mu}_{14}(4) \cong 3.05 \text{ years}. \quad (3.432)$$

At the system operation states z_{15}, the ferry technical system is composed of two subsystems S_1 and S_2 forming a series structure shown in Fig. 2.16. The system safety structure and its components safety functions are the same as at the operation state z_3. Thus, considering (3.287)–(3.291), its conditional safety function is given by

$$[\boldsymbol{s}(t,\cdot)]^{(15)} = \left[1, [\boldsymbol{s}(t,1)]^{(15)}, [\boldsymbol{s}(t,2)]^{(15)}, [\boldsymbol{s}(t,3)]^{(15)}, [\boldsymbol{s}(t,4)]^{(15)}\right], \quad t \in \langle 0, \infty), \quad (3.433)$$

where

$$[\boldsymbol{s}(t,1)]^{(15)} = 12\exp[-0.291t] + 8\exp[-0.363t] + 6\exp[-0.357t] - 16\exp[-0.324t] - 6\exp[-0.33t] - 3\exp[-0.396t], \quad (3.434)$$

$$[\boldsymbol{s}(t,2)]^{(15)} = 12\exp[-0.36t] + 14\exp[-0.44t] - 22\exp[-0.4t] - 3\exp[-0.48t], \quad (3.435)$$

$$[\boldsymbol{s}(t,3)]^{(15)} = 12\exp[-0.415t] + 8\exp[-0.51t] + 6\exp[-0.515t] - 16\exp[-0.465t] - 6\exp[-0.46t] - 3\exp[-0.56t], \quad (3.436)$$

$$[\boldsymbol{s}(t,4)]^{(15)} = 12\exp[-0.46t] + 8\exp[-0.565t] + 6\exp[-0.57t] - 16\exp[-0.515t] - 6\exp[-0.51t] - 3\exp[-0.62t]. \quad (3.437)$$

The expected values and standard deviations of the ferry technical system conditional lifetimes in the safety state subsets $\{1,2,3,4\}$, $\{2,3,4\}$, $\{3,4\}$, $\{4\}$ at the operation state z_{15}, calculated from the results given by (3.434)–(3.437), according to (3.8)–(3.9), respectively are:

$$\mu_{15}(1) \cong 4.94, \quad \mu_{15}(2) \cong 3.9, \quad \mu_{15}(3) \cong 3.44, \quad \mu_{15}(4) \cong 3.1 \text{ years}, \quad (3.438)$$

$$\sigma_{15}(1) \cong 4.6, \quad \sigma_{15}(2) \cong 3.6, \quad \sigma_{15}(3) \cong 3.2, \quad \sigma_{15}(4) \cong 2.9 \text{ years}, \quad (3.439)$$

and further, using (3.10) and (3.438), the mean values of the ferry technical system conditional lifetimes in the particular safety states 1, 2, 3, 4 at the operation state z_{15}, respectively are:

$$\bar{\mu}_{15}(1) \cong 1.04, \quad \bar{\mu}_{15}(2) \cong 0.46, \quad \bar{\mu}_{15}(3) \cong 0.34, \quad \bar{\mu}_{15}(4) \cong 3.1 \text{ years}. \quad (3.440)$$

At the system operation state z_{16}, the ferry technical system is composed of two subsystems S_1 and S_2 forming a series structure shown in Fig. 2.16. The system safety structure and its components safety functions are the same as at the operation state z_{11}. Thus, considering (3.401)–(3.405), its conditional safety function is given by

$$[\boldsymbol{s}(t,\cdot)]^{(16)} = \left[1, [\boldsymbol{s}(t,1)]^{(16)}, [\boldsymbol{s}(t,2)]^{(16)}, [\boldsymbol{s}(t,3)]^{(16)}, [\boldsymbol{s}(t,4)]^{(16)}\right], \quad t \in \langle 0, \infty), \quad (3.441)$$

where

$$[\boldsymbol{s}(t,1)]^{16} = 12\exp[-0.363t] + 8\exp[-0.462t] - 16\exp[-0.396t] - 3\exp[-0.495t], \quad (3.442)$$

$$[\boldsymbol{s}(t,2)]^{16} = 12\exp[-0.42t] + 8\exp[-0.53t] + 6\exp[-0.5t] - 16\exp[-0.46t] - 6\exp[-0.49t] - 3\exp[-0.57t], \quad (3.443)$$

$$[\boldsymbol{s}(t,3)]^{16} = 12\exp[-0.475t] + 8\exp[-0.6t] + 6\exp[-0.575t] - 16\exp[-0.525t] - 6\exp[-0.55t] - 3\exp[-0.65t], \quad (3.444)$$

$$[\boldsymbol{s}(t,4)]^{16} = 12\exp[-0.52t] + 8\exp[-0.655t] + 6\exp[-0.63t] - 16\exp[-0.575t] - 6\exp[-0.6t] - 3\exp[-0.71t]. \quad (3.445)$$

The expected values and standard deviations of the ferry technical system conditional lifetimes in the safety state subsets $\{1,2,3,4\}$, $\{2,3,4\}$, $\{3,4\}$, $\{4\}$ at the operation state z_{16}, calculated from the results given by (3.442)–(3.445), according to (3.8)–(3.9), respectively are:

$$\mu_{16}(1) \cong 3.91, \quad \mu_{16}(2) \cong 3.37, \quad \mu_{16}(3) \cong 3.07, \quad \mu_{16}(4) \cong 2.76 \text{ years}, \quad (3.446)$$

$$\sigma_{16}(1) \cong 3.64, \quad \sigma_{16}(2) \cong 3.16, \quad \sigma_{16}(3) \cong 2.82, \quad \sigma_{16}(4) \cong 2.57 \text{ years}, \quad (3.447)$$

and further, using (3.10) and (3.446), the mean values of the ferry technical system conditional lifetimes in the particular safety states 1, 2, 3, 4 at the operation state z_{16}, respectively are:

$$\bar{\mu}_{16}(1) \cong 0.54, \quad \bar{\mu}_{16}(2) \cong 0.34, \quad \bar{\mu}_{16}(3) \cong 0.27, \quad \bar{\mu}_{16}(4) \cong 2.76 \text{ years.} \tag{3.448}$$

At the system operation state z_{17}, the ferry technical system is composed of three subsystems S_1, S_2 and S_5 forming a series structure shown in Fig. 2.15. The system safety structure and its components safety functions are the same as at the operation state z_2. Thus, considering (3.269)–(3.273), its conditional safety function is given by

$$[\boldsymbol{s}(t,\cdot)]^{(17)} = \left[1, [\boldsymbol{s}(t,1)]^{(17)}, [\boldsymbol{s}(t,2)]^{(17)}, [\boldsymbol{s}(t,3)]^{(17)}, [\boldsymbol{s}(t,4)]^{(17)}\right], \quad t \geq 0, \tag{3.449}$$

where

$$[\boldsymbol{s}(t,1)]^{(17)} = 12\exp[-0.462t] + 8\exp[-0.561t] - 16\exp[-0.495t] - 3\exp[-0.594t]. \tag{3.450}$$

$$[\boldsymbol{s}(t,2)]^{(17)} = 12\exp[-0.54t] + 8\exp[-0.65t] + 6\exp[-0.62t] - 16\exp[-0.58t] - 6\exp[-0.61t] - 3\exp[-0.69t], \tag{3.451}$$

$$[\boldsymbol{s}(t,3)]^{(17)} = 12\exp[-0.62t] + 8\exp[-0.745t] + 6\exp[-0.72t] - 16\exp[-0.67t] - 6\exp[-0.695t] - 3\exp[-0.795t], \tag{3.452}$$

$$[\boldsymbol{s}(t,4)]^{(17)} = 12\exp[-0.685t] + 8\exp[-0.82t] + 6\exp[-0.795t] - 16\exp[-0.74t] - 6\exp[-0.765t] - 3\exp[-0.875t]. \tag{3.453}$$

The expected values and standard deviations of the ferry technical system conditional lifetimes in the safety state subsets $\{1,2,3,4\}$, $\{2,3,4\}$, $\{3,4\}$, $\{4\}$ at the operation state z_{17}, calculated from the results given by (3.450)–(3.453), according to (3.8)–(3.9), respectively are:

$$\mu_{17}(1) \cong 2.86, \quad \mu_{17}(2) \cong 2.43, \quad \mu_{17}(3) \cong 2.14, \quad \mu_{17}(4) \cong 1.93 \text{ years,} \tag{3.454}$$

$$\sigma_{17}(1) \cong 2.74, \quad \sigma_{17}(2) \cong 2.35, \quad \sigma_{17}(3) \cong 2.05, \quad \sigma_{17}(4) \cong 1.85 \text{ years,} \tag{3.455}$$

and further, using (3.10) and (3.454), the mean values of the ferry technical system conditional lifetimes in the particular safety states 1, 2, 3, 4 at the operation state z_{17}, respectively are:

$$\bar{\mu}_{17}(1) \cong 0.43, \quad \bar{\mu}_{17}(2) \cong 0.29, \quad \bar{\mu}_{17}(3) \cong 0.21, \quad \bar{\mu}_{17}(4) \cong 1.93 \text{ years.} \tag{3.456}$$

At the system operation state z_{18}, the ferry technical system is composed of two subsystems S_3 and S_4 forming a series structure shown in Fig. 2.14. The system safety structure and its components safety functions are the same as at the operation state z_1. Thus, considering (3.246)–(3.250), its conditional safety function is given by

$$[\boldsymbol{s}(t,\cdot)]^{(18)} = \Big[1, [\boldsymbol{s}(t,1)]^{(18)}, [\boldsymbol{s}(t,2)]^{(18)}, [\boldsymbol{s}(t,3)]^{(18)}, [\boldsymbol{s}(t,4)]^{(18)}\Big], \quad t \in \langle 0, \infty), \tag{3.457}$$

where

$$[\boldsymbol{s}(t,1)]^{(18)} = \exp[-0.123t] \tag{3.458}$$

$$[\boldsymbol{s}(t,2)]^{(18)} = \exp[-0.155t] \tag{3.459}$$

$$[\boldsymbol{s}(t,3)]^{(18)} = \exp[-0.175t], \tag{3.460}$$

$$[\boldsymbol{s}(t,4)]^{(18)} = \exp[-0.20t]. \tag{3.461}$$

The expected values and standard deviations of the ferry technical system conditional lifetimes in the safety state subsets $\{1,2,3,4\}$, $\{2,3,4\}$, $\{3,4\}$, $\{4\}$ at the operation state z_{18}, calculated from the results given by (3.458)–(3.461), according to (3.8)–(3.9), respectively are:

$$\mu_{18}(1) \cong 8.13, \quad \mu_{18}(2) \cong 6.45, \quad \mu_{18}(3) \cong 5.71, \quad \mu_{18}(4) \cong 5.00 \text{ years}, \tag{3.462}$$

$$\sigma_{18}(1) \cong 8.13, \quad \sigma_{18}(2) \cong 6.45, \quad \sigma_{18}(3) \cong 5.71, \quad \sigma_{18}(4) \cong 5.00 \text{ years}, \tag{3.463}$$

and further, using (3.10) and (3.462) the mean values of the ferry technical system conditional lifetimes in the particular safety states 1, 2, 3, 4 at the operation state z_{18}, respectively are:

$$\bar{\mu}_{18}(1) \cong 1.68, \quad \bar{\mu}_{18}(2) \cong 0.74, \quad \bar{\mu}_{18}(3) \cong 0.71, \quad \bar{\mu}_{18}(4) \cong 5.00 \text{ years}. \tag{3.464}$$

In the case when the ferry technical system operation time is large enough, its unconditional safety function is given by the vector

$$\boldsymbol{s}(t,\cdot) = [1, \boldsymbol{s}(t,1), \boldsymbol{s}(t,2), \boldsymbol{s}(t,3), \boldsymbol{s}(t,4)], \quad t \geq 0, \tag{3.465}$$

where according to (3.101) and considering the ferry technical system operation process transient probabilities at the operation states determined by (2.48), the vector coordinates are given respectively by

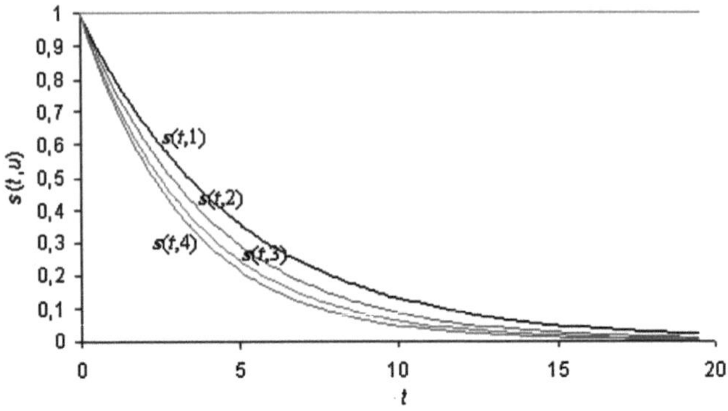

Fig. 3.5 The graph of the ferry technical system safety function $s(t,\cdot)$ coordinates

$$s(t,1) = 0.038 \times [s(t,1)]^{(1)} + 0.002 \times [s(t,1)]^{(2)} + 0.026 \times [s(t,1)]^{(3)} + 0.036 \times [s(t,1)]^{(4)} + 0.363 \times [s(t,1)]^{(5)} + 0.026 \times [s(t,1)]^{(6)} + 0.005 \times [s(t,1)]^{(7)} + 0.016 \times [s(t,1)]^{(8)} + 0.037 \times [s(t,1)]^{(9)} + 0.002 \times [s(t,1)]^{(10)} + 0.003 \times [s(t,1)]^{(11)} + 0.016 \times [s(t,1)]^{(12)} + 0.351 \times [s(t,1)]^{(13)} + 0.034 \times [s(t,1)]^{(14)} + 0.024 \times [s(t,1)]^{(15)} + 0.003 \times [s(t,1)]^{(16)} + 0.005 \times [\bar{s}(t,1)]^{(17)} + 0.013 \times [s(t,1)]^{(18)}, \quad (3.466)$$

$$s(t,2) = 0.038 \times [s(t,2)]^{(1)} + 0.002 \times [s(t,2)]^{(2)} + 0.026 \times [s(t,2)]^{(3)} + 0.036 \times [s(t,2)]^{(4)} + 0.363 \times [s(t,2)]^{(5)} + 0.026 \times [s(t,2)]^{(6)} + 0.005 \times [s(t,2)]^{(7)} + 0.016 \times [s(t,2)]^{(8)} + 0.037 \times [s(t,2)]^{(9)} + 0.002 \times [s(t,2)]^{(10)} + 0.003 \times [s(t,2)]^{(11)} + 0.016 \times [s(t,2)]^{(12)} + 0.351 \times [s(t,2)]^{(13)} + 0.034 \times [s(t,2)]^{(14)} + 0.024 \times [s(t,2)]^{(15)} + 0.003 \times [s(t,2)]^{(16)} + 0.005 \times \left[\bar{s}(t,2)\right]^{(17)} + 0.013 \times [s(t,2)]^{(18)}, \quad (3.467)$$

$$s(t,3) = 0.038 \times [s(t,3)]^{(1)} + 0.002 \times [s(t,3)]^{(2)} + 0.026 \times [s(t,3)]^{(3)} + 0.036 \times [s(t,3)]^{(4)} + 0.363 \times [s(t,3)]^{(5)} + 0.026 \times [s(t,3)]^{(6)} + 0.005 \times [s(t,3)]^{(7)} + 0.016 \times [s(t,3)]^{(8)} + 0.037 \times [s(t,3)]^{(9)} + 0.002 \times [s(t,3)]^{(10)} + 0.003 \times [s(t,3)]^{(11)} + 0.016 \times [s(t,3)]^{(12)} + 0.351 \times [s(t,3)]^{(13)} + 0.034 \times [s(t,3)]^{(14)} + 0.024 \times [s(t,3)]^{(15)} + 0.003 \times [s(t,3)]^{(16)} + 0.005 \times [\bar{s}(t,3)]^{(17)} + 0.013 \times [s(t,3)]^{(18)}, \quad (3.468)$$

$$
\begin{aligned}
s(t,4) = {} & 0.038 \times [s(t,4)]^{(1)} + 0.002 \times [s(t,4)]^{(2)} + 0.026 \times [s(t,4)]^{(3)} \\
& + 0.036 \times [s(t,4)]^{(4)} + 0.363 \times [s(t,4)]^{(5)} + 0.026 \times [s(t,4)]^{(6)} \\
& + 0.005 \times [s(t,4)]^{(7)} + 0.016 \times [s(t,4)]^{(8)} + 0.037 \times [s(t,4)]^{(9)} \\
& + 0.002 \times [s(t,4)]^{(10)} + 0.003 \times [s(t,4)]^{(11)} + 0.016 \times [s(t,4)]^{(12)} \\
& + 0.351 \times [s(t,4)]^{(13)} + 0.034 \times [s(t,4)]^{(14)} + 0.024 \times [s(t,4)]^{(15)} \\
& + 0.003 \times [s(t,4)]^{(16)} + 0.005 \times [\bar{s}(t,4)]^{(17)} + 0.013 \times [s(t,4)]^{(18)},
\end{aligned} \tag{3.469}
$$

where $[s(t,u)]^{(b)}$, $u = 1,2,3,4$, $b = 1,2,\ldots,18$, are given by (3.247)–(3.250), (3.270)–(3.273), (3.288)–(3.291), (3.311)–(3.314), (3.318)–(3.322), (3.342)–(3.345), (3.350)–(3.353), (3.368)–(3.371), (3.376)–(3.379), (3.384)-(3.387), (3.402)–(3.405), (3.410)–(3.413), (3.418)–(3.421), (3.426)–(3.429), (3.434)–(3.437), (3.442)–(3.445), (3.450)–(3.453), (3.458)–(3.461).

The graph of the five-state ferry technical system safety function is shown in Fig. 3.5.

The expected values and the standard deviations of the ferry unconditional lifetimes in the safety state subsets, $\{1,2,3,4\}$, $\{2,3,4\}$, $\{3,4\}$, $\{4\}$, calculated from the results given by (3.466)–(3.469), according to (3.7)–(3.9), and considering (2.48), (3.251), (3.274), (3.292), (3.315), (3.323), (3.346), (3.354), (3.372), (3.380), (3.388), (3.406), (3.414), (3.422), (3.430), (3.438), (3.446), (3.454), (3.462), respectively are:

$$
\begin{aligned}
\mu(1) \cong {} & 0.038 \times 8.13 + 0.002 \times 2.86 + 0.026 \times 4.94 + 0.036 \times 4.2 + 0.363 \times 4.2 \\
& + 0.026 \times 4.01 + 0.005 \times 2.86 + 0.016 \times 9.71 + 0.037 \times 9.71 + 0.002 \times 2.86 \\
& + 0.003 \times 3.91 + 0.016 \times 4.2 + 0.351 \times 4.2 + 0.034 \times 4.2 + 0.024 \times 4.94 \\
& + 0.003 \times 3.91 + 0.005 \times 2.86 + 0.013 \times 8.1 \cong 4.70 \text{ years},
\end{aligned} \tag{3.470}
$$

$$
\sigma(1) \cong 4.3 \text{ years}, \tag{3.471}
$$

$$
\begin{aligned}
\mu(2) \cong {} & 0.038 \times 6.45 + 0.002 \times 2.43 + 0.026 \times 3.90 + 0.036 \times 3.80 + 0.363 \times 3.80 \\
& + 0.026 \times 3.24 + 0.005 \times 2.43 + 0.016 \times 7.69 + 0.037 \times 7.69 + 0.002 \times 2.43 \\
& + 0.003 \times 3.37 + 0.016 \times 3.80 + 0.351 \times 3.80 + 0.034 \times 3.80 + 0.024 \times 3.90 \\
& + 0.003 \times 3.37 + 0.005 \times 2.43 + 0.013 \times 6.45 \cong 4.11 \text{ years},
\end{aligned} \tag{3.472}
$$

$$
\sigma(2) \cong 3.64 \text{ years}, \tag{3.473}
$$

$$
\begin{aligned}
\mu(3) \cong {} & 0.037 \times 5.71 + 0.002 \times 2.14 + 0.026 \times 3.44 + 0.036 \times 3.38 + 0.363 \times 3.38 \\
& + 0.026 \times 2.88 + 0.005 \times 2.14 + 0.016 \times 6.89 + 0.037 \times 6.89 + 0.002 \times 2.14 \\
& + 0.003 \times 3.07 + 0.016 \times 3.38 + 0.351 \times 3.38 + 0.034 \times 3.38 + 0.024 \times 3.44 \\
& + 0.003 \times 3.07 + 0.005 \times 2.14 + 0.013 \times 5.71 \cong 3.66 \text{ years},
\end{aligned} \tag{3.474}
$$

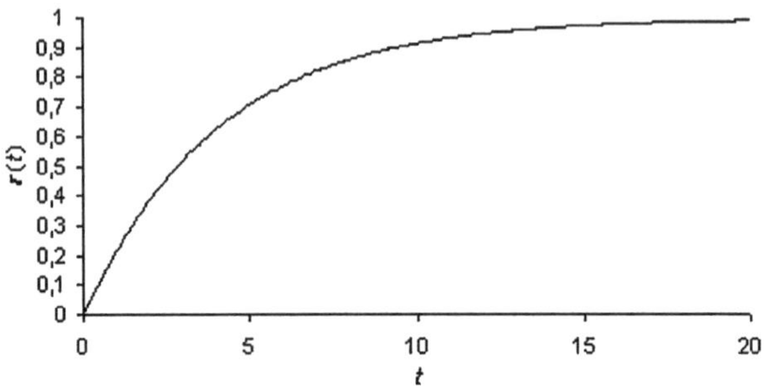

Fig. 3.6 The graph of the ferry technical system risk function $\boldsymbol{r}(t)$

$$\sigma(3) \cong 3.34 \text{ years}, \tag{3.475}$$

$$\begin{aligned}\mu(4) \cong\ & 0.038 \times 5.00 + 0.002 \times 1.93 + 0.026 \times 3.1 + 0.036 \times 3.05 + 0.363 \times 3.05 \\ & + 0.026 \times 2.61 + 0.005 \times 1.93 + 0.016 \times 6.06 + 0.037 \times 6.25 + 0.002 \times 1.93 \\ & + 0.003 \times 2.76 + 0.016 \times 3.05 + 0.351 \times 3.05 + 0.034 \times 3.05 + 0.024 \times 3.10 \\ & + 0.003 \times 2.76 + 0.005 \times 1.93 + 0.013 \times 5.00 \cong 3.29 \text{ years},\end{aligned} \tag{3.476}$$

$$\sigma(2) \cong 2.89 \text{ years}, \tag{3.477}$$

And further, considering (3.10) and (3.470), (3.472), (3.474) and (3.476), the unconditional lifetimes in the particular safety states 1, 2, 3, 4 respectively are:

$$\bar{\mu}(1) = \mu(1) - \mu(2) = 0.59, \quad \bar{\mu}(2) = \mu(2) - \mu(3) = 0.77 \text{ year}, \tag{3.478}$$

$$\bar{\mu}(3) = \mu(3) - \mu(4) = 0.45, \quad \bar{\mu}(4) = \mu(4) = 2.29 \text{ years}. \tag{3.479}$$

Since the critical safety state is $r = 2$, the system risk function, according to (3.11), is given by

$$\boldsymbol{r}(t) = 1 - \boldsymbol{s}(t, 2), \tag{3.480}$$

where $\boldsymbol{s}(t, 2)$ is given by (3.467).

The graph of the ferry technical system risk function $\boldsymbol{r}(t)$ is presented in Fig. 3.6.

From (3.480), according to (3.12), the moment when the system risk function exceeds a permitted level, for instance $\delta = 0.05$, is

$$\tau = \boldsymbol{r}^{-1}(\delta) \cong 0.21 \text{ year}. \tag{3.481}$$

The safety characteristics of the ferry technical system operating at the variable conditions predicted in this section are different from those determined in Sect. 1.4.3 at the system constant operation conditions. This fact suggests the necessity of considering the system at the variable operation conditions, which upon the improved accuracy of the system operation process identification, makes its system safety prediction more precise.

Using the results of the maritime ferry technical system safety prediction given by (3.472)–(3.473) and the results of the classical renewal theory presented in Sect. 3.4, we may predict the renewal and availability characteristics of this technical system in the case when it is repairable and its time of renovation is either ignored or non-ignored.

First, assuming that the maritime ferry technical system is repaired after the exceeding of its safety critical state $r = 2$ and that the time of the system renovation is ignored and applying *Proposition 3.4*, we obtain the following results:

(a) the time $S_N(2)$ until the *N*th exceeding by the system the safety critical state $r = 2$, for sufficiently large *N*, has approximately normal distribution $N(4.11N, 3.64\sqrt{N})$, i.e.,

$$F^{(N)}(t,2) = P(S_N(2) < t) \cong F_{N(0,1)}\left(\frac{t - 4.11N}{3.64\sqrt{N}}\right), \quad t \in (-\infty, \infty);$$

(b) the expected value and the variance of the time $S_N(2)$ until the *N*th exceeding by the system the safety critical state $r = 2$ are respectively given by

$$E[S_N(2)] \cong 4.11N, \quad D[S_N(2)] \cong 13.250N;$$

(c) the number $N(t,2)$ of exceeding by the system the safety critical state $r = 2$ up to the moment t, $t \geq 0$, for sufficiently large t, approximately has the distribution of the form

$$P(N(t,2) = N) \cong F_{N(0,1)}\left(\frac{4.11(N+1) - t}{1.795\sqrt{t}}\right) - F_{N(0,1)}\left(\frac{4.11N - t}{1.795\sqrt{t}}\right),$$
$$N = 0, 1, \ldots;$$

(d) the expected value and the variance of the number $N(t,2)$ of exceeding by the system the safety critical state $r = 2$ up to the moment t, $t \geq 0$, for sufficiently large t, approximately are respectively given by

$$H(t,2) = 0.243t, \quad D(t,2) = 0.191t.$$

Assuming that the maritime ferry technical system is repaired after the exceeding its safety critical state $r = 2$ and that the time of the system renovation is non-ignored and it has the mean value $\mu_0(2) = 0.019$ year and the standard deviation $\sigma_0(2) = 0.0095$ year and applying *Proposition 3.5*, we obtain the following results:

(a) the time $\bar{S}_N(2)$ until the *N*th exceeding by the system the safety critical state $r = 2$, for sufficiently large *N*, has approximately normal distribution $N(4.11N + 0.019(N-1), \sqrt{13.250N - 0.0000903(N-1)})$, i.e.,

$$\overline{F}^{(N)}(t,2) = P(\overline{S}_N(2) < t) \cong F_{N(0,1)}\left(\frac{t - 4.129N + 0.019}{\sqrt{13.25009N - 0.0000903}}\right),$$
$$t \in (-\infty, \infty);$$

(b) the expected value and the variance of the time $\overline{S}_N(2)$ until the Nth exceeding by the system the safety critical state $r = 2$, for sufficiently large N, are respectively given by

$$E[\overline{S}_N(2)] \cong 4.11N + 0.019(N - 1), \quad D[\overline{S}_N(2)] \cong 13.25N + 0.0000903(N - 1);$$

(c) the number $\overline{N}(t,2)$ of exceeding by the system the safety critical state $r = 2$ up to the moment t, $t \geq 0$, for sufficiently large t, has approximately distribution of the form

$$P(\overline{N}(t,2) = N) \cong F_{N(0,1)}\left(\frac{4.129(N+1) - t - 0.019}{1.791\sqrt{t + 0.019}}\right) - F_{N(0,1)}\left(\frac{4.129N - t - 0.019}{1.791\sqrt{t + 0.019}}\right), \quad N = 0, 1, \ldots;$$

(d) the expected value and the variance of the number $\overline{N}(t,2)$ of exceeding by the system the safety critical state $r = 2$ up to the moment t, $t \geq 0$, for sufficiently large t, are respectively given by

$$\overline{H}(t,2) \cong 0.242(t + 0.019), \quad \overline{D}(t,2) \cong 0.188(t + 0.019);$$

(e) the time $\overline{\overline{S}}_N(2)$ until the Nth system's renovation, for sufficiently large N, has approximately normal distribution $N(4.129N, 3.640\sqrt{N})$, i.e.,

$$\overline{\overline{F}}^{(N)}(t,2) = P(\overline{\overline{S}}_N(2) < t) \cong F_{N(0,1)}\left(\frac{t - 4.129N}{3.640\sqrt{N}}\right), \quad t \in (-\infty, \infty);$$

(f) the expected value and the variance of the time $\overline{\overline{S}}_N(2)$ until the Nth system's renovation, for sufficiently large N, are respectively given by

$$E[\overline{\overline{S}}_N(2)] \cong 4.129N, \quad D[\overline{\overline{S}}_N(2)] \cong 13.25009N;$$

(g) the number $\overline{\overline{N}}(t,2)$ of the system's renovations up to the moment t, $t \geq 0$, for sufficiently large t, has approximately distribution of the form

$$P(\overline{\overline{N}}(t,2) = N) \cong F_{N(0,1)}\left(\frac{4.129(N+1) - t}{1.791\sqrt{t}}\right) - F_{N(0,1)}\left(\frac{4.129N - t}{1.791\sqrt{t}}\right),$$
$$N = 0, 1, \ldots;$$

(h) the expected value and the variance of the number $\overline{\overline{N}}(t,2)$ of system's renovations up to the moment t, $t \geq 0$, for sufficiently large t, are respectively given by

$$\overline{\overline{H}}(t,2) \cong 0.242t, \quad \overline{\overline{D}}(t,2) \cong 0.188t;$$

(i) the steady availability coefficient of the system at the moment t, $t \geq 0$, for sufficiently large t, is given by

$$A(t,2) \cong 0.995, \quad t \geq 0;$$

(j) the steady availability coefficient of the system in the time interval $\langle t, t+\tau)$, $\tau > 0$, for sufficiently large t, is given by

$$A(t,\tau,2) \cong 0.242 \int_{\tau}^{\infty} s(t,2)\,\mathrm{d}t, \quad t \geq 0, \ \tau > 0,$$

where $s(t,2)$ is given by (7.378).

The ferry technical system operating at the varying in time operation conditions safety and risk evaluation and prediction is approximate because of non-sufficiently exact input data. Although the ferry operation process parameters are sufficiently exact because of the collected very good and sufficient statistical data necessary for their estimation, the ferry technical system safety structures are fixed on a very high level of generality and without high detailed accuracy. The input safety parameters of the ferry technical system components also are not sufficiently accurate because of the lack exact statistical data necessary for their estimation. Therefore, the achieved results may only be considered as an illustration of the proposed methods, possibilities of applications in the ferry technical system safety and risk analysis and prediction. However, the obtained evaluation may be a very useful example in technical systems of ships operating at restricted and open sea waters, safety and risk characteristics prediction, especially during the design and when planning and improving their operation processes, safety and effectiveness. To improve the quality of the achieved results it is supposed that the statistical data should be collected for the additional period of time and after that the full identification of at least the ferry technical system operation process would be performed and this process, unknown parameters and main characteristics could be determined and used in its safety and risk for more precise analysis and prediction. The final characteristics of the ferry technical system safety and risk are used in the appliqué part of Chap. 6 for its operation and safety optimization.

3.7 Summary

In this chapter the final results coming from the general analytical models of complex technical multistate systems, reliability, availability and safety are based on the results of the previous chapters of the book. The material given in this chapter delivers the procedures and algorithms that allow to find the main and practically important reliability and safety and renewal and availability characteristics of the complex technical systems at the variable operation condition. The reliability characteristics of the exemplary system and the port oil transportation system and the safety characteristics of the ferry technical system operating at the variable conditions predicted in this chapter are different from those determined in Chap. 1 for these systems operating at constant conditions. This fact justifies the sensibility of considering complex technical systems at the variable operation conditions, that is appearing out in a natural way from practice. This approach, upon the good accuracy of systems' operation processes and their components reliability and safety parameters identification, makes their reliability, safety, renewal and availability characteristics prediction more precise.

The results of this chapter, together with the results of their applications, and the linear programming are used in Chap. 6 for complex technical systems, reliability, availability and safety improvement, optimization and operation cost analysis.

References

1. Amari SV, Misra RB (1997) Comment on: Dynamic reliability analysis of coherent multistate systems. IEEE Trans Reliab 46:460–461
2. Aven T (1985) Reliability evaluation of multistate systems with multistate components. IEEE Trans Reliab 34:473–479
3. Aven T (1993) On performance measures for multistate monotone systems. Reliab Eng Syst Saf 41:259–266
4. Aven T, Jensen U (1999) Stochastic models in reliability. Springer, New York
5. Barlow RE, Wu AS (1978) Coherent systems with multi-state components. Math Oper Res 4:275–281
6. Brunelle RD, Kapur KC (1999) Review and classification of reliability measures for multistate and continuum models. IEEE Trans 31:1117–1180
7. Ferreira F, Pacheco A (2007) Comparison of level-crossing times for Markov and semi-Markov processes. Stat Probab Lett 77(2):151–157
8. Glynn PW, Haas PJ (2006) Laws of large numbers and functional central limit theorems for generalized semi-Markov processes. Stoch Model 22(2):201–231
9. Grabski F (2002) Semi-Markov models of systems reliability and operations analysis. System Research Institute, Polish Academy of Science (in Polish)
10. Guze S, Kołowrocki K (2008) Reliability analysis of multi-state ageing consecutive "k out of n: F" systems. Int J Mater Struct Reliab 6(1):47–60
11. Guze S, Kołowrocki K, Soszyńska J (2008) Modeling environment and infrastructure influence on reliability and operation processes of port transportation systems. Summer safety & reliability seminars. J Pol Saf Reliab Assoc 2(1):179–188
12. Habibullah MS, Lumanpauw E, Kolowrocki K, Soszynska J, Ming NG (2009) A computational tool for general model of operation processes in industrial systems. Operation processes. Electron J Reliab Risk Anal Theory Appl 2(4):181–191
13. Huang J, Zuo MJ, Wu Y (2000) Generalized multi-state k-out-of-n:G systems. IEEE Trans Reliab 49:105–111
14. Hudson JC, Kapur KC (1982) Reliability theory for multistate systems with multistate components. Microelectron Reliab 22:1–7

15. Hudson JC, Kapur KC (1983) Reliability analysis of multistate systems with multistate components. Trans Inst Ind Eng 15:127–135
16. Hudson J, Kapur K (1985) Reliability bounds for multistate systems with multistate components. Oper Res 33:735–744
17. Kołowrocki K (1998) On applications of asymptotic reliability functions to the reliability and risk evaluation of pipelines. Int J Press Vessel Pip 75:545–558
18. Kołowrocki K (2003) An asymptotic approach to reliability evaluation of large multi-state systems with applications to piping transportation systems. Int J Press Vessel Pip 80:59–73
19. Kołowrocki K (2004) Reliability of large systems. Elsevier, Amsterdam
20. Kołowrocki K (2006) Reliability and risk evaluation of complex systems in their operation processes. Int J Mater Struct Reliab 4(2):129–147
21. Kołowrocki K (2007) Reliability modelling of complex systems—part 1. Int J Gnedenko e-Forum Reliab Theory Appl 2(3–4):116–127
22. Kołowrocki K (2007) Reliability modelling of complex systems—part 2. Int J Gnedenko e-Forum Reliab Theory Appl 2(3–4):128–139
23. Kołowrocki K (2008) Reliability and risk analysis of multi-state systems with degrading components. Summer safety & reliability seminars. J Pol Saf Reliab Assoc 2(2):205–216
24. Kołowrocki K (2008) Reliability of large systems. In: Encyclopedia of quantitative risk analysis and assessment, vol 4. Wiley, New York, pp 1466–1471
25. Kołowrocki K, Soszyńska J (2006) Reliability and availability of complex systems. Qual Reliab Eng Int 22(1):79–99
26. Kołowrocki K, Soszyńska J (2008) A general model of industrial systems operation processes related to their environment and infrastructure. Summer safety & reliability seminars. J Pol Saf Reliab Assoc 2(2):223–226
27. Kolowrocki K, Soszynska J (2009) Modeling environment and infrastructure influence on reliability and operation process of port oil transportation system. Electron J Reliab Risk Anal Theory Appl 2(3):131–142
28. Kolowrocki K, Soszynska J (2009) Safety and risk evaluation of Stena Baltica ferry in variable operation conditions. Electron J Reliab Risk Anal Theory Appl 2(4):168–180
29. Kołowrocki K, Soszyńska J (2009) Statistical identification and prediction of the port oil pipeline system's operation process and its reliability and risk evaluation. Summer safety & reliability seminars. J Pol Saf Reliab Assoc 3(2):241–250
30. Kołowrocki K, Soszyńska J (2009) Methods and algorithms for evaluating unknown parameters of components reliability of complex technical systems. Summer safety & reliability seminars. J Pol Saf Reliab Assoc 4(2):223–230
31. Kolowrocki K, Soszynska J (2010) Reliability, availability and safety of complex technical systems: modelling – identification – prediction – optimization. Summer safety & reliability seminars. J Pol Saf Reliab Assoc 4(1):133–158
32. Kolowrocki K, Soszynska J (2010) Reliability modeling of a port oil transportation system's operation processes. Int J Perform Eng 6(1):77–87
33. Kołowrocki K, Soszyńska J (2010) Safety and risk evaluation of a ferry technical system. Summer safety & reliability seminars. J Pol Saf Reliab Assoc 4(1):159–172
34. Kołowrocki K, Soszyńska J, Judziński M, Dziula P (2007) On multi-state safety analysis in shipping. Int J Reliab Qual Saf Eng Syst Reliab Saf 146:547–567
35. Kołowrocki K, Soszyńska J, Xie M, Kien M, Salahudin M (2008) Safety and reliability of complex industrial systems and process. Summer safety & reliability seminars. J Pol Saf Reliab Assoc 4(2):227–234
36. Kołowrocki K, Soszyńska J (2011) On safety analysis of complex technical maritime transportation system. J Risk Reliab 225(1):1–10
37. Kossow A, Preuss W (1995) Reliability of linear consecutively-connected systems with multistate components. IEEE Trans Reliab 44:518–522
38. Limnios N, Oprisan G (2005) Semi-Markov processes and reliability. Birkhauser, Boston
39. Lisnianski A, Levitin G (2003) Multi-state system reliability assessment, optimisation and applications. World Scientific Publishing Co. Pte. Ltd., Singapore

40. Meng F (1993) Component-relevancy and characterisation in multi-state systems. IEEE Trans Reliab 42:478–483
41. Mercier S (2008) Numerical bounds for semi-Markovian quantities and application to reliability. Methodol Comput Appl Probab 10(2):179–198
42. Natvig B (1982) Two suggestions of how to define a multi-state coherent system. Adv Appl Probab 14:434–455
43. Natvig B (1984) Multi-state coherent systems. In: Encyclopaedia of statistical sciences. Wiley, New York
44. Natvig B, Streller A (1984) The steady-state behaviour of multistate monotone systems. J Appl Probab 21:826–835
45. Ohio F, Nishida T (1984) On multistate coherent systems. IEEE Trans Reliab 33:284–287
46. Soszyńska J (2004) Reliability of large series system in variable operation conditions. In: Jt Proc 17. Gdynia Maritime University Press, Gdynia, pp 36–43
47. Soszyńska J (2004) Reliability of large parallel systems in variable operation conditions. In Fac of Navig Res Work 16. Gdynia, pp 168–180
48. Soszyńska J (2006) Reliability of large series–parallel system in variable operation conditions. Int J Autom Comput 3(2):199–206
49. Soszyńska J (2006) Reliability evaluation of a port oil transportation system in variable operation conditions. Int J Press Vessel Pip 83(4):304–310
50. Soszyńska J (2006) Safety analysis of multistate systems in variable operations conditions (in Polish). Diagnostyka 3(39):25–34
51. Soszyńska J (2007) Systems reliability analysis in variable operation conditions. PhD thesis, Gdynia Maritime University-System Research Institute Warsaw (in Polish)
52. Soszyńska J (2008) Asymptotic approach to reliability evaluation of large "*m* out of *l*"-series system in variable operation conditions. Summer safety & reliability seminars. J Pol Saf Reliab Assoc 2(2):323–346
53. Soszyńska J (2010) Reliability and risk evaluation of a port oil pipeline transportation system in variable operation conditions. Int J Press Vessel Pip 87(2–3):81–87
54. Soszyńska J, Kołowrocki K, Blokus-Roszkowska A, Guze S (2010) Prediction of complex technical systems operation processes. Summer safety & reliability seminars. J Pol Saf Reliab Assoc 4(2):79–510
55. Xue J (1985) On multi-state system analysis. IEEE Trans Reliab 34:329–337
56. Xue J, Yang K (1995) Dynamic reliability analysis of coherent multi-state systems. IEEE Trans Reliab 4(44):683–688
57. Xue J, Yang K (1995) Symmetric relations in multi-state systems. IEEE Trans Reliab 4(44):689–693
58. Yu K, Koren I, Guo Y (1994) Generalised multistate monotone coherent systems. IEEE Trans Reliab 43:242–250

Chapter 4
Complex Technical System Operation Processes Identification

4.1 Introduction

The general model of complex technical systems operation processes is proposed in Chap. 2. The reliability and safety models of various multistate complex technical systems are considered in Chap. 1. Consequently, the general joint models linking these system reliability models with the model of their operation processes, allowing us for the reliability and safety analysis of complex technical systems at the variable operation conditions, are constructed in Chap. 3. To be able to apply these general models practically in the evaluation and prediction of the reliability and safety of real complex technical systems it is necessary to have the statistical methods concerned with determining the unknown parameters of the proposed models [1–4, 6, 7, 10, 11, 13, 14, 17, 25]. Particularly, concerning the system operation process, the probabilities of the system operation process stay at the operation states at the initial moment, the probabilities of the system operation process transitions between the system operation states and the distributions of the conditional sojourn times of the system operation process at the particular operation states should be identified [5, 10–13, 15, 16]. It is also necessary to have the methods of testing the hypotheses concerned with the conditional sojourn times of the system operation process at the operation states and the procedures of testing the uniformity of their realizations coming from different sets of empirical data [13, 18].

4.2 Identification of Complex Technical Systems Operation Processes

We assume, as in Chaps. 2 and 3, that a system during its operation at the fixed moment t, $t \in \langle 0, +\infty \rangle$, may be at one of v, $v \in N$, different operations states z_b, $b = 1, 2, \ldots, v$. Next, we mark by $Z(t)$, $t \in \langle 0, +\infty \rangle$, the system operation

K. Kołowrocki and J. Soszyńska-Budny, *Reliability and Safety of Complex Technical Systems and Processes*, Springer Series in Reliability Engineering, DOI: 10.1007/978-0-85729-694-8_4,

process, which is a function of a continuous variable t, taking discrete values in the set $\{z_1, z_2, \ldots, z_v\}$ of the system operation states. We assume a semi-Markov model [8, 9, 13, 17, 19–22, 24] of the system operation process $Z(t)$ and we mark by θ_{bl} its random conditional sojourn times at the operation states z_b, when its next operation state is $z_l, b, l = 1, 2, \ldots, v$, $b \neq l$.

Under these assumption, the operation process may be described by the vector $[p_b(0)]_{1\times v}$ of probabilities of the system operation process stay at the particular operations states at the initial moment $t = 0$, the matrix $[p_{bl}(t)]_{v\times v}$ of the probabilities of the system operation process transitions between the operation states and the matrix $[H_{bl}(t)]_{v\times v}$ of the distribution functions of the conditional sojourn times θ_{bl} of the system operation process at the operation states or equivalently by the matrix $[h_{bl}(t)]_{v\times v}$ of the density functions of the conditional sojourn times θ_{bl}, $b, l = 1, 2, \ldots, v, b \neq l$, of the system operation process at the operation states. All these parameters of the complex technical system operation process are unknown and before their use for the prognosis of this process their characteristics have to be estimated on the basis of statistical data coming from practice.

4.2.1 *Defining Unknown Parameters of System Operation Process and Data Collection*

To make the estimation of the unknown parameters of the system operations process, the experiment delivering the necessary statistical data should be precisely planned.

First, before the experiment, we should perform the following preliminary steps:

(i) to analyze the system operation process;
(ii) to fix or to define the system operation process following general parameters:

- the number of the operation states of the system operation process v,
- the operation states of the system operation process $z_1, z_2, \ldots, z_v$;

(iii) to fix the possible transitions between the system operation states;
(iv) to fix the set of the unknown parameters of the system operation process semi-Markov model.

Next, to estimate the unknown parameters of the system operation process, based on the experiment, we should collect the necessary statistical data performing the following steps:

(i) to fix and collect the following statistical data necessary for evaluating the probabilities $p_b(0)$ of the system operation process stay at the operation states at the initial moment $t = 0$:

- the duration time of the experiment Θ,
- the number of the investigated (observed) realizations of the system operation process $n(0)$,

- the vector of the realizations $n_b(0)$, $b = 1, 2, \ldots, v$, of the number of stays of the operation process respectively at the operation states $z_1, z_2, \ldots, z_v$, at the initial moments $t = 0$ of all $n(0)$ observed realizations of the system operation process

$$[n_b(0)] = [n_1(0), n_2(0), \ldots, n_v(0)],$$

where

$$n_1(0) + n_2(0) + n_v(0) = n(0);$$

(ii) to fix and collect the following statistical data necessary for evaluating the probabilities p_{bl} of the system operation process transitions between the system operation states:

- the matrix of the realization of the numbers n_{bl}, $b, l = 1, 2, \ldots, v$, $b \neq l$, of the transitions of the system operation process from the operation state z_b into the operation state z_l at all observed realizations of the system operation process

$$[n_{bl}] = \begin{bmatrix} n_{11}\ n_{12} \ldots n_{1v} \\ n_{21}\ n_{22} \ldots n_{2v} \\ \ldots \\ n_{v1}\ n_{v2} \ldots n_{vv} \end{bmatrix},$$

where

$$n_{bb} = 0 \quad \text{for } b = 1, 2, \ldots, v,$$

- the vector of the realizations of the numbers n_b, $b = 1, 2, \ldots, v$, of departures of the system operation process from the operation states z_b (the sums of the numbers of the bth rows of the matrix $[n_{bl}]$)

$$[n_b] = [n_1, n_2, \ldots, n_v],$$

where

$$n_1 = n_{11} + n_{12} + \cdots + n_{1v},$$
$$n_2 = n_{21} + n_{22} + \cdots + n_{2v},$$
$$\ldots$$
$$n_v = n_{v1} + n_{v2} + \cdots + n_{vv};$$

(iii) to fix and collect the following statistical data necessary for evaluating the unknown parameters of the distributions $H_{bl}(t)$ of the conditional sojourn times θ_{bl} of the system operation process at the particular operation states:

- the numbers n_{bl}, $b, l = 1, 2, \ldots, v, b \neq l$, of realizations of the conditional sojourn times θ_{bl}, $b, l = 1, 2, \ldots, v, b \neq l$, of the system operation process at the operation state z_b when the next transition is to the operation state z_l during the observation time Θ,
- the realizations θ_{bl}^k, $k = 1, 2, \ldots, n_{bl}$, of the conditional sojourn times θ_{bl} of the system operation process at the operation state z_b when the next transition is to the operation state z_l during the observation time Θ for each $b, l = 1, 2, \ldots, v, b \neq l$.

4.2.2 Estimating Basic Parameters of System Operation Process

After collecting the statistical data, it is possible to estimate the unknown parameters of the system operation process performing the following steps:

(i) to determine the vector

$$[p(0)] = [p_1(0), p_2(0), \ldots, p_v(0)], \tag{4.1}$$

of the realizations of the probabilities $p_b(0)$, $b = 1, 2, \ldots, v$, of the system operation process stay at the operation states at the initial moment $t = 0$, according to the formula

$$p_b(0) = \frac{n_b(0)}{n(0)} \quad \text{for } b = 1, 2, \ldots, v, \tag{4.2}$$

where

$$n(0) = \sum_{b=1}^{v} n_b(0), \tag{4.3}$$

is the number of realizations of the system operation process starting at the initial moment $t = 0$;

(ii) to determine the matrix

$$[p_{bl}] = \begin{bmatrix} p_{11} \; p_{12} & \cdots & p_{1v} \\ p_{21} \; p_{22} & \cdots & p_{2v} \\ & \cdots & \\ p_{v1} \; p_{v2} & \cdots & p_{vv} \end{bmatrix}, \tag{4.4}$$

of realizations of the probabilities p_{bl}, $b, l = 1, 2, \ldots, v$, of system operation process transitions from the operation state z_b to the operation state z_l according to the formula

$$p_{bl} = \frac{n_{bl}}{n_b} \quad \text{for } b, l = 1, 2, \ldots, v, \quad b \neq l, \quad p_{bb} = 0 \quad \text{for } b = 1, 2, \ldots, v, \tag{4.5}$$

where

$$n_b = \sum_{b \neq l}^{v} n_{bl}, \quad b = 1, 2, \ldots, v, \tag{4.6}$$

is the realization of the total number of the system operation process departures from the operation state z_b during the experiment time Θ.

4.2.3 Estimating Parameters of Distributions of System Conditional Sojourn Times at Operation States

Prior to estimating the parameters of the distributions of the conditional sojourn times of the system operation process at the particular operation states, we have to determine the following empirical characteristics of the realizations of the conditional sojourn time of the system operation process at the particular operation states:

- the realizations of the empirical mean values $\bar{\theta}_{bl}$ of the conditional sojourn times θ_{bl} of the system operation process at the operation state z_b when the next transition is to the operation state z_l, according to the formula

$$\bar{\theta}_{bl} = \frac{1}{n_{bl}} \sum_{k=1}^{n_{bl}} \theta_{bl}^k, \quad b, l = 1, 2, \ldots, v, \quad b \neq l, \tag{4.7}$$

- the number $\bar{r}_{bl}$ of the disjoint intervals $I_j = \langle a_{bl}^j, b_{bl}^j), j = 1, 2, \ldots, \bar{r}_{bl}$ which include the realizations θ_{bl}^k, $k = 1, 2, \ldots, n_{bl}$, of the conditional sojourn times θ_{bl} at the operation state z_b when the next transition is to the operation state z_l, according to the formula

$$\bar{r}_{bl} \cong \sqrt{n_{bl}},$$

- the length d_{bl} of the intervals $I_j = \langle a_{bl}^j, b_{bl}^j), j = 1, 2, \ldots, \bar{r}_{bl}$, according to the formula

$$d_{bl} = \frac{\bar{R}_{bl}}{\bar{r}_{bl} - 1},$$

where

$$\bar{R}_{bl} = \max_{1 \leq k \leq n_{bl}} \theta_{bl}^k - \min_{1 \leq k \leq n_{bl}} \theta_{bl}^k,$$

- the ends a_{bl}^j, b_{bl}^j, of the intervals $I_j = \langle a_{bl}^j, b_{bl}^j)$, $j = 1, 2, \ldots, \bar{r}_{bl}$, according to the formulae

$$a_{bl}^1 = \max\{\min_{1 \le k \le n_{bl}} \theta_{bl}^k - \frac{d_{bl}}{2}, 0\},$$

$$b_{bl}^j = a_{bl}^1 + jd_{bl}, \quad j = 1, 2, \ldots, \bar{r}_{bl},$$

$$a_{bl}^j = b_{bl}^{j-1}, \quad j = 2, 3 \ldots, \bar{r}_{bl},$$

in such a way that

$$I_1 \cup I_2 \cup \ldots \cup I_{\bar{r}_{bl}} = \langle a_{bl}^1, b_{bl}^{\bar{r}_{bl}})$$

and

$$I_i \cap I_j = \emptyset \quad \text{for all } i \ne j, \quad i, j \in \{1, 2, \ldots, \bar{r}_{bl}\},$$

- the number n_{bl}^j of the realizations θ_{bl}^k in the intervals $I_j = \langle a_{bl}^j, b_{bl}^j)$, $j = 1, 2, \ldots, \bar{r}_{bl}$, according to the formula

$$n_{bl}^j = \#\{k : \theta_{bl}^k \in I_j, \ k \in \{1, 2, \ldots, n_{bl}\}\}, \quad j = 1, 2, \ldots, \bar{r}_{bl},$$

where

$$\sum_{j=1}^{\bar{r}_{bl}} n_{bl}^j = n_{bl},$$

whereas the symbol # means the number of elements of the set.

To estimate the parameters of the distributions of the conditional sojourn times of the system operation process at the particular operation states distinguished in Chap. 2, we proceed respectively in the following way:

- for the uniform distribution with the density function given by (2.5), the estimates of the unknown parameters are:

$$x_{bl} = a_{bl}^1 \quad y_{bl} = x_{bl} + \bar{r}_{bl} d_{bl}; \tag{4.8}$$

- for the triangular distribution with the density function given by (2.6), the estimates of the unknown parameters are:

$$x_{bl} = a_{bl}^1, \quad y_{bl} = x_{bl} + \bar{r}_{bl} d_{bl}, \quad z_{bl} = \bar{\theta}_{bl}; \tag{4.9}$$

- for the double trapezium distribution with the density function given by (2.7), the estimates of the unknown parameters are:

$$x_{bl} = a_{bl}^1, \quad y_{bl} = x_{bl} + \bar{r}_{bl} d_{bl}, \quad q_{bl} = \frac{n_{bl}^1}{n_{bl} d_{bl}}, \quad w_{bl} = \frac{n_{bl}^{\bar{r}_{bl}}}{n_{bl} d_{bl}}, \quad z_{bl} = \bar{\theta}_{bl}; \tag{4.10}$$

- for the quasi-trapezium distribution with the density function given by (2.8), the estimates of the unknown parameters are:

$$x_{bl} = a_{bl}^1, \quad y_{bl} = x_{bl} + \bar{r}_{bl} d_{bl}, \quad q_{bl} = \frac{n_{bl}^1}{n_{bl} d_{bl}}, \quad w_{bl} = \frac{n_{bl}^{\bar{r}_{bl}}}{n_{bl} d_{bl}}, \quad z_{bl}^1 = \bar{\theta}_{bl}^1, \quad z_{bl}^2 = \bar{\theta}_{bl}^2, \tag{4.11}$$

where

$$\bar{\theta}_{bl}^1 = \frac{1}{n_{(me)}} \sum_{k=1}^{n_{(me)}} \theta_{bl}^k, \quad \bar{\theta}_{bl}^2 = \frac{1}{n_{bl} - n_{(me)}} \sum_{k=n_{(me)}+1}^{n_{bl}} \theta_{bl}^k, \quad n_{(me)} = \left[\frac{n_{bl} + 1}{2} \right], \tag{4.12}$$

and $[x]$ denotes the entire part of x;

- for the exponential distribution with the density function given by (2.9), the estimates of the unknown parameters are:

$$x_{bl} = a_{bl}^1, \quad \alpha_{bl} = \frac{1}{\bar{\theta}_{bl} - x_{bl}}; \tag{4.13}$$

- for the Weibull's distribution with the density function given by (2.10), the estimates of the unknown parameters are (the expressions for estimates of parameters α_{bl} and β_{bl} are not explicit):

$$x_{bl} = a_{bl}^1, \quad \alpha_{bl} = \frac{n_{bl}}{\sum_{k=1}^{n_{bl}} (\theta_{bl}^k)^{\beta_{bl}}}, \quad \alpha_{bl} = \frac{\frac{n_{bl}}{\beta_{bl}} + \sum_{k=1}^{n_{bl}} \ln(\theta_{bl}^k - x_{bl})}{\sum_{k=1}^{n_{bl}} (\theta_{bl}^k)^{\beta_{bl}} \ln(\theta_{bl}^k - x_{bl})}; \tag{4.14}$$

- for the chimney distribution with the density function given by (2.11), the estimates of the unknown parameters are:

$$x_{bl} = a_{bl}^1, \quad y_{bl} = x_{bl} + \bar{r}_{bl} d_{bl}, \tag{4.15}$$

and moreover, if

$$\hat{n}_{bl} = \max_{1 \le j \le \bar{r}_{bl}} \{ n_{bl}^j \} \tag{4.16}$$

and i, $i \in \{1, 2, \ldots, \bar{r}_{bl}\}$, is the number of the interval including the largest number of realizations i.e. such as that

$$n_{bl}^i = \hat{n}_{bl}, \tag{4.17}$$

then:

- for $i = 1$

either

$$z_{bl}^1 = x_{bl} + (i-1)\, d_{bl}, \quad z_{bl}^2 = x_{bl} + i d_{bl}, \quad A_{bl} = 0, \quad C_{bl} = \frac{n_{bl}^i}{n_{bl}},$$
$$D_{bl} = \frac{n_{bl}^{i+1} + \cdots + n_{bl}^{\bar{r}_{bl}}}{n_{bl}}, \tag{4.18}$$

while

$$n_{bl}^{i+1} = 0 \quad \text{or} \quad n_{bl}^{i+1} \neq 0 \quad \text{and} \quad \frac{n_{bl}^{i}}{n_{bl}^{i+1}} \geq 3, \tag{4.19}$$

or

$$z_{bl}^{1} = x_{bl} + (i-1)d_{bl}, \quad z_{bl}^{2} = x_{bl} + (i+1)d_{bl}, \quad A_{bl} = 0, \tag{4.20}$$

$$C_{bl} = \frac{n_{bl}^{i} + n_{bl}^{i+1}}{n_{bl}}, \quad D_{bl} = \frac{n_{bl}^{i+2} + \cdots + n_{bl}^{\bar{r}_{bl}}}{n_{bl}}, \tag{4.21}$$

while

$$n_{bl}^{i+1} \neq 0 \quad \text{and} \quad \frac{n_{bl}^{i}}{n_{bl}^{i+1}} < 3; \tag{4.22}$$

- for $i = 2, 3, \ldots, \bar{r}_{bl} - 1$
  either

$$z_{bl}^{1} = x_{bl} + (i-1)d_{bl}, \quad z_{bl}^{2} = x_{bl} + id_{bl}, \quad A_{bl} = \frac{n_{bl}^{1} + \cdots + n_{bl}^{i-1}}{n_{bl}}, \quad C_{bl} = \frac{n_{bl}^{i}}{n_{bl}}, \tag{4.23}$$

$$D_{bl} = \frac{n_{bl}^{i+1} + \cdots + n_{bl}^{\bar{r}_{bl}}}{n_{bl}}, \tag{4.24}$$

while

$$n_{bl}^{i-1} = 0 \quad \text{or} \; n_{bl}^{i-1} \neq 0 \quad \text{and} \quad \frac{n_{bl}^{i}}{n_{bl}^{i-1}} \geq 3 \tag{4.25}$$

and while

$$n_{bl}^{i+1} = 0 \quad \text{or} \; n_{bl}^{i+1} \neq 0 \quad \text{and} \quad \frac{n_{bl}^{i}}{n_{bl}^{i+1}} \geq 3, \tag{4.26}$$

or

$$z_{bl}^{1} = x_{bl} + (i-1)d_{bl}, \quad z_{bl}^{2} = x_{bl} + (i+1)d, \quad A_{bl} = \frac{n_{bl}^{1} + \cdots + n_{bl}^{i-1}}{n_{bl}}, \tag{4.27}$$

$$C_{bl} = \frac{n_{bl}^{i} + n_{bl}^{i+1}}{n_{bl}}, \quad D_{bl} = \frac{n_{bl}^{i+2} + \cdots + n_{bl}^{\bar{r}_{bl}}}{n_{bl}}, \tag{4.28}$$

while

$$n_{bl}^{i-1} = 0 \quad \text{or} \; n_{bl}^{i-1} \neq 0 \quad \text{and} \quad \frac{n_{bl}^{i}}{n_{bl}^{i-1}} \geq 3 \tag{4.29}$$

and while

$$n_{bl}^{i+1} \neq 0 \quad \text{and} \quad \frac{n_{bl}^{i}}{n_{bl}^{i+1}} < 3, \tag{4.30}$$

or

$$z_{bl}^{1} = x_{bl} + (i-2)d_{bl}, \quad z_{bl}^{2} = x_{bl} + id_{bl}, \quad A_{bl} = \frac{n_{bl}^{1} + \cdots + n_{bl}^{i-2}}{n_{bl}}, \quad C_{bl} = \frac{n_{bl}^{i-1} + n_{bl}^{i}}{n_{bl}}, \tag{4.31}$$

$$D_{bl} = \frac{n_{bl}^{i+1} + \cdots + n_{bl}^{\bar{r}_{bl}}}{n_{bl}}, \tag{4.32}$$

while

$$n_{bl}^{i-1} \neq 0 \quad \text{and} \quad \frac{n_{bl}^{i}}{n_{bl}^{i-1}} < 3 \tag{4.33}$$

and while

$$n_{bl}^{i+1} = 0 \quad \text{or } n_{bl}^{i+1} \neq 0 \quad \text{and} \quad \frac{n_{bl}^{i}}{n_{bl}^{i+1}} \geq 3, \tag{4.34}$$

or

$$z_{bl}^{1} = x_{bl} + (i-2)d_{bl}, \quad z_{bl}^{2} = x_{bl} + (i+1)d_{bl}, \quad A_{bl} = \frac{n_{bl}^{1} + \cdots + n_{bl}^{i-2}}{n_{bl}}, \tag{4.35}$$

$$C_{bl} = \frac{n_{bl}^{i-1} + n_{bl}^{i} + n_{bl}^{i+1}}{n_{bl}}, \quad D_{bl} = \frac{n_{bl}^{i+2} + \cdots + n_{bl}^{\bar{r}_{bl}}}{n_{bl}}, \tag{4.36}$$

while

$$n_{bl}^{i-1} \neq 0 \quad \text{and} \quad \frac{n_{bl}^{i}}{n_{bl}^{i-1}} < 3 \tag{4.37}$$

and while

$$n_{bl}^{i+1} \neq 0 \quad \text{and} \quad \frac{n_{bl}^{i}}{n_{bl}^{i+1}} < 3; \tag{4.38}$$

- for $i = \bar{r}_{bl}$
  either

$$z_{bl}^{1} = x_{bl} + (i-1)d_{bl}, \quad z_{bl}^{2} = x_{bl} + id_{bl}, \quad A_{bl} = \frac{n_{bl}^{1} + \cdots + n_{bl}^{i-1}}{n_{bl}}, \tag{4.39}$$

$$C_{bl} = \frac{n_{bl}^{i}}{n_{bl}}, \quad D_{bl} = 0, \tag{4.40}$$

while

$$n_{bl}^{i-1} = 0 \text{ or } n_{bl}^{i-1} \neq 0 \text{ and } \frac{n_{bl}^{i}}{n_{bl}^{i-1}} \geq 3, \tag{4.41}$$

or

$$z_{bl}^{1} = x_{bl} + (i-2)d_{bl}, \quad z_{bl}^{2} = x_{bl} + id_{bl}, \quad A_{bl} = \frac{n_{bl}^{1} + \cdots + n_{bl}^{i-2}}{n_{bl}}, \tag{4.42}$$

$$C_{bl} = \frac{n_{bl}^{i-1} + n_{bl}^{i}}{n_{bl}}, \quad D_{bl} = 0, \tag{4.43}$$

while

$$n_{bl}^{i-1} \neq 0 \quad \text{and} \quad \frac{n_{bl}^{i}}{n_{bl}^{i-1}} < 3. \tag{4.44}$$

4.2.4 Identification of Distribution Functions of System Conditional Sojourn Times at Operation States

To formulate and then to verify the non-parametric hypothesis concerning the form of the distribution of the system operation process conditional sojourn time θ_{bl} at the operation state z_b when the next transition is to the operation state z_l, on the basis of at least 30 of its realizations θ_{bl}^{k}, $k = 1, 2, \ldots, n_{bl}$, it is due to proceed according to the following (Fig. 4.1) scheme:

- to construct and plot the realization of the histogram of the system operation process conditional sojourn time θ_{bl} at the operation state z_b, defined by the following formula

$$\bar{h}_{n_{bl}}(t) = \frac{n_{bl}^{j}}{n_{bl}} \quad \text{for } t \in I_j, \tag{4.45}$$

- to analyze the realization of the histogram $\bar{h}_{n_{bl}}(t)$, comparing it with the graphs of the density functions $h_{bl}(t)$ of the previously distinguished distributions in Chap. 2, to select one of them and to formulate the null hypothesis H_0, concerning the unknown form of the distribution of the conditional sojourn time θ_{bl} in the following form:
  H_0: The system operation process conditional sojourn time θ_{bl} at the operation state z_b when the next transition is to the operation state z_l, has the distribution with the density function $h_{bl}(t)$;

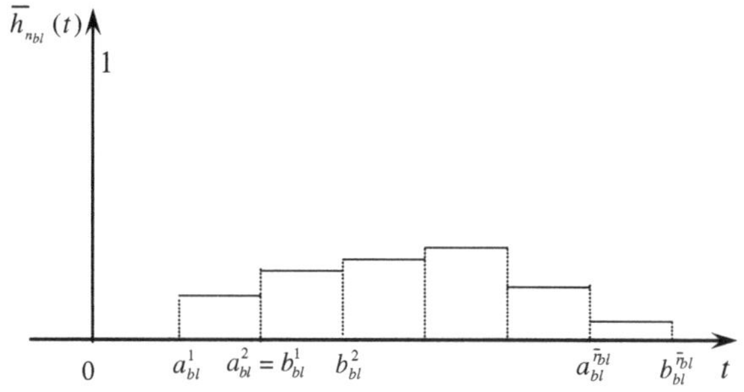

Fig. 4.1 The graph of the realization of the histogram of the system operation process conditional sojourn time θ_{bl} at the operation state z_b

- to join each of the intervals I_j that has the number n_{bl}^j of realizations less than 4 either with the neighbor interval I_{j+1} or with the neighbor interval I_{j-1} in such a way that the number of realizations in all the intervals is not less than 4;
- to fix a new number of intervals $\bar{\bar{r}}_{bl}$;
- to determine new intervals

$$\bar{I}_j = \langle \bar{a}_{bl}^j, \bar{b}_{bl}^j), \quad j = 1,\ 2,\ ..,\bar{\bar{r}}_{bl};$$

- to fix the number $\bar{n}_{bl}^j$ of realizations in new intervals $\bar{I}_j$, $j = 1,\ 2,\ ..,\ \bar{\bar{r}}_{bl}$;
- to calculate the hypothetical probabilities so that the variable θ_{bl} takes values from the interval $\bar{I}_j$, under the assumption that the hypothesis H_0 is true, i.e. the probabilities

$$p_j = P(\theta_{bl} \in \bar{I}_j) = P(\bar{a}_{bl}^j \le \theta_{bl} < \bar{b}_{bl}^j) = H_{bl}(\bar{b}_{bl}^j) - H_{bl}(\bar{a}_{bl}^j), \quad j = 1,\ 2,\ldots,\bar{\bar{r}}_{bl}, \tag{4.46}$$

where $H_{bl}(\bar{b}_{bl}^j)$ and $H_{bl}(\bar{a}_{bl}^j)$ are the values of the distribution function $H_{bl}(t)$ of the random variable θ_{bl} corresponding to the density function $h_{bl}(t)$ assumed in the null hypothesis H_0;

- to calculate the realization of the χ^2(chi-square)-Pearson's statistics $U_{n_{bl}}$, according to the formula

$$u_{n_{bl}} = \sum_{j=1}^{\bar{\bar{r}}_{bl}} \frac{(\bar{n}_{bl}^j - n_{bl}p_j)^2}{n_{bl}p_j}; \tag{4.47}$$

- to assume the significance level α ($\alpha = 0.01$, $\alpha = 0.02$, $\alpha = 0.05$ or $\alpha = 0.10$) of the test;
- to fix the number $\bar{\bar{r}}_{bl} - l - 1$ of degrees of freedom, substituting for l for the distinguished in Chap. 2 distributions respectively the following values: $l = 0$ for the uniform, triangular, double trapezium, quasi-trapezium and chimney distributions, $l = 1$ for the exponential distribution and $l = 2$ for the Weibull's distribution;
- to read from the tables of the χ^2-Pearson's distribution the value u_α for the fixed values of the significance level α and the number of degrees of freedom $\bar{\bar{r}}_{bl} - l - 1$ such that the following equality holds

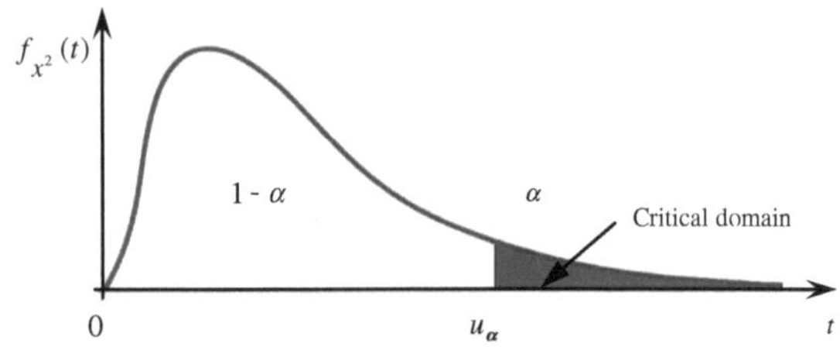

Fig. 4.2 The graphical interpretation of the critical interval and the acceptance interval for the chi-square goodness-of-fit test

$$P(U_{n_{bl}} > u_\alpha) = \alpha, \tag{4.48}$$

and then to determine the critical domain in the form of the interval (Fig. 4.2) $(u_\alpha, +\infty)$ and the acceptance domain in the form of the interval $\langle 0, u_\alpha \rangle$,

- to compare the obtained value $u_{n_{bl}}$ of the realization of the statistics $U_{n_{bl}}$ with at read from the tables, critical value u_α of the chi-square random variable and to decide on the previously formulated null hypothesis H_0 in the following way: if the value $u_{n_{bl}}$ does not belong to the critical domain, i.e. when $u_{n_{bl}} \le u_\alpha$, then we do not reject the hypothesis H_0, if the value $u_{n_{bl}}$ belongs to the critical domain, i.e. when $u_{n_{bl}} > u_\alpha$, then we reject the hypothesis H_0.

4.2.5 Testing Uniformity of Statistical Data of Complex Technical System Operation Processes

The statistical data needed in Sect. 4.2.4 for estimating the unknown parameters of the complex technical system operation process very often come from different experiments of the same operation process and are collected into separate data sets. Before combining them into one set of data in order to perform the unknown parameters, evaluation with the methods and procedures described in Sect. 4.2.4, we have to carryout the uniformity testing of these statistical data sets.

4.2.5.1 Procedure of System Operation Process Data Collection

To carryout the uniformity testing of the statistical data collected in two separate data sets coming from the same system operation process realizations in two different experiments, we should collect the necessary statistical data performing the following steps:

(i) to fix two independent experiments of the system operation process data collection and their following basic parameters:

- the duration times of the experiments Θ_1, Θ_2,
- the system operation processes, single realizations,
- the number of the investigated (observed) realizations of the system operation process $n_1(0)$, $n_2(0)$;

(ii) to fix and collect the following statistical data concerned with the empirical distributions of the conditional sojourn times θ_{bl}^1 and θ_{bl}^2, $b, l \in \{1, 2, \ldots, v\}$, $b \neq l$, of the system operation process at the particular operation states, respectively in the first and second experiments:

- the number of realizations

$$n_{bl}^1, \quad b, l \in \{1, 2, \ldots, v\}, \quad b \neq l,$$

of the sojourn time θ_{bl}^1, $b, l \in \{1, 2, \ldots, v\}$, in the first experiment,

- the sample of non-decreasing ordered realizations

$$\theta_{bl}^{1k}, \quad k = 1, 2, \ldots, n_{bl}^1, \quad b \neq l, \tag{4.49}$$

of the sojourn time $\theta_{bl}^1, b, l \in \{1, 2, \ldots, v\}$, in the first experiment,

- the number of realizations

$$n_{bl}^2, \quad b, l \in \{1, 2, \ldots, v\}, \quad b \neq l,$$

of the sojourn time $\theta_{bl}^2, b, l \in \{1, 2, \ldots, v\}$, in the second experiment,

- the sample of non-decreasing ordered realizations

$$\theta_{bl}^{2k}, \quad k = 1, 2, \ldots, n_{bl}^2, \quad b \neq l, \tag{4.50}$$

of the sojourn time θ_{bl}^2, $b, l \in \{1, 2, \ldots, v\}$, in the second experiment.

4.2.5.2 Procedure of Testing Uniformity of Distributions of System Operation Process Conditional Sojourn Times at Operation States

We consider test λ based on the Kolmogorov-Smirnov theorem [18] that can be used for testing whether two independent samples of realizations of the conditional sojourn time θ_{bl}, $b, l \in \{1, 2, \ldots, v\}$, $b \neq l$, at the particular operation states of the system operation process are drawn from the population with the same distribution.

We assume that we have defined in the previous section two independent samples of non-decreasing ordered realizations (4.49) and (4.50) of the sojourn times θ_{bl}^1 and θ_{bl}^2, $b, l \in \{1, 2, \ldots, v\}$, $b \neq l$, coming from two different experiments, respectively composed of n_{bl}^1 and n_{bl}^2 realizations and we define their corresponding empirical distribution functions

$$H_{bl}^1(t) = \frac{1}{n_{bl}^1} \#\{k : \theta_{bl}^{1k} < t, k \in \{1, 2, \ldots, n_{bl}^1\}\}, \quad t \geq 0, \quad b, l \in \{1, 2, \ldots, v\}, \quad b \neq l, \tag{4.51}$$

and

$$H_{bl}^{2}(t) = \frac{1}{n_{bl}^{2}} \#\{k: \theta_{bl}^{2k} < t,\ k \in \{1, 2, \ldots, n_{bl}^{2}\}\},\quad t \geq 0,\quad b, l \in \{1, 2, \ldots, v\},\quad b \neq l. \tag{4.52}$$

Then, according to the Kolmogorov-Smirnov theorem [18], the sequence of distribution functions given by the equation

$$Q_{n_1 n_2}(\lambda) = P\left(D_{n_1 n_2} < \frac{\lambda}{\sqrt{n}}\right) \tag{4.53}$$

defined for $\lambda > 0$, where

$$n_1 = n_{bl}^{1},\quad n_2 = n_{bl}^{2},\quad n = \frac{n_1 n_2}{n_1 + n_2}, \tag{4.54}$$

and

$$D_{n_1 n_2} = \max_{-\infty < t < +\infty} \left| H_{bl}^{1}(t) - H_{bl}^{2}(t) \right|, \tag{4.55}$$

is convergent, as $n \to \infty$, to the limit distribution function

$$Q(\lambda) = \sum_{k=-\infty}^{+\infty} (-1)^k e^{-2k^2 \lambda^2},\quad \lambda > 0. \tag{4.56}$$

The distribution function $Q(\lambda)$ given by (4.56) is called λ distribution and its table of values are available.

The convergence of the sequence $Q_{n_1 n_2}(\lambda)$ to the λ distribution $Q(\lambda)$ means that for sufficiently large n_1 and n_2 we may use the following approximate formula

$$Q_{n_1 n_2}(\lambda) \cong Q(\lambda). \tag{4.57}$$

Hence, it follows that if we define the statistic

$$U_n = D_{n_1 n_2} \sqrt{n}, \tag{4.58}$$

where $D_{n_1 n_2}$ is defined by (4.55), then by (4.53) and (4.57), we have

$$P(U_n < \lambda) = P(D_{n_1 n_2} \sqrt{n} < \lambda) = P\left(D_{n_1 n_2} < \frac{\lambda}{\sqrt{n}}\right) = Q_{n_1 n_2}(\lambda) \cong Q(\lambda) \quad \text{for } \lambda > 0. \tag{4.59}$$

This result means that in order to formulate and then verify the hypothesis that the two independent samples of the realizations of the system operation process conditional sojourn times θ_{bl}^{1} and θ_{bl}^{2}, $b, l \in \{1, 2, \ldots, v\}$, $b \neq l$, at the operation state z_b when the next transition is to the operation state z_l are coming from the population with the same distribution, it is necessary to proceed according to the following scheme:

- to fix the number of realizations n_{bl}^{1} and n_{bl}^{2} in the samples,
- to collect the realizations (4.49) and (4.50) of the conditional sojourn times θ_{bl}^{1} and θ_{bl}^{2} of the system operation process in the samples,
- to find the realization of the empirical distribution functions $H_{bl}^{1}(t)$ and $H_{bl}^{2}(t)$ defined by (4.51) and (4.52) respectively, in the following forms:

$$H_{bl}^{1}(t) = \begin{cases} \frac{n_{bl}^{11}}{n_{bl}^{1}} = 0, & t \le \theta_{bl}^{11} \\ \frac{n_{bl}^{12}}{n_{bl}^{1}}, & \theta_{bl}^{11} < t \le \theta_{bl}^{12} \\ \frac{n_{bl}^{13}}{n_{bl}^{1}}, & \theta_{bl}^{12} < t \le \theta_{bl}^{13} \\ \cdot \quad \cdot \quad \cdot & \\ \frac{n_{bl}^{1k}}{n_{bl}^{1}}, & \theta_{bl}^{1k-1} < t \le \theta_{bl}^{1k} \\ \cdot \quad \cdot \quad \cdot & \\ \frac{n_{bl}^{1n_{bl}^{1}}}{n_{bl}^{1}}, & \theta_{bl}^{1n_{bl}^{1}-1} < t \le \theta_{bl}^{1n_{bl}^{1}} \\ \frac{n_{bl}^{1n_{bl}^{1}+1}}{n_{bl}^{1}} = 1, & t \ge \theta_{bl}^{1n_{bl}^{1}} \end{cases}, \tag{4.60}$$

$$H_{bl}^{2}(t) = \begin{cases} \frac{n_{bl}^{21}}{n_{bl}^{2}} = 0, & t \le \theta_{bl}^{21} \\ \frac{n_{bl}^{22}}{n_{bl}^{2}}, & \theta_{bl}^{21} < t \le \theta_{bl}^{22} \\ \frac{n_{bl}^{23}}{n_{bl}^{2}}, & \theta_{bl}^{22} < t \le \theta_{bl}^{23} \\ \cdot \quad \cdot \quad \cdot & \\ \frac{n_{bl}^{2k}}{n_{bl}^{2}}, & \theta_{bl}^{2k-1} < t \le \theta_{bl}^{2k} \\ \cdot \quad \cdot \quad \cdot & \\ \frac{n_{bl}^{2n_{bl}^{2}}}{n_{bl}^{2}}, & \theta_{bl}^{2n_{bl}^{2}-1} < t \le \theta_{bl}^{2n_{bl}^{1}} \\ \frac{n_{bl}^{2n_{bl}^{2}+1}}{n_{bl}^{2}} = 1, & t \ge \theta_{bl}^{2n_{bl}^{2}} \end{cases}, \tag{4.61}$$

where

$$n_{bl}^{11} = 0, \quad n_{bl}^{1\,n_{bl}^{1}+1} = n_{bl}^{1}, \tag{4.62}$$

and

$$n_{bl}^{1k} = \#\{j : \theta_{bl}^{1j} < \theta_{bl}^{1k}, \quad j \in \{1, 2, \ldots, n_{bl}^{1}\}\}, \quad k = 2, 3, \ldots, n_{bl}^{1}, \tag{4.63}$$

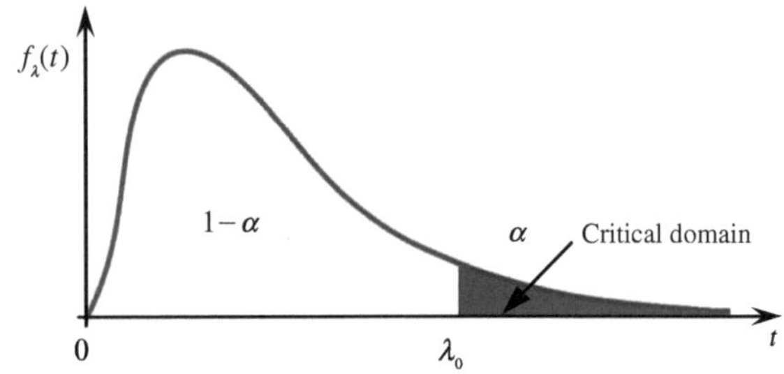

Fig. 4.3 The graphical interpretation of the critical domain and the acceptance domain for the two-sample Smirnov-Kolmogorov test

is the number of the sojourn time θ_{bl}^1 realizations less than its realization θ_{bl}^{1k}, $k = 2, 3, \ldots, n_{bl}^1$, and respectively

$$n_{bl}^{21} = 0, \quad n_{bl}^{2\,n_{bl}^2+1} = n_{bl}^2, \tag{4.64}$$

and

$$n_{bl}^{2k} = \#\{j : \theta_{bl}^{2j} < \theta_{bl}^{2k}, \quad j \in \{1, 2, \ldots, n_{bl}^2\}\}, \quad k = 2, 3, \ldots, n_{bl}^2, \tag{4.65}$$

is the number of the sojourn time θ_{bl}^2 realizations less than its realization θ_{bl}^{2k}, $k = 2, 3, \ldots, n_{bl}^2$,

- to calculate the realization of the statistic u_n defined by (4.58) according to the formula

$$u_n = d_{n_{bl}^1 n_{bl}^2} \sqrt{n}, \tag{4.66}$$

where

$$d_{n_{bl}^1 n_{bl}^2} = \max\{d_{n_{bl}^1 n_{bl}^2}^1, d_{n_{bl}^1 n_{bl}^2}^2\}, \tag{4.67}$$

$$d_{n_{bl}^1 n_{bl}^2}^1 = \max\{|H_{bl}^1(\theta_{bl}^{1k}) - H_{bl}^2(\theta_{bl}^{1k})|, \quad k \in \{1, 2, \ldots, n_{bl}^1\}\}, \tag{4.68}$$

$$d_{n_{bl}^1 n_{bl}^2}^2 = \max\{|H_{bl}^1(\theta_{bl}^{2k}) - H_{bl}^2(\theta_{bl}^{2k})|, \quad k \in \{1, 2, \ldots, n_{bl}^2\}\}, \tag{4.69}$$

$$n = \frac{n_{bl}^1 n_{bl}^2}{n_{bl}^1 + n_{bl}^2}, \tag{4.70}$$

- to formulate the null hypothesis H_0 in the following form:

H_0 : The samples of realizations (4.49) and (4.50) come from the populations with the same distributions,

- to fix the significance level α ($\alpha = 0.01$, $\alpha = 0.02$, $\alpha = 0.05$ or $\alpha = 0.10$) of the test,
- to read from the tables of λ distribution, corresponding to $1 - \alpha$, the value λ_0 such that the following equality holds

$$P(U_n < \lambda_0) = Q(\lambda_0) = 1 - \alpha, \tag{4.71}$$

- to determine the critical domain in the form of the interval (Fig. 4.3) $(\lambda_0, +\infty)$ and the acceptance domain in the form of the interval $(0, \lambda_0\rangle$,
- to compare the obtained value u_n of the realization of the statistics U_n with that read from the tables critical value λ_0,
- to decide on the previously formulated null hypothesis H_0 in the following way:
- if the value u_n does not belong to the critical domain, i.e. when $u_n \leq \lambda_0$, then we do not reject the hypothesis H_0, if the value u_n belongs to the critical domain, i.e. when $u_n > \lambda_0$, then we reject the hypothesis H_0.

In the case when the null hypothesis H_0 is not rejected we may join the statistical data from the considered two separate sets into one new set of data and if there are no other sets of statistical data including the realizations of the sojourn time θ_{bl}, we proceed with the data of this new set in the way described in Sects. 4.2.1–4.2.4. Conversely, if there are other sets of statistical data including the realizations of the sojourn time θ_{bl}, we select the next one and perform the procedure of this section for data from this set and data from the previously formed new set. We continue this procedure up to the moment when the store of the statistical data sets including the realizations of the sojourn time θ_{bl}, is exhausted.

4.3 Applications

4.3.1 Statistical Identification of Exemplary System Operation Process

The operation process of the considered exemplary system is described in Sect. 2.3.1, where the number of system operation process states $v = 4$ is fixed and the operation states z_b, $b = 1, 2, 3, 4$ are arbitrarily defined. Moreover, it is fixed that there are possible transitions between all system operation states.

4.3.1.1 Defining Parameters and Data Collection of Exemplary System Operation Process

The unknown parameters of the system operation process semi-markov model are:

- the initial probabilities $p_b(0)$, $b = 1, 2, 3, 4$, of the exemplary system operation process stay at the particular states z_b at the moment $t = 0$,
- the probabilities p_{bl}, $b, l = 1, 2, 3, 4$, $b \neq l$, of the exemplary system operation process transitions from the operation state z_b into the operation state z_l,
- the distributions of the exemplary system operation process conditional sojourn times θ_{bl}, $b, l = 1, 2, 3, 4$, $b \neq l$, at the particular operation states and their mean values $M_{bl} = E[\theta_{bl}]$, $b, l = 1, 2, 3, 4$, $b \neq l$.

To identify all these parameters of the exemplary system operation process the statistical data about this process are needed. As the considered system is an exemplary one, we do not have the statistical data collected that are needed for estimating the parameters. Therefore, only the values of the unknown parameters of the exemplary system operation process we need in our considerations are arbitrarily fixed in the two following sections.

4.3.1.2 Evaluating Basic Parameters of Exemplary System Operation Process

As we do not have the statistical data collected that are needed for estimating the unknown basic parameters of the exemplary system operation process, we assume their values arbitrarily as in Sect. 2.3.1, where the matrix $[p_{bl}]$ of the probabilities p_{bl}, b, $l = 1, 2, 3, 4$, of the system operation process transitions from the operation state z_b into the operation state z_l is arbitrarily defined.

4.3.1.3 Evaluating Parameters of Distributions of Exemplary System Conditional Sojourn Times at Operation States

We do not have the realizations of the conditional sojourn times θ_{bl}, b, $l = 1, 2, 3, 4$, of the exemplary system operation process at the particular operation states that could allow to identify their distributions. Therefore, we assume arbitrarily their mean values $M_{bl} = E[\theta_{bl}]$, b, $l = 1, 2, 3, 4$, that are given in Sect. 2.3.1.

4.3.1.4 Identification of Distribution Functions of Exemplary System Conditional Sojourn Times at Operation States

As we do not have the realizations of the conditional sojourn times θ_{bl}, b, $l = 1, 2, 3, 4$, of the exemplary system operation process at the particular operation states, it is not possible to identify their distributions.

4.3.2 Statistical Identification of Port Oil Piping Transportation System Operation Process

On the basis of the expert opinions concerning the operation process of the considered port oil pipeline transportation system, in Sect. 2.3.2, the number of the pipeline system operation process states $v = 7$ is fixed and the operation states z_b, $b = 1, 2, \ldots, 7$, are defined. Moreover, it is fixed that there are possible transitions between all system operation states.

4.3.2.1 Defining Parameters and Data Collection of Port Oil Piping Transportation System Operation Process

The unknown parameters of the system operation process semi-Markov model are:

- the initial probabilities $p_b(0)$, $b = 1, 2, \ldots, 7$, of the pipeline system operation process stay at the particular states z_b at the moment $t = 0$,
- the probabilities p_{bl}, $b, l = 1, 2, \ldots, 7$, $b \neq l$, of the pipeline system operation process transitions from the operation state z_b into the operation state z_l,
- the distributions of the pipeline system conditional sojourn times θ_{bl}, $b, l = 1, 2, \ldots, 7$, $b \neq l$, at the particular operation states and their mean values $M_{bl} = E[\theta_{bl}]$, $b, l = 1, 2, \ldots, 7$, $b, l = 1, 2, \ldots, 7$, $b \neq l$.

To identify all these parameters of the pipeline system operation process the statistical data about this process is needed.

The statistical data collected by the system operators necessary for evaluating the initial transient probabilities of the piping system operation process at the particular states are:

- the pipeline system operation process observation/experiment time

$$\Theta = 329 \text{ days} = 47 \text{ weeks},$$

- the number of pipeline system operation process realizations

$$n(0) = 41,$$

- the vector of realizations

$$n_1(0) = 14, \; n_2(0) = 2, \; n_3(0) = 0, \; n_4(0) = 0, \; n_5(0) = 9, \; n_6(0) = 8, \; n_7(0) = 8,$$

of the number of pipeline system operation processes stay at the particular operation states z_b at the initial moment $t = 0$

$$[n_b(0)] = [14, 2, 0, 0, 9, 8, 8].$$

The collected statistical data necessary for evaluating the probabilities of transitions of the pipeline system operation process between the operation states are:

- the matrix of realizations

$$\begin{array}{lllllll}
n_{11} = 0, & n_{12} = 1, & n_{13} = 1, & n_{14} = 0, & n_{15} = 24, & n_{16} = 5, & n_{17} = 14, \\
n_{21} = 1, & n_{22} = 0, & n_{23} = 0, & n_{24} = 0, & n_{25} = 0, & n_{26} = 0, & n_{27} = 4, \\
n_{31} = 1, & n_{32} = 0, & n_{33} = 0, & n_{34} = 0, & n_{35} = 0, & n_{36} = 0, & n_{37} = 0, \\
n_{41} = 0, & n_{42} = 0, & n_{43} = 0, & n_{44} = 0, & n_{45} = 0, & n_{46} = 0, & n_{47} = 1, \\
n_{51} = 21, & n_{52} = 1, & n_{53} = 0, & n_{54} = 1, & n_{55} = 0, & n_{56} = 10, & n_{57} = 10, \\
n_{61} = 2, & n_{62} = 0, & n_{63} = 0, & n_{64} = 0, & n_{65} = 14, & n_{66} = 0, & n_{67} = 5, \\
n_{71} = 17, & n_{72} = 2, & n_{73} = 0, & n_{74} = 0, & n_{75} = 7, & n_{76} = 6, & n_{77} = 0,
\end{array}$$

of the number of pipeline system operation process transitions from the state z_b into the state z_l during the experiment time $\Theta = 329$ days

$$[n_{bl}] = \begin{bmatrix} 0 & 1 & 1 & 0 & 24 & 5 & 14 \\ 1 & 0 & 0 & 0 & 0 & 0 & 4 \\ 1 & 0 & 0 & 0 & 0 & 0 & 0 \\ 0 & 0 & 0 & 0 & 0 & 0 & 1 \\ 21 & 1 & 0 & 1 & 0 & 10 & 10 \\ 2 & 0 & 0 & 0 & 14 & 0 & 5 \\ 17 & 2 & 0 & 0 & 7 & 6 & 0 \end{bmatrix},$$

- the vector of realizations

$$n_1 = n_{11} + n_{12} + n_{13} + n_{14} + n_{15} + n_{16} + n_{17} = 45,$$
$$n_2 = n_{21} + n_{22} + n_{23} + n_{24} + n_{25} + n_{26} + n_{27} = 5,$$
$$n_3 = n_{31} + n_{32} + n_{33} + n_{34} + n_{35} + n_{36} + n_{37} = 1,$$
$$n_4 = n_{41} + n_{42} + n_{43} + n_{44} + n_{45} + n_{46} + n_{47} = 1,$$
$$n_5 = n_{51} + n_{52} + n_{53} + n_{54} + n_{55} + n_{56} + n_{57} = 43,$$
$$n_6 = n_{61} + n_{62} + n_{63} + n_{64} + n_{65} + n_{66} + n_{67} = 21,$$
$$n_7 = n_{71} + n_{72} + n_{73} + n_{74} + n_{75} + n_{76} + n_{77} = 32,$$

(the sums of the number of matrix $[n_{bl}]$ rows) of the total numbers of the pipeline system operation process transitions from the operation state z_b during the experiment time $\Theta = 329$ days

$$[n_b] = [45, 5, 1, 1, 43, 21, 32].$$

The collected statistical data necessary for evaluating the unknown parameters of the distributions of the conditional sojourn times of the port oil piping transportation system operation process at the particular operation states are:

- the numbers n_{bl}, $b, l = 1, 2, \ldots, 7$, of the realizations θ_{bl}^k of the conditional sojourn times θ_{bl} of the port oil piping transportation system operation process at the operation state z_b when the next transition is to the operation state z_l;
- the realizations θ_{bl}^k, $k = 1,2, \ldots, n_{bl}$, of the conditional sojourn times θ_{bl} of the port oil piping transportation system operation process at the operation state z_b when the next transition is to the operation state z_l.

For instance, the statistical data for the sojourn time θ_{15} collected by the experts operating the piping system are as follows:

- the number of realizations $n_{15} = 24$,
- the realizations: $\theta_{15}^1 = 930$, $\theta_{15}^2 = 3840$, $\theta_{15}^3 = 1290$, $\theta_{15}^4 = 480$, $\theta_{15}^5 = 5575$, $\theta_{15}^6 = 4680$, $\theta_{15}^7 = 4350$, $\theta_{15}^8 = 2100$, $\theta_{15}^9 = 840$, $\theta_{15}^{10} = 2460$, $\theta_{15}^{11} = 1560$, $\theta_{15}^{12} = 1020$, $\theta_{15}^{13} = 1860$, $\theta_{15}^{14} = 960$, $\theta_{15}^{15} = 930$, $\theta_{15}^{16} = 910$, $\theta_{15}^{17} = 480$, $\theta_{15}^{18} = 410$, $\theta_{15}^{19} = 960$, $\theta_{15}^{20} = 480$, $\theta_{15}^{21} = 1440$, $\theta_{15}^{22} = 4710$, $\theta_{15}^{23} = 540$, $\theta_{15}^{24} = 5180$.

4.3.2.2 Evaluating Basic Parameters of Port Oil Piping Transportation System Operation Process

On the basis of the statistical data from Sect. 4.3.2.1, using respectively the formulae (4.1–4.3) and (4.4–4.6) given in Sect. 4.2.2, it is possible to evaluate the following unknown basic parameters of the port oil piping transportation system operation process:

- the vector

$$[p(0)] = [0.34, 0.05, 0, 0, 0.23, 0.19, 0.19]$$

  of the initial probabilities $p_b(0)$, $b = 1, 2, \ldots 7$, of the pipeline system operation process stay at the particular states z_b at the $t = 0$,
- the matrix

$$[p_{bl}] = \begin{bmatrix} 0 & 0.022 & 0.022 & 0 & 0.534 & 0.111 & 0.311 \\ 0.2 & 0 & 0 & 0 & 0 & 0 & 0.8 \\ 1 & 0 & 0 & 0 & 0 & 0 & 0 \\ 0 & 0 & 0 & 0 & 0 & 0 & 1 \\ 0.488 & 0.023 & 0 & 0.023 & 0 & 0.233 & 0.233 \\ 0.095 & 0 & 0 & 0 & 0.667 & 0 & 0.238 \\ 0.531 & 0.062 & 0 & 0 & 0.219 & 0.188 & 0 \end{bmatrix}$$

  of the probabilities p_{bl}, $b, l = 1, 2, \ldots, 7$, of transitions of the pipeline system operation process from the operation state z_b into the operation state z_l.

The values of some probabilities existing in the vector $[p(0)]$ and in the matrix $[p_{bl}]$, besides of those standing on the main diagonal, equal to zero do not mean that the events they are concerned with, cannot appear. They are evaluated on the basis of real statistical data and their values may change and become more precise if the duration of the experiment is longer.

4.3.2.3 Evaluating Parameters of Distributions of Port Oil Piping Transportation System Conditional Sojourn Times at Operation States

On the basis of the statistical data partly presented in Sect. 4.3.2.1, using the procedure and the formulae given in Sect. 4.2.3, it is possible to determine the empirical parameters of the conditional sojourn times of the pipeline system operation process at the particular operation states. To illustrate the application of this procedure and these formulae, we perform it for θ_{15}, that is one of the conditional sojourn times having the most populous set of realizations listed at the end of Sect. 4.3.2.1.

The results for the conditional sojourn time θ_{15} are:

- the realization $\bar{\theta}_{15}$ of the defined by (4.7) mean value of the conditional sojourn time θ_{15} of the pipeline system operation process at the operation state z_1 when the next transition is to the operation state z_5

$$\bar{\theta}_{15} = \frac{1}{24}\sum_{k=1}^{24}\theta_{15}^k = 1999.4,$$

- the number $\bar{r}_{15}$ of the disjoint intervals $I_j = \langle a_{15}^j, b_{15}^j), \ j = 1, 2, \ldots, \bar{r}_{15}$, that include the realizations $\theta_{15}^k, \ k = 1, 2, \ldots, 24$, of the conditional sojourn times θ_{15} at the operation state z_1 when the next transition is to the operation state z_5

$$\bar{r}_{15} \cong \sqrt{24} \cong 5,$$

- the length d_{15} of the intervals $I_j = \langle a_{15}^j, b_{15}^j), \ j = 1, 2, \ldots, 5$, which after considering

$$\bar{R}_{15} = \max_{1 \le k \le 24} \theta_{15}^k - \min_{1 \le k \le 24} \theta_{15}^k = 5575 - 410 = 5165,$$

is

$$d_{15} = \frac{\bar{R}_{15}}{\bar{r}_{15} - 1} = \frac{5165}{4} = 1291,$$

- the ends $a_{15}^j, \ b_{15}^j$, of the intervals $I_j = \langle a_{15}^j, b_{15}^j), \ j = 1, 2, \ldots, 5$, which after considering

$$\min_{1 \le k \le 24} \theta_{15}^k - \frac{d_{15}}{2} = 410 - \frac{1291}{2} = -235.5,$$

are

$$\begin{aligned}
a_{15}^1 = \max\{-235.5, \ 0\} = 0, \quad & b_{15}^1 = a_{15}^1 + 1291 = 0 + 1291 = 1291, \\
a_{15}^2 = b_{15}^1 = 1291, \quad & b_{15}^2 = a_{15}^1 + 2 \cdot 1291 = 0 + 2582 = 2582, \\
a_{15}^3 = b_{15}^2 = 2582, \quad & b_{15}^3 = a_{15}^1 + 3 \cdot 1291 = 0 + 3873 = 3873, \\
a_{15}^4 = b_{15}^3 = 3873, \quad & b_{15}^4 = a_{15}^1 + 4 \cdot 1291 = 0 + 5164 = 5164, \\
a_{15}^5 = b_{15}^4 = 5164, \quad & b_{15}^5 = a_{15}^1 + 5 \cdot 1291 = 0 + 6455 = 6455,
\end{aligned}$$

- the number n_{15}^j of the realizations θ_{15}^k in particular intervals $I_j = \langle a_{15}^j, b_{15}^j)$, $j = 1, 2, \ldots, 5$,

$$n_{15}^1 = 13, \quad n_{15}^2 = 5, \quad n_{15}^3 = 1, \quad n_{15}^4 = 4, \quad n_{15}^5 = 1.$$

4.3.2.4 Identification of Distribution Functions of Port Oil Piping Transportation System Conditional Sojourn Times at Operation States

Using the procedure given in Sect. 4.2.4 and the statistical data from Sect. 4.3.2.1 and the results from Sect. 4.3.2.3, we may verify the hypotheses on the distributions of the pipeline system operation process conditional sojourn times θ_{bl}, $b, l = 1, 2, \ldots, 7, b \neq l$, at the particular operation states. To do this, we need a sufficient number of realizations of these variables [2, 4, 23, 26, 27], namely, the sets of their realizations should contain at least 30 realizations coming from the experiment. This condition is not satisfied for the statistical data we have in disposal and that are presented in Sect. 4.3.2.1. However, to make the procedure familiar to the reader, we perform it for the conditional sojourn time θ_{15}, the one having the most numerous set of realizations and preliminarily analyzed in Sect. 4.3.2.3.

The realization $\bar{h}_{15}(t)$ of the histogram of the pipeline system operation process conditional sojourn time θ_{15}, defined by (4.45), is presented in Table 4.1 and illustrated in Fig. 4.4.

After analyzing and comparing the realization $\bar{h}_{15}(t)$ of the histogram with the graphs of the density functions $h_{bl}(t)$ of the previously distinguished distributions in Chap. 2, we formulate the null hypothesis H_0 in the following form:

H_0 : The pipeline system operation process conditional sojourn time θ_{15} at the operation state z_1 when the next transition is to the operation state z_5, has the chimney distribution with the density function defined by (2.11) of the form

$$h_{15}(t) = \begin{cases} 0, & t < x_{15} \\ \dfrac{A_{15}}{z_{15}^1 - x_{15}} & x_{15} \le t \le z_{15}^1 \\ \dfrac{C_{15}}{z_{15}^2 - z_{15}^1}, & z_{15}^1 \le t \le z_{15}^2 \\ \dfrac{D_{15}}{y_{15} - z_{15}^2}, & z_{15}^2 \le t \le y_{15} \\ 0, & t > y_{15}. \end{cases} \tag{4.72}$$

Since, according to (4.16–4.17), we have

$$\hat{n}_{15} = \max_{1 \le j \le 5}\{ n_{15}^j\} = 13 \quad \text{and} \quad n_{15}^1 = \hat{n}_{15} = 13,$$

Table 4.1 The realization of the histogram of the pipeline system operation process conditional sojourn time θ_{15}

Histogram of the conditional sojourn time θ_{15}					
$I_j = \langle a_{15}^j, b_{15}^j)$	0–1291	1291–2582	2582–3873	3873–5164	5164–6455
n_{15}^j	13	5	1	4	1
$\bar{h}_{15}(t) = n_{15}^j / n_{15}$	13/24	5/24	1/24	4/24	1/24

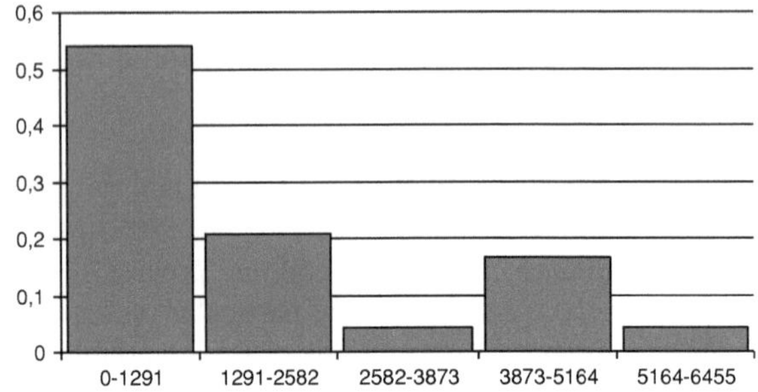

Fig. 4.4 The graph of the histogram of the pipeline system operation process conditional sojourn time θ_{15}

then $i = 1$. Moreover, by (4.22), we get

$$n_{15}^2 = 5 \neq 0 \quad \text{and} \quad \frac{n_{15}^1}{n_{15}^2} = \frac{13}{5} < 3.$$

Therefore, we estimate the unknown parameters of the density function of the hypothetical chimney distribution using the formulae (4.15), (4.20) and (4.21) and obtain the following results

$$x_{15} = a_{15}^1 = 0, \quad y_{15} = x_{15} + \bar{r}_{15} d_{15} = 0 + 5 \times 1291 = 6455,$$
$$z_{15}^1 = x_{15} + (1-1)d_{15} = 0, \quad z_{15}^2 = x_{15} + (1+1)d_{15} = 0 + 2 \times 1291 = 2582,$$
$$A_{15} = 0, \; C_{15} = \frac{n_{15}^1 + n_{15}^2}{n_{15}} = \frac{13+5}{24} = 0.75, \quad D_{15} = \frac{n_{15}^3 + n_{15}^4 + n_{15}^5}{n_{15}} = \frac{1+4+1}{24} = 0.25.$$

Substituting the above results into (4.72), we get completely defined the hypothetical density function in the form

$$h_{15}(t) = \begin{cases} 0, & t < 0 \\ \dfrac{0.75}{2582-0}, & 0 \le t < 2582 \\ \dfrac{0.25}{6455-2582}, & 2582 \le t < 6455 \\ 0, & t \ge 6455 \end{cases}$$

$$= \begin{cases} 0, & t < 0 \\ 0.000290473, & 0 \le t < 2582 \\ 0.000064549, & 2582 \le t < 6455 \\ 1, & t \ge 6455. \end{cases} \tag{4.73}$$

Hence the hypothetical distribution function $H_{15}(t)$ of the conditional sojourn time θ_{15}, after taking the integral of the hypothetical density function $h_{15}(t)$ given by (4.73), takes the following form

$$H_{15}(t) = \int_0^t h_{15}(t)dt = \begin{cases} 0, & t<0 \\ 0.000290473t, & 0 \le t < 2582 \\ 0.000064549t + 0.583336205, & 2582 \le t < 6455 \\ 1, & t \ge 6455. \end{cases} \tag{4.74}$$

Next, we join the intervals defined in the realization of the histogram $h_{15}(t)$ that have the number n_{15}^j, of realizations less than 4 into new intervals and we perform the following steps:

- fix the new number of intervals $\bar{\bar{r}}_{15} = 3$,
- determine the new intervals

$$\bar{I}_1 = \langle 0,\ 1291), \quad \bar{I}_2 = \langle 1291,\ 2582), \quad \bar{I}_3 = \langle 2582,\ 6455),$$

- fix the number of realizations in the new intervals

$$\bar{n}_{15}^1 = 13, \quad \bar{n}_{15}^2 = 5, \quad \bar{n}_{15}^3 = 6,$$

- calculate, using (4.46), the hypothetical probabilities that the variable θ_{15} takes values from the new intervals

$$p_1 = P(\theta_{15} \in \bar{I}_1) = P(0 \le \theta_{15} < 1291) = H_{15}(1291) - H_{15}(0) = 0.375 - 0 = 0.375,$$

$$p_2 = P(\theta_{15} \in \bar{I}_2) = P(1291 \le \theta_{15} < 2582) = H_{15}(2582) - H_{15}(1291) = 0.75 - 0.375 = 0.375,$$

$$p_3 = P(\theta_{15} \in \bar{I}_3) = P(2582 \le \theta_{15} < 6455) = H_{15}(6455) - H_{15}(2582) = 1 - 0.75 = 0.25,$$

- calculate, using (4.47), the realization of the χ^2(chi-square)-Pearson's statistics

$$u_{24} = \sum_{j=1}^{3} \frac{(\bar{n}_{15}^j - n_{15}p_j)^2}{n_{15}p_j} = \frac{(13 - 24 \times 0.375)^2}{24 \times 0.375} + \frac{(5 - 24 \times 0.375)^2}{24 \times 0.375} + \frac{(6 - 24 \times 0.25)^2}{24 \times 0.25}$$
$$\cong 1.78 + 1.78 + 0 = 3.56,$$

- assume the significance level $\alpha = 0.05$,
- fix the number of degrees of freedom

$$\bar{\bar{r}}_{15} - l - 1 = 3 - 0 - 1 = 2,$$

- read from the tables of the χ^2-Pearson's distribution the value u_α for the fixed values of the significance level $\alpha = 0.05$ and the number of degrees of freedom $\bar{\bar{r}}_{15} - l = 1 = 2$, such that, according to (4.48), the following equality holds

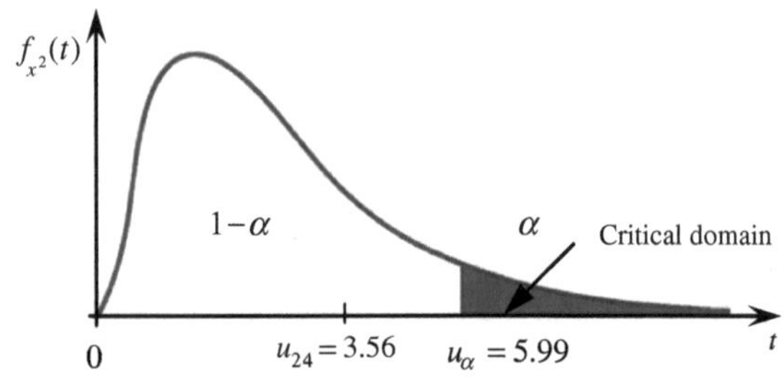

Fig. 4.5 The graphical interpretation of the critical interval and the acceptance interval for the chi-square goodness-of-fit test

$$P(U_{24} > u_\alpha) = \alpha = 0.05,$$

this amounts to $u_\alpha = 5.99$ and we determine the critical domain in the form of the interval $(5.99, +\infty)$and the acceptance domain in the form of the interval (Fig. 4.5) $\langle 0, 5.99\rangle$

- we compare the obtained value $u_{24} = 3.56$ of the realization of the statistics U_{24} with that read from the tables, critical value $u_\alpha = 5.99$ of the chi-square random variable and since the value $u_{24} = 3.56$ does not belong to the critical domain, i.e.

$$u_{24} = 3.56 \leq u_\alpha = 5.99,$$

we do not reject the hypothesis H_0.

After getting such a result, in the case that have enough statistical data, we may assume that the sojourn time θ_{15} has the chimney distribution with the density function given by (4.73). Otherwise, if the null hypothesis H_0 is rejected, we should select other density function from the distinguished distributions and repeat the procedure of testing.

If we decide to accept the density function $h_{15}(t)$ of the conditional sojourn time θ_{15} given by (4.73), then either applying the general formula (2.12) for mean value or applying the particular formula (2.11) for chimney distribution, it is possible to find its mean value evaluation:

$$M_{15} = \int_0^\infty t h_{15}(t)dt = \frac{1}{2}[A_{15}(x_{15} + z_{15}^1) + C_{15}(z_{15}^1 + z_{15}^2) + D_{15}(z_{15}^2 + y_{15})]$$

$$= \frac{1}{2}[0(0+0) + 0.75(0 + 2582) + 0.25(2582 + 6455)] \cong 2098.$$

Because of the lack of sufficient number of realizations of the piping system operation process conditional sojourn times at the operation states, it is not possible to identify statistically their distributions. In those cases of not identified distributions, using formula (4.7), it is possible to find the approximate empirical values of the mean values $M_{bl} = E[\theta_{bl}]$ of the conditional sojourn times at the particular operation states that are as follow:

$$M_{12}=1920,\quad M_{13}=480,\quad M_{15}=1999.4,\quad M_{16}=1250,\quad M_{17}=1129.6,$$
$$M_{21}=9960,\quad M_{27}=810,\quad M_{31}=575,\quad M_{47}=380,\quad M_{51}=874.7,\quad M_{52}=480,$$
$$M_{54}=300,\quad M_{56}=436.3,\quad M_{57}=1042.5,\quad M_{61}=325,\quad M_{65}=510.7,$$
$$M_{67}=438,\quad M_{71}=874.1,\quad M_{72}=510,\quad M_{75}=2585.7,\quad M_{76}=2380.$$

These approximate values are used in Sect. 2.3.2 in the evaluation and prediction of the characteristics of the port oil piping transportation system operation process.

As there are no realizations of the port oil piping transportation system operation process conditional sojourn times at the operation states

$$\theta_{14},\ \theta_{23},\ \theta_{24},\ \theta_{25},\ \theta_{26},\ \theta_{32},\ \theta_{34},\ \theta_{35},\ \theta_{36},\ \theta_{37},\ \theta_{41},\ \theta_{42},\ \theta_{43},\ \theta_{45},$$
$$\theta_{46},\ \theta_{53},\ \theta_{62},\ \theta_{63},\ \theta_{64},\ \theta_{73},\ \theta_{74},$$

it is impossible to estimate their empirical conditional mean values

$$M_{14},\quad M_{23},\quad M_{24},\quad M_{25},\quad M_{26},\quad M_{32},\quad M_{34},\quad M_{35},\quad M_{36},\quad M_{37},\quad M_{41},$$
$$M_{42},\quad M_{43}, M_{45},\quad M_{46},\quad M_{53},\quad M_{62},\quad M_{63},\quad M_{64},\quad M_{73},\quad M_{74},$$

4.3.3 Statistical Identification of Maritime Ferry Technical System Operation Process

On the basis of expert opinion concerning the operation process of the considered maritime ferry technical system, in Sect. 2.3.3, the number of system operation process states $v = 18$ is fixed and the operation states z_b, $b = 1, 2, \ldots, 18$, are defined. Moreover, it is fixed that there are possible only transitions between the neighboring system operation states, i.e., from the operation states z_b to the operation states z_{b+1}, $b = 1, 2, \ldots, 17$, and from the operation state z_{18} to the operation state z_1.

4.3.3.1 Defining Parameters and Data Collection of Ferry Technical System Operation Process

The unknown parameters of the system operation process semi-Markov model are:

- the initial probabilities $p_b(0)$, $b = 1, 2, \ldots, 18$, of the ferry technical system operation process stay at the particular states z_b at the moment $t = 0$,
- the probabilities $p_{b\,b+1}$, $b = 1, 2, \ldots, 17$, and p_{181} of the ferry technical system operation process transitions from the operation state z_b into the operation state z_{b+1} and from the operation state z_{18} into the operation state z_1,

- the distributions of the ferry technical system operation process conditional sojourn times $\theta_{b\,b+1}$, $b = 1, 2, \ldots, 17$, and θ_{181} at the particular operation states and their mean values $M_{b\,b+1} = E[\theta_{b\,b+1}]$, $b = 1, 2, \ldots, 17$, and $M_{181} = E[\theta_{181}]$.

To identify all these parameters of the ferry technical system operation process the statistical data about this process is needed. The statistical data collected during the ferry spring voyages are partly given below. The collected spring statistical data necessary to evaluate the probabilities of straying of the ferry technical system operation process at the particular states at the initial moment $t = 0$, are:

- the ferry technical system operation process observation/experiment time

$$\Theta = 42 \text{ days},$$

- the number the ferry operation process realizations

$$n(0) = 42,$$

- the vector of realizations

$$n_1(0) = 42,\ n_2(0) = 0,\ \ldots,\ n_{18}(0) = 0,$$

of the number of ferry technical system operation process transients at the particular operation states z_b at the initial moment $t = 0$

$$[n_b(0)] = [42, 0, \ldots, 0].$$

The collected statistical data necessary to evaluate the probabilities of transitions of the ferry technical system operation process between the operation states are:

- the matrix of realizations

$$n_{11} = 0,\ n_{12} = 42,\ n_{13} = 0,\ \ldots,\ n_{117} = 0,\ n_{118} = 0,$$
$$n_{21} = 0,\ n_{22} = 0,\ n_{23} = 42, \ldots,\ n_{217} = 0,\ n_{218} = 0,$$
$$\ldots$$
$$n_{171} = 0,\ n_{172} = 0,\ n_{173} = 0,\ \ldots,\ n_{1717} = 0,\ n_{1718} = 42,$$
$$n_{181} = 42,\ n_{182} = 0,\ n_{183} = 0,\ \ldots,\ n_{1817} = 0,\ n_{1818} = 0,$$

of the number of ferry technical system operation process transitions from the state z_b into the state z_l during the experiment time $\Theta = 42$ days

$$[n_{bl}] = \begin{bmatrix} 0\ 42\ \ 0 \ldots 0\ \ 0 \\ 0\ \ 0\ \ 42 \ldots 0\ \ 0 \\ \ldots \\ 0\ \ 0\ \ 0 \ldots 0\ 42 \\ 42\ \ 0\ \ 0 \ldots 0\ \ 0 \end{bmatrix},$$

- the matrix of realizations

$$n_1 = n_{11} + n_{12} + \cdots + n_{118} = 42,$$
$$n_2 = n_{21} + n_{22} + \cdots + n_{218} = 42,$$
$$\cdots$$
$$n_{17} = n_{171} + n_{172} + \cdots + n_{1718} = 42,$$
$$n_{18} = n_{181} + n_{182} + \cdots + n_{1818} = 42,$$

(the sums of the number of matrix $[n_{bl}]$ rows) of the total number of ferry technical system operation process transitions from the operation state z_b during the experiment time $\Theta = 42$ days

$$[n_b] = [42,\ 42,\ \ldots,\ 42].$$

The statistical data necessary collected by experts to evaluate the unknown parameters of the distributions of the conditional sojourn times of the ferry technical system operation process at the particular operation states are:

- the numbers $n_{b\,b+1}$, $b = 1, 2, \ldots, 17$, of the realizations $\theta^k_{b\,b+1}$ of the conditional sojourn times $\theta_{b\,b+1}$ of the ferry technical system operation process at the operation state z_b when the next transition is to the operation state z_{b+1} and the number n_{181} of the realizations θ^k_{181} of the conditional sojourn times θ_{181} of the ferry technical system operation process at the operation state z_{18} when the next transition is to the operation state z_1;
- the realizations $\theta^k_{b\,b+1}$, $k = 1,2, \ldots, n_{b\,b+1}$, $b = 1, 2, \ldots, 17$, of the conditional sojourn times $\theta_{b\,b+1}$ of the ferry technical system operation process at the operation state z_b when the next transition is to the operation state z_{b+1} and the realizations θ^k_{181}, $k = 1,2, \ldots, n_{181}$, of the conditional sojourn times θ_{181} of the ferry technical system operation process at the operation state z_{18} when the next transition is to the operation state z_1.

The exemplary statistical data collected by experts for two of the conditional sojourn times, namely for θ_{12} and θ_{23}, are presented below.

The statistical data for the sojourn time θ_{12} collected by the experts operating the ferry technical system are as follows:

- the number of realizations $n_{12} = 42$;
- the realizations: 55, 52, 47, 75, 60, 60, 62, 43, 50, 61, 65, 63, 45, 45, 40, 20, 33, 50, 43, 15, 45, 57, 97, 68, 58, 35, 45, 75, 72, 62, 37, 44, 46, 78, 59, 65, 53, 25, 55, 84, 71, 67.

The statistical data for the sojourn time θ_{23} collected by the experts operating the ferry technical system are as follows:

- the number of realizations $n_{23} = 42$;
- the realizations: 4, 3, 3, 2, 2, 2, 2, 3, 3, 4, 3, 2, 2, 2, 2, 2, 2, 3, 2, 2, 3, 2, 2, 3, 3, 4, 3, 3, 2, 3, 6, 3, 2, 2, 2, 2, 2, 2, 3, 2, 2, 2.

4.3.3.2 Evaluating Basic Parameters of Ferry Technical System Operation Process

On the basis of the statistical data from Sect. 4.3.3.1, using the formulae (4.1–4.3) and respectively (4.4–4.6) given in Sect. 4.2.2, it is possible to evaluate the following basic parameters of the ferry technical system operation process:

- the vector

$$[p(0)] = [1,\ 0,\ 0,\ \ldots,\ 0],$$

of the initial probabilities $p_b(0)$, $b = 1,\ 2,\ \ldots,\ 18$, of the ferry technical system operation process stay at the particular operation states z_b at the moment $t = 0$,
- the matrix

$$[p_{bl}] = \begin{bmatrix} 0\,1\,0 \ldots 0\,0 \\ 0\,0\,1 \ldots 0\,0 \\ \ldots \\ 0\,0\,0 \ldots 0\,1 \\ 1\,0\,0 \ldots 0\,0 \end{bmatrix},$$

of the probabilities p_{bl}, $b, l = 1,\ 2,\ \ldots, 18$, of the ferry technical system operation process transitions from the operation state z_b into the operation state z_l.

4.3.3.3 Evaluating Parameters of Distributions of Ferry Technical System Conditional Sojourn Times at Operation States

On the basis of the statistical data partly presented in Sect. 4.3.3.1, using the formulae given in Sect. 4.2.3, it is possible to determine the empirical parameters of the conditional sojourn times of the ferry technical system operation process at the particular operation states. To illustrate this procedure and the formulae application, we perform it for two of the conditional sojourn times, namely θ_{12} and θ_{23}.

The results for the ferry technical system operation process conditional sojourn time θ_{12} are:

- the realization $\bar{\theta}_{12}$ of the defined by (4.7) mean value of the conditional sojourn times θ_{12} of the ferry technical system operation process at the operation state z_1 when the next transition is to the operation state z_2

$$\bar{\theta}_{12} = \frac{1}{42} \sum_{k=1}^{42} \theta_{12}^k = 54.3, \tag{4.75}$$

- the number $\bar{r}_{12}$ of disjoint intervals $I_j = \langle a_{12}^j, b_{12}^j), \ j = 1, 2, \ldots, \bar{r}_{12}$, that include the realizations θ_{12}^k, $k = 1, 2, \ldots, 42$, of the conditional sojourn times θ_{12} at the operation state z_1 when the next transition is to the operation state z_2

$$\bar{r}_{12} = \sqrt{n_{12}} = \sqrt{42} \cong 6,$$

- the length d_{12} of the intervals $I_j = \langle a_{12}^j, b_{12}^j), j = 1, 2, \ldots, 6$, which after considering

$$\bar{R}_{12} = \max_{1 \le j \le 42} \theta_{12}^k - \min_{1 \le j \le 42} \theta_{12}^k = 97 - 15 = 82,$$

is

$$d_{12} = \frac{\bar{R}}{\bar{r}_{12} - 1} = \frac{82}{5} \cong 16,$$

- the ends a_{12}^j, b_{12}^j, of the intervals $I_j = \langle a_{12}^j, b_{12}^j), \ j = 1, 2, \ldots, 6$, which after considering

$$\min_{1 \le k \le 42} \theta_{12}^k - \frac{d_{12}}{2} = 15 - \frac{16}{2} = 15 - 8 = 7,$$

are

$$\begin{aligned}
a_{12}^1 &= \max\{7, 0\} = 7, \quad b_{12}^1 = a_{12}^1 + d = 7 + 16 = 23,\\
a_{12}^2 &= b_{12}^1 = 23, \quad b_{12}^2 = a_{12}^1 + 2 \times 16 = 7 + 32 = 39,\\
a_{12}^3 &= b_{12}^2 = 39, \quad b_{12}^3 = a_{12}^1 + 3 \times 16 = 7 + 48 = 55,\\
a_{12}^4 &= b_{12}^3 = 55, \quad b_{12}^3 = a_{12}^1 + 3 \times 16 = 7 + 48 = 55,\\
a_{12}^5 &= b_{12}^4 = 71, \quad b_{12}^5 = a_{12}^1 + 5 \times 16 = 7 + 80 = 87,\\
a_{12}^6 &= b_{12}^5 = 87, \quad b_{12}^6 = a_{12}^1 + 6 \times 16 = 7 + 96 = 103,
\end{aligned}$$

- the number n_{12}^j of the realizations θ_{12}^k in particular intervals $I_j = \langle a_{12}^j, b_{12}^j), j = 1, 2, \ldots, 6$,

$$n_{12}^1 = 2, \quad n_{12}^2 = 4, \quad n_{12}^3 = 14, \quad n_{12}^4 = 15, \quad n_{12}^5 = 6, \quad n_{12}^6 = 1.$$

The results for the ferry technical system operation process conditional sojourn time θ_{23} are:

- the realization of the mean value $\bar{\theta}_{23}$ of the conditional sojourn times θ_{23} of the ferry technical system operation process at the operation state z_2 when the next transition is to the operation state z_3

$$\bar{\theta}_{23} = \frac{1}{42} \sum_{k=1}^{42} \theta_{23}^k = 2.6, \quad b, l = 1, 2, \ldots, 18, \quad b \neq l, \tag{4.76}$$

- the number $\bar{r}_{23}$ of the disjoint intervals $I_j = \langle a_{23}^j, b_{23}^j), j = 1, 2, \ldots, \bar{r}_{23}$, which include the realizations θ_{23}^k, $k = 1, 2, \ldots, 42$, of the conditional sojourn times θ_{23} at the operation state z_2 when the next transition is to the operation state z_3

$$\bar{r}_{23} = \sqrt{42} \cong 6,$$

- the length d_{23} of the intervals $I_j = \langle a_{23}^j, b_{23}^j)$, $j = 1, 2, \ldots, 6$, which after considering

$$\bar{R}_{23} = \max_{l \le k \le 42} \theta_{12}^k - \min_{l \le k \le 42} \theta_{23}^k = 6 - 2 = 4,$$

is

$$d_{23} = \frac{\bar{R}_{23}}{\bar{r}_{23} - 1} = \frac{4}{5} = 0.8,$$

- the ends a_{23}^j, b_{23}^j, of the intervals $I_j = \langle a_{23}^j, b_{23}^j)$, $j = 1, 2, \ldots, 6$, which after considering

$$\min_{1 \le k \le 42} \theta_{23}^k - \frac{d_{23}}{2} = 2 - \frac{0.8}{2} = 2 - 0.4 = 1.6,$$

are

$$a_{23}^1 = \max\{1.6, 0\} = 1.6, \quad b_{23}^1 = a_{23}^1 + 0.8 = 1.6 + 0.8 = 2.4,$$
$$a_{23}^2 = b_{23}^1 = 2.4, \quad b_{23}^2 = a_{23}^1 + 2 \cdot 0.8 = 1.6 + 1.6 = 3.2,$$
$$a_{23}^3 = b_{23}^2 = 3.2, \quad b_{23}^3 = a_{23}^1 + 3 \cdot 0.8 = 1.6 + 2.4 = 4.0,$$
$$a_{23}^4 = b_{23}^3 = 4.0, \quad b_{23}^4 = a_{23}^1 + 4 \cdot 0.8 = 1.6 + 3.2 = 4.8,$$
$$a_{23}^5 = b_{23}^4 = 4.8, \quad b_{23}^5 = a_{23}^5 + 5 \cdot 0.8 = 1.6 + 4.0 = 5.6,$$
$$a_{23}^6 = b_{23}^5 = 5.6, \quad b_{23}^6 = a_{23}^6 + 6 \cdot 0.8 = 1.6 + 4.8 = 6.4,$$

- the number n_{23}^j of the realizations θ_{23}^k in particular intervals $I_j = \langle a_{23}^j, b_{23}^j)$, $j = 1, 2, \ldots, 6$,

$$n_{23}^1 = 24, \quad n_{23}^2 = 14, \quad n_{23}^3 = 0, \quad n_{23}^4 = 3, \quad n_{23}^5 = 0, \quad n_{23}^6 = 1.$$

4.3.3.4 Identification of Distribution Functions of Ferry Technical System Conditional Sojourn Times at Operation States

Using the procedure given in Sect. 4.2.4 and data from Sect. 4.3.3.1 and the results from Sect. 4.3.3.3, we may verify the hypotheses on the distributions of the ferry technical system conditional sojourn times θ_{bl}, $b, l = 1, 2, \ldots, 18$, $b \neq l$, in the particular operation states. To do this, we need a sufficient number of realizations of these variables [2, 4, 23, 26–27], namely the sets of their realizations should

Table 4.2 The realization of the histogram of the ferry technical system operation process conditional sojourn time θ_{12}

Histogram of the conditional sojourn time θ_{12}						
$I_j = \langle a_{bl}^j, b_{bl}^j)$	7–23	23–39	39–55	55–71	71–87	87–103
n_{12}^j	2	4	14	15	6	1
$\bar{h}_{12}(t) = n_{12}^j/n_{12}$	2/42	4/42	14/42	15/42	6/42	1/42

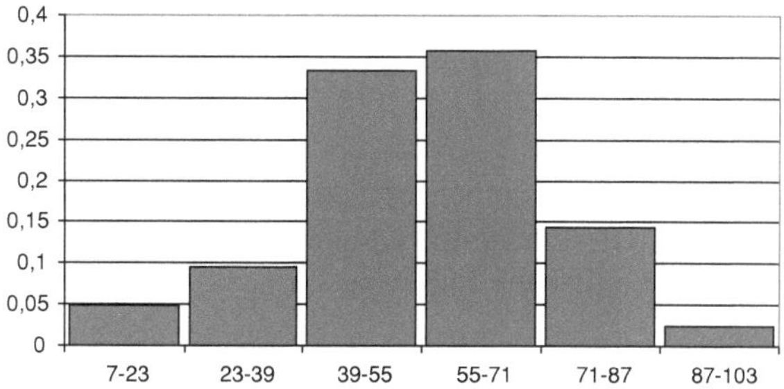

Fig. 4.6 The histogram of the ferry technical system operation process conditional sojourn time θ_{12}

contain at least 30 realizations coming from the experiment. This condition is satisfied for the statistical data we have at disposal and that are partly presented in Sect. 4.3.3.1. To make the procedure familiar to the reader, we perform it for the conditional sojourn times θ_{12} and θ_{23} preliminarily analyzed in Sect. 4.3.3.3.

The realization $\bar{h}_{12}(t)$ defined by (4.45) of the we get histogram of the ferry technical system operation process conditional sojourn time θ_{12} is presented in Table 4.2 and illustrated in Fig. 4.6.

After analyzing and comparing the realization $\bar{h}_{12}(t)$ of the histogram with the graphs of the density functions $h_{bl}(t)$ of the previously distinguished distributions in Chap. 2, we formulate the null hypothesis H_0 in the following form:

H_0: The ferry technical system operation time conditional sojourn time θ_{12} at the operation state z_1 when the next transition is to the operation state z_2, has the triangular distribution with the density function defined by (2.6) of the form

$$h_{12}(t) = \begin{cases} 0, & t < x_{12} \\ \frac{2}{y_{12} - x_{12}} \frac{t - x_{12}}{z_{12} - x_{12}}, & x_{12} \le t \le z_{12} \\ \frac{2}{y_{12} - x_{12}} \frac{y_{12} - t}{y_{12} - z_{12}}, & z_{12} \le t \le y_{12} \\ 0, & t > y_{12}. \end{cases} \tag{4.77}$$

We estimate the unknown parameters of the density function of the hypothetical triangular distribution using the formulac (4.9) and the evaluation (4.75) and we obtain the following results

$$x_{12} = a_{12}^1 = 7, \quad y_{12} = x_{12} + \bar{r}_{12} d_{12} = 7 + 6.16 = 103, \quad z_{12} = \bar{\theta}_{12} = 54.3.$$

Substituting the above results into (4.77), we completely define the hypothetical density function in the form

$$h_{12}(t) = \begin{cases} 0, & t<7 \\ \dfrac{2}{103-7}\dfrac{t-7}{54.3-7}, & 7 \le t < 54.3 \\ \dfrac{2}{103-7}\dfrac{103-t}{103-54.3}, & 54.3 \le t < 103 \\ 0, & t \ge 103 \end{cases}$$

$$= \begin{cases} 0, & t<7 \\ 0.000440451t - 0.003083157, & 7 \le t < 54.3 \\ -0.000427789t + 0.044062286, & 54.3 \le t < 103 \\ 0, & t \ge 103. \end{cases} \tag{4.78}$$

Hence, the hypothetical distribution function $H_{12}(t)$ of the conditional sojourn time θ_{12}, after taking the integral of the hypothetical density function $h_{12}(t)$ given by (4.78), takes the following form

$$H_{12}(t) = \int_0^t h_{12}(t)dt$$

$$= \begin{cases} 0, & t<7 \\ 0.0002202255t^2 - 0.003083157t + 0.0107910495, & 7 \le t < 54.3 \\ -0.0002138945t^2 + 0.044062286t - 1.2692080266 & 54.3 \le t < 103 \\ 1, & t \ge 103. \end{cases} \tag{4.79}$$

Next, we join the intervals defined in the realization of the histogram $h_{12}(t)$ which have the numbers n_{12}^j, of realizations less than 4 into new intervals and we perform the following steps:

- fix the new number of intervals $\bar{\bar{r}}_{12} = 4$,
- determine the new intervals

$$\bar{I}_1 = \langle 7, 39), \quad \bar{I}_2 = \langle 39, 55), \quad \bar{I}_3 = \langle 55, 71), \quad \bar{I}_4 = \langle 71, 103),$$

- fix the number of realizations in the new intervals

$$\bar{n}_{12}^1 = 6, \quad \bar{n}_{12}^2 = 14, \quad \bar{n}_{12}^3 = 15, \quad \bar{n}_{12}^4 = 7,$$

- calculate, using (4.46), the hypothetical probabilities that the variable θ_{12} takes values from the new intervals

$$p_1 = P(\theta_{12} \in \bar{I}_1) = P(7 \le \theta_{12} < 39) = H_{12}(39) - H_{12}(7)$$
$$= 0.2255 - 0 = 0.2255,$$

$$p_2 = P(\theta_{12} \in \bar{I}_2) = P(39 \le \theta_{12} < 55) = H_{12}(55) - H_{12}(39)$$
$$= 0.5072 - 0.2255 = 0.2817,$$

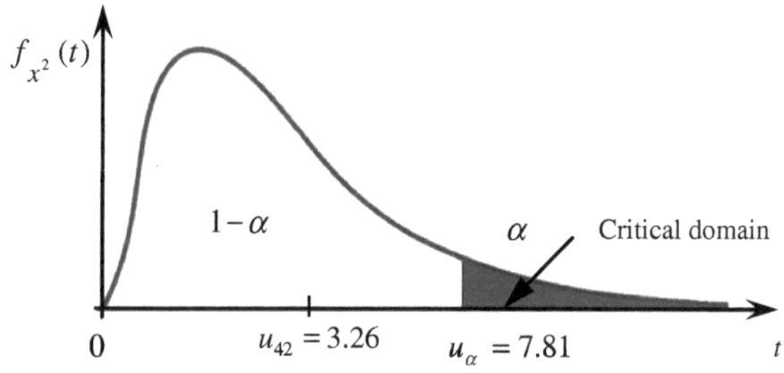

Fig. 4.7 The graphical interpretation of the critical interval and the acceptance interval for the chi-square goodness-of-fit test

$$p_3 = P(\theta_{12} \in \bar{I}_3) = P(55 \leq \theta_{12} < 71) = H_{12}(71) - H_{12}(55) = 0.7810 - 0.5072 = 0.2738$$

$$p_4 = P(\theta_{12} \in \bar{I}_4) = P(71 \leq \theta_{12} < 103) = H_{12}(103) - H_{12}(71) = 1 - 0.7810 = 0.2190,$$

- calculate, using (4.47), the realization of the χ^2(chi-square)-Pearson's statistics

$$u_{42} = \sum_{j=1}^{4} \frac{(\bar{n}_{12}^{j} - n_{12}p_j)^2}{n_{12}p_j} = \frac{(6 - 42 \times 0.2255)^2}{42 \times 0.2255} + \frac{(14 - 42 \times 0.2817)^2}{42 \times 0.2817} + \frac{(15 - 42 \times 0.2738)^2}{42 \times 0.2738} + \frac{(7 - 42 \times 0.2190)^2}{42 \times 0.2190} \cong 1.27 + 0.39 + 1.07 + 0.53 = 3.26,$$

- assume the significance level $\alpha = 0.05$,
- fix the number of degrees of freedom

$$\bar{\bar{r}}_{12} - l - 1 = 4 - 0 - 1 = 3,$$

- read from the tables of the χ^2-Pearson's distribution the value u_α for the fixed values of the significance level $\alpha = 0.05$ and the number of degrees of freedom $\bar{\bar{r}}_{12} - l = 1 = 3$, such that, according to (4.49), the following equality holds

$$P(U_{42} > u_\alpha) = \alpha = 0.05$$

that amounts $u_\alpha = 7.81$ and determine the critical domain in the form of the interval $(7.81, +\infty)$ and the acceptance domain in the form of the interval (Fig. 4.7) $\langle 0, 7.81 \rangle$
- we compare the obtained value $u_{42} = 3.26$ of the realization of the statistics U_{42} with the read from the tables, critical value $u_\alpha = 7.81$ of the chi-square

Table 4.3 The realization of the histogram of the ferry technical system operation process conditional sojourn time θ_{23}

Histogram of the conditional sojourn time θ_{23}						
$I_j = \langle a_{bl}^j, b_{bl}^j)$	1.6–2.4	2.4–3.2	3.2–4.0	4.0–4.8	4.8–5.6	5.6–6.4
n_{23}^j	24	14	0	3	0	1
$\bar{h}_{23}(t) = n_{23}^j/n_{23}$	24/42	14/42	0/42	3/42	0/42	1/42

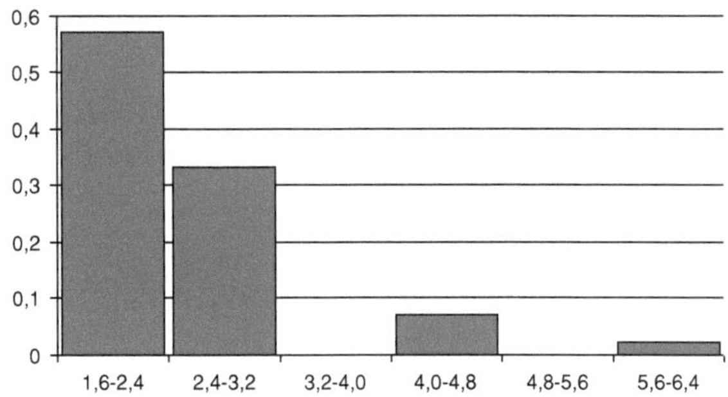

Fig. 4.8 The histogram of the ferry technical system operation process conditional sojourn time θ_{23}

random variable and since the value $u_{42} = 3.26$ does not belong to the critical domain, i.e.

$$u_{42} = 3.26 \leq u_{\alpha} = 7.81,$$

we do not reject the hypothesis H_0.

The realization $\bar{h}_{23}(t)$ of the histogram defined by (4.45) of the ferry technical system operation process conditional sojourn time θ_{23} is presented in Table 4.3 and illustrated in Fig. 4.8.

After analyzing and comparing the realization $\bar{h}_{23}(t)$ of the histogram with the graphs of the density functions $h_{bl}(t)$ of the previously distinguished distributions in Chap. 2, we formulate the null hypothesis H_0 in the following form:

H_0: The ferry technical system operation process conditional sojourn time θ_{23} at the operation state z_2 when the next transition is to the operation state z_3, has the exponential distribution with the density function defined by (2.9) of the form

$$h_{23}(t) = \begin{cases} 0, & t < x_{23} \\ \alpha_{23} \exp[-\alpha_{23}(t - x_{23})], & t \geq x_{23}, \end{cases} \tag{4.80}$$

We estimate the unknown parameters of the density function of the hypothetical exponential distribution using the formulae (4.13) and the evaluation (4.76) and obtain the following results

$$x_{23} = a_{23}^1 = 1.6, \quad \alpha_{23} = \frac{1}{\bar{\theta}_{23} - x_{23}} = \frac{1}{2.6 - 1.6} = 1$$

Substituting the above results into (4.80), we completely define the hypothetical density function in the form

$$h_{23}(t) = \begin{cases} 0, & t < 1.6 \\ 1\exp[-1(t-1.6)]. & t \geq 1.6 \end{cases} = \begin{cases} 0, & t < 1.6 \\ \exp[-t+1.6], & t \geq 1.6. \end{cases} \tag{4.81}$$

Hence, the hypothetical distribution function $H_{23}(t)$ of the conditional sojourn time θ_{23}, after taking the integral of the hypothetical density function $h_{23}(t)$ given by (4.81), takes the following form

$$H_{23}(t) = \int_0^t h_{23}(t)dt = \begin{cases} 0, & t < 1.6 \\ 1 - \exp[-t+1.6], & t \geq 1.6. \end{cases} \tag{4.82}$$

Next, we join the intervals defined in the realization of the histogram $h_{23}(t)$ which have the numbers n_{23}^j, of realizations less than 4 into new intervals and we perform the following steps:

- fix the new number of intervals $\bar{\bar{r}}_{23} = 3$,
- determine the new intervals

$$\bar{I}_1 = \langle 1.6, 2.4), \quad \bar{I}_2 = \langle 2.4, 3.2), \quad \bar{I}_3 = \langle 3.2, \infty),$$

- fix the numbers of realizations in the new intervals

$$\bar{n}_{23}^1 = 24, \quad \bar{n}_{23}^2 = 14, \quad \bar{n}_{23}^3 = 4,$$

- calculate, using (4.46), the hypothetical probabilities that the variable θ_{23} takes values from the new intervals

$$\begin{aligned} p_1 &= P(\theta_{23} \in \bar{I}_1) = P(1.6 \leq \theta_{23} < 2.4) = H_{23}(2.4) - H_{23}(1.6) \\ &= (1 - \exp[-2.4+1.6]) - (1 - \exp[-1.6+1.6]) = 1 - \exp[-0.8] \\ &= 1 - 0.4493 = 0.5507, \end{aligned}$$

$$\begin{aligned} p_2 &= P(\theta_{23} \in \bar{I}_2) = P(2.4 \leq \theta_{23} < 3.2) = H_{23}(3.2) - H_{23}(2.4) \\ &= (1 - \exp[-3.2+1.6]) - (1 - \exp[-2.4+1.6]) = -\exp[-1.6] + \exp[-0.8] \\ &= -0.2019 + 0.4493 = 0.2474, \end{aligned}$$

$$\begin{aligned} p_3 &= P(\theta_{23} \in \bar{I}_3) = P(2.4 \leq \theta_{23} < \infty) = 1 - (p_1 + p_2) \\ &= 1 - (0.5507 + 0.2474) = 0.2019, \end{aligned}$$

- calculate, using (4.47), the realization of the χ^2(chi-square)-Pearson's statistics

Fig. 4.9 The graphical interpretation of the critical interval and the acceptance interval for the chi-square goodness-of-fit test

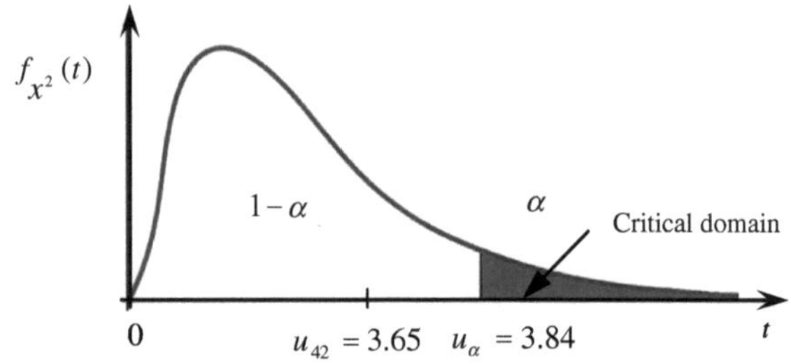

$$u_{42} = \sum_{j=1}^{3} \frac{(\bar{n}_{23}^{j} - n_{23}p_j)^2}{n_{23}p_j} = \frac{(24 - 42 \times 0.5507)^2}{42 \times 0.5507} + \frac{(14 - 42 \times 0.2474)^2}{42 \times 0.2474}$$

$$+ \frac{(4 - 42 \times 0.2019)^2}{42 \times 0.2019} \cong 0.03 + 1.25 + 2.37 = 3.65,$$

- assume the significance level $\alpha = 0.05$,
- fix the number of degrees of freedom

$$\bar{\bar{r}}_{23} - l - 1 = 3 - 1 - 1 = 1,$$

- read from the tables of the χ^2-Pearson's distribution the value u_α for the fixed values of the significance level $\alpha = 0.05$ and the number of degrees of freedom $\bar{\bar{r}}_{23} - l = 1 = 1$, such that, according to (4.49), the following equality holds

$$P(U_{42} > u_\alpha) = \alpha = 0.05$$

  this amounts to $u_\alpha = 3.84$ and we determine the critical domain in the form of the interval (Fig. 4.9) $(3.84, +\infty)$ and the acceptance domain in the form of the interval $\langle 0, 3.84 \rangle$
- we compare the obtained value $u_{42} = 3.65$ of the realization of the statistics U_{42} with that read from the table, critical value $u_\alpha = 3.84$ of the chi-square random variable and since the value $u_{42} = 3.65$ does not belong to the critical domain, i.e.

$$u_{42} = 3.65 \le u_\alpha = 3.84,$$

  we do not reject the hypothesis H_0.

Proceeding afterwards in an analogous way as in the case of the conditional sojourn times θ_{12} and θ_{23}, we identify the density functions of the remaining sojourn times. Their forms are given in Chap. 2. For the identified distributions, by application of either the general formulae for the mean value given by (2.12) or the particular formulae (2.13–2.19), the mean values $M_{b\,b+1} = E[\theta_{b\,b+1}]$, $b = 1, 2, \ldots, 17$, and $M_{181} = E[\theta_{181}]$ of the ferry technical system operation process conditional sojourn times at the particular operation states are determined in

Sect. 2.3.3 and applied for the evaluation and prediction of the characteristics of the ferry technical system operation process.

4.3.4 Testing Uniformity of Statistical Data of Ferry Technical System Operation Process

The ferry technical system operation process is described in Sect. 2.3.3. Its semi-Markov model unknown parameters are evaluated in Sect. 4.3.3 on the basis of the statistical data coming from its spring realizations. Additionally, we have at disposal the statistical data coming from this winter operation process realizations given below in this section. Thus, we may consider the results of two different experiments of the statistical data collection from the ferry technical system operation process, one performed in spring and another in winter. From these experiments two sets of the realizations of the ferry technical system operation process conditional sojourn times at particular operation states $\theta_{b\,b+1}$, $b = 1, 2, \ldots, 17$, and θ_{181} are collected. Joining these two sets of data into one new set of data containing all data from the two experiments and using it for the evaluation of the ferry technical system operation process may lead to a much better estimation of these parameters. But, before the joining of these data into one new set it is necessary to be sure that they are the realizations of the same operation process. Otherwise, the evaluations may be very far from these parameters, real values. Testing the uniformity of the statistical data of these two separate sets before joining them into one set of data may give the reason for doing this.

4.3.4.1 Ferry Technical System Operation Process Data Collection

The statistical data of the spring experiment on the ferry technical system operation process given in Sect. 4.3.3.1 are characterized by the following parameters:

- the ferry technical system operation process experiment time $\Theta_1 = 42$ days,
- the number of ferry technical system operation process realizations

$$n_1(0) = 42,$$

- the number of realizations $\theta^{1k}_{b\,b+1}$, $b = 1, 2, \ldots, 17$, and θ^{1k}_{181} of the ferry technical system operation process conditional sojourn times at the particular operation states

$$n^1_{bl} = 42.$$

The realizations $\theta^{1k}_{b\,b+1} = \theta^{k}_{b\,b+1}$, $b = 1, 2, \ldots, 17$, $k = 1, 2, \ldots, 42$, of the ferry technical system operation process conditional sojourn times $\theta^{1}_{b\,b+1}$ for

$b = 1, 2, \ldots, 17$, and the realizations $\theta_{181}^{1k} = \theta_{181}^{k}$, $k = 1, 2, \ldots, 42$, of the technical system operation process conditional sojourn time θ_{181}^{1}, are partly given in Sect. 4.3.3.1.

The winter experiment on the ferry technical system operation process is characterized by the following parameters:

- the ferry technical system operation process experiment time $\Theta_2 = 40$ days,
- the number of ferry technical system operation process realizations

$$n_2(0) = 40,$$

- the number of realizations $\theta_{b\,b+1}^{2k}$, $b = 1, 2, \ldots, 17$, and θ_{181}^{2k} of the ferry technical system operation process conditional sojourn times at particular operation states

$$n_{bl}^{2} = 40.$$

The realizations $\theta_{b\,b+1}^{2k}$, $b = 1, 2, \ldots, 17$, $k = 1, 2, \ldots, 40$, of the ferry technical system operation process conditional sojourn time $\theta_{b\,b+1}^{2}$ for $b = 1, 2, \ldots, 17$, and the realizations θ_{181}^{2k}, $k = 1, 2, \ldots, 40$, of the ferry technical system operation process conditional sojourn time θ_{181}^{2}, are collected by experts and partly presented in Sect. 4.3.4.2.

4.3.4.2 Spring and Winter Statistical Data of Ferry Technical System Operation Process Uniformity Analysis

We use the two-sample Smirnov-Kolmogorov test described in Sect. 4.2.5.2 to verify the hypotheses that the spring and the winter data sets consisting of the ferry technical system operation process conditional sojourn times at the particular operation states come from the population with the same distribution.

The procedure of testing the uniformity of data sets collected by the experts in the spring and in winter is illustrated on the example of the spring realizations θ_{12}^{1k}, $k = 1, 2, \ldots, 42$, and the winter realizations θ_{12}^{2k}, $k = 1, 2, \ldots, 40$, of the ferry technical system operation process conditional sojourn time θ_{12}^{1} and respectively the ferry technical system operation process conditional sojourn time θ_{12}^{2} at the operation state z_1 while the next operation state is z_2.

For the spring data, the ordered sample of realizations θ_{12}^{1k} presented in Sect. 4.3.3.1 is: 15, 20, 25, 33, 35, 37, 40, 43, 43, 44, 45, 45, 45, 45, 46, 47, 50, 50, 52, 53, 55, 55, 57, 58, 59, 60, 60, 61, 62, 62, 63, 65, 65, 67, 68, 71, 72, 75, 75, 78, 84, 97, and after applying (4.60–4.62), the conditional sojourn time θ_{12}^{1} has the empirical distribution function of the form

$$H_{12}^{1}(t) = \begin{cases} 0 & t \le 15, \\ 1/42, & 15 < t \le 20, \\ 2/42, & 20 < t \le 25, \\ 3/42, & 25 < t \le 33, \\ 4/42 & 33 < t \le 35, \\ 5/42, & 35 < t \le 37, \\ 6/42, & 37 < t \le 40, \\ 8/42, & 40 < t \le 43, \\ 9/42, & 43 < t \le 44, \\ 13/42, & 44 < t \le 45, \\ 14/42, & 45 < t \le 46, \\ 15/42, & 46 < t \le 47, \\ 17/42, & 47 < t \le 50, \\ 18/42, & 50 < t \le 52, \\ 19/42, & 52 < t \le 53, \\ 21/42, & 53 < t \le 55, \\ 22/42, & 55 < t \le 57, \\ 23/42, & 57 < t \le 58, \\ 24/42, & 58 < t \le 59, \\ 26/42, & 59 < t \le 60, \\ 27/42, & 60 < t \le 61, \\ 29/42, & 61 < t \le 62, \\ 30/42, & 62 < t \le 63, \\ 32/42, & 63 < t \le 65, \\ 33/42, & 65 < t \le 67, \\ 34/42, & 67 < t \le 68, \\ 35/42, & 68 < t \le 71, \\ 36/42, & 71 < t \le 72, \\ 38/42, & 72 < t \le 75, \\ 39/42, & 75 < t \le 78, \\ 40/42, & 78 < t \le 84, \\ 41/42, & 84 < t \le 97, \\ 1, & t > 97; \end{cases}$$

For the winter data collected by the experts, the ordered sample of realizations θ_{12}^{2k} is: 12, 15, 15, 18, 19, 20, 25, 33, 33, 34, 36, 37, 37, 37, 40, 41, 44, 46, 48, 48, 50, 53, 55, 57, 59, 60, 60, 61, 62, 63, 65, 65, 65, 67, 69, 75, 75, 75, 80, 90, and after applying (4.63–4.65), the conditional sojourn time θ_{12}^{2} has the empirical distribution function of the form

$$H_{12}^2(t) = \begin{cases} 0 & t \le 12, \\ 1/40, & 12 < t \le 15, \\ 3/40, & 15 < t \le 18, \\ 4/40, & 18 < t \le 19, \\ 5/40, & 19 < t \le 20, \\ 6/40, & 20 < t \le 25, \\ 7/40, & 25 < t \le 33, \\ 9/40, & 33 < t \le 34, \\ 10/40, & 34 < t \le 36, \\ 11/40, & 36 < t \le 37, \\ 14/40, & 37 < t \le 40, \\ 15/40, & 40 < t \le 41, \\ 16/40, & 41 < t \le 44, \\ 17/40, & 44 < t \le 46, \\ 18/40, & 46 < t \le 48, \\ 20/40, & 48 < t \le 50, \\ 21/40, & 50 < t \le 53, \\ 22/40, & 53 < t \le 55, \\ 23/40, & 55 < t \le 57, \\ 24/40, & 57 < t \le 59, \\ 25/40, & 59 < t \le 60, \\ 27/40, & 60 < t \le 61, \\ 28/40, & 61 < t \le 62, \\ 29/40, & 62 < t \le 63, \\ 30/40, & 63 < t \le 65, \\ 34/40, & 65 < t \le 67, \\ 35/40, & 67 < t \le 69, \\ 36/40, & 69 < t \le 75, \\ 38/40, & 75 < t \le 80, \\ 39/40, & 80 < t \le 90, \\ 1, & t > 90. \end{cases}$$

Consequently, the null hypothesis is as follows:

H_0: The winter and spring sets of realizations of the ferry technical system operation process conditional sojourn times θ_{12}^1 and θ_{12}^2 come from populations with the same distribution.

Using the empirical distributions $H_{12}^1(t)$ and $H_{12}^2(t)$, we form a common Table 4.4 composed of all their values. In the Table 4.4, the values t_k assume all realizations θ_{12}^{1k}, $k = 1, 2, \ldots, 42$, and θ_{12}^{2k}, $k = 1, 2, \ldots, 40$, of the conditional sojourn times θ_{12}^1 and θ_{12}^2 i.e. they represent all discontinuity points of the empirical distribution functions $H_{12}^1(t)$ and $H_{12}^2(t)$ where they have jumps in their values $H_{12}^1(t_k)$ and $H_{12}^2(t_k)$ respectively.

Further, according to (4.68–4.70), from Table 4.4, we get

$$d_{42\,40} = \max_{t_k} \left| H_{12}^1(t_k) - H_{12}^2(t_k) \right| \cong 0.209,$$

and according to (4.70)

$$n = \frac{42 \times 40}{42 + 40} = 20.49.$$

Thus, considering (4.66), the realization of the statistics U_n defined by (4.59), is

$$u_n = d_{42\,40}\sqrt{n} = 0.209\sqrt{20.49} \cong 0.946.$$

To verify the hypothesis H_0, we use the two-sample Smirnov-Kolmogorov test λ at the significance level $\alpha = 0.05$.

From the table of the λ distribution for $1 - \alpha = 1 - 0.05 = 0.95$, we find the critical value λ_0 such that according to (4.71) the condition

$$P(U_n < \lambda_0) = Q(\lambda_0) = 1 - 0.05 = 0.95$$

is satisfied. This value amounts to $\lambda_0 \cong 1.36$. Since

$$u_n \cong 0.946 < \lambda_0 = 1.36,$$

we do not have arguments to reject the null hypothesis H_0.

Proceeding in an analogous way with the statistical data collected at the remaining operation states, we may test positively the uniformity of the spring sets of the realizations of the ferry technical system operation process conditional sojourn times $\theta_{b\,b+1}^1$, $b = 2, 3, \ldots, 17$, and θ_{181}^1 and respectively the winter sets of the realizations of the ferry technical system operation process conditional sojourn times $\theta_{b\,b+1}^2$, $b = 2, 3, \ldots, 17$, and θ_{181}^2. After that, we may join the statistical data collected in spring and winter and create new statistical data sets of realizations of the ferry technical system operation process conditional sojourn times $\theta_{b\,b+1}$, $b = 1, 2, \ldots, 17$, and θ_{181} with the following new statistical data:

- the ferry technical system operation process experiment time

$$\Theta = 82 \text{ days},$$

- the number of ferry technical system operation process realizations

$$n(0) = 82,$$

Table 4.4 The values and differences of the distribution functions $H_{12}^{1}(t_k)$ and $H_{12}^{2}(t_k)$ at all their discontinuity points

$t_k = \theta_{12}^{1k} \vee \theta_{12}^{2k}$	$H_{12}^{1}(t_k)$	$H_{12}^{2}(t_k)$	$\lvert H_{12}^{1}(t_k) - H_{12}^{2}(t_k) \rvert$
12	0	0	0
15	0	1/40	0.025
18	1/42	3/40	0.051
19	1/42	4/40	0.076
20	1/42	5/40	0.101
25	2/42	6/40	0.102
33	3/42	7/40	0.104
34	4/42	9/40	0.129
35	4/42	10/40	0.155
36	5/42	10/40	0.131
37	5/42	11/40	0.156
40	6/42	14/40	0.207
41	8/42	15/40	0.185
43	8/42	16/40	0.209
44	9/42	16/40	0.186
45	13/42	17/40	0.115
46	14/42	17/40	0.092
47	15/42	18/40	0.093
48	17/42	18/40	0.045
50	17/42	20/40	0.095
52	18/42	21/40	0.096
53	19/42	21/40	0.073
55	21/42	22/40	0.05
57	22/42	23/40	0.051
58	23/42	24/40	0.052
59	24/42	24/40	0.029
60	26/42	25/40	0.006
61	27/42	24/40	0.043
62	29/42	28/40	0.009
63	30/42	29/40	0.011
65	32/42	30/40	0.012
67	33/42	34/40	0.064
68	34/42	35/40	0.065
69	35/42	35/40	0.042
71	35/42	36/40	0.067
72	36/42	36/40	0.043
75	38/42	36/40	0.005
78	39/42	38/40	0.021
80	40/42	38/40	0.002
84	40/42	39/40	0.023
90	41/42	39/40	0.001
97	41/42	1	0.024
>97	1	1	0

- the number of realizations $\theta^{k}_{b\,b+1}$, $b = 1, 2, \ldots, 17$, and θ^{k}_{181} of the ferry technical system operation process conditional sojourn times $\theta_{b\,b+1}$, $b = 1, 2, \ldots, 17$, and θ_{181} at the particular operation states

$$n_{b\,b+1} = 82,\ n_{181} = 82,$$

- the realizations $\theta^{k}_{b\,b+1}$, $b = 1, 2, \ldots, 17$, and θ^{k}_{181}, $k = 1, 2, \ldots, 82$, of the ferry technical system operation process conditional sojourn times $\theta_{b\,b+1}$, $b = 1, 2, \ldots, 17$, and θ_{181}.

After these join the statistical data of two experiments, we may go to the ferry technical system operation process identification and proceeding accordingly to the procedures proposed in Sects. 4.2.1–4.2.4, we may improve the accuracy of the results obtained in Sect. 4.3.3.

4.4 Summary

The proposed statistical methods of identification of the unknown parameters of the system operation processes allow us for the identification of the models discussed in Chaps. 3 and 6 and also their practical applications in the evaluation, prediction and optimization of reliability, availability and safety of real complex technical systems. This possibility is illustrated in the appliqué sections of these chapters, where on the basis of these methods and the statistical data presented in this chapter, the identification and then the evaluation, prediction and optimization of reliability, availability and safety of the exemplary system, the port oil pipeline system and the ferry technical system are done.

References

1. Barbu V, Limnios N (2006) Empirical estimation for discrete-time semi-Markov processes with applications in reliability. J Nonparametric Stat 18(7–8):7–8 483-498
2. Collet J (1996) Some remarks on rare-event approximation. IEEE Trans Reliab 45:106–108
3. Gamiz ML, Roman Y (2008) Non-parametric estimation of the availability in a general repairable. Reliab Eng Syst Saf 93(8):1188–1196
4. Giudici P, Figini S (2009) Applied data mining for business and industry. Wiley, London
5. Habibullah MS, Lumanpauw E, Kolowrocki K, Soszynska J, Ming NG (2009) A computational tool for general model of operation processes in industrial systems operation processes. Electron J Reliab Risk Anal: Theory Appl 2(4):181–191
6. Helvacioglu S, Insel M (2008) Expert system applications in marine technologies. Ocean Eng 35(11–12):1067–1074
7. Hryniewicz O (1995) Lifetime tests for imprecise data and fuzzy reliability requirements. In: Onisawa T, Kacprzyk J (eds) Reliability and safety analyses under fuzziness. Physica Verlag, Heidelberg, pp 169–182
8. Kołowrocki K (2004) Reliability of large systems. Elsevier, Amsterdam

9. Kołowrocki K, Soszyńska J (2008) A general model of industrial systems operation processes related to their environment and infrastructure. Summer Safety & Reliability Seminars. J Pol Saf Reliab Assoc 2(2):223–226
10. Kolowrocki K, Soszynska J (2009) Modeling environment and infrastructure influence on reliability and operation process of port oil transportation system. Electron J Reliab Risk Anal: Theory Appl 2(3):131–142
11. Kolowrocki K, Soszynska J (2009) Safety and risk evaluation of Stena Baltica ferry in variable operation conditions. Electron J Reliab Risk Anal: Theory Appl 2(4):68–180
12. Kołowrocki K, Soszyńska J (2009) Statistical identification and prediction of the port oil pipeline system's operation process and its reliability and risk evaluation. Summer Safety & Reliability Seminars. J Pol Saf Reliab Assoc 3(2):241–250
13. Kołowrocki K, Soszyńska J (2009) Methods and algorithms for evaluating unknown parameters of operation processes of complex technical systems. Summer Safety & Reliability Seminars. J Pol Saf Reliab Assoc 3(1–2):211–222
14. Kołowrocki K, Soszyńska J (2009) Methods and algorithms for evaluating unknown parameters of components reliability of complex technical systems. Summer Safety & Reliability Seminars. J Pol Saf Reliab Assoc 4(2):223–230
15. Kołowrocki K, Soszyńska J (2009) Statistical identification and prediction of the port oil pipeline system's operation process and its reliability and risk evaluation. Summer Safety & Reliability Seminars. J Pol Saf Reliab Assoc 4(2):241–250
16. Kolowrocki K, Soszynska J (2010) Reliability modeling of a port oil transportation system's operation processes. Int J of Perform Eng 6(1):77–87
17. Kolowrocki K, Soszynska J (2010) Reliability, availability and safety of complex technical systems: modelling–identification–prediction–optimization. Summer Safety & Reliability Seminars. J Pol Saf Reliab Assoc 4(1):133–158
18. Kolowrocki K, Soszynska J (2010) Testing uniformity of statistical data sets coming from complex systems operation processes. Summer Safety & Reliability Seminars. J Pol Saf Reliab Assoc 4(1):123–132
19. Limnios N, Oprisan G (2005) Semi-Markov processes and reliability. Birkhauser, Boston
20. Limnios N, Ouhbi B, Sadek A (2005) Empirical estimator of stationary distribution for semi-Markov processes. Commun Stat-Theory Methods 34(4):987–995 12
21. Macci C (2008) Large deviations for empirical estimators of the stationary distribution of a semi-Markov process with finite state space. Commun Stat-Theory Methods 37(19):3077–3089
22. Mercier S (2008) Numerical bounds for semi-Markovian quantities and application to reliability. Methodol Comput Appl Probab 10(2):179–198
23. Rice JA (2007) Mathematical statistics and data analysis. Duxbury. Thomson Brooks/Cole. University of California, Berkeley
24. Soszyńska J (2007) Systems reliability analysis in variable operation conditions. PhD thesis, Gdynia Maritime University-System Research Institute Warsaw (in Polish)
25. Soszyńska J, Kołowrocki K, Blokus-Roszkowska A, Guze S (2010) Identification of complex technical system components safety models. Summer Safety & Reliability Seminars. J Pol Saf Reliab Assoc 4(2):399–496
26. Vercellis S (2009) Data mining and optimization for decision making. Wiley, Indianapolis
27. Wilson AG, Graves TL, Hamada MS et al (2006) Advances in data combination, analysis and collection for system reliability assessment. Stat Sci 21(4):514–531

Chapter 5
Complex Technical System Components Reliability and Safety Identification

5.1 Introduction

The general joint models linking the multistate systems' reliability and safety models considered in Chap. 1 with the models of their operation processes proposed in Chap. 2, giving the possibility of the reliability and safety analysis of complex technical systems at the variable operating conditions, are constructed in Chap. 3. To be able to apply these general models practically in the evaluation and prediction of the reliability and safety of real complex technical systems it is necessary to elaborate the statistical methods concerned with determining unknown parameters of the proposed models [2–6, 8–13, 19–22, 24]. Particularly, concerning the multistate systems' safety and reliability models, the unknown parameters of the conditional multistate reliability and safety functions of the system components at the various operation states should be identified [1, 7, 14–17, 25–28]. It is also necessary to have the methods of testing the hypotheses concerned with the conditional multistate reliability functions of the system components at the system various operation states.

5.2 Identification of Complex Technical System Components Reliability and Safety Models

We assume, as in Sects. 3.2 and 3.5, that the changes of the system operation process $Z(t)$ have an influence on the system components' reliability and safety characteristics. We mark the conditional multistate reliability function of the system component while the system is at the operation state z_b, $b = 1, 2, \ldots, v$, by the vector

K. Kołowrocki and J. Soszyńska-Budny, *Reliability and Safety of Complex Technical Systems and Processes*, Springer Series in Reliability Engineering,
DOI: 10.1007/978-0-85729-694-8_5,

$$[R(t,\ \cdot)]^{(b)} = [1, [R(t,\ 1)]^{(b)}, \ldots, [R(t, z)]^{(b)}], \tag{5.1}$$

where

$$[R(t, u)]^{(b)} = P\left(T^{(b)}(u) > t|Z(t) = z_b\right) \quad \text{for } t \in \langle 0, \infty), \\ u = 1, 2, \ldots, z,\ b = 1, 2, \ldots, v, \tag{5.2}$$

is the conditional reliability function standing the probability that the conditional lifetime $T^{(b)}(u)$ of the system component at the reliability states subset $\{u, u + 1, \ldots, z\}$ is greater than t, while the system operation process $Z(t)$ is in the operation state z_b, $b = 1, 2, \ldots, v$.

Further, we assume that the coordinates (5.2) of the vector of the conditional multistate reliability function (5.1) are exponential reliability functions of the form

$$[R(t, u)]^{(b)} = \exp\left[-[\lambda(u)]^{(b)} t\right] \quad \text{for } t \in \langle 0, \infty), \quad u = 1, 2, \ldots, z,\ b = 1, 2, \ldots, v. \tag{5.3}$$

Similarly, we mark the conditional multistate safety function of the system component while the system is at the operation state z_b, $b = 1, 2, \ldots, v$, by the vector

$$[s(t,\ \cdot)]^{(b)} = \left[1, [s(t,\ 1)]^{(b)}, \ldots, [s(t, z)]^{(b)}\right], \tag{5.4}$$

where

$$[s(t, u)]^{(b)} = P(T^{(b)}(u) > t|Z(t) = z_b) \quad \text{for } t \in \langle 0, \infty), \\ u = 1, 2, \ldots, z,\ b = 1, 2, \ldots, v, \tag{5.5}$$

is the conditional safety function standing the probability that the conditional lifetime $T^{(b)}(u)$ of the system component in the safety state subset $\{u, u + 1, \ldots, z\}$ is greater than t, while the system operation process $Z(t)$ is at the operation state $z_b, b = 1, 2, \ldots, v$.

Further, we assume that the coordinates (5.5) of the vector of the conditional multistate safety function (5.4) are exponential safety functions of the form

$$[s(t, u)]^{(b)} = \exp\left[-[\lambda(u)]^{(b)} t\right] \quad \text{for } t \in \langle 0, \infty), \quad u = 1, 2, \ldots, z,\ b = 1, 2, \ldots, v. \tag{5.6}$$

The above assumptions mean that the density function of the system component conditional lifetime $T^{(b)}(u)$ in the reliability or safety state subset $\{u, u + 1, \ldots, z\}$, $u = 1, 2, \ldots, z$, at the operation state z_b, $b = 1, 2, \ldots, v$, is exponential of the form

$$[f(t, u)]^{(b)} = [\lambda(u)]^{(b)} \exp[-[\lambda(u)]^{(b)} t] \quad \text{for } t \in \langle 0, \infty), \tag{5.7}$$

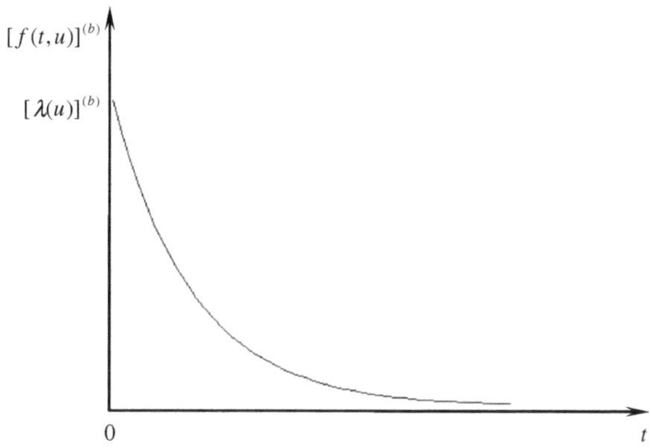

Fig. 5.1 The graph of the conditional density function $[f(t,u)]^{(b)}$

where $[\lambda(u)]^{(b)}$, $[\lambda(u)]^{(b)} \geq 0$, is an unknown intensity of the system component departure from this subset of the reliability or safety states.

The exemplary graph of the conditional density function $[f(t,u)]^{(b)}$ defined by (5.7) is illustrated in Fig. 5.1.

The assumptions on the system components' reliability and safety models expressed respectively in the formulae (5.1–5.3) and (5.4–5.6) are identical. Thus, the methods of estimating the unknown intensities of departures from the reliability state subsets or respectively from the safety state subsets, existing in the formula (5.7) for the density function, are the same. Therefore, in this section we present these methods in detail mainly for the system components' reliability models and after that we apply them for the identification of the system components' reliability models and the system components' safety models as well.

5.2.1 Defining Unknown Parameters of System Components Reliability and Safety Models and Data Collection

First, before fixing the subsystems and components of the system that are intended to be observed at the various operation states, we should analyze the system operation process and to fix or to define its following general parameters:

- the number of the operation states of the system operation process v,
- the operation states of the system operation process $z_1, z_2, \ldots, z_v$.

Next, we should perform the following steps:

(i) fix the subsystems of the system operating at the particular operation states;
(ii) fix, describe and mark the components of the subsystems operating at the particular operation states.

To make the estimation of the unknown parameters of the system components conditional multistate reliability or safety functions the experiment delivering the necessary statistical data should be precisely planned.

Before the experiment, we should perform the following preliminary steps:

(i) analyze the processes of reliability or safety states changing of all system components at the particular operation states;
(ii) fix or define its following general parameters:

- the number of the reliability or safety states of the system components $z + 1$,
- the reliability or safety states of the system components $0, 1, \ldots, z$;

(iii) fix the possible transitions between the system components reliability or safety states;
(iv) fix the set of the unknown parameters of the system components reliability models.

5.2.1.1 Collecting Data Coming from Components Reliability and Safety States Changing Processes

To estimate the unknown parameters of the system components' multistate reliability or safety models, during the experiment, we should collect the necessary statistical data performing the following steps:

(i) fix the experiment kind subjected to the defined below *Cases 1–6;*
(ii) fix and to collect, in *Cases 1–6*, the following statistical data necessary to evaluate the component unknown intensities of departure from the reliability or safety state subsets:

- the number of experiment posts,
- the experiments duration times,
- the number of the observed realizations of the component lifetimes up to the first departure from the reliability or safety state subsets,
- the realizations of the component lifetimes up to the first departure from the reliability or safety state subsets.

The fixed kinds of experiments and the collected statistical data sets are described below. They are only defined in the reliability aspect as their definitions in the safety aspect are analogous.

Case 1 The estimation of the component intensity of departure from the reliability state subset on the basis of the realizations of the component lifetimes up to the first departure from the reliability state subset on several experimental posts—Completed investigations, the same observation time on all experimental posts.

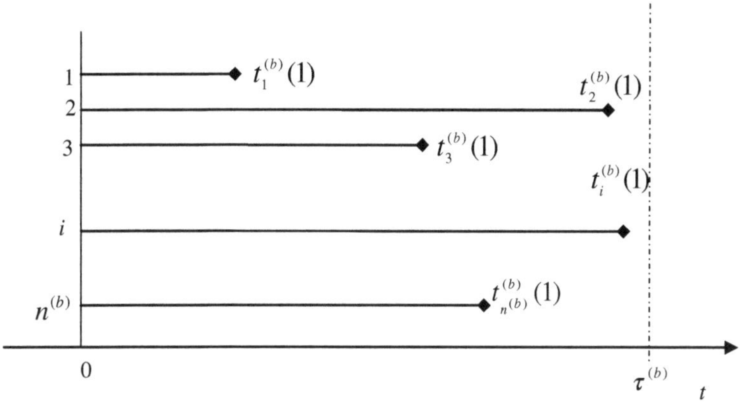

Fig. 5.2 The scheme of the realizations of the component lifetimes up to the first departure from the reliability state subset on $n^{(b)}$ observational posts (completed investigations, the same observation time on all experimental posts)

We assume that during the time $\tau^{(b)}$, $\tau^{(b)} > 0$, we have been observing the realizations of the component lifetime $T^{(b)}(u)$ in the reliability state subset $\{u, u+1, \ldots, z\}$, $u = 1, 2, \ldots, z$, at the operation state z_b, $b = 1, 2, \ldots, v$, on $n^{(b)}$ identical experimental posts. We assume that at the beginning of the experiment all components are new identical components straying at the best reliability state z and during the fixed observation time $\tau^{(b)}$ all components have left the reliability state subset $\{1, 2, \ldots, z\}$, i.e. all observed components reached the worst reliability state 0 (Fig. 5.2). It means that the number $m^{(b)}(u)$ of components that have left the reliability state subset $\{u, u+1, \ldots, z\}$, $u = 1, 2, \ldots, z$, is equal to $n^{(b)}$, i.e. $m^{(b)}(u) = n^{(b)}$, $u = 1, 2, \ldots, z$.

We mark by

$$A^{(b)}(u) = \{t_i^{(b)}(u) : i = 1, 2, \ldots, m^{(b)}(u)\}, \; u = 1, 2, \ldots, z,$$

the set of the moments $t_i^{(b)}(u)$, $i = 1, 2, \ldots, m^{(b)}(u)$, $u = 1, 2, \ldots, z$, of departures from the reliability state subset $\{u, u+1, \ldots, z\}$, $u = 1, 2, \ldots, z$, of the component on the ith observational post, i.e. the realizations of the component lifetimes $T_i^{(b)}(u)$ to the first departure from the reliability state subset, that are the independent random variables with the exponential distribution defined by the density function (5.1).

Case 2 The estimation of the component intensity of departure from the reliability state subset on the basis of the realizations of the component lifetimes up to the first departure from the reliability state subset on several experimental posts—Non-completed investigations, the same observation time on all experimental posts.

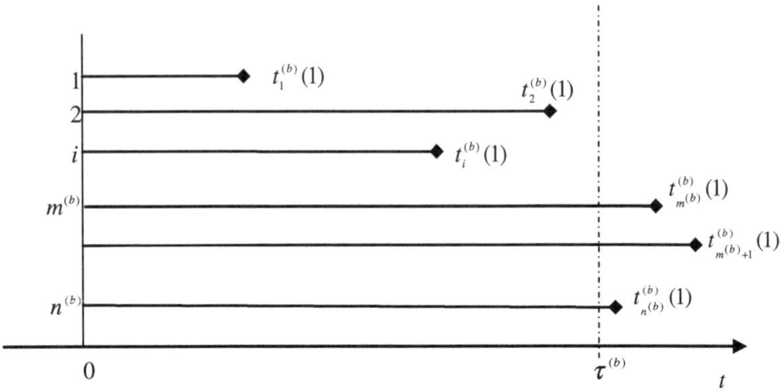

Fig. 5.3 The scheme of the realizations of the component lifetimes up to the first departure from the reliability state subset on $n^{(b)}$ observational posts (non-completed investigations, the same observation time on all experimental posts)

We assume that during the time $\tau^{(b)}$, $\tau^{(b)} > 0$, we have been observing the realizations of the component lifetimes $T^{(b)}(u)$ in the reliability state subset $\{u, u+1, \ldots, z\}$, $u = 1, 2, \ldots, z$, at the operation state z_b, $b = 1, 2, \ldots, v$, on $n^{(b)}$ identical experimental posts. We assume that at the beginning of the experiment all components are new identical components straying at the best reliability state z and that during the fixed observation time $\tau^{(b)}$not all components have left the reliability state subset $\{1, 2, \ldots, z\}$, i.e. $m^{(b)}$, $m^{(b)} < n^{(b)}$, observed components reached the worst reliability state 0 (Fig. 5.3). It means that the number $m^{(b)}(u)$ of components that have left the reliability state subset $\{u, u+1, \ldots, z\}$, $u = 1, 2, \ldots, z$, is less or equal to $n^{(b)}$, i.e. $(m^{(b)}(u) \le n^{(b)}$, $u = 1, 2, \ldots, z$.

We mark by

$$A^{(b)}(u) = \{t_i^{(b)}(u) : i = 1, 2, \ldots, m^{(b)}(u)\}, \; u = 1, 2, \ldots, z,$$

the set of the moments $t_i^{(b)}(u)$, $i = 1, 2, \ldots, m^{(b)}(u)$, $u = 1, 2, \ldots, z$, of departures from the reliability state subset $\{u, u+1, \ldots, z\}$, $u = 1, 2, \ldots, z$, of the component on the ith observational post, i.e. the realizations of the component lifetimes $T_i^{(b)}(u)$ to the first departure from the reliability state subset, that are the independent random variables with the exponential distribution defined by the density function (5.1).

Case 3 The estimation of the component intensity of departure from the reliability state subset on the basis of the realizations of the component lifetimes up to the first departure from the reliability state subset on several experimental posts—Non-completed investigations, different observation times on particular experimental posts.

We assume that we have been observing the realizations of the component lifetimes $T^{(b)}(u)$ in the reliability state subset $\{u, u+1, \ldots, z\}$, $u = 1, 2, \ldots, z$, at

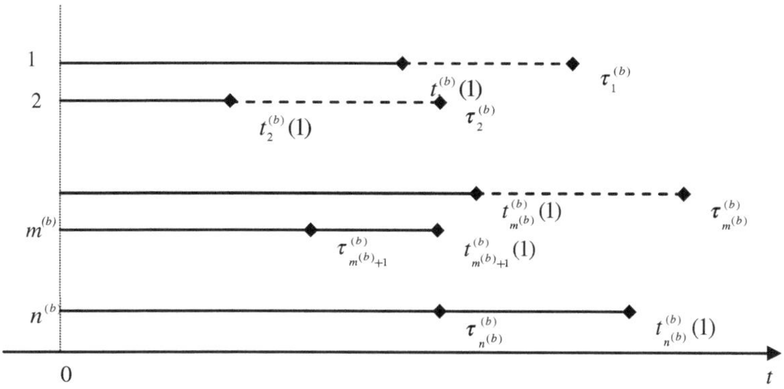

Fig. 5.4 The scheme of the realizations of the component lifetimes up to the first departure from the reliability state subset on $n^{(b)}$ observational posts (non-completed investigations, different observation times on all experimental posts)

the operation state z_b, $b = 1, 2, \ldots, v$, on $n^{(b)}$ identical experimental posts. We assume that the observation times on particular experimental posts are different and we mark by $\tau_i^{(b)}$, $\tau_i^{(b)} > 0$, $i = 1, 2, \ldots, n^{(b)}$, the observation time respectively on the ith experimental post. We assume that at the beginning of the experiment all components are new identical components straying at the best reliability state z and that during the fixed observation times $\tau_i^{(b)}$ not all components have left the reliability state subset $\{1, 2, \ldots, z\}$, i.e. $m^{(b)}$, $m^{(b)} < n^{(b)}$, observed components reached the worst reliability state 0 (Fig. 5.4). It means that the number $m^{(b)}(u)$ of components that have left the reliability state subset $\{u, u+1, \ldots, z\}$, $u = 1, 2, \ldots, z$, is less or equal to $n^{(b)}$, i.e. $m^{(b)}(u) \le n^{(b)}$, $u = 1, 2, \ldots, z$.

We mark by

$$A^{(b)}(u) = \{t_i^{(b)}(u) : i = 1, 2, \ldots, m^{(b)}(u)\}, \; u = 1, 2, \ldots, z,$$

the set of the moments $t_i^{(b)}(u)$, $i = 1, 2, \ldots, m^{(b)}(u)$, $u = 1, 2, \ldots, z$, of departures from the reliability state subset $\{u, u+1, \ldots, z\}$, $u = 1, 2, \ldots, z$, of the component on the ith observational post, i.e. the realizations of the component lifetimes $T_i^{(b)}(u)$ to the first departure from the reliability state subset, that are the independent random variables with the exponential distribution defined by the density function (5.1).

Case 4 The estimation of the component intensity of departure from the reliability state subset on the basis of the realizations of the component simple renewal flow (stream) on one experimental post.

We assume that during the time $\tau^{(b)}$, $\tau^{(b)} > 0$, we have been observing the realizations of the component lifetime $T^{(b)}(u)$ in the reliability state subset $\{u, u+1, \ldots, z\}$, $u = 1, 2, \ldots, z$, at the operation state z_b, $b = 1, 2, \ldots, v$, on one

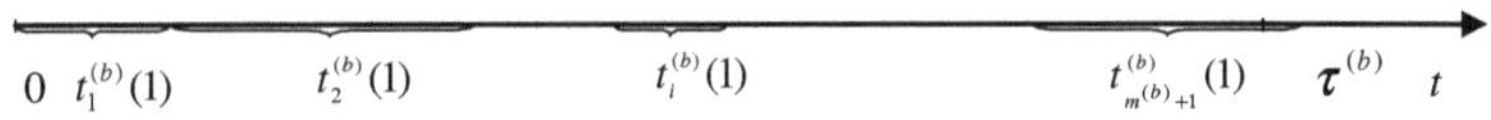

Fig. 5.5 The scheme of the realizations of the component simple renewal flow (stream) on one experimental post

experimental post. We assume that at the moment when the component is leaving the reliability state subset $\{1,2,\ldots,z\}$, i.e. the observed component reached the worst reliability state 0, it is replaced at once by the same new component straying at the reliability state z (Fig. 5.5). It means that at the beginning all components are new identical components straying at the best reliability state z. We assume that during the fixed observation time $m^{(b)}$ components have left the reliability state subset $\{1,2,\ldots,z\}$, i.e. $m^{(b)}$ observed components reached the worst reliability state 0. It means that the number $m^{(b)}(u)$ of components that have left the reliability state subset $\{u,u+1,\ldots,z\}$, $u=1,2,\ldots,z$, is equal either to $m^{(b)}$ or to $m^{(b)}+1$, i.e. $m^{(b)}(u)=m^{(b)}$ or $m^{(b)}(u)=m^{(b)}+1$, $u=1,2,\ldots,z$.

We mark by

$$A^{(b)}(u)=\{t_i^{(b)}(u):\ i=1,2,\ldots,m^{(b)}(u)\},\ u=1,2,\ldots,z,$$

the set of the moments $t_i^{(b)}(u)$, $i=1,2,\ldots,m^{(b)}(u)$, $u=1,2,\ldots,z$, of departures from the reliability state subset $\{u,u+1,\ldots,z\}$, $u=1,2,\ldots,z$, of the component on the ith observational post, i.e. the realizations of the component lifetime $T_i^{(b)}(u)$ to the first departure from the reliability state subset, that are the independent random variables with the exponential distribution defined by the density function (5.1).

Case 5 The estimation of the component intensity of departure from the reliability state subset on the basis of the realizations of the component simple renewal flows (streams) on several experimental posts—The same observation time on all experimental posts.

We assume that during the time $\tau_i^{(b)}$, $\tau^{(b)}>0$, we have been observing the realizations of the component lifetime $T^{(b)}(u)$ in the reliability state subset $\{u,u+1,\ldots,z\}$, $u=1,2,\ldots,z$, at the operation state z_b, $b=1,2,\ldots,v$, on $n^{(b)}$ experimental posts. We assume that, at each observation post, at the moment when the component is leaving the reliability state subset $\{1,2,\ldots,z\}$, i.e. the observed component reached the worst reliability state 0, it is replaced at once by the same new component straying at the reliability state z (Fig. 5.6). It means that, at each experiment post, at the beginning all components are new identical components straying at the best reliability state z. We assume that, at the jth, $j=1,2,\ldots,n^{(b)}$ experimental post, during the fixed observation time $m_j^{(b)}$ components have left the reliability state subset $\{1,2,\ldots,z\}$, i.e. $m_j^{(b)}$ observed components reached the worst reliability state 0. It means that the number $m_j^{(b)}(u)$ of components that have

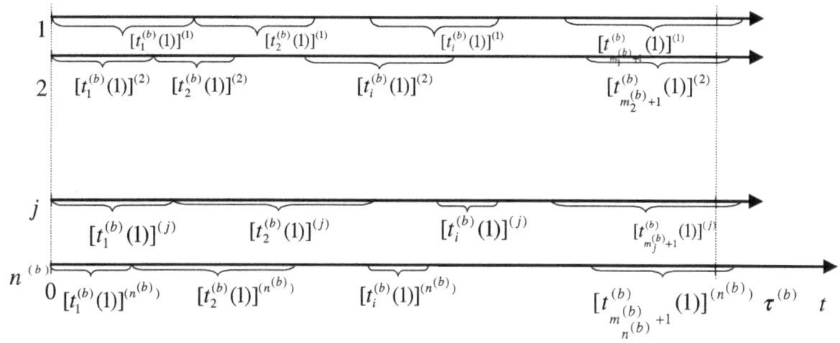

Fig. 5.6 The scheme of the realizations of the component simple renewal flows (streams) on several experimental posts (the same observation time on all experimental posts)

left the reliability state subset $\{u, u+1, \ldots, z\}$, $u = 1, 2, \ldots, z$, is equal either to $m_j^{(b)}$ or to $m_j^{(b)} + 1$, i.e. $m_j^{(b)}(u) = m_j^{(b)}$ or $m_j^{(b)}(u) = m_j^{(b)} + 1,\ u = 1, 2, \ldots, z$

We mark by

$$A_j^{(b)}(u) = \{[t_i^{(b)}(u)]^{(j)} : i = 1, 2, \ldots, m_j^{(b)}(u)\}, \quad u = 1, 2, \ldots, z,\ j = 1, 2, \ldots, n^{(b)},$$

the sets of the times $[t_i^{(b)}(u)]^{(j)}$, $i = 1, 2, \ldots, m_j^{(b)}(u)$, to the components departures from the reliability state subset $\{u, u+1, \ldots, z\}$, $u = 1, 2, \ldots, z$, at the jth, $j = 1, 2, \ldots, n^{(b)}$, experimental post, i.e. the realizations of the component lifetimes $T_i^{(b)}(u)$ to the first departure from the reliability state subset, that are random variable with the exponential distribution defined by the density function (5.1).

Case 6 The estimation of the component intensity of departure from the reliability state subset on the basis of the realizations of the component simple renewal flows (streams) on several experimental posts—Different observation times on experimental posts.

We assume that we have been observing the realizations of the component lifetime $T^{(b)}(u)$ in the reliability state subset $\{u, u+1, \ldots, z\}$, $u = 1, 2, \ldots, z$, at the operation state z_b, $b = 1, 2, \ldots, v$, on $n^{(b)}$ experimental posts. We assume that the observation times on particular experimental posts are different and we mark by $\tau_j^{(b)}$, $\tau_j^{(b)} > 0$, $i = 1, 2, \ldots, n^{(b)}$, the observation time respectively on the ith experimental post. We assume that, at each observation post, at the moment when the component is leaving the reliability state subset $\{1, 2, \ldots, z\}$, i.e. the observed component reached the worst reliability state 0, it is replaced at once by the same new component straying at the reliability state z (Fig. 5.7). It means that, at each experiment post, at the beginning all components are new identical components straying at the best reliability state z. We assume that, at the jth, $j = 1, 2, \ldots, n^{(b)}$, experimental post, during the fixed observation time $m_j^{(b)}$ components have left the reliability state subset $\{1, 2, \ldots, z\}$, i.e. $m_j^{(b)}$ observed components reached the

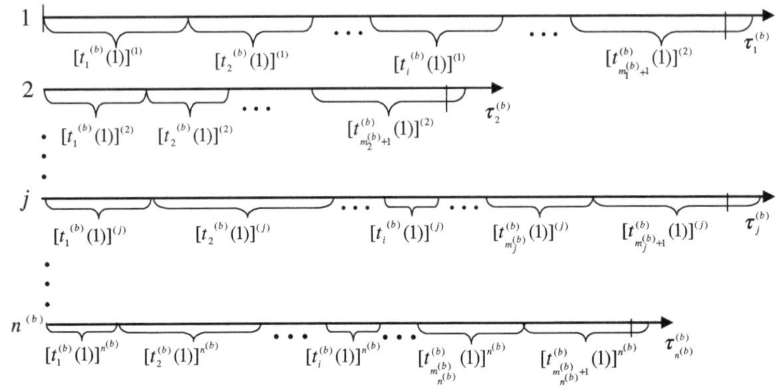

Fig. 5.7 The scheme of the realizations of the component simple renewal flows (streams) on several experimental posts (different observation times on experimental posts)

worst reliability state 0. It means that the number $m_j^{(b)}(u)$ of components that have left the reliability state subset $\{u, u+1, \ldots, z\}$, $u = 1, 2, \ldots, z$, is equal either to $m_j^{(b)}$ or to $m_j^{(b)} + 1$, i.e. $m_j^{(b)}(u) = m_j^{(b)}$ or $m_j^{(b)}(u) = m_j^{(b)} + 1$, $u = 1, 2, \ldots, z$.

We mark by

$$A_j^{(b)}(u) = \left\{ \left[t_i^{(b)}(u)\right]^{(j)} :\ i = 1, 2, \ldots, m_j^{(b)}(u) \right\},$$
$$u = 1, 2, \ldots, z, \ j = 1, 2, \ldots, n^{(b)},$$

the sets of the times $\left[t_i^{(b)}(u)\right]^{(j)}$, $i = 1, 2, \ldots, m_j^{(b)}(u)$, to the components departures from the reliability state subset $\{u, u+1, \ldots, z\}$, $u = 1, 2, \ldots, z$, at the jth, $j = 1, 2, \ldots, n^{(b)}$, experimental post, i.e. the realizations of the component lifetimes $T_i^{(b)}(u)$ to the first departure from the reliability state subset, that are random variable with the exponential distribution defined by the density function (5.1).

To make clear the way of data collection, it has to be expressed that the realizations of the component lifetimes to the first departure from the reliability state subset $\{1, 2, \ldots, z\}$, i.e. the times up to the moment when the component reached the worst reliability state 0 presented in Figs. 5.2–5.7 include the realizations of the component lifetimes to the first departure from the reliability state subsets $\{u, u+1, \ldots, z\}$, $u = 2, 3, \ldots, z.$, For instance, the realization $t_1^{(b)}(1)$ presented in Fig. 5.2 includes the realizations $t_1^{(b)}(z)$, $t_1^{(b)}(z-1), \ldots, t_1^{(b)}(2)$, which because of ageing, fulfill the following condition

$$t_1^{(b)}(z) \le t_1^{(b)}(z-1) \le \cdots \le t_1^{(b)}(2) \le t_1^{(b)}(1)$$

illustrated in Fig. 5.8.

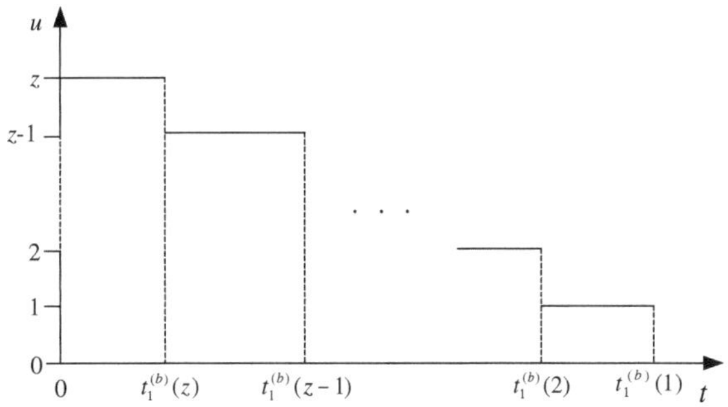

Fig. 5.8 The relationship between the realizations of the component lifetimes in the reliability state subsets

5.2.1.2 Collecting Reliability and Safety Data Coming from Experts

In this case, as we do not have statistical data coming from the system components' reliability or safety states' changing processes, we try to get approximate data on components' reliability and safety from experts. To make the estimation of the unknown parameters of the system components' multistate reliability or safety models on the basis of the expert opinions the approximate values

$$[\hat{\mu}(u)]^{(b)}, \quad u = 1, 2, \ldots, z, \; b = 1, 2, \ldots, v,$$

of the mean values

$$[\mu(u)]^{(b)} = E[T^{(b)}(u)], \quad u = 1, 2, \ldots, z, \; b = 1, 2, \ldots, v,$$

of the system component lifetimes $T^{(b)}(u)$, $u = 1, 2, \ldots, z$, $b = 1, 2, \ldots, v$, in the reliability or safety state subset $\{u, u+1, \ldots, z\}$, $u = 1, 2, \ldots, z$, while the system is operating at the operation state z_b, $b = 1, 2, \ldots, v$ should be fixed.

5.2.2 Estimating Parameters of Conditional Multistate Exponential Reliability and Safety Functions of System Components

Prior to the estimation of the system components unknown intensities of departures from the reliability or safety states subsets, we have to decide whether we have statistical data coming from components reliability or safety states changing processes or from experts and after that we have to select one of two procedures of estimating presented below, respectively in Sects. 5.2.2.1 and 5.2.2.2.

5.2.2.1 Evaluating System Components Intensities of Departures from Reliability and Safety State Subsets on Basis of Data Coming from Components Reliability and Safety States Changing Processes

On the basis of statistical data described in Sect. 5.2.1.1, we can find the estimate $[\hat{\lambda}(u)]^{(b)}$ of the system component unknown intensity $[\lambda(u)]^{(b)}$of departure from the reliability or safety state subset $\{u, u+1, \ldots, z\}$, $u = 1, 2, \ldots, z$. The formulae for all kinds of experiments, considered in Sect. 5.2.1.1 are presented below.

Case 1 The maximum likelihood evaluation of the unknown component intensity of departure $[\lambda(u)]^{(b)}$ from the reliability or safety state subset $\{u, u+1, \ldots, z\}$, $u = 1, 2, \ldots, z$, is

$$[\hat{\lambda}(u)]^{(b)} = \frac{n^{(b)}}{\sum_{i=1}^{n^{(b)}} t_i^{(b)}(u)}, \quad u = 1, 2, \ldots, z. \tag{5.8}$$

Case 2 The maximum likelihood evaluation of the unknown component intensity of departure $[\lambda(u)]^{(b)}$ from the reliability or safety state subset $\{u, u+1, \ldots, z\}$, $u = 1, 2, \ldots, z$, is

$$[\hat{\lambda}(u)]^{(b)} = \frac{m^{(b)}(u)}{\sum_{i=1}^{m^{(b)}(u)} t_i^{(b)}(u) + \tau^{(b)}[n^{(b)} - m^{(b)}(u)]}, \quad u = 1, 2, \ldots, z. \tag{5.9}$$

Assuming the observation time $\tau^{(b)}$ as the moment of departure from the reliability or safety state subset $\{u, u+1, \ldots, z\}$, $u = 1, 2, \ldots, z$, of the components that have not left this reliability or safety state subset we get so-called pessimistic evaluation of the intensity of departure $[\lambda(u)]^{(b)}$ from the reliability or safety state subset $\{u, u+1, \ldots, z\}$, $u = 1, 2, \ldots, z$, of the form

$$[\hat{\lambda}(u)]^{(b)} = \frac{n^{(b)}}{\sum_{i=1}^{m^{(b)}(u)} t_i^{(b)}(u) + \tau^{(b)}[n^{(b)} - m^{(b)}(u)]}, \quad u = 1, 2, \ldots, z. \tag{5.10}$$

Case 3 The maximum likelihood evaluation of the unknown component intensity of departure $[\lambda(u)]^{(b)}$ from the reliability or safety state subset $\{u, u+1, \ldots, z\}$, $u = 1, 2, \ldots, z$, is

$$[\hat{\lambda}(u)]^{(b)} = \frac{m^{(b)}(u)}{\sum_{i=1}^{m^{(b)}(u)} t_i^{(b)}(u) + \sum_{i=m^{(b)}(u)+1}^{n^{(b)}} \tau_i^{(b)}}, \quad u = 1, 2, \ldots, z. \tag{5.11}$$

Assuming the observation times $\tau_i^{(b)}$, $i = m^{(b)}(u), m^{(b)}(u)+1, \ldots, n^{(b)}$, as the moment of departure from the reliability or safety state subset $\{u, u+1, \ldots, z\}$, $u = 1, 2, \ldots, z$, of the components that have not left this reliability or safety state subset we get a so-called pessimistic evaluation of the intensity of departure $\lambda^{(b)}(u)$ from the reliability or safety state subset of the form

$$[\hat{\lambda}(u)]^{(b)} = \frac{n^{(b)}}{\sum_{i=1}^{m^{(b)}(u)} t_i^{(b)}(u) + \sum_{i=m^{(b)}(u)+1}^{n^{(b)}} \tau_i^{(b)}}, \quad u = 1, 2, \ldots, z. \tag{5.12}$$

Case 4 The maximum likelihood evaluation of the unknown component intensity of departure $[\lambda(u)]^{(b)}$ from the reliability or safety state subset $\{u, u+1, \ldots, z\}$, $u = 1, 2, \ldots, z$, is

$$[\hat{\lambda}(u)]^{(b)} = \frac{m^{(b)}(u)}{\sum_{i=1}^{m^{(b)}(u)} t_i^{(b)}(u) + d^{(b)}(u)}, \quad u = 1, 2, \ldots, z, \tag{5.13}$$

where

$$d^{(b)}(u) = \begin{cases} \tau^{(b)} - \sum\limits_{i=1}^{m^{(b)}(u)} t_i^{(b)}(1), & \text{if } m^{(b)}(u) = m^{(b)} \\ 0, & \text{if } m^{(b)}(u) = m^{(b)} + 1, \quad u = 1, 2, \ldots, z. \end{cases} \tag{5.14}$$

In the case when $m^{(b)}(u) = m^{(b)}$, $u = 1, 2, \ldots, z$, after assuming the observation time $\tau^{(b)}$ as the moment of departure from the reliability or safety state subset $\{u, u+1, \ldots, z\}$, $u = 1, 2, \ldots, z$, of the last component that has not left this reliability or safety state subset, we get a so-called pessimistic evaluation of the intensity of departure $[\lambda(u)]^{(b)}$ from the reliability or safety state subset $\{u, u+1, \ldots, z\}$, $u = 1, 2, \ldots, z$, of the form

$$[\hat{\lambda}(u)]^{(b)} = \frac{m^{(b)} + 1}{\sum_{i=1}^{m^{(b)}(u)} t_i^{(b)}(u) + d^{(b)}(u)}, \quad u = 1, 2, \ldots, z. \tag{5.15}$$

Case 5 The maximum likelihood evaluation of the unknown component intensity of departure $[\lambda(u)]^{(b)}$ from the reliability or safety state subset $\{u, u+1, \ldots, z\}$, $u = 1, 2, \ldots, z$, is either

$$[\hat{\lambda}(u)]^{(b)} = \frac{\sum_{j=1}^{n^{(b)}} m_j^{(b)}(u)}{\sum_{j=1}^{n^{(b)}} \sum_{i=1}^{m_j^{(b)}(u)} \left[t_i^{(b)}(u)\right]^j + \sum_{j=1}^{n^{(b)}} d_j^{(b)}(u)}, \quad u = 1, 2, \ldots, z, \tag{5.16}$$

where

$$d_j^{(b)}(u) = \begin{cases} \tau^{(b)} - \sum_{i=1}^{m_j^{(b)}(u)} \left[t_i^{(b)}(1)\right]^{(j)}, & \text{if } m_j^{(b)}(u) = m_j^{(b)} \\ 0, & \text{if } m_j^{(b)}(u) = m_j^{(b)} + 1, \quad u = 1,2,\ldots,z, \end{cases} \tag{5.17}$$

for $j = 1, 2, \ldots, n^{(b)}$.

In the case if there exist j, $j \in \{1, 2, \ldots, n^{(b)}\}$, such that $m_j^{(b)}(u) = m_j^{(b)}$, $u = 1, 2, \ldots, z$, assuming the observation time $\tau^{(b)}$ as the moment of departures from the reliability or safety state subset $\{u, u+1, \ldots, z\}$, $u = 1, 2, \ldots, z$, of the last components on all experimental posts that have not left this reliability or safety state subset we get a so-called pessimistic evaluation of the intensity of departure $[\lambda(u)]^{(b)}$ from the reliability or safety state subset $\{u, u+1, \ldots, z\}$, $u = 1, 2, \ldots, z$, of the form

$$[\hat{\lambda}(u)]^{(b)} = \frac{\sum_{j=1}^{n^{(b)}} m_j^{(b)} + n^{(b)}}{\sum_{j=1}^{n^{(b)}} \sum_{i=1}^{m_j^{(b)}(u)} \left[t_i^{(b)}(u)\right]^j + \sum_{j=1}^{n^{(b)}} d_j^{(b)}(u)}, \quad u = 1, 2, \ldots, z. \tag{5.18}$$

Case 6 The maximum likelihood evaluation of the unknown component intensity of departure $[\lambda(u)]^{(b)}$ from the reliability or safety state subset $\{u, u+1, \ldots, z\}$, $u = 1, 2, \ldots, z$, is either

$$[\hat{\lambda}(u)]^{(b)} = \frac{\sum_{j=1}^{n^{(b)}} m_j^{(b)}(u)}{\sum_{j=1}^{n^{(b)}} \sum_{i=1}^{m_j^{(b)}(u)} \left[t_i^{(b)}(u)\right]^j + \sum_{j=1}^{n^{(b)}} \bar{d}_j^{(b)}(u)}, \quad u = 1, 2, \ldots, z, \tag{5.19}$$

where

$$\bar{d}_j^{(b)}(u) = \begin{cases} \tau_j^{(b)} - \sum_{i=1}^{m_j^{(b)}(u)} \left[t_i^{(b)}(1)\right]^{(j)}, & \text{if } m_j^{(b)}(u) = m_j^{(b)} \\ 0, & \text{if } m_j^{(b)}(u) = m_j^{(b)} + 1, \quad u = 1,2,\ldots,z, \end{cases} \tag{5.20}$$

for $j = 1, 2, \ldots, n^{(b)}$.

In the case when there exist j, $j \in \{1, 2, \ldots, n^{(b)}\}$, such that $m_j^{(b)}(u) = m_j^{(b)}$, $u = 1, 2, \ldots, z$, assuming the observation times $\tau_j^{(b)}$, $j = 1, 2, \ldots, n^{(b)}$, as the moments of departures from the reliability or safety state subset $\{u, u+1, \ldots, z\}$, $u = 1, 2, \ldots, z$, of the last components on experimental posts that have not left this reliability or safety state subset, we get a so-called pessimistic evaluation of the intensity of departure $[\lambda(u)]^{(b)}$ from the reliability or safety state subset $\{u, u+1, \ldots, z\}$, $u = 1, 2, \ldots, z$, of the form

$$[\hat{\lambda}(u)]^{(b)} = \frac{\sum_{j=1}^{n^{(b)}} m_j^{(b)} + n^{(b)}}{\sum_{j=1}^{n^{(b)}} \sum_{i=1}^{m_j^{(b)}(u)} \left[t_i^{(b)}(u)\right]^j + \sum_{j=1}^{n^{(b)}} \bar{d}_j^{(b)}(u)}, \quad u = 1, 2, \ldots, z. \tag{5.21}$$

5.2.2.2 Evaluating System Components Intensities of Departure from Reliability and Safety State Subsets on the Basis of Data Coming from Experts

On the basis of the approximate values

$$[\hat{\mu}(u)]^{(b)}, \quad u = 1, 2, \ldots, z, \ b = 1, 2, \ldots, v,$$

of the mean values

$$[\mu(u)]^{(b)} = E[T(u)]^{(b)}, \quad u = 1, 2, \ldots, z, \ b = 1, 2, \ldots, v,$$

of the system component lifetimes $[T(u)]^{(b)}$, $u = 1, 2, \ldots, z$, $b = 1, 2, \ldots, v$, in the reliability or safety state subset $\{u, u+1, \ldots, z\}$, $u = 1, 2, \ldots, z$, while the system is operating at the operation state z_b, $b = 1, 2, \ldots, v$ coming from experts and described in Sect. 5.2.1.2, we want to estimate the values $[\hat{\lambda}(u)]^{(b)}$ of the components unknown intensities $[\lambda(u)]^{(b)}$ of departure from the reliability or safety state subset $\{u, u+1, \ldots, z\}$, $u = 1, 2, \ldots, z$, while the system is operating at the operation state z_b, $b = 1, 2, \ldots, v$

The formula for all system components is given by the following approximate equation

$$[\lambda(u)]^{(b)} \cong [\hat{\lambda}(u)]^{(b)} = \frac{1}{[\hat{\mu}(u)]^{(b)}}, \quad u = 1, 2, \ldots, z, \ b = 1, 2, \ldots, v. \tag{5.22}$$

5.2.3 Identification of Conditional Multistate Exponential Reliability and Safety Functions of System Components

Prior to the identification of the system components conditional multistate exponential reliability or safety functions, we have to decide whether we have statistical data coming from components reliability states changing processes or from experts and after that we have to select one of two procedures of estimation presented below, in Sects. 5.2.3.1 and 5.2.3.2, respectively.

5.2.3.1 Identifying System Components Conditional Multistate Exponential Reliability and Safety Functions on the Basis of Data Coming from Components Reliability and Safety States Changing Processes

Having the evaluation of the unknown intensity $[\lambda(u)]^{(b)}$ of the component departure from the reliability or safety state subset, performed according to one of the formulae given in Sect. 5.2.2.1 on the basis of the realizations of the system component lifetime $T^{(b)}(u)$ in the reliability or safety state subset $\{u, u+1, \ldots, z\}$ at the operation state z_b, described in Sect. 5.2.1.1, we may formulate and then verify the non-parametric hypothesis concerning the exponential form of the coordinates $[R(t,u)]^{(b)}$, $u = 1, 2, \ldots, z$, $b = 1, 2, \ldots, v$, of the vector of its conditional multistate reliability function $[R(t,\ \cdot)]^{(b)}$ or the coordinates $[s(t,u)]^{(b)}$, $u = 1, 2, \ldots, z$, $b = 1, 2, \ldots, v$, of the vector of its conditional multistate safety function $[s(t,\ \cdot)]^{(b)}$. The procedures for the identification of the system components' reliability functions and safety functions are analogous. Therefore, we present the way of the identification of the reliability functions only.

To perform this task in the reliability aspect, it is due to act according to the scheme below:

- to formulate the null hypothesis H_0, concerned with the form of the component multistate reliability $[R(t,\ \cdot)]^{(b)}$in the following form:
  H_0: The conditional multistate reliability function of the system component

$$[R(t,\ \cdot)]^{(b)} = [1,\ [R(t,\ 1)]^{(b)}, \ldots,\ [R(t,z)]^{(b)}], \tag{5.23}$$

  has the exponential reliability function coordinates of the forms

$$[R(t,u)]^{(b)} = \exp[-[\lambda(u)]^{(b)} t] \quad \text{for } t \in \langle 0, \infty), \quad u = 1, 2, \ldots, z; \tag{5.24}$$

- to fix the number $\bar{n}^{(b)}$, $b = 1, 2, \ldots, v$of observed components' realizations of the system component conditional lifetime $T^{(b)}(1)$, $b = 1, 2, \ldots, v$ in the reliability state subset $\{1, 2, \ldots, z\}$, i.e., the number of observed components that reached the worst reliability state 0, according to the formula

$$\bar{n}^{(b)} = \begin{cases} n^{(b)} & \text{in Cases } 1, 2, 3 \\ m^{(b)} + 1 & \text{in Case } 4 \\ \sum_{j=1}^{n^{(b)}} m_j^{(b)} + n^{(b)} & \text{in Cases } 5, 6; \end{cases}$$

- to fix the number $\bar{m}^{(b)}(u)$, $u = 1, 2, \ldots, z$, $b = 1, 2, \ldots, v$, of realizations of the system component conditional lifetime $T^{(b)}(u)$, $b = 1, 2, \ldots, v$, in the reliability state subset $\{u, u+1, \ldots, z\}$, $u = 1, 2, \ldots, z$,

$$\bar{m}^{(b)}(u) = \begin{cases} m^{(b)}(u) & \text{in Cases } 1,2,3,4 \\ = \sum_{j=1}^{n^{(b)}} m_j^{(b)}(u) & \text{in Cases } 5,6; \end{cases}$$

- to fix the realizations $t_1^{(b)}(u),\ t_2^{(b)}(u), \ldots, t_{\bar{m}^{(b)}(u)}^{(b)}(u),\ u = 1,2,\ldots,z$, of the system component conditional lifetime $T^{(b)}(u),\ b = 1,2,\ldots,v$, in the reliability state subset $\{u, u+1, \ldots, z\},\ u = 1,2,\ldots,z$,
- to determine the number $\bar{r}^{(b)}(u)$ of disjoint intervals $I_j^{(b)}(u) = \langle x_j^{(b)}(u), y_j^{(b)}(u)),$ $j = 1,2,\ldots,\bar{r}^{(b)}(u)$, that include the realizations $t_2^{(b)}(u),\ t_1^{(b)}(u), \ldots, t_{\bar{m}^{(b)}(u)}^{(b)}(u)$, $u = 1,2,\ldots,z$, of the system component conditional lifetime $T^{(b)}(u)$ in the reliability state subset $\{u, u+1, \ldots, z\},\ u = 1,2,\ldots,z$, according to the formula

$$\bar{r}^{(b)}(u) \cong \sqrt{\bar{m}^{(b)}(u)};$$

- to determine the length $d^{(b)}(u)$ of the intervals $I_j^{(b)}(u) = \langle x_j^{(b)}(u), y_j^{(b)}(u)),$ $j = 1,2,\ldots,\bar{r}^{(b)}(u)$, according to the formula

$$d^{(b)}(u) = \frac{\bar{R}^{(b)}(u)}{\bar{r}^{(b)}(u) - 1},$$

where

$$\bar{R}^{(b)}(u) = \max_{1 \le i \le \bar{m}^{(b)}(u)} \left\{t_i^{(b)}(u)\right\} - \min_{1 \le i \le m^{(b)}(u)} \left\{t_i^{(b)}(u)\right\};$$

- to determine the ends $x_j^{(b)}(u),\ y_j^{(b)}(u)$, of the intervals $I_j^{(b)}(u) = \langle x_j^{(b)}(u), y_j^{(b)}(u)),$ $j = 1,2,\ldots,\bar{r}^{(b)}(u)$, according to the formulae

$$x_1^{(b)}(u) = \max\left\{\min_{1 \le i \le \bar{m}^{(b)}(u)} \left\{t_i^{(b)}(u)\right\} - \frac{d^{(b)}(u)}{2},\ 0\right\}$$

$$y_j^{(b)}(u) = x_1^{(b)}(u) + jd^{(b)}(u), \quad j = 1,2,\ldots,\bar{r}^{(b)}(u),$$

$$x_j^{(b)}(u) = y_{j-1}^{b)}(u), \quad j = 2,3,\ldots,\bar{r}^{(b)}(u),$$

in a way such that

$$I_1^{(b)}(u) \cup I_2^{(b)}(u) \cup \ldots \cup I_{\bar{r}^{(b)}(u)}^{(b)}(u) = \langle x_1^{(b)}(u), y_{\bar{r}^{(b)}(u)}^{(b)}(u)),$$

and

$$I_i^{(b)}(u) \cap I_j^{(b)}(u) = \varnothing \quad \text{for all } i \neq j, \quad i,j \in \left\{1, 2, \ldots, \bar{r}^{(b)}(u)\right\}$$

and the ends of the interval $I^{(b)}_{\bar{r}^{(b)}(u)+1}(u) = \langle x^{(b)}_{\bar{r}^{(b)}(u)+1}(u), y^{(b)}_{\bar{r}^{(b)}+1}(u)\rangle$, including the remaining unknown realizations, assuming

$$x^{(b)}_{\bar{r}^{(b)}(u)+1}(u) = y^{(b)}_{\bar{r}^{(b)}}(u), \quad y^{(b)}_{\bar{r}^{(b)}+1}(u) = +\infty;$$

- to determine the number of realizations $n_j^{(b)}(u)$ in particular intervals $I_j^{(b)}(u), j = 1, 2, \ldots, \bar{r}^{(b)}(u) + 1$,according to the formula

$$n_j^{(b)}(u) = \#\left\{i : t_i^{(b)}(u) \in I_j^{(b)}(u), \quad i \in \left\{1, 2, \ldots, \bar{m}^{(b)}(u)\right\}\right\}, \quad j = 1, 2, \ldots, \bar{r}^{(b)}(u),$$

$$n^{(b)}_{\bar{r}^{(b)}(u)+1}(u) = \bar{n}^{(b)} - \bar{m}^{(b)}(u)$$

where

$$\sum_{j=1}^{\bar{r}^{(b)}(u)+1} n_j^{(b)}(u) = n^{(b)},$$

whereas the symbol # means the number of elements of a set;

- to construct the realization of the histogram of the conditional system component lifetime $T^{(b)}(u)$, $b = 1, 2, \ldots, v$, in the reliability state subset $\{u, u + 1, \ldots, z\}$, $u = 1, 2, \ldots, z$, at the system operation state z_b, $b = 1, 2, \ldots, v$,

$$\bar{f}^{(b)}_{n^{(b)}}(t, u) = \frac{n_j^{(b)}(u)}{\bar{n}^{(b)}} \quad \text{for } t \in I_j^{(b)}(u);$$

- to join each of the intervals $I_j^{(b)}(u)$, $j = 1, 2, \ldots, \bar{r}^{(b)}(u)$ that has the number $n_j^{(b)}(u)$ of realizations <4 either with the neighbor interval $I_{j+1}^{(b)}(u)$ or with the neighbor interval $I_{j-1}^{(b)}(u)$, in such a way that the number of realizations in all intervals are not <4, and to join the interval $I^{(b)}_{\bar{r}^{(b)}(u)+1}(u)$ including unknown realizations with its neighboring new interval;
- to fix a new number of intervals

$$\bar{\bar{r}}^{(b)}(u),$$

- to determine new intervals

$$\bar{I}_j^{(b)}(u) = \langle \bar{x}_j^{(b)}(u), \bar{y}_j^{(b)}(u) \rangle, \quad j = 1, 2, \ldots, \bar{\bar{r}}^{(b)}(u);$$

- to fix the number $\bar{n}_j^{(b)}(u)$ of realizations in new intervals $\bar{I}_j^{(b)}(u)$, $j = 1, 2, \ldots, \bar{\bar{r}}^{(b)}(u)$;
- to calculate the hypothetical probabilities so that the lifetime $T^{(b)}(u)$ takes values from the interval $\bar{I}_j^{(b)}(u)$, under the assumption that the hypothesis H_0 is true, i.e. the probabilities

$$\begin{aligned} p_j^{(b)}(u) &= P\left(T^{(b)}(u) \in \bar{I}_j^{(b)}(u)\right) = P\left(\bar{x}_j^{(b)}(u) \le T^{(b)}(u) < \bar{y}_j^{(b)}(u)\right) \\ &= \left[R(\bar{x}_j^{(b)}(u), u)\right]^{(b)} - \left[R(\bar{y}_j^{(b)}(u), u)\right]^{(b)}, \ j = 1, 2, \ldots, \bar{\bar{r}}^{(b)}(u), \end{aligned} \tag{5.25}$$

where $[R(\bar{x}_j^{(b)}(u), u)]^{(b)}$ and $[R(\bar{y}_j^{(b)}(u), u)]^{(b)}$ are the values of the coordinate reliability function $[R(t, u)]^{(b)}$ of the multistate reliability function defined by (5.23–5.24) in the null hypothesis H_0;

- to calculate the realization u_n of the χ^2 (chi-square)-Pearson's statistics U_n, according to the formula

$$u_n = \sum_{j=1}^{\bar{\bar{r}}^{(b)}(u)} \frac{(\bar{n}_j^{(b)}(u) - n^{(b)} p_j^{(b)}(u))^2}{n^{(b)} p_j^{(b)}(u)}; \tag{5.26}$$

- to assume the significance level α ($\alpha = 0.01$, $\alpha = 0.02$, $\alpha = 0.05$ or $\alpha = 0.10$) of the test;
- to fix the number $\bar{\bar{r}}^{(b)}(u) - l - 1$ of degrees of freedom, substituting $l = 1$;
- to read from the Tables of the χ^2-Pearson's distribution the value u_α for the fixed values of the significance level α and the number of degrees of freedom $\bar{\bar{r}}(u) - l - 1$ such that the following equality holds

$$P(U_n > u_\alpha) = \alpha, \tag{5.27}$$

and then to determine the critical domain in the form of the interval $(u_\alpha, +\infty)$ and the acceptance domain in the form of the interval $\langle 0, u_\alpha \rangle$ (see: Fig. 5.9);

- to compare the obtained value u_n of the realization of the statistics U_n with that read from the Tables' critical value u_α of the chi-square random variable and to verify the previously formulated null hypothesis H_0 in the following way: if the value u_n does not belong to the critical domain, i.e. when $u_n \le u_\alpha$, we do not reject the hypothesis H_0, if the value u_n belongs to the critical domain, i.e. when $u_n > u_\alpha$, we reject the hypothesis H_0.

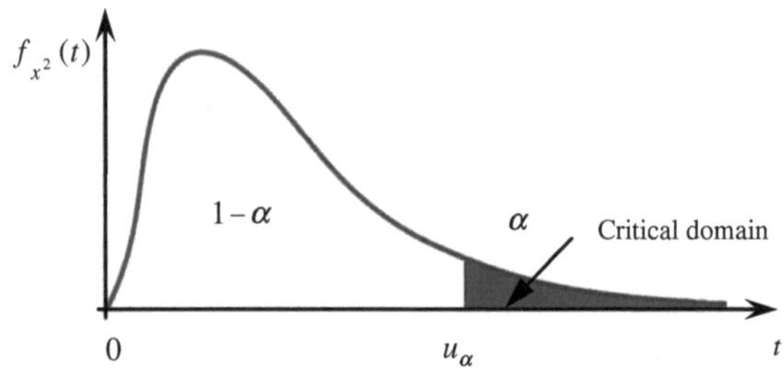

Fig. 5.9 The graphical interpretation of the critical interval and the acceptance interval for the chi-square goodness-of-fit test

5.2.3.2 Identifying System Components Conditional Multistate Exponential Reliability and Safety Functions on Basis of Data Coming from Experts

In this case, as we do not have statistical data from the system components' reliability or safety states changing processes, we arbitrarily assume the conditional multistate exponential reliability or safety functions with the approximately estimated intensities of departures from the reliability or safety states according to the formula (5.22) given in Sect. 5.2.2.2.

5.3 Applications

5.3.1 Statistical Identification of Exemplary System Components Reliability

The considered exemplary system reliability structure changing at the various operation states and its components and their unknown reliability parameters are described in Sects. 2.3.1 and 3.6.1.

5.3.1.1 Defining Parameters of Exemplary System Components Reliability Models and Data Collection

At all the system operation process states z_b, $b = 1, 2, 3, 4$, defined in Sect. 2.3.1, we distinguish the following four reliability states 0, 1, 2, 3, ($z = 3$) of the system and its components, defined in Sect. 1.4.1. Moreover, we fix the possible transitions between the components' reliability states only from better to worse.

From the above, the subsystems S_υ, $\upsilon = 1, 2$, are composed of four-state, i.e. $z = 3$, components $E_{ij}^{(\upsilon)}$, $\upsilon = 1, 2$, with the conditional four-state reliability functions given by the vector

$$\left[R_{ij}^{(\upsilon)}(t,\cdot)\right]^{(b)} = 1\left[, \left[R_{ij}^{(v)}(t,1)\right]^{(b)}, \left[R_{ij}^{(\upsilon)}(t,2)\right]^{(b)}, \left[R_{ij}^{(\upsilon)}(t,3)\right]^{(b)}\right], \quad b = 1,2,3,4, \tag{5.28}$$

with the exponential co-ordinates

$$\left[R_{ij}^{(\upsilon)}(t,1)\right]^{(b)} = \exp\left[-\left[\lambda_{ij}^{(\upsilon)}(1)\right]^{(b)}\right], \left[R_{ij}^{(\upsilon)}(t,2)\right]^{(b)} = \exp\left[-\left[\lambda_{ij}^{(\upsilon)}(2)\right]^{(b)}\right],$$
$$\left[R_{ij}^{(\upsilon)}(t,3)\right]^{(b)} = \exp\left[-\left[\lambda_{ij}^{(\upsilon)}(3)\right]^{(b)}\right], \tag{5.29}$$

different at various operation states z_b, $b = 1,2,3,4$, and with the intensities of departure from the reliability state subsets $\{1,2,3\}$, $\{2,3\}$, $\{3\}$, respectively

$$\left[\lambda_{ij}^{(\upsilon)}(1)\right]^{(b)}, \left[\lambda_{ij}^{(\upsilon)}(1)\right]^{(b)}, \left[\lambda_{ij}^{(\upsilon)}(3)\right]^{(b)}, \ b = 1,2,3,4.$$

5.3.1.2 Collecting Data Coming from Exemplary System Components Reliability States Changing Processes

We arbitrarily suppose that we have at disposal data collected from the exemplary system components' reliability states changing processes due to *Case 2* described in Sect. 5.2.1.1. Namely, we have at disposal the following data for particular components $E_{ij}^{(\upsilon)}$, $\upsilon = 1,2$, of the exemplary system:

- the number of identical experiment posts $n^{(b)} = n_{ij}^{(b)}$,
- the observation times $\tau^{(b)} = \tau_{ij}^{(b)}$,
- the numbers $m^{(b)}(u) = m_{ij}^{(b)}(u)$ of components that have left the reliability state subsets $\{u, u+1, \ldots, 3\}$, $u = 1,2,3$,
- the sets $A^{(b)}(u) = A_{ij}^{(b)}(u) = \{t_i^{(b)}(u) : i = 1,2,\ldots,m^{(b)}(u)\}$ of realizations $t_i^{(b)}(u) = t_{ij}^{(b)}(u)$ of the component lifetimes $T_{ij}^{(b)}(u)$ in the reliability state subsets $\{u, u+1, \ldots, 3\}$, $u = 1,2,3$, at the operation state z_b, $b = 1,2,3,4$.

For instance, we suppose that the collected data for the component $E_{11}^{(1)}$ of the subsystem S_1 at the operation state z_1 are as follows:

$$n^{(1)} = 40, \quad \tau^{(1)} = 2600, \ m^{(1)}(1) = 32,$$

$$A^{(1)}(1) = \{30, 44, 209, 240, 263, 265, 280, 285, 288, 289, 289, 302, 307, 350, 381, 400, 430, 441, 452, 490, 490, 790, 837, 852, 856, 869, 1176, 1191, 1253, 1697, 1700, 2454\}, \tag{5.30}$$

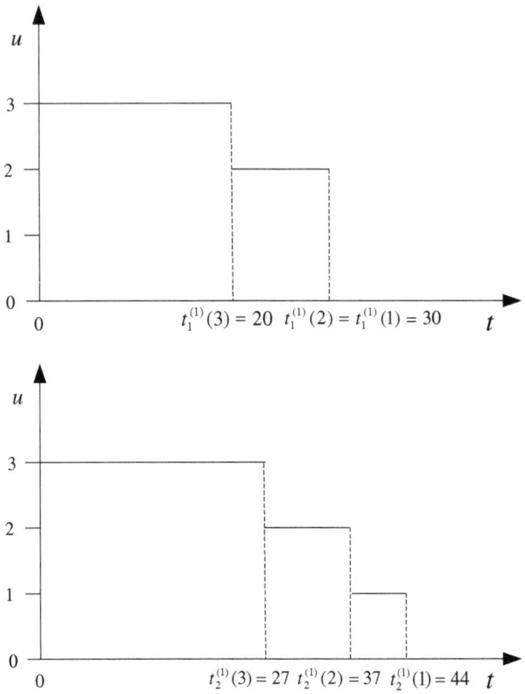

Fig. 5.10 The realizations of the components $E_{11}^{(1)}$ lifetimes $T_{11}^{(1)}(1)$, $T_{11}^{(1)}(2)$ and $T_{11}^{(1)}(3)$ in the reliability state subsets $\{1,2,3\}$, $\{2,3\}$, and $\{3\}$

$$n^{(1)} = 40, \quad \tau^{(1)} = 2600, \ m^{(1)}(2) = 32,$$

$$A^{(1)}(2) = \{30,\ 37,\ 37,\ 60,\ 63,\ 65,\ 69,\ 69,\ 80,\ 85,\ 88,\ 302,\ 307,\ 350,\ 352,\ 381,\ 400,\ 430,\ 441,\ 462,\ 470,\ 490,\ 637,\ 652,\ 656,\ 669,\ 776,\ 891,\ 1053,\ 1597,\ 1600,\ 2254\} \tag{5.31}$$

$$n^{(1)} = 40, \quad \tau^{(1)} = 2600, \ m^{(1)}(3) = 32,$$

$$A^{(1)}(3) = \{20,\ 27,\ 37,\ 60,\ 63,\ 65,\ 69,\ 69,\ 80,\ 85,\ 88,\ 302,\ 307,\ 350,\ 352,\ 381,\ 400,\ 430,\ 441,\ 462,\ 470,\ 490,\ 637,\ 652,\ 656,\ 669,\ 776,\ 891,\ 1053,\ 1597,\ 1600,\ 2054\}. \tag{5.32}$$

The first realizations $t_1^{(1)}(1) = 30$, $t_1^{(1)}(2) = 30$, $t_1^{(1)}(3) = 20$ and the second realizations $t_2^{(1)}(1) = 44$, $t_2^{(1)}(2) = 37$, $t_2^{(1)}(3) = 27$ of the component $E_{11}^{(1)}$ lifetimes $T_{11}^{(1)}(1)$, $T_{11}^{(1)}(2)$ and $T_{11}^{(1)}(3)$ in the reliability state subsets $\{1,2,3\}$, $\{2,3\}$, and $\{3\}$ taken from the sets $A_1^{(1)}(1)$, $A_1^{(1)}(2)$ and $A_1^{(1)}(3)$ and are presented in Fig. 5.10.

5.3.1.3 Evaluating Exemplary System Components Intensities of Departures from Reliability State Subsets on Basis of Data Coming from Components Reliability States Changing Processes

As by the arbitrary assumption, there are data collected from the exemplary system components' reliability states changing processes, their reliability models identification using the methods of Sect. 5.2.2.1 is possible.

To identify the intensities of departures from the reliability state subsets, we can use statistical data included in Sect. 5.3.1.2 and formula (5.9) in order to find the approximate values $[\hat{\lambda}_{ij}^{(\upsilon)}(1)]^{(b)}$, $[\hat{\lambda}_{ij}^{(\upsilon)}(2)]^{(b)}$ and $[\hat{\lambda}_{ij}^{(\upsilon)}(3)]^{(b)}$ of the subsystems S_υ, $\upsilon = 1, 2$, components unknown intensities $[\lambda_{ij}^{(\upsilon)}(1)]^{(b)}$, $[\lambda_{ij}^{(\upsilon)}(2)]^{(b)}$ and $[\lambda_{ij}^{(\upsilon)}(3)]^{(b)}$ of departure respectively from the reliability states subsets $\{1, 2, 3\}$, $\{2, 3\}$, $\{3\}$, while the system is operating at the operation state z_b, $b = 1, 2, 3, 4$, and we can use the formula (5.10) to get their pessimistic evaluations.

To illustrate this procedure, we find the evaluations $[\hat{\lambda}_{11}^{(1)}(1)]^{(1)}$, $[\hat{\lambda}_{11}^{(1)}(2)]^{(1)}$ and $[\hat{\lambda}_{11}^{(1)}(3)]^{(1)}$ of the intensities $[\lambda_{11}^{(1)}(1)]^{(1)}$, $[\lambda_{11}^{(1)}(2)]^{(1)}$ and $[\lambda_{11}^{(1)}(3)]^{(1)}$ of departures respectively from the reliability state subsets $\{1, 2, 3\}, \{2, 3\},$ and $\{3\}$ of the component $E_{11}^{(1)}$ of the subsystem S_1, while the system is operating in the operation state z_1.

We proceed as follows:

- from data (5.30), we have

$$n^{(1)} = 40, \quad \tau^{(1)} = 2600, \; m^{(1)}(1) = 32,$$

$$\sum_{i=1}^{m^{(1)}(1)} t_1^{(1)}(1) = 30 + 44 + \cdots + 1700 + 2454 = 20200,$$

then, according to (5.9), the evaluations $[\hat{\lambda}_{11}^{(1)}(1)]^{(1)}$ of the intensity $[\lambda_{11}^{(1)}(1)]^{(1)}$ of departure from the reliability state subset $\{1, 2, 3\}$ is

$$\left[\hat{\lambda}_{11}(1)\right]^{(1)} = \frac{m^{(1)}(1)}{\sum_{i=1}^{m^{(1)}(1)} t_i^{(1)}(1) + \tau^{(1)}[n^{(1)} - m^{(1)}(1)]} = \frac{32}{20200 + 2600[40 - 32]} \cong 0.0008 \tag{5.33}$$

and according to (5.10), its pessimistic evaluation is

$$[\hat{\lambda}_{11}(1)]^{(1)} = \frac{n^{(1)}}{\sum_{i=1}^{m^{(1)}(1)} t_i^{(1)}(1) + \tau^{(1)}[n^{(1)} - m^{(1)}(1)]} = \frac{40}{20200 + 2600[40 - 32]} \cong 0.0010.$$

- from data (5.31), we have

$$n^{(1)} = 40, \quad \tau^{(1)} = 2600, \ m^{(1)}(2) = 32,$$

$$\sum_{i=1}^{m^{(1)}(2)} t_1^{(1)}(2) = 30 + 37 + \cdots + 1600 + 2254 = 15853,$$

then, according to (5.9), the evaluations $[\hat{\lambda}_{11}^{(1)}(2)]^{(1)}$ of the intensity $[\lambda_{11}^{(1)}(2)]^{(1)}$ of departure from the reliability state subset $\{2, 3\}$ is

$$\left[\hat{\lambda}_{11}(2)\right]^{(1)} = \frac{m^{(1)}(2)}{\sum_{i=1}^{m^{(1)}(2)} t_i^{(1)}(2) + \tau^{(1)}[n^{(1)} - m^{(1)}(2)]} = \frac{32}{15853 + 2600[40 - 32]} \cong 0.0009 \qquad (5.34)$$

and according to (5.10), its pessimistic evaluation is

$$[\hat{\lambda}_{11}(2)]^{(1)} = \frac{n^{(1)}}{\sum_{i=1}^{m^{(1)}(2)} t_i^{(1)}(2) + \tau^{(1)}[n^{(1)} - m^{(1)}(2)]} = \frac{40}{15853 + 2600[40 - 32]} \cong 0.0011.$$

- from data (5.32), we have

$$n^{(1)} = 40, \quad \tau^{(1)} = 2600, \ m^{(1)}(3) = 32,$$

$$\sum_{i=1}^{m^{(1)}(3)} t_1^{(1)}(3) = 20 + 27 + \cdots + 1600 + 2054 = 15633,$$

then, according to (5.9), the evaluations $[\hat{\lambda}_{11}^{(1)}(3)]^{(1)}$ of the intensity $[\lambda_{11}^{(1)}(3)]^{(1)}$ of departure from the reliability state subset $\{3\}$ is

$$\left[\hat{\lambda}_{11}(3)\right]^{(1)} = \frac{m^{(1)}(3)}{\sum_{i=1}^{m^{(1)}(3)} t_i^{(1)}(3) + \tau^{(1)}[n^{(1)} - m^{(1)}(3)]} = \frac{32}{15633 + 2600[40 - 32]} \cong 0.0009 \qquad (5.35)$$

and according to (5.10), its pessimistic evaluation is

$$\left[\hat{\lambda}_{11}(3)\right]^{(1)} = \frac{m^{(1)}(3)}{\sum_{i=1}^{m^{(1)}(3)} t_i^{(1)}(3) + \tau^{(1)}[n^{(1)} - m^{(1)}(3)]} = \frac{40}{15633 + 2600[40 - 32]}$$
$$\cong 0.0011.$$

The evaluations of the intensities of departure $[\lambda_{21}^{(1)}(u)]^{(1)}$, $u = 1, 2, 3$, of the component $E_{21}^{(1)}$ are the same as that determined by (5.33–5.35) as this component is identical with the component $E_{11}^{(1)}$.

In this way, we may obtain the evaluations of the unknown intensities of departure for all the remaining system components.

Substituting the evaluations of the intensities of departures respectively into the formulae (5.29), we get the exponential coordinates of the exemplary system components' reliability functions (5.28) which after testing, partly presented in Sect. 5.3.1.3, are used in Sect. 3.6.1 of Chap. 3 for evaluation and prediction of this system reliability.

5.3.1.4 Identifying Exemplary System Components Conditional Multistate Exponential Reliability Functions on the Basis of Data Coming from System Components Reliability States Changing Processes

As by the arbitrary assumption, there are data collected from the system components reliability states changing processes, it is possible to verify the hypotheses on the exponential forms of the system components' conditional reliability functions. To this end, we use the procedure given in Sect. 5.2.3.1.

Applying this procedure and using the statistical data from Sect. 5.3.1.2 and the results from Sect. 5.3.1.3, we may verify the hypotheses on the conditional exponential four-state exemplary system components' reliability functions $[R_{ij}^{(\upsilon)}(t, \cdot)]^{(b)}$, $\upsilon = 1, 2$, $b = 1, 2, 3, 4$, at the particular operation states z_b, $b = 1, 2, 3, 4$. To do this, we need a sufficient number of realizations of the system components lifetime in the reliability state subsets [2, 4, 18, 23, 24], namely the sets of their realizations should contain at least 30 realizations coming from the experiment. This condition is satisfied for the statistical data we have at disposal which are partly presented in Sect. 5.3.1.2. To make the procedure familiar to the reader, we perform it for the conditional reliability function of the subsystem S_1 component $E_{11}^{(1)}$ at the system operation state z_1.

Considering the evaluated values of the unknown intensities of the component $E_{11}^{(1)}$ departure from the reliability state subsets given by (5.33–5.35), we formulate the null hypothesis H_0 and the alternative hypothesis H_A, concerned with the form of its multistate reliability $[R(t, \cdot)]^{(1)}$ in the following form:

H_0: The conditional multistate reliability function of the system component $E_{11}^{(1)}$ at the operation state z_1

$$\left[R_{11}^{(1)}(t, \cdot)\right]^{(1)} = \left[1, \left[R_{11}^{(1)}(t, 1)\right]^{(1)}, \left[R_{11}^{(1)}(t, 2)\right]^{(1)}, \left[R_{11}^{(1)}(t, 3)\right]^{(1)}\right],$$

has the exponential reliability function coordinates of the form

$$\left[R_{11}^{(1)}(t,1)\right]^{(1)} = \exp[-0.0008t], \ \left[R_{11}^{(1)}(t,2)\right]^{(1)} = \exp[-0.0009t],$$

$$[R_{11}^{(1)}(t,3)]^{(1)} = \exp[-0.0009t] \quad \text{for } t \in \langle 0, \infty). \tag{5.36}$$

To verify the hypothesis concerning the exponential form of the coordinate $[R_{11}^{(1)}(t, 1)]^{(1)}$ defined by (5.36), it is due to act according to the scheme below:

- we fix the number of observed components and the number of realizations of the exemplary system component conditional lifetime $T_{11}^{(1)}(1)$ in the reliability state subset $\{1, 2, 3\}$ which according to (5.30) are

$$\bar{n}^{(1)} = n^{(1)} = 40, \ \bar{m}^{(1)} = m^{(1)}(1) = 32,$$

- we fix the realizations $t_i^{(1)}(1)$, $i = 1, 2, \ldots, 32$, of the exemplary system component conditional lifetime $T_{11}^{(1)}(1)$ in the reliability state subset $\{1, 2, 3\}$ that are given by (5.30),
- we determine the number $\bar{r}^{(1)}(1)$ of the disjoint intervals $I_j^{(1)}(1) = \langle x_j^{(1)}(1), y_j^{(1)}(1))$ $j = 1, 2, \ldots, \bar{r}^{(1)}(1)$, which include the realizations $t_i^{(1)}(1)$, $i = 1, 2, \ldots, 32$, of the system component conditional lifetime $T_{11}^{(1)}(1)$ in the reliability state subset $\{1, 2, 3\}$

$$\bar{r}^{(1)}(1) \cong \sqrt{\bar{m}^{(1)}(1)} = \sqrt{32} \cong 6,$$

- we determine the length $d^{(1)}(1)$ of the intervals $I_j^{(1)}(1) = \langle x_j^{(1)}(1), y_j^{(1)}(1))$, $j = 1, 2, \ldots, 6$, which after considering

$$\bar{R}^{(1)}(1) = \max_{1 \le i \le 32} \{t_i^{(1)}(1)\} - \min_{1 \le i \le 32} \{t_i^{(1)}(1)\} = 2454 - 30 = 2424,$$

is

$$d^{(1)}(1) = \frac{2424}{6 - 1} \cong 485,$$

- we determine the ends $x_j^{(1)}(1)$, $y_j^{(1)}(1)$, of the intervals $I_j^{(1)}(1) = \langle x_j^{(1)}(1), y_j^{(1)}(1))$, $j = 1, 2, \ldots, 6$, which after considering

$$\min_{1 \le i \le 32}\{t_i^{(1)}(1)\} - \frac{d^{(1)}}{2} = 30 - 242.5 = -212.5,$$

are

$$x_1^{(1)}(1) = \max\{-212.5,\ 0\} = 0,\quad y_1^{(1)}(1) = x_1^{(1)}(1) + 1d^{(1)}(1) = 0 + 1 \cdot 485 = 485,$$

$$x_2^{(1)}(1) = y_1^{(1)}(1) = 485,\quad y_2^{(1)}(1) = x_1^{(1)}(1) + 2d^{(1)}(1) = 0 + 2 \cdot 485 = 970,$$

$$x_3^{(1)}(1) = y_2^{(1)}(1) = 970,\quad y_3^{(1)}(1) = x_1^{(1)}(1) + 3d^{(1)}(1) = 0 + 3 \cdot 485 = 1455,$$

$$x_4^{(1)}(1) = y_3^{(1)}(1) = 1455,\quad y_4^{(1)}(1) = x_1^{(1)}(1) + 4d^{(1)}(1) = 0 + 4 \cdot 485 = 1940,$$

$$x_5^{(1)}(1) = y_4^{(1)}(1) = 1940,\quad y_5^{(1)}(1) = x_1^{(1)}(1) + 5d^{(1)}(1) = 0 + 5 \cdot 485 = 2425,$$

$$x_6^{(1)}(1) = y_5^{(1)}(1) = 2425,\quad y_6^{(1)}(1) = x_1^{(1)}(1) + 6d^{(1)}(1) = 0 + 6 \cdot 485 = 2910,$$

and the ends of the interval $I_7^{(1)}(1) = \langle x_7^{(1)}(1), y_7^{(1)}(1))$, including the remaining unknown realizations

$$x_7^{(1)}(1) = y_6^{(1)}(1) = 2910,\quad y_7^{(1)}(1) = +\infty,$$

- we determine the numbers of realizations $n_j^{(1)}(1)$ in particular intervals $I_j^{(1)}(1) = \langle x_j^{(1)}(1), y_j^{(1)}(1)),\ j = 1, 2, \ldots, 7,$

$$n_1^{(1)}(1) = 19,\ n_2^{(1)}(1) = 7,\ n_3^{(1)}(1) = 3,\ n_4^{(1)}(1) = 2,\ n_5^{(1)}(1) = 0,$$
$$n_6^{(1)}(1) = 1, n_7^{(1)}(1) = 40 - 32 = 8,$$

- we find the realization $\bar{f}_{40}(t, 1)$ of the histogram of the exemplary system component conditional lifetime $T_{11}^{(1)}(1)$ which is presented in Table 5.1,
- we join the intervals defined in the realization of the histogram $\bar{f}_{40}(t, 1)$, which have the numbers $n_j^{(1)}(1)$, of realizations <4 into new intervals and the interval $I_7^{(1)}(1)$with its new neighboring interval,
- we fix the new number of intervals $\bar{\bar{r}}^{(1)}(1) = 3$,
- we determine the new intervals

$$\bar{I}_1^{(1)}(1) = \langle 0,\ 485),\ \bar{I}_2^{(1)}(1) = \langle 485,\ 970),\ \bar{I}_3^{(1)}(1) = \langle 970,\ +\infty),$$

- we fix the number of realizations in the new intervals

$$\bar{n}_1^{(1)}(1) = 19,\ \bar{n}_2^{(1)}(1) = 7,\ \vec{n}_3^{(1)}(1) = 14,$$

Table 5.1 The realization of the histogram of the exemplary system conditional lifetime $T_{11}^{(1)}(1)$

Histogram of the conditional lifetime $T_{11}^{(1)}(1)$							
$I_j^{(1)}(1) = \langle x_j^{(1)}(1), y_j^{(1)}(1)\rangle$	0–485	485–970	970–1455	1455–1940	1940–2425	2425–2910	2910–$+\infty$
$n_j^{(1)}(1)$	19	7	3	2	0	1	8
$\bar{f}_{40}(t,1) = n_j^{(1)}(1)/\bar{n}^{(1)}$	19/40	7/40	3/40	2/40	0/40	1/40	8/40

- we calculate, using (5.25), the hypothetical probabilities that the conditional lifetime $T_{11}^{(1)}(1)$ takes values from the new intervals

$$\begin{aligned} p_1^{(1)}(1) &= P(T_{11}^{(1)}(1) \in \bar{I}_1^{(1)}(1)) = P(0 \le T_{11}^{(1)}(1) < 485) \\ &= [R_{11}^{(1)}(0,1)]^{(1)} - [R_{11}^{(1)}(485,1)]^{(1)} \\ &= \exp[-0.0008 \cdot 0] - \exp[-0.0008 \cdot 485]) = 1 - \exp[-0.388] \\ &= 1 - 0.6784 = 0.3216, \end{aligned}$$

$$\begin{aligned} p_2^{(1)}(1) &= P(T_{11}^{(1)}(1) \in \bar{I}_2^{(1)}(1)) = P(485 \le T_{11}^{(1)}(1) < 970) \\ &= [R_{11}^{(1)}(485,1)]^{(1)} - [R_{11}^{(1)}(970,1)]^{(1)} \\ &= \exp[-0.0008 \cdot 485] - \exp[-0.0008 \cdot 970]) = \exp[-0.388] - \exp[-0.776] \\ &= 0.6784 - 0.4602 = 0.2182, \end{aligned}$$

$$\begin{aligned} p_3^{(1)}(1) &= P(T_{11}^{(1)}(1) \in \bar{I}_3^{(1)}(1)) = P(970 \le T_{11}^{(1)}(1)\langle\infty) \\ &= 1 - [p_1^{(1)}(1) + p_2^{(1)}(1)] = 1 - [0.3216 + 0.2182] = 0.4602, \end{aligned}$$

- we calculate, using (5.26), the realization of the χ^2(chi-square)-Pearson's statistics

$$u_{40} = \sum_{j=1}^{3} \frac{(\bar{n}_j^{(1)}(1) - n^{(1)} p_j^{(1)}(1))^2}{n^{(1)} p_j^{(1)}(1)} = \frac{(19 - 40 \cdot 0.3216)^2}{40 \cdot 0.3216}$$

$$+ \frac{(7 - 40 \cdot 0.2182)^2}{40 \cdot 0.2182} + \frac{(14 - 40 \cdot 0.4602)^2}{40 \cdot 0.4602}$$

$$\cong 2.93 + 0.34 + 1.06 = 4.33,$$

- we assume the significance level $\alpha = 0.01$,
- we fix the number of degrees of freedom

$$\bar{\bar{r}}^{(1)}(1) - l - 1 = 3 - 1 - 1 = 1,$$

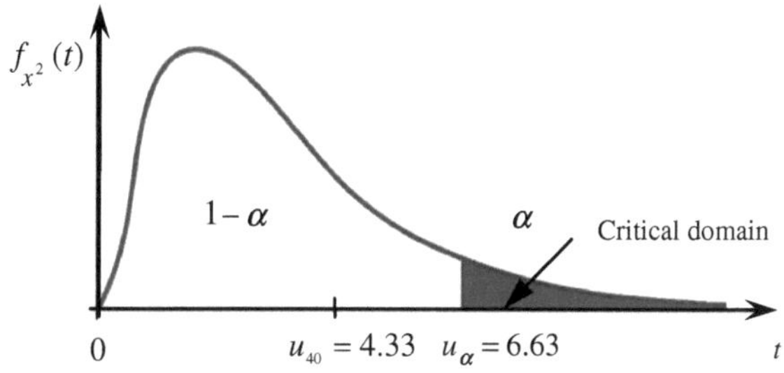

Fig. 5.11 The graphical interpretation of the critical interval and the acceptance interval for the chi-square goodness-of-fit test

- we read from the Tables of the χ^2-Pearson's distribution the value u_α for the fixed values of the significance level $\alpha = 0.01$ and the number of degrees of freedom $\bar{\bar{r}}^{(1)}(1) - l = 1 = 1$, such that, according to (5.27), the following equality holds

$$P(U_{40} > u_\alpha) = \alpha = 0.01$$

which amounts to $u_\alpha = 6.63$ and determine the critical domain in the form of the interval $(6.63, +\infty)$ and the acceptance domain in the form of the interval $\langle 0, 6.63 \rangle$ (Fig. 5.11),

- we compare the obtained value $u_{40} = 4.33$ of the realization of the statistics U_{40} read from the Tables of the chi-square distribution critical value $u_\alpha = 6.63$ and since the value $u_{40} = 4.33$ does not belong to the critical domain, i.e.

$$u_{40} = 4.33 \le u_\alpha = 6.63,$$

then we do not reject the hypothesis H_0 in its part concerned with the coordinate $[R_{11}^{(1)}(t,\ 1)]^{(1)}$.

To verify the hypothesis concerning the exponential form of the coordinate $[R_{11}^{(1)}(t,\ 2)]^{(1)}$ defined by (5.36), it is due to act according to the scheme below:

- we fix the number of observed components and the number of realizations of the exemplary system component conditional lifetime $T_{11}^{(1)}(2)$ in the reliability state subset $\{2, 3\}$ which according to (5.31) are

$$\bar{n}^{(1)} = n^{(1)} = 40, \quad \bar{m}^{(1)}(2) = m^{(1)}(2) = 32,$$

- we fix the realizations $t_i^{(1)}(2)$, $i = 1, 2, \ldots, 32$, of the exemplary system component conditional lifetime $T_{11}^{(1)}(2)$ in the reliability state subset $\{2, 3\}$ that are given by (5.31),

- we determine the number $\bar{r}^{(1)}(2)$ of the disjoint intervals $I_j^{(1)}(2) = \langle x_j^{(1)}(2), y_j^{(1)}(2))$, $j = 1, 2, \ldots, \bar{r}^{(1)}(2)$, which include the realizations $t_i^{(1)}(2)$, $i = 1, 2, \ldots, 32$, of the system component conditional lifetime $T_{11}^{(1)}(2)$ in the reliability states subsets $\{2, 3\}$

$$\bar{r}^{(1)}(2) \cong \sqrt{\bar{m}^{(1)}(2)} = \sqrt{32} \cong 6,$$

- we determine the length $d^{(1)}(2)$ of the intervals $I_j^{(1)}(2) = \langle x_j^{(1)}(2),\ y_j^{(1)}(2)), j = 1, 2, \ldots, 6$, which after considering

$$\bar{R}^{(1)}(2) = \max_{1 \le i \le 32} \left\{ t_i^{(1)}(2) \right\} - \min_{1 \le i \le 32} \left\{ t_i^{(1)}(2) \right\} = 2254 - 30 = 2224,$$

is

$$d^{(1)}(2) = \frac{2224}{6 - 1} \cong 445,$$

- we determine the ends $x_j^{(1)}(2)$, $y_j^{(1)}(2)$, of the intervals $I_j^{(1)}(2) = \langle x_j^{(1)}(2), y_j^{(1)}(2))$, $j = 1, 2, \ldots, 6$, which after considering

$$\min_{1 \le i \le 32} \left\{ t_i^{(1)}(2) \right\} - \frac{d^{(1)}(2)}{2} = 30 - 222.5 = -192.5,$$

are

$$x_1^{(1)}(2) = \max\{-192.5,\ 0\} = 0, \quad y_1^{(1)}(2) = x_1^{(1)}(2) + 1d^{(1)}(2) = 0 + 1 \cdot 445 = 445$$

$$x_2^{(1)}(2) = y_1^{(1)}(2) = 445, \quad y_2^{(1)}(2) = x_1^{(1)}(2) + 2d^{(1)}(2) = 0 + 2 \cdot 445 = 890,$$

$$x_3^{(1)}(2) = y_2^{(1)}(2) = 890, \quad y_3^{(1)}(2) = x_1^{(1)}(2) + 3d^{(1)}(2) = 0 + 3 \cdot 445 = 1335,$$

$$x_4^{(1)}(2) = y_3^{(1)}(2) = 1335, \quad y_4^{(1)}(2) = x_1^{(1)}(2) + 4d^{(1)}(2) = 0 + 4 \cdot 445 = 1780,$$

$$x_5^{(1)}(2) = y_4^{(1)}(2) = 1780, \quad y_5^{(1)}(2) = x_1^{(1)}(2) + 5d^{(1)}(2) = 0 + 5 \cdot 445 = 2225,$$

$$x_6^{(1)}(2) = y_5^{(1)}(2) = 2225, \quad y_6^{(1)}(2) = x_1^{(1)}(2) + 6d^{(1)}(2) = 0 + 6 \cdot 445 = 2670,$$

Table 5.2 The realization of the histogram of the exemplary system conditional lifetime $T_{11}^{(1)}(2)$

Histogram of the conditional lifetime $T_{11}^{(1)}(2)$							
$I_j^{(1)}(2) = \langle x_j^{(1)}(2), y_j^{(1)}(2))$	0 – 445	445– 890	890– 1335	1335– 1780	1780– 2225	2225– 2670	2670– $+\infty$
$n_j^{(1)}(2)$	19	8	2	2	0	1	8
$\bar{f}_{40}(t,2) = n_j^{(1)}(2)/\bar{n}^{(1)}$	19/40	8/40	2/40	2/40	0/40	1/40	8/40

and the ends of the interval $I_7^{(1)}(2) = \langle x_7^{(1)}(2), y_7^{(1)}(2))$, including the remaining unknown realizations

$$x_7^{(1)}(2) = y_6^{(1)}(2) = 2670, \quad y_7^{(1)}(2) = +\infty,$$

- we determine the numbers of realizations $m_j^{(1)}(2)$ in particular intervals $I_j^{(1)}(2) = \langle x_j^{(1)}(2), y_j^{(1)}(2)), j = 1, 2, \ldots, 7,$

$$n_1^{(1)}(2) = 19, \quad n_2^{(1)}(2) = 8, \quad n_3^{(1)}(2) = 2, \quad n_4^{(1)}(2) = 2, \quad n_5^{(1)}(2) = 0,$$
$$n_6^{(1)}(2) = 1, n_7^{(1)}(2) = 40 - 32 = 8,$$

- we find the realization $\bar{f}_{40}(t,2)$ of the histogram of the exemplary system component conditional lifetime $T_{11}^{(1)}(2)$ presented in Table 5.2,
- we join the intervals defined in the realization of the histogram $\bar{f}_{40}(t,2)$ that have the numbers $n_j^{(1)}(2)$, of realizations <4 into new intervals and the interval $I_7^{(1)}(2)$ with its new neighboring interval,
- we fix the new number of intervals $\bar{\bar{r}}^{(1)}(2) = 3$,
- we determine the new intervals

$$\bar{I}_1^{(1)}(2) = \langle 0, \ 445), \quad \bar{I}_2^{(1)}(2) = \langle 445, \ 890), \quad \bar{I}_3^{(1)}(2) = \langle 890, \ +\infty),$$

- we fix the number of realizations in the new intervals

$$\bar{n}_1^{(1)}(2) = 19, \quad \bar{n}_2^{(1)}(2) = 8, \quad \bar{n}_3^{(1)}(2) = 13,$$

- we calculate, using (5.25), the hypothetical probabilities that the conditional lifetime $T_{11}^{(1)}(2)$ takes values from the new intervals

$$\begin{aligned}p_1^{(1)}(2) &= P(T_{11}^{(1)}(2) \in \bar{I}_1^{(1)}(2)) = P(0 \le T_{11}^{(1)}(2) < 445)\\ &= [R_{11}^{(1)}(0,2)]^{(1)} - [R_{11}^{(1)}(445,2)]^{(1)}\\ &= \exp[-0.0009 \cdot 0] - \exp[-0.0009 \cdot 445]) = 1 - \exp[-0.4005]\\ &= 1 - 0.6700 = 0.3300,\end{aligned}$$

$$\begin{aligned}p_2^{(1)}(2) &= P(T_{11}^{(1)}(2) \in \bar{I}_2^{(1)}(2)) = P(445 \le T_{11}^{(1)}(2) < 890)\\ &= [R_{11}^{(1)}(445,2)]^{(1)} - [R_{11}^{(1)}(890,2)]^{(1)}\\ &= \exp[-0.0009 \cdot 445] - \exp[-0.0009 \cdot 890])\\ &= \exp[-0.4005] - \exp[-0.8010]\\ &= 0.6700 - 0.4489 = 0.2211,\end{aligned}$$

$$\begin{aligned}p_3^{(1)}(2) &= P(T_{11}^{(1)}(2) \in \bar{I}_3^{(1)}(2)) = P(890 \le T_{11}^{(1)}(2) \langle 0, \infty)\\ &= 1 - [p_1^{(1)}(2) + p_2^{(1)}(2)] = 1 - [0.3300 + 0.2211] = 0.4489,\end{aligned}$$

- we calculate, using (5.26), the realization of the χ^2 (chi-square)-Pearson's statistics

$$u_{40} = \sum_{j=1}^{3} \frac{\left(\bar{n}_j^{(1)}(2) - \bar{n}^{(1)} p_j^{(1)}(2)\right)^2}{\bar{n}^{(1)} p_j^{(1)}(2)} = \frac{(19 - 40 \cdot 0.3300)^2}{40 \cdot 0.3300} + \frac{(8 - 40 \cdot 0.2211)^2}{40 \cdot 0.2211}$$
$$+ \frac{(13 - 40 \cdot 0.4489)^2}{40 \cdot 0.4489} \cong 2.55 + 0.08 + 1.37 = 4.00,$$

- we assume the significance level $\alpha = 0.01$,
- we fix the number of degrees of freedom

$$\bar{\bar{r}}^{(1)}(2) - l - 1 = 3 - 1 - 1 = 1,$$

- we read from the Tables of the χ^2-Pearson's distribution the value u_α for the fixed values of the significance level $\alpha = 0.01$ and the number of degrees of freedom $\bar{\bar{r}}^{(1)}(2) - l = 1 = 1$, such that, according to (5.27), the following equality holds

$$P(U_{40} > u_\alpha) = \alpha = 0.01$$

which amounts to $u_\alpha = 6.63$ and determine the critical domain in the form of the interval $(6.63, +\infty)$ and the acceptance domain in the form of the interval $\langle 0, 6.63 \rangle$, (Fig. 5.12),

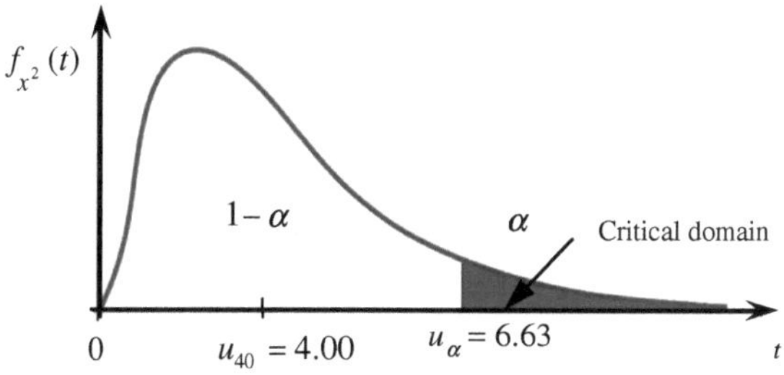

Fig. 5.12 The graphical interpretation of the critical interval and the acceptance interval for the chi-square goodness-of-fit test

- we compare the obtained value $u_{40} = 4.00$ of the realization of the statistics U_{40} read from the Tables of the chi-square distribution critical value $u_\alpha = 6.63$ and since the value $u_{40} = 4.00$ does not belong to the critical domain, i.e.

$$u_{40} = 4.00 \leq u_\alpha = 6.63,$$

we do not reject the hypothesis H_0 in its part concerned with the coordinate $[R_{11}^{(1)}(t,\ 2)]^{(1)}$.
To verify the hypothesis concerning the exponential form of the coordinate $[R_{11}^{(1)}(t,\ 3)]^{(1)}$, defined by (5.36), it is due to act according to the scheme below:

- we fix the number of observed components and the number of realizations of the system component conditional lifetime $T_{11}^{(1)}(3)$ in the reliability state subset $\{3\}$ which according to (5.33) are

$$\bar{n}^{(1)} = n^{(1)} = 40,\ \bar{m}^{(1)}(3) = m^{(1)}(3) = 32,$$

- we fix the realizations $t_i^{(1)}(3)$, $i = 1, 2, \ldots, 32$, of the exemplary system component conditional lifetime $T_{11}^{(1)}(3)$ in the reliability state subset $\{3\}$ which are given by (5.33),
- we determine the number $\bar{r}^{(1)}(3)$ of the disjoint intervals $I_j^{(1)}(3) = \langle x_j^{(1)}(3), y_j^{(1)}(3))$, $j = 1, 2, \ldots, \bar{r}^{(1)}(3)$, which include the realizations $t_i^{(1)}(3)$, $i = 1, 2, \ldots, 32$, of the system component conditional lifetime $T_{11}^{(1)}(3)$ in the reliability state subset $\{3\}$

$$\bar{r}^{(1)}(3) \cong \sqrt{\bar{m}^{(1)}(3)} = \sqrt{32} \cong 6,$$

- we determine the length $d^{(1)}(3)$ of the intervals $I_j^{(1)}(3) = \langle x_j^{(1)}(3), y_j^{(1)}(3))$, $j = 1, 2, \ldots, 6$, which after considering

$$\bar{R}^{(1)} = \max_{1 \le i \le 32} \left\{ t_i^{(1)}(3) \right\} - \min_{1 \le i \le 32} \{ t_i^{(1)}(3) \} = 2054 - 20 = 2034,$$

is

$$d^{(1)}(3) = \frac{2034}{6-1} \cong 407,$$

- we determine the ends $x_j^{(1)}(3)$, $y_j^{(1)}(3)$, of the intervals $I_j^{(1)}(3) = \langle x_j^{(1)}(3), y_j^{(1)}(3))$, $j = 1, 2, \ldots, 6$, which after considering

$$\min_{1 \le i \le 32} \{ t_i^{(1)}(3) \} - \frac{d^{(1)}(3)}{2} = 20 - 203.5 = -183.5,$$

are

$$x_1^{(1)}(3) = \max\{-183.5,\ 0\} = 0,$$
$$y_1^{(1)}(3) = x_1^{(1)}(3) + 1d^{(1)}(3) = 0 + 1 \cdot 407 = 407,$$

$$x_2^{(1)}(3) = y_1^{(1)}(3) = 407, \quad y_2^{(1)}(3) = x_1^{(1)}(3) + 2d^{(1)}(3) = 0 + 2 \cdot 407 = 814,$$

$$x_3^{(1)}(3) = y_2^{(1)}(3) = 814, \quad y_3^{(1)}(3) = x_1^{(1)}(3) + 3d^{(1)}(3) = 0 + 3 \cdot 407 = 1221,$$

$$x_4^{(1)}(3) = y_3^{(1)}(3) = 1221, \quad y_4^{(1)}(3) = x_1^{(1)}(3) + 4d^{(1)}(3) = 0 + 4 \cdot 407 = 1628,$$

$$x_5^{(1)}(3) = y_4^{(1)}(3) = 1628, \quad y_5^{(1)}(3) = x_1^{(1)}(3) + 5d^{(1)}(3) = 0 + 5 \cdot 407 = 2035,$$

$$x_6^{(1)}(3) = y_5^{(1)}(3) = 2035, \quad y_6^{(1)}(3) = x_1^{(1)}(3) + 6d^{(1)}(3) = 0 + 6 \cdot 407 = 2442,$$

and the ends of the interval $I_7^{(1)}(3) = \langle x_7^{(1)}(3), y_7^{(1)}(3))$, including the remaining unknown realizations

$$x_7^{(1)}(3) = y_6^{(1)}(3) = 2442, \ y_7^{(1)}(3) = +\infty,$$

- we determine the number of realizations $n_j^{(1)}(3)$ in particular intervals $I_j^{(1)}(3) = \langle x_j^{(1)}(3), y_j^{(1)}(3))$, $j = 1, 2, \ldots, 6$, according

$$n_1^{(1)}(3) = 17, \quad n_2^{(1)}(3) = 10, \quad n_3^{(1)}(3) = 2, \quad n_4^{(1)}(3) = 2, \quad n_5^{(1)}(3) = 0,$$
$$n_6^{(1)}(3) = 1, \quad n_7^{(1)}(3) = 8,$$

- we find the realization of the histogram of the exemplary system component conditional lifetimes $T_{11}^{(1)}(3)$ as presented in Table 5.3,

Table 5.3 The realization of the histogram of the exemplary system conditional lifetime $T_{11}^{(1)}(3)$

Histogram of the conditional lifetime $T_{11}^{(1)}(3)$							
$I_j^{(1)}(3) = \langle x_j^{(1)}(3), y_j^{(1)}(3))$	0 – 407	407– 814	814– 1221	1221– 1628	1628– 2035	2035– 2442	2442– $+\infty$
$n_j^{(1)}(3)$	17	10	2	2	0	1	8
$\bar{f}_{40}(t,3) = n_j^{(1)}(3)/\bar{n}^{(1)}$	17/40	10/40	2/40	2/40	0/40	1/40	8/40

- we join the intervals defined in the realization of the histogram $\bar{f}_{40}(t,3)$ that have the numbers $n_j^{(1)}(3)$, of realizations <4 into new intervals and the interval $I_7^{(1)}(2)$ with its new neighboring interval,
- we fix the new number of intervals $\bar{\bar{r}}^{(1)}(3) = 3$,
- we determine the new intervals

$$\bar{I}_1^{(1)}(3) = \langle 0,\ 407), \quad \bar{I}_2^{(1)}(3) = \langle 407,\ 814), \quad \bar{I}_3^{(1)}(3) = \langle 814,\ +\infty),$$

- we fix the number of realizations in the new intervals

$$\bar{n}_1^{(1)}(3) = 17,\ \bar{n}_2^{(1)}(3) = 10,\ \bar{n}_3^{(1)}(3) = 13,$$

- we calculate, using (5.25), the hypothetical probabilities that the conditional lifetime $T_{11}^{(1)}(3)$ takes values from the new intervals

$$\begin{aligned} p_1^{(1)}(3) &= P(T_{11}^{(1)}(3) \in \bar{I}_1^{(1)}(3)) = P(0 \le T_{11}^{(1)}(3) < 407) \\ &= [R_{11}^{(1)}(0,3)]^{(1)} - [R_{11}^{(1)}(407,3)]^{(1)} \\ &= \exp[-0.0009 \cdot 0] - \exp[-0.0009 \cdot 407]) = 1 - \exp[-0.3663] \\ &= 1 - 0.6933 = 0.3067, \end{aligned}$$

$$\begin{aligned} p_2^{(1)}(3) &= P(T_{11}^{(1)}(3) \in \bar{I}_2^{(1)}(3)) = P(407 \le T_{11}^{(1)}(3) < 814) \\ &= [R_{11}^{(1)}(407,3)]^{(1)} - [R_{11}^{(1)}(814,3)]^{(1)} \\ &= \exp[-0.0009 \cdot 407] - \exp[-0.0009 \cdot 814]) = \exp[-0.3663] - \exp[-0.7326] \\ &= 0.6933 - 0.4807 = 0.2126, \end{aligned}$$

$$\begin{aligned} p_3^{(1)}(3) &= P(T_{11}^{(1)}(3) \in \bar{I}_3^{(1)}(3)) = P(814 \le T_{11}^{(1)}(3) \langle \infty) \\ &= 1 - [p_1^{(1)}(3) + p_2^{(1)}(3)] = 1 - [0.3067 + 0.2126] = 0.4807, \end{aligned}$$

- we calculate, using (5.26), the realization of the χ^2 (chi-square)-Pearson's statistics

$$u_{40} = \sum_{j=1}^{3} \frac{(\bar{n}_j^{(1)}(3) - \bar{n}^{(1)} p_j^{(1)}(3))^2}{\bar{n}^{(1)} p_j^{(1)}(3)} = \frac{(17 - 40 \cdot 0.3067)^2}{40 \cdot 0.3067} + \frac{(10 - 40 \cdot 0.2126)^2}{40 \cdot 0.2126}$$
$$+ \frac{(13 - 40 \cdot 0.4807^2}{40 \cdot 0.4807} \cong 1.83 + 0.26 + 2.02 = 4.11,$$

- we assume the significance level $\alpha = 0.01$,
- we fix the number of degrees of freedom

$$\bar{\bar{r}}^{(1)}(3) - l - 1 = 3 - 1 - 1 = 1,$$

- we read from the Tables of the χ^2-Pearson's distribution the value u_α for the fixed values of the significance level $\alpha = 0.01$ and the number of degrees of freedom $\bar{\bar{r}}^{(1)}(3) - l = 1 = 1$, such that, according to (5.27), the following equality holds

$$P(U_{40} > u_\alpha) = \alpha = 0.01$$

which amounts to $u_\alpha = 6.63$ and determine the critical domain in the form of the interval $(6.63, +\infty)$and the acceptance domain in the form of the interval $(0, 6.63)$(Fig. 5.13),

- we compare the obtained value $u_{40} = 4.11$ of the realization of the statistics U_{40} read from the Tables of the chi-square distribution critical value $u_\alpha = 6.63$ and since the value $u_{40} = 4.11$ does not belong to the critical domain, i.e.

$$u_{40} = 4.11 \le u_\alpha = 6.63,$$

we do not reject the hypothesis H_0 in its part concerned with the coordinate $\left[R_{11}^{(1)}(t,\ 3)\right]^{(1)}$.

In conclusion, we accept in full the hypothesis H_0 that the reliability function of the component $E_{11}^{(1)}$ at the operation state z_1 is exponential with the coordinates given by (5.36).

For the remaining cases, proceeding afterwards in an analogous way as in the case of the conditional reliability function $\left[R_{11}^{(1)}(t,\ \cdot)\right]^{(1)}$ of the component $E_{11}^{(1)}$ at the operation state z_1, we can get'all the results of testing the hypotheses on the forms of the reliability functions defined by (5.28–5.29) given in Sect. 3.6.1 and used for the evaluation and prognosis of the reliability characteristics of the exemplary system operating at the variable operation conditions.

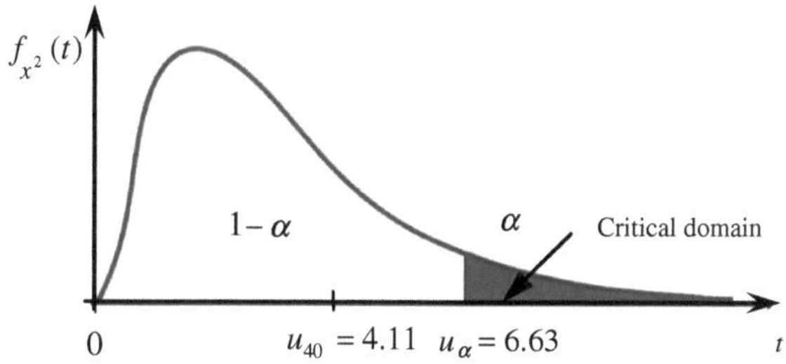

Fig. 5.13 The graphical interpretation of the critical interval and the acceptance interval for the chi-square goodness-of-fit test

5.3.2 Statistical Identification of Port Oil Piping Transportation System Components Reliability

The considered port oil piping transportation system reliability structure changing at the various operation states and its components and their unknown reliability parameters are described in Sects. 2.3.2 and 3.6.2.

5.3.2.1 Defining Parameters of Port Oil Piping Transportation System Components Reliability Models and Data Collection

At all the system operation states z_b, $b = 1, 2, \ldots, 7,$, defined in Sect. 2.3.2, we distinguish the following three reliability states 0, 1, 2 ($z = 2$) of the system and its components, defined in Sect. 1.4.2. Moreover, we fix the possible transitions between the components reliability states only from better to worse.

From the above, the port oil pipeline transportation subsystems S_υ, $\upsilon = 1, 2, 3$, are composed of three-state, i.e. $z = 2$, components $E_{ij}^{(\upsilon)}$, $\upsilon = 1, 2, 3$, having the conditional three-state reliability functions

$$[R_{ij}^{(\upsilon)}(t, \cdot)]^{(b)} = [1, [R_{ij}^{(\upsilon)}(t, 1)]^{(b)}, [R_{ij}^{(\upsilon)}(t, 2)]^{(b)}], \quad b = 1, 2, \ldots, 7, \tag{5.37}$$

with the exponential coordinates

$$[R_{ij}^{(\upsilon)}(t, 1)]^{(b)} = \exp[-[\lambda_{ij}^{(\upsilon)}(1)]^{(b)}], \; [R_{ij}^{(\upsilon)}(t, 2)]^{(b)} = \exp[-[\lambda_{ij}^{(\upsilon)}(2)]^{(b)}], \tag{5.38}$$

and with the intensities of departure from the reliability state subsets $\{1, 2\}$, $\{2\}$, respectively

$$[\lambda_{ij}^{(\upsilon)}(1)]^{(b)}, \; [\lambda_{ij}^{(\upsilon)}(2)]^{(b)}, \quad b = 1, 2, \ldots, 7,$$

different at the particular operation states z_b, $b = 1, 2, \ldots, 7$, where $i = 1, 2$ and $j = 1, 2, \ldots, 178$ for $\upsilon = 1$, $i = 1, 2$ and $j = 1, 2, \ldots, 719$ for $\upsilon = 2$, $i = 1, 2, 3$ and $j = 1, 2, \ldots, 362$ for $\upsilon = 3$.

Table 5.4 The approximate mean values $[\hat{\mu}_{ij}^{(1)}(u)]^{(b)}$ of the piping subsystem components' conditional lifetimes $[T_{ij}^{(1)}(u)]^{(b)}$ at the operation states z_b

Component of subsystem S_1		Mean value $[\hat{\mu}_{ij}^{(1)}(u)]^{(b)}$ (*in years*)	
Operation state z_b	Reliability state subset$\{u, u+1, \ldots, 2\}$	$E_{ij}^{(1)}, i = 1, 2,$ $j = 1, 2, \ldots, 176$	$E_{ij}^{(1)}$ $i = 1, 2,$ $j = 177,\ 178$
z_1			
z_2			
z_3	{1,2}	161	60
	{2}	114	55
z_4	{1,2}	161	60
	{2}	114	55
z_5	{1,2}	161	60
	{2}	114	55
z_6	{1,2}	161	60
	{2}	114	55
z_7			

5.3.2.2 Collecting Data on Port Oil Piping Transportation System Components Reliability Coming from Experts

We have the approximate realizations

$$[\hat{\mu}_{ij}^{(\upsilon)}(u)]^{(b)}, \quad \upsilon = 1, 2, 3,\ u = 1, 2,\ b = 1, 2, \ldots, 7,$$

of the mean values

$$[\mu_{ij}^{(\upsilon)}(u)]^{(b)} = E[[T_{ij}^{(\upsilon)}(u)]^{(b)}], \quad \upsilon = 1, 2, 3,\ u = 1, 2,\ b = 1, 2, \ldots, 7,$$

of the conditional lifetimes $[T_{ij}^{(\upsilon)}(u)]^{(b)}$, $\upsilon = 1, 2, 3,\ u = 1, 2,\ b = 1, 2, \ldots, 7$, at the reliability state subsets $\{u, u+1, \ldots, 2\}$, $u = 1, 2$, of the components $E_{ij}^{(\upsilon)}$, $\upsilon = 1, 2, 3$, of the oil pipeline subsystems S_υ, $\upsilon = 1, 2, 3$, at the particular operation states z_b, $b = 1, 2, \ldots, 7$ estimated on the basis of expert opinion.

In Table 5.4, as an example of data coming from experts, there is given the approximate realizations of the conditional lifetime at the reliability state subsets of the components of the oil pipeline subsystems S_1.

5.3.2.3 Evaluating Port Oil Piping Transportation System Components Intensities of Departures from Reliability State Subsets on the Basis of Data Coming from Experts

To evaluate approximately the parameters of multistate reliability functions of the port oil pipeline system components the statistical data from experts presented in

Sect. 5.3.2.2 can be used. The collected statistical data partly presented in Table 5.4 and the resulting from (5.22) formula

$$[\hat{\lambda}_{ij}^{(\upsilon)}(u)]^{(b)} = \frac{1}{[\hat{\lambda}_{ij}^{(\upsilon)}(u)]^{(b)}}, \quad \upsilon = 1,2,3, \ u = 1,2, \ b = 1,2,\ldots,7, \qquad (5.39)$$

application yield the approximate values $[\hat{\lambda}_{ij}^{(\upsilon)}(u)]^{(b)}$of subsystems S_υ, $\upsilon = 1,2,3$, components' unknown intensities $[\lambda_{ij}^{(\upsilon)}(u)]^{(b)}$ of departure from the reliability state subsets, {1, 2}, {2}, while the system is operating at the operation state z_b, $b = 1,2,\ldots,7$.

For instance, substituting into (5.39) the values of the mean lifetimes

$$\left[\hat{\mu}_{ij}^{(1)}(1)\right]^{(3)} = 161, \quad \left[\hat{\mu}_{ij}^{(1)}(2)\right]^{(3)} = 114,$$

taken from the fifth and sixth rows of the third column of Table 5.4, we obtain the approximate evaluations of the unknown intensities of departure of components $E_{ij}^{(1)}$, $i = 1,2\, j = 1,2,\ldots,176$, of the subsystem from the reliability state subsets$\{1,2\}$, $\{2\}$ while the piping system is operating at the operation state z_3 that respectively amount to

$$\left[\hat{\lambda}_{ij}^{(1)}(1)\right]^{(3)} = \frac{1}{\left[\hat{\mu}_{ij}^{(1)}(1)\right]^{(3)}} = \frac{1}{161} \cong 0.0062,$$
$$\left[\hat{\lambda}_{ij}^{(1)}(2)\right]^{(3)} = \frac{1}{\left[\hat{\mu}_{ij}^{(1)}(2)\right]^{(3)}} = \frac{1}{114} \cong 0.0088, \quad i = 1,2, \ j = 1,2,\ldots,176. \qquad (5.40)$$

Similarly, substituting into (5.39) the values of the mean lifetimes

$$[\hat{\mu}_{ij}^{(1)}(1)]^{(3)} = 60, \quad [\hat{\mu}_{ij}^{(1)}(2)]^{(3)} = 55,$$

taken from the fifth and sixth rows of the fourth column of Table 5.4, we obtain the approximate evaluations of the unknown intensities of departure of components $E_{ij}^{(1)}$, $i = 3$, $j = 177, 178$, of subsystem S_1 from the reliability state subsets $\{1,2\}$, $\{2\}$ while the piping system is operating at the operation state z_3 that respectively amount to

$$\left[\hat{\lambda}_{ij}^{(1)}(1)\right]^{(3)} = \frac{1}{\left[\hat{\mu}_{ij}^{(\upsilon)}(1)\right]^{(3)}} = \frac{1}{610} \cong 0.0167,$$
$$\left[\hat{\lambda}_{ij}^{(1)}(2)\right]^{(3)} = \frac{1}{\left[\hat{\mu}_{ij}^{(\upsilon)}(2)\right]^{(3)}} = \frac{1}{55} \cong 0.0182, \quad i = 3, \ j = 177, 178. \qquad (5.41)$$

The evaluations of unknown intensities of departure from the reliability state subsets {1, 2}, {2} of all components of the port oil piping transportation system operating at the various operation states, can be obtained in the same way.

Substituting the above and the remaining evaluations of the intensities of departures respectively into the formulae (5.37–5.38), we get the exponential coordinates of the port oil piping transportation system components three-state reliability functions which after arbitrary acceptance in Sect. 5.3.2.4 are used in Sect. 3.6.2 for the evaluation and prediction of the system reliability characteristics.

5.3.2.4 Identifying Port Oil Piping Transportation System Components Conditional Multistate Exponential Reliability Functions on the Basis of Data Coming from Experts

As there are no data collected from the port oil piping transportation system components' reliability states changing processes, it is not possible to verify the hypotheses on the exponential forms of the port oil piping transportation system components' conditional reliability functions. We arbitrarily assume that these reliability functions are exponential and using the results of the previous section and the relationships (5.37–5.38) given in Sect. 5.3.2.1 we fix their forms.

For instance, using the evaluations (5.40–5.41) of the previous section and the formulae (5.37–5.38), we conclude that at the system operation state, the subsystem S_1 consists of two identical pipelines, each composed of 176 pipe segments with the conditional reliability functions coordinates

$$\left[R_{ij}^{(1)}(t,1)\right]^{(3)} = \exp[-0.0062t], \ \left[R_{ij}^{(1)}(t,2)\right]^{(3)} = \exp[-0.0088t],$$
$$i = 1,2, \ j = 1,2,\ldots,176$$

and two valves with the conditional reliability functions co-ordinates

$$\left[R_{ij}^{(1)}(t,1)\right]^{(3)} = \exp[-0.0167t], \ \left[R_{ij}^{(1)}(t,2)\right]^{(3)} = \exp[-0.0182t],$$
$$i = 1,2, \ j = 177,178.$$

The exponential coordinates of the port oil piping transportation system all components three-state reliability functions arbitrarily fixed in this way are given and used in Sect. 3.6.2 for the evaluation and prediction of the system reliability characteristics.

5.3.3 Statistical Identification of Maritime Ferry Technical System Components Safety

The considered ferry technical system safety structure changing at the various operation states and its components and their unknown safety parameters are described in Sects. 2.3.3 and 3.6.3.

5.3.3.1 Defining Parameters of Ferry Technical System Components Safety Models and Data Collection

At all the system operation states z_b, $b = 1, 2\ldots, 18$, defined in Sect. 2.3.3, we distinguish the following five safety states $0, 1, 2, 3, 4 (z = 4)$ of the system and its components, defined in Sect. 1.4.3. Moreover, we fix the possible transitions between the components safety states only from better to worse.

From the above, the ferry technical subsystems S_υ, $\upsilon = 1, 2\ldots, 5$ are composed of five-state, i.e. $z = 4$, components $E_{ij}^{(\upsilon)}$, $\upsilon = 1, 2, \ldots, 5$ having the conditional five-state safety functions

$$\left[s_{ij}^{(\upsilon)}(t, \cdot)\right]^{(b)} = \left[1, \left[s_{ij}^{(\upsilon)}(t, 1)\right]^{(b)}, \left[s_{ij}^{(\upsilon)}(t, 2)\right]^{(b)}, \left[s_{ij}^{(\upsilon)}(t, 3)\right]^{(b)}, \left[s_{ij}^{(\upsilon)}(t, 4)\right]^{(b)}, \right.$$
$$b = 1, 2, \ldots, 18, \tag{5.42}$$

with the exponential coordinates

$$\left[s_{ij}^{(\upsilon)}(t, 1)\right]^{(b)} = \exp\left[-\left[\lambda_{ij}^{(\upsilon)}(1)\right]^{(b)}\right], \quad \left[s_{ij}^{(\upsilon)}(t, 2)\right]^{(b)} = \exp\left[-\left[\lambda_{ij}^{(\upsilon)}(2)\right]^{(b)}\right],$$
$$\left[s_{ij}^{(\upsilon)}(t, 3)\right]^{(b)} = \exp\left[-\left[\lambda_{ij}^{(\upsilon)}(3)\right]^{(b)}\right], \quad \left[s_{ij}^{(\upsilon)}(t, 4)\right]^{(b)} = \exp\left[-\left[\lambda_{ij}^{(\upsilon)}(4)\right]^{(b)}\right], \tag{5.43}$$

and with the intensities of departure from the safety states subsets $\{1, 2, 3, 4\}$, $\{2, 3, 4\}$, $\{3, 4\}$, $\{4\}$, respectively

$$\left[\lambda_{ij}^{(\upsilon)}(1)\right]^{(b)} \left[\lambda_{ij}^{(\upsilon)}(2)\right]^{(b)}, \left[\lambda_{ij}^{(\upsilon)}(3)\right]^{(b)}, \left[\lambda_{ij}^{(\upsilon)}(4)\right]^{(b)}, \quad b = 1, 2, \ldots, 18,$$

different at the particular operation states z_b, $b = 1, 2, \ldots, 18$, where $i = 1$ and $j = 1$ for $\upsilon = 1$, $i = 1$ and $j = 1, 2, 3, 4$, $i = 2$ and $j = 1, 2$, $i = 3, 4, 5, 6, 7$ and $j = 1$ for $\upsilon = 2$, $i = 1, 2, 3, 4, 5$ and $j = 1$ for $\upsilon = 3$, $i = 1, 2$ and $j = 1$ for $\upsilon = 4$, $i = 1, 2, 3$ and $j = 1$ for $\upsilon = 5$.

5.3.3.2 Collecting Data on Ferry Technical System Components Safety Models Coming from Experts

We have the approximate realizations

$$[\hat{\mu}_{ij}^{(\upsilon)}(u)^{(b)}, \quad \upsilon = 1, 2, 3, 4, 5, \ u = 1, 2, 3, 4, \ b = 1, 2, \ldots, 18,$$

of the mean values

$$[\mu_{ij}^{(\upsilon)}(u)]^{(b)} = E[[T_{ij}^{(\upsilon)}(u)^{(b)}] \quad \upsilon = 1,2,3,4,5,\ u = 1,2,3,4,\ b = 1,2,\ldots,18,$$

of the conditional lifetimes $[T_{ij}^{(\upsilon)}(u)^{(b)}$, $\upsilon = 1,2,3,4,5$, $u = 1,2,3,4$, $b = 1,2,\ldots,18$, in the safety state subsets $\{u, u+1,\ldots,4\}$, $u = 1,2,3,4$, of the components $E_{ij}^{(\upsilon)}$ of the ferry technical subsystems S_b, $\upsilon = 1,2,3,4,5$, at the particular operation states z_b, $b = 1,2,\ldots,18$, estimated on the basis of expert opinions.

In the Table 5.5, as an example of data coming from experts, there are given the approximate realizations of the conditional lifetime at the safety state subsets of the components of the ferry technical subsystems S_1.

5.3.3.3 Evaluating Ferry Technical System Components Intensities of Departures from Safety State Subsets on the Basis of Data Coming from Experts

To evaluate approximately the parameters of multistate safety functions of the ferry technical system components the statistical data coming from experts partly presented in Sect. 5.3.3.2 can be used. The statistical data collected in Table 5.5 and the resulting from (5.22) formula

$$[\hat{\lambda}_{ij}^{(\upsilon)}(u)]^{(b)} = \frac{1}{[\hat{\mu}_{ij}^{(\upsilon)}(u)]^{(b)}}, \quad \upsilon = 1,2,3,4,5,\ u = 1,2,3,4, \quad b = 1,2,\ldots,18, \tag{5.44}$$

application yield the approximate values $[\hat{\lambda}_{ij}^{(\upsilon)}(u)]^{(b)}$ of subsystems S_υ, $\upsilon = 1,2,3$, components unknown intensities $[\lambda_{ij}^{(\upsilon)}(u)]^{(b)}$ of departure from the safety state subsets $\{1,2,3,4\}$, $\{2,3,4\}$, $\{3,4\}$, $\{4\}$, while the system is operating at the operation state z_b, $b = 1,2,\ldots,18$.

For instance, substituting into (5.44) the values the mean lifetimes

$$[\hat{\mu}_{ij}^{(1)}(1)]^{(2)} = 30,\ [\hat{\mu}_{ij}^{(1)}(2)]^{(2)} = 25,\ [\hat{\mu}_{ij}^{(1)}(3)]^{(2)} = 22,\ [\hat{\mu}_{ij}^{(1)}(4)]^{(2)} = 20,$$

taken from the fourth, fifth, sixth and seventh rows of the third column of Table 5.5, we obtain the approximate evaluations of the unknown intensities of departure of component $E_{11}^{(1)}$ of subsystem S_1 from the safety state subsets $\{1,2,3,4\}$, $\{2,3,4\}$, $\{3,4\}$, $\{4\}$, while the ferry technical system is operating at the operation state z_2 that respectively amount to

$$\left[\hat{\lambda}_{ij}^{(1)}(1)\right]^{(2)} = \frac{1}{\left[\hat{\mu}_{ij}^{(1)}(1)\right]^{(2)}} = \frac{1}{30} \cong 0.033,\ \left[\hat{\lambda}_{ij}^{(1)}(2)\right]^{(2)} = \frac{1}{\left[\hat{\mu}_{ij}^{(1)}(2)\right]^{(2)}} = \frac{1}{25} \cong 0.040,$$

Table 5.5 The approximate mean values $[\hat{\mu}_{11}^{(1)}(u)]^{(b)}$ of the ferry subsystem components' conditional lifetimes $[T_{11}^{(1)}(u)]^{(b)}$ at the operation states z_b

Component of subsystem S_1		$E_{11}^{(1)}$
Operation state z_b	Safety state subset $\{u, u+1, \ldots, 4\}$	Mean value $[\hat{\mu}_{11}^{(1)}(1)]^{(b)}$ (*in years*)
z_1		
z_2	{1,2,3,4}	30
	{2,3,4}	25
	{3,4}	22
	{4}	20
z_3	{1,2,3,4}	30
	{2,3,4}	25
	{3,4}	22
	{4}	20
z_4	{1,2,3,4}	30
	{2,3,4}	25
	{3,4}	22
	{4}	20
z_5	{1,2,3,4}	30
	{2,3,4}	25
	{3,4}	22
	{4}	20
z_6	{1,2,3,4}	30
	{2,3,4}	25
	{3,4}	22
	{4}	20
z_7	{1,2,3,4}	30
	{2,3,4}	25
	{3,4}	22
	{4}	20
z_8		
z_9		
z_{10}	{1,2,3,4}	30
	{2,3,4}	25
	{3,4}	22
	{4}	20
z_{11}	{1,2,3,4}	30
	{2,3,4}	25
	{3,4}	22
	{4}	20
z_{12}	{1,2,3,4}	30
	{2,3,4}	25
	{3,4}	22
	{4}	20

(continued)

Table 5.5 (continued)

Component of subsystem S_1		$E_{11}^{(1)}$
Operation state z_b	Safety state subset$\{u, u+1, \ldots, 4\}$	Mean value $[\hat{\mu}_{11}^{(1)}(1)]^{(b)}$ *(in years)*
z_{13}	{1,2,3,4}	30
	{2,3,4}	25
	{3,4}	22
	{4}	20
z_{14}	{1,2,3,4}	30
	{2,3,4}	25
	{3,4}	22
	{4}	20
z_{15}	{1,2,3,4}	30
	{2,3,4}	25
	{3,4}	22
	{4}	20
z_{16}	{1,2,3,4}	30
	{2,3,4}	25
	{3,4}	22
	{4}	20
z_{17}	{1,2,3,4}	30
	{2,3,4}	25
	{3,4}	22
	{4}	20
z_{18}		

$$\left[\hat{\lambda}_{ij}^{(1)}(3)\right]^{(2)} = \frac{1}{\left[\hat{\mu}_{ij}^{(1)}(3)\right]^{(2)}} = \frac{1}{22} \cong 0.045, \; \left[\hat{\lambda}_{ij}^{(1)}(4)\right]^{(2)} = \frac{1}{\left[\hat{\mu}_{ij}^{(1)}(4)\right]^{(2)}} = \frac{1}{20} \cong 0.050. \tag{5.45}$$

The evaluations of all unknown intensities of departure from the reliability state subsets $\{1, 2, 3, 4\}$, $\{2, 3, 4\}$, $\{3, 4\}$, $\{4\}$, of components of the ferry technical system operating at various operation states, can be obtained in the same way.

Substituting the above evaluations of the intensities of departures respectively into the formulae (5.42–5.43), we get the exponential coordinates of the ferry technical system components' five-state safety functions which after arbitrary acceptance in Sect. 5.3.3.4 are used in Sect. 3.6.2 for the evaluation and prediction of the system safety characteristics.

5.3.3.4 Identifying Ferry Technical System Components Conditional Multistate Exponential Safety Functions on the Basis of Data Coming from Experts

As there are no data collected from the ferry technical system components safety states changing processes, it is not possible to verify the hypotheses on the exponential forms of this system components' conditional safety functions. We arbitrarily assume that these safety functions are exponential and using the results of the previous section and the relationships (5.42–5.43) given in Sect. 5.3.3.1 we fix their forms.

For instance, using the evaluations (5.45) of the previous section and (5.42–5.43), we conclude that at the system operation state z_2, subsystem S_1 consists of a component $E_{11}^{(1)}$ with the conditional multistate safety function co-ordinates

$$[s_{11}^{(1)}(t,1)]^{(2)} = \exp[-0.033t], \quad [s_{11}^{(1)}(t,2)]^{(2)} = \exp[-0.040t],$$
$$[s_{11}^{(1)}(t,3)]^{(2)} = \exp[-0.045t], \quad [s_{11}^{(1)}(t,4)]^{(2)} = \exp[-0.05t].$$

The exponential coordinates of the ferry technical system components safety functions arbitrarily fixed in this way are used in Sect. 3.6.3 for the evaluation and prediction of the system safety characteristics.

5.4 Summary

The proposed statistical methods of identification of the unknown parameters of the system reliability and safety of the components of complex technical systems allow for the models discussed in Chap. 3 and their unknown parameters identification and practical applications in evaluation and prediction of the reliability, availability and safety of real complex technical systems. The methods and the results of these models' identification allow for the optimization of the reliability, availability and safety of the complex technical systems presented in Chap. 6.

References

1. Aven T (1985) Reliability evaluation of multistate systems with multistate components. IEEE Trans Reliab 34:473–479
2. Collet J (1996) Some remarks on rare-event approximation. IEEE Trans Reliab 45:106–108
3. Gamiz ML, Roman Y (2008) Non-parametric estimation of the availability in a general repairable. Reliab Eng Sys Saf 93(8):1188–1196
4. Giudici P, Figini S (2009) Applied data mining for business and industry. Wiley, Chichester
5. Helvacioglu S, Insel M (2008) Expert system applications in marine technologies. Ocean Eng 35(11–12):1067–1074
6. Hryniewicz O (1995) Lifetime tests for imprecise data and fuzzy reliability requirements. In: Onisawa T, Kacprzyk J (eds) Reliability and safety analyses under fuzziness. Physica Verlag, Heidelberg, pp 169–182
7. Kołowrocki K (2004) Reliability of large systems. Elsevier, Amsterdam
8. Kołowrocki K (2007) Reliability modelling of complex systems–Part 1. Int J Gnedenko e-Forum Reliab Theory Appl 2(3–4):116–127
9. Kołowrocki K (2007) Reliability modelling of complex systems–Part 2. Int J Gnedenko e-Forum Reliab Theory Appl 2(3–4):128–139

10. Kolowrocki K, Soszynska J (2009) Modeling environment and infrastructure influence on reliability and operation process of port oil transportation system. Electron J Reliab Risk Anal: Theory Appl 2(3):131–142
11. Kolowrocki K, Soszynska J (2009) Safety and risk evaluation of Stena Baltica ferry in variable operation conditions. Electron J Reliab Risk Anal Theory Appl 2(4):168–180
12. Kołowrocki K, Soszyńska J (2009) Methods and algorithms for evaluating unknown parameters of components reliability of complex technical systems Summer Safety & Reliability Seminars. J Pol Saf Reliab Assoc 4(2):223–230
13. Kolowrocki K, Soszynska J (2010) Reliability, availability and safety of complex technical systems modelling identification prediction optimization Summer Safety & Reliability Seminars. J Pol Saf Reliab Assoc 4(1):133–158
14. Kołowrocki K, Soszyńska J (2010) Safety and risk evaluation of a ferry technical system Summer Safety & Reliability Seminars. J Pol Saf Reliab Assoc 4(1):159–172
15. Kołowrocki K, Soszyńska J (2011) On safety analysis of complex technical maritime transportation system. J Risk Reliab 225(1):1–10
16. Kołowrocki K, Soszyńska J, Judziński M, Dziula P (2007) On multi-state safety analysis in shipping. Int J Reliab Qual Saf Eng Sys Reliab Saf 14(6):547–567
17. Lisnianski A, Levitin G (2003) Multi-state system reliability. Assessment, optimisation and applications. World Scientific Publishing Co. Pte. Ltd, London
18. Rice JA (2007) Mathematical statistics and data analysis. Duxbury. Thomson Brooks/Cole, University of California, Berkeley
19. Soszyńska J (2006) Reliability evaluation of a port oil transportation system in variable operation conditions. Int J Press Vessel Pip 83(4):304–310
20. Soszyńska J (2007) Systems reliability analysis in variable operation conditions. PhD Thesis, Gdynia Maritime University-System Research Institute Warsaw (in Polish)
21. Soszyńska J (2010) Reliability and risk evaluation of a port oil pipeline transportation system in variable operation conditions. Int J Press Vessel Pip 87(2–3):81–87
22. Soszyńska J, Kołowrocki K, Blokus-Roszkowska A, Guze S (2010) Identification of complex technical system components safety models Summer Safety & Reliability Seminars. J Pol Saf Reliab Assoc 4(2):399–496
23. Vercellis S (2009) Data mining and optimization for decision making. Wiley, Hoboken
24. Wilson AG, Graves TL, Hamada MS et'al (2006) Advances in data combination, analysis and collection for system reliability assessment. Stat Sci 21(4):514–531
25. Xue J (1985) On multi-state system analysis. IEEE Trans Reliab 34:329–337
26. Xue J, Yang K (1995) Dynamic reliability analysis of coherent multi-state systems. IEEE Trans Reliab 4(44):683–688
27. Xue J, Yang K (1995) Symmetric relations in multi-state systems. IEEE Trans Reliab 4(44):689–693
28. Yu K, Koren I, Guo Y (1994) Generalized multistate monotone coherent systems. IEEE Trans Reliab 43:242–250

Chapter 6
Complex Technical Systems Operation, Reliability, Availability, Safety and Cost Optimization

6.1 Introduction

Complex technical systems' reliability and safety improvement and decreasing their operation costs and risk are of great value in the industrial practice [1, 4, 6–9, 15, 16]. In everyday practice, the tools that could be applied to improve the reliability and safety characteristics of the considered multistate non-repairable systems are needed [3, 5]. Tools are needed for allowing to find the distributions, the expected values and the variances of the optimal times until the successive exceeding of the reliability or the safety critical state and the distributions, the expected values and the variances of the optimal number of exceedings of the reliability or the safety critical state at a fixed moment of the operation time for these systems when they are repairable and when their renovation is ignored, and to compare the values of these characteristics before and after their operation processes' optimization [2, 3, 5, 10–12, 14]. Tools can also be applied to find the distribution functions, the expected values and the variances of the optimal times until the successive renovations and distribution functions, the expected values and variances of the optimal times until the successive exceeding of the reliability or safety critical state and distributions, the expected values and variances of the optimal number of renovations up to a fixed moment of the operation time and distributions, the expected values and variances of the optimal number of exceedings of the reliability or safety critical state up to a fixed moment of the operation time and steady optimal availability coefficients and optimal availability coefficients in a fixed operation time interval for these systems when they are repairable and the time of their renovation is non-ignored and to compare the values of these characteristics before and after their operation processes' optimization [2, 3, 5, 10–12, 14]. These tools may be applied to perform the effective cost analysis [13] of complex technical systems at the variable operating conditions and its application to the evaluation the cost before and after the system operation process optimization [3, 5]. The tools based on the methods of corrective and

K. Kołowrocki and J. Soszyńska-Budny, *Reliability and Safety of Complex Technical Systems and Processes*, Springer Series in Reliability Engineering, DOI: 10.1007/978-0-85729-694-8_6, © Springer-Verlag London Limited 2011

preventive maintenance analysis [17] may be applied to fix the optimal maintenance policy maximizing the availability and minimizing the renovation cost of the complex technical systems at the variable operating conditions [3, 5].

6.2 Optimization of Operation, Reliability and Safety and Cost Analysis of Complex Technical Systems

6.2.1 *Optimal Transient Probabilities of Complex Technical System Operation Process at Operation States*

Considering Eq. 3.6, it is natural to assume that the system operation process has significant influence on the system reliability. This influence is also clearly expressed in Eq. 3.7 for the mean values of the system unconditional lifetimes in the reliability state subsets.

From the linear Eq. 3.7, we can see that the mean value of the system unconditional lifetime $\mu(u)$, $u = 1, 2, \ldots, z$, is determined by the limit values of transient probabilities p_b, $b = 1, 2, \ldots, v$, of the system operation process at the operation states given by (2.22) and the mean values $\mu_b(u)$, $b = 1, 2, \ldots, v$, $u = 1, 2, \ldots, z$, of the system conditional lifetimes in the reliability state subsets $\{u, u+1, \ldots, z\}$, $u = 1, 2, \ldots, z$, given by (3.8). Therefore, the system lifetime optimization approach based on the linear programming [2, 3, 5] can be proposed. Namely, we may look for the corresponding optimal values $\dot{p}_b$, $b = 1, 2, \ldots, v$, of the transient probabilities p_b, $b = 1, 2, \ldots, v$, of the system operation process at the operation states to maximize the mean value $\mu(u)$ of the unconditional system lifetimes in the reliability state subsets $\{u, u+1, \ldots, z\}$, $u = 1, 2, \ldots, z$, under the assumption that the mean values $\mu_b(u)$, $b = 1, 2, \ldots, v$, $u = 1, 2, \ldots, z$, of the system conditional lifetimes in the reliability state subsets are fixed. As a special and practically important case of the above formulated system lifetime optimization problem, if r, $r = 1, 2, \ldots, z$, is a system critical reliability state, we may look for the optimal values $\dot{p}_b$, $b = 1, 2, \ldots, v$, of the transient probabilities p_b, $b = 1, 2, \ldots, v$, of the system operation process at the system operation states to maximize the mean value $\mu(r)$ of the unconditional system lifetime in the reliability state subset $\{r, r+1, \ldots, z\}$, $r = 1, 2, \ldots, z$, under the assumption that the mean values $\mu_b(r)$, $b = 1, 2, \ldots, v$, $r = 1, 2, \ldots, z$, of the system conditional lifetimes in this reliability state subset are fixed. More exactly, we may formulate the optimization problem as a linear programming model with the objective function of the following form

$$\mu(r) = \sum_{b=1}^{v} p_b \mu_b(r) \tag{6.1}$$

for a fixed $r \in \{1, 2, \ldots, z\}$ and with the following bound constraints

$$\breve{p}_b \le p_b \le \hat{p}_b, \quad b = 1, 2, \ldots, v, \tag{6.2}$$

$$\sum_{b=1}^{v} p_b = 1, \tag{6.3}$$

where

$$\mu_b(r), \ \mu_b(r) \ge 0, \quad b = 1, 2, \ldots, v, \tag{6.4}$$

are fixed mean values of the system conditional lifetimes in the reliability state subset $\{r, r+1, \ldots, z\}$ and

$$\breve{p}_b, \ 0 \le \breve{p}_b \le 1 \quad \text{and} \quad \hat{p}_b, \ 0 \le \hat{p}_b \le 1, \ \breve{p}_b \le \hat{p}_b, \quad b = 1, 2, \ldots, v, \tag{6.5}$$

are lower and upper bounds of the unknown transient probabilities p_b, $b = 1, 2, \ldots, v$, respectively.

Now, we can obtain the optimal solution of the formulated by (6.1–6.5) the linear programming problem, i.e. we can find the optimal values $\dot{p}_b$ of the transient probabilities p_b, $b = 1, 2, \ldots, v$, which maximize the objective function given by (6.1).

First, we arrange the system conditional lifetime mean values $\mu_b(r)$, $b = 1, 2, \ldots, v$, in non-increasing order

$$\mu_{b_1}(r) \ge \mu_{b_2}(r) \ge \cdots \ge \mu_{b_v}(r),$$

where $b_i \in \{1, 2, \ldots, v\}$ for $i = 1, 2, \ldots, v$.

Next, we substitute

$$x_i = p_{b_i}, \ \breve{x}_i = \breve{p}_{b_i}, \ \hat{x}_i = \hat{p}_{b_i} \quad \text{for } i = 1, 2, \ldots, v \tag{6.6}$$

and we maximize with respect to x_i, $i = 1, 2, \ldots, v$, the linear form (6.1) which after this transformation takes the form

$$\mu(r) = \sum_{i=1}^{v} x_i \mu_{b_i}(r) \tag{6.7}$$

for a fixed $r \in \{1, 2, \ldots, z\}$ with the following bound constraints

$$\breve{x}_i \le x_i \le \hat{x}_i, \quad i = 1, 2, \ldots, v, \tag{6.8}$$

$$\sum_{i=1}^{v} x_i = 1, \tag{6.9}$$

where

$$\mu_{b_i}(r), \ \mu_{b_i}(r) \ge 0, \quad i = 1, 2, \ldots, v,$$

are fixed mean values of the system conditional lifetimes in the reliability state subset $\{r, r+1, \ldots, z\}$ arranged in non-increasing order and

$$\breve{x}_i,\ 0 \le \breve{x}_i \le 1 \quad \text{and} \quad \widehat{x}_i,\ 0 \le \widehat{x}_i \le 1,\ \breve{x}_i \le \widehat{x}_i, \quad i = 1, 2, \ldots, v, \tag{6.10}$$

are lower and upper bounds of the unknown probabilities x_i, $i = 1, 2, \ldots, v$, respectively.

To find the optimal values of x_i, $i = 1, 2, \ldots, v$, we define

$$\breve{x} = \sum_{i=1}^{v} \breve{x}_i, \quad \hat{y} = 1 - \breve{x} \tag{6.11}$$

and

$$\breve{x}^{0} = 0,\ \widehat{x}^{0} = 0 \quad \text{and} \quad \breve{x}^{I} = \sum_{i=1}^{I} \breve{x}_i,\ \widehat{x}^{I} = \sum_{i=1}^{I} \widehat{x}_i \quad \text{for } I = 1, 2, \ldots, v. \tag{6.12}$$

Next, we find the largest value $I \in \{0, 1, \ldots, v\}$ such that

$$\widehat{x}^{I} - \breve{x}^{I} < \hat{y} \tag{6.13}$$

and we fix the optimal solution that maximizes (6.7) in the following way:

(i) if $I = 0$, the optimal solution is

$$\dot{x}_1 = \hat{y} + \breve{x}_1 \quad \text{and} \quad \dot{x}_i = \breve{x}_i \quad \text{for } i = 2, 3, \ldots, v; \tag{6.14}$$

(ii) if $0 < I < v$, the optimal solution is

$$\dot{x}_i = \widehat{x}_i \quad \text{for } i = 1, 2, \ldots, I,\ \dot{x}_{I+1} = \hat{y} - \widehat{x}^{I} + \breve{x}^{I} + \breve{x}_{I+1} \quad \text{and}$$
$$\dot{x}_i = \breve{x}_i \quad \text{for } i = I + 2, I + 3, \ldots, v; \tag{6.15}$$

(iii) if $I = v$, the optimal solution is

$$\dot{x}_i = \widehat{x}_i \quad \text{for } i = 1, 2, \ldots, v. \tag{6.16}$$

Finally, after making the inverse substitution to (6.6), we get the optimal limit transient probabilities

$$\dot{p}_{b_i} = \dot{x}_i \quad \text{for } i = 1, 2, \ldots, v, \tag{6.17}$$

that maximize the system mean lifetime in the reliability state subset $\{r, r + 1, \ldots, z\}$, defined by the linear form (6.1), giving its maximum value in the following form

$$\dot{\mu}(r) = \sum_{b=1}^{v} \dot{p}_b \mu_b(r) \tag{6.18}$$

for a fixed $r \in \{1, 2, \ldots, z\}$.

6.2.2 Optimal Reliability and Safety Characteristics of Complex Technical Systems

From expression (6.18) for the maximum mean value $\dot{\mu}(r)$ of the system unconditional lifetime in the reliability state subset $\{r, r+1, \ldots, z\}$, replacing in it the critical reliability state r by the reliability state u, $u = 1, 2, \ldots, z$, we obtain the corresponding optimal solutions for the mean values of the system unconditional lifetimes in the reliability state subsets $\{u, u+1, \ldots, z\}$ of the form

$$\dot{\mu}(u) = \sum_{b=1}^{v} \dot{p}_b \mu_b(u) \quad \text{for } u = 1, 2, \ldots, z. \tag{6.19}$$

Further, according to (3.5–3.6), the corresponding optimal unconditional multistate reliability function of the system is the vector

$$\dot{\boldsymbol{R}}(t, \cdot) = [1, \ \dot{\boldsymbol{R}}(t, 1), \ldots, \ \dot{\boldsymbol{R}}(t, z)], \tag{6.20}$$

with the coordinates given by

$$\dot{\boldsymbol{R}}(t, u) \cong \sum_{b=1}^{v} \dot{p}_b [\boldsymbol{R}(t, u)]^{(b)} \quad \text{for } t \geq 0, \quad u = 1, 2, \ldots, z. \tag{6.21}$$

By applying (3.9), the corresponding optimal values of the variances of the system unconditional lifetimes in the system reliability state subsets are

$$\dot{\sigma}^2(u) = 2 \int_0^{\infty} t \, \dot{\boldsymbol{R}}(t, u) \mathrm{d}t - [\dot{\mu}(u)]^2, \quad u = 1, 2, \ldots, z, \tag{6.22}$$

where $\dot{\mu}(u)$ is given by (6.19) and $\dot{\boldsymbol{R}}(t, u)$ is given by (6.21).

And, by (3.10), the optimal solutions for the mean values of the system unconditional lifetimes in the particular reliability states are

$$\dot{\bar{\mu}}(u) = \dot{\mu}(u) - \dot{\mu}(u+1), \quad u = 1, \ldots, z-1, \ \ \dot{\bar{\mu}}(z) = \dot{\mu}(z). \tag{6.23}$$

Moreover, considering (3.11) and (3.12), the corresponding optimal system risk function and the optimal moment when the risk exceeds a permitted level δ, respectively are given by

$$\dot{\boldsymbol{r}}(t) = 1 - \dot{\boldsymbol{R}}(t, r), \quad t \geq 0, \tag{6.24}$$

and

$$\dot{\tau} = \dot{\boldsymbol{r}}^{-1}(\delta), \tag{6.25}$$

where $\dot{\boldsymbol{R}}(t, r)$ is given by (6.21) for $u = r$ and $\dot{\boldsymbol{r}}^{-1}(t)$, if it exists, is the inverse function of the optimal risk function $\dot{\boldsymbol{r}}(t)$.

All the considered reliability characteristics and presented results in this section may be transferred to the complex technical systems safety analysis and formulated by simply replacing their reliability function $\boldsymbol{R}(t,\cdot)$ and its optimal form $\dot{\boldsymbol{R}}(t,\cdot)$ respectively by their safety function $\boldsymbol{s}(t,\cdot)$ and its optimal form $\dot{\boldsymbol{s}}(t,\cdot)$.

6.2.3 *Optimal Renewal and Availability Characteristics of Complex Technical Systems*

The direct replacing $\mu(r)$ by $\dot{\mu}(r)$ and $\sigma(r)$ by $\dot{\sigma}(r)$ in the expressions for the repairable systems renewal and availability characteristics pointed in *Propositions 3.4* and *3.5*, we get their corresponding optimal values pointed out in the following two propositions [3, 5].

Proposition 6.1 *If components of the multistate repairable system with ignored time of renovation have the exponential reliability functions at the operation states* z_b, $b = 1, 2, \ldots, v$, *given by* (3.13–3.14) *or respectively by* (3.15–3.16) *and the system reliability critical state is* r, $r \in \{1, 2, \ldots, z\}$, *then after the system operation process optimization:*

(a) *the optimal time* $\dot{S}_N(r)$ *until the Nth exceeding by the system, the reliability critical state r, for sufficiently large N, has approximately normal distribution* $N(N\dot{\mu}(r), \sqrt{N}\dot{\sigma}(r))$, *i.e.,*

$$\dot{F}^{(N)}(t,r) = P(\dot{S}_N(r) < t) \cong F_{N(0,1)}\left(\frac{t - N\dot{\mu}(r)}{\sqrt{N}\dot{\sigma}(r)}\right),$$
$$t \in (-\infty, \infty), \quad r \in \{1, 2, \ldots, z\}; \tag{6.26}$$

(b) *the expected value and the variance of the optimal time* $\dot{S}_N(r)$ *until the Nth exceeding by the system the reliability critical state r, for sufficiently large N, are respectively given by*

$$E[\dot{S}_N(r)] = N\dot{\mu}(r), \quad D[\dot{S}_N(r)] = N\dot{\sigma}^2(r), \qquad r \in \{1, 2, \ldots, z\}; \tag{6.27}$$

(c) *the optimal number* $\dot{N}(t,r)$ *of exceedings by the system the reliability critical state r up to the moment t,* $t \geq 0$, *for sufficiently large t, has approximately distribution of the form*

$$P(\dot{N}(t,r) = N) \cong F_{N(0,1)}\left(\frac{(N+1)\dot{\mu}(r) - t}{\dot{\sigma}(r)\sqrt{\frac{t}{\dot{\mu}(r)}}}\right) - F_{N(0,1)}\left(\frac{N\dot{\mu}(r) - t}{\dot{\sigma}(r)\sqrt{\frac{t}{\dot{\mu}(r)}}}\right),$$
$$N = 0, 1, \ldots, \quad r \in \{1, 2, \ldots, z\}; \tag{6.28}$$

(d) *the expected value and the variance of the optimal number* $N(t,r)$ *of exceeding by the system the reliability critical state r up to the moment* t, $t \geq 0$, *for sufficiently large t, are respectively given by*

$$\dot{H}(t,r) = \frac{t}{\dot{\mu}(r)}, \quad \dot{D}(t,r) = \frac{t}{[\dot{\mu}(r)]^3}[\dot{\sigma}(r)]^2, \qquad r \in \{1,2,\ldots,z\}; \tag{6.29}$$

where $\dot{\mu}(r)$ *and* $\dot{\sigma}(r)$ *are given by* (6.19 *and* 6.22) *for* $u = r$.

Proposition 6.2 *If components of the multistate repairable system with non-ignored time of renovation at the operational states* z_b, $b = 1,2,\ldots,v$, *have the exponential reliability functions at the operation states* z_b, $b = 1,2,\ldots,v$, *given by* (3.13–3.14) *or respectively by* (3.15–3.16), *the system reliability critical state is* r, $r \in \{1,2,\ldots,z\}$, *and the successive times of system's renovations are independent and have an identical distribution function with the expected value* $\mu_o(r)$ *and the variance* $\sigma_o^2(r)$, *then after the system operation process optimization:*

(a) *the optimal time* $\dot{\bar{S}}_N(r)$ *until the Nth exceeding by the system the reliability critical state r, for sufficiently large N, has approximately normal distribution* $N(N\dot{\mu}(r) + (N-1)\mu_o(r)), \sqrt{N\dot{\sigma}^2(r) + (N-1)\sigma_o^2(r)})$, *i.e.,*

$$\dot{\bar{F}}^{(N)}(t,r) = P(\dot{\bar{S}}_N(r) < t) \cong F_{N(0,1)}\left(\frac{t - N(\dot{\mu}(r) + \mu_o(r)) + \mu_o(r)}{\sqrt{N(\dot{\sigma}^2(r) + \sigma_o^2(r)) - \sigma_o^2(r)}}\right),$$
$$t \in (-\infty,\infty), \quad r \in \{1,2,\ldots,z\}; \tag{6.30}$$

(b) *the expected value and the variance of the optimal time* $\dot{\bar{S}}_N(r)$ *until the Nth exceeding by the system the reliability critical state r, for sufficiently large N, are respectively given by*

$$E[\dot{\bar{S}}_N(r)] \cong N\dot{\mu}(r) + (N-1)\mu_o(r), \quad D[\dot{\bar{S}}_N(r)] \cong N\dot{\sigma}^2(r) + (N-1)\sigma_o^2(r),$$
$$r \in \{1,2,\ldots,z\}; \tag{6.31}$$

(c) *the optimal number* $\dot{\bar{N}}(t,r)$ *of exceedings by the system the reliability critical state r up to the moment t,* $t \geq 0$, *for sufficiently large t, has approximately distribution of the form*

$$P(\dot{\bar{N}}(t,r) = N) \cong F_{N(0,1)}\left(\frac{(N+1)(\dot{\mu}(r) + \mu_o(r)) - t - \mu_0(r)}{\sqrt{\frac{t + \mu_0(r)}{\dot{\mu}(r) + \mu_0(r)}(\dot{\sigma}^2(r) + \sigma_o^2(r))}}\right)$$
$$-F_{N(0,1)}\left(\frac{N(\dot{\mu}(r) + \mu_o(r)) - t - \mu_0(r)}{\sqrt{\frac{t + \mu_0(r)}{\dot{\mu}(r) + \mu_0(r)}(\dot{\sigma}^2(r) + \sigma_o^2(r))}}\right),$$
$$N = 0,1,\ldots, \quad r \in \{1,2,\ldots,z\}; \tag{6.32}$$

(d) *the expected value and the variance of the optimal number* $\dot{\bar{\bar{N}}}(t,r)$ *of exceedings by the system the reliability critical state r up to the moment t,* $t \geq 0$, *for sufficiently large t, are respectively given by*

$$\dot{\bar{\bar{H}}}(t,r) \cong \frac{t+\mu_0(r)}{\dot{\mu}(r)+\mu_o(r)}, \quad \dot{\bar{D}}(t,r) \cong \frac{t+\mu_0(r)}{(\dot{\mu}(r)+\mu_o(r))^3}(\dot{\sigma}^2(r)+\sigma_o^2(r)),$$
$$r \in \{1,2,\ldots,z\}; \tag{6.33}$$

(e) *the optimal time* $\dot{\bar{\bar{S}}}_N(r)$ *until the Nth system's renovation, for sufficiently large N, has approximately normal distribution* $N(N(\dot{\mu}(r)+\mu_o(r)),$ $\sqrt{N(\dot{\sigma}^2(r)+\sigma_o^2(r))})$, *i.e.,*

$$\dot{\bar{\bar{F}}}^{(N)}(t,r) = P(\dot{\bar{\bar{S}}}_N(r) < t) \cong F_{N(0,1)}\left(\frac{t-N(\dot{\mu}(r)+\mu_o(r))}{\sqrt{N(\dot{\sigma}^2(r)+\sigma_o^2(r))}}\right), \quad t \in (-\infty,\infty),$$
$$r \in \{1,2,\ldots,z\}; \tag{6.34}$$

(f) *the expected value and the variance of the optimal time* $\dot{\bar{\bar{S}}}_N(r)$ *until the Nth system's renovation, for sufficiently large N, are respectively given by*

$$E[\dot{\bar{\bar{S}}}_N(r)] \cong N(\dot{\mu}(r)+\mu_o(r)), \quad D[\dot{\bar{\bar{S}}}_N(r)] \cong N(\dot{\sigma}^2(r)+\sigma_o^2(r)),$$
$$r \in \{1,2,\ldots,z\}; \tag{6.35}$$

(g) *the optimal number* $\dot{\bar{\bar{N}}}(t,r)$ *of system's renovations up to the moment t,* $t \geq 0$, *for sufficiently large t, has approximately distribution of the form*

$$P(\dot{\bar{\bar{N}}}(t,r) = N) \cong F_{N(0,1)}\left(\frac{(N+1)(\dot{\mu}(r)+\mu_o(r))-t}{\sqrt{\frac{t}{\dot{\mu}(r)+\mu_0(r)}(\dot{\sigma}^2(r)+\sigma_o^2(r))}}\right)$$
$$-F_{N(0,1)}\left(\frac{N(\dot{\mu}(r)+\mu_o(r))-t}{\sqrt{\frac{t}{\dot{\mu}(r)+\mu_0(r)}(\dot{\sigma}^2(r)+\sigma_o^2(r))}}\right),$$
$$N = 0,1,\ldots, \quad r \in \{1,2,\ldots,z\}; \tag{6.36}$$

(h) *the expected value and the variance of the optimal number* $\dot{\bar{\bar{N}}}(t,r)$ *of system's renovations up to the moment t,* $t \geq 0$, *for sufficiently large t, are respectively given by*

$$\dot{\bar{\bar{H}}}(t,r) \cong \frac{t}{\dot{\mu}(r)+\mu_o(r)}, \quad \dot{\bar{\bar{D}}}(t,r) \cong \frac{t}{(\dot{\mu}(r)+\mu_o(r))^3}(\dot{\sigma}^2(r)+\sigma_o^2(r)),$$
$$r \in \{1,2,\ldots,z\}, \tag{6.37}$$

(i) *the optimal steady availability coefficient of the system at the moment* $t,\ t \geq 0$, *for sufficiently large t, is given by*

$$\dot{A}(t,r) \cong \frac{\dot{\mu}(r)}{\dot{\mu}(r) + \mu_o(r)}, \quad t \geq 0, \quad r \in \{1,2,\ldots,z\}; \tag{6.38}$$

(j) *the optimal steady availability coefficient of the system in the time interval* $\langle t, t+\tau),\ \tau > 0$, *for sufficiently large t, is given by*

$$\dot{A}(t,\tau,r) \cong \frac{1}{\dot{\mu}(r) + \mu_o(r)} \int_{\tau}^{\infty} \dot{\boldsymbol{R}}(t,r)\mathrm{d}t, \quad t \geq 0, \quad \tau > 0, \quad r \in \{1,2,\ldots,z\}; \tag{6.39}$$

where $\dot{\boldsymbol{R}}(t,r)$ *is given by the formula* (6.21) *and* $\dot{\mu}(r)$ *and* $\dot{\sigma}(r)$ *are given by* (6.19 *and* 6.22) *for* $u = r$.

6.2.4 Optimal Sojourn Times of Complex Technical System Operation Processes at Operation States and Operation Strategy

Replacing in (2.22) the limit transient probabilities p_b of the system operation process at the operation states by their optimal values $\dot{p}_b$ found in Sect. 6.2.1 and the mean values M_b of the unconditional sojourn times at the operation states by their corresponding unknown optimal values $\dot{M}_b$ maximizing the mean value of the system lifetime in the reliability states subset $\{r, r+1,\ldots,z\}$ defined by (6.1), we get the system of equations

$$\dot{p}_b = \frac{\pi_b \dot{M}_b}{\sum_{l=1}^{v} \pi_l \dot{M}_l}, \quad b = 1,2,\ldots,v. \tag{6.40}$$

After simple transformations the above system takes the form

$$\begin{aligned} (\dot{p}_1 - 1)\pi_1\dot{M}_1 + \dot{p}_1\pi_2\dot{M}_2 + \cdots + \dot{p}_1\pi_v\dot{M}_v &= 0 \\ \dot{p}_2\pi_1\dot{M}_1 + (\dot{p}_2 - 1)\pi_2\dot{M}_2 + \cdots + \dot{p}_2\pi_v\dot{M}_v &= 0 \\ &\vdots \\ \dot{p}_v\pi_1\dot{M}_1 + \dot{p}_v\pi_2\dot{M}_2 + \cdots + (\dot{p}_v - 1)\pi_v\dot{M}_v &= 0, \end{aligned} \tag{6.41}$$

where $\dot{M}_b$ are unknown variables that we want to find, $\dot{p}_b$ are optimal transient probabilities determined by (6.17) and π_b are steady probabilities determined by (2.23).

Since the system of equations (6.41) is homogeneous and it can be proved that the determinant of its main matrix is equal to zero, then it has nonzero solutions and moreover, these solutions are ambiguous. Thus, if we fix some of the optimal values

$\dot{M}_b$ of the mean values M_b of the unconditional sojourn times at the operation states, for instance by arbitrarily fixing one or a few of them, we may find the values of the remaining ones and in this way get the solution of this equation.

Having this solution, it is also possible to look for the optimal values $\dot{M}_{bl}$ of the mean values M_{bl} of the conditional sojourn times at the operation states using the following system of equations

$$\sum_{l=1}^{v} p_{bl}\dot{M}_{bl} = \dot{M}_b, \quad b = 1,2,\ldots,v, \tag{6.42}$$

obtained from (2.21) by replacing M_b by $\dot{M}_b$ and M_{bl} by $\dot{M}_{bl}$, where p_{bl} are known probabilities of the system operation process transitions between the operation states z_b i z_l, $b,l = 1,2,\ldots,v$, $b \neq l$, defined by (2.22).

Another very useful and much easier to be applied in practice tool that can help in planning the operation processes of complex technical systems are the system operation process optimal mean values of the total system operation process sojourn times $\hat{\theta}_b$ at the particular operation states z_b, $b = 1,2,\ldots,v$, during the fixed system operation time θ, which can be obtained by replacing in the formula (2.24) the transient probabilities p_b at the operation states z_b by their optimal values $\dot{p}_b$ and resulting in the following expression

$$\dot{E}[\hat{\theta}_b] = \dot{p}_b\theta, \quad b = 1,2,\ldots,v. \tag{6.43}$$

The knowledge of the optimal values $\dot{M}_b$ of the mean values of the unconditional sojourn times and the optimal values $\dot{M}_{bl}$ of the mean values of the conditional sojourn times at the operation states and the optimal mean values $\dot{E}[\hat{\theta}_b]$ of the total sojourn times at the particular operation states during the fixed system operation time may be the basis for changing the complex technical systems' operation processes in order to ensure that these systems operates more reliably and safely. Their knowledge may also be useful in these systems' operation cost analysis.

6.2.5 Cost Analysis of Complex Technical System Operations

We consider the complex technical multistate system consisted of n components and we assume that the operation costs of its single basic components at the operation state z_b, $b = 1,2,\ldots,v$, during the system operation time θ, $\theta \geq 0$, amount to

$$c_i(\theta, b), \quad i = 1,2,\ldots,n.$$

First, we suppose that the system is non-repairable, i.e. the system during the operation has not exceeded the critical reliability state r. In this case, the total cost of the non-repairable system during the operation time θ, $\theta \geq 0$, is given by

$$C(\theta) \cong \sum_{b=1}^{v} p_b \sum_{i=1}^{n} c_i(\theta, b), \quad \theta \geq 0, \tag{6.44}$$

where p_b, $b = 1, 2, \ldots, v$, are transient probabilities defined by (2.22).

Further, we additionally assume that the system is repairable after exceeding the critical reliability state r, its renovation time is ignored and the cost of the system singular renovation is c_{ig}.

Then, the total operation cost of the repairable system to ignoring its renovation time during the operation time θ, $\theta \geq 0$, amounts to

$$C_{ig}(\theta) \cong \sum_{b=1}^{v} p_b \sum_{i=1}^{n} c_i(\theta, b) + c_{ig} H(\theta, r), \quad \theta \geq 0, \tag{6.45}$$

where p_b, $b = 1, 2, \ldots, v$, are transient probabilities defined by (2.22) and $H(\theta, r)$ is the mean value of the number of exceedings of the critical reliability state r by the system operating at the variable conditions during the operation time θ defined by (3.58).

Now, we assume that the system is repairable after exceeding the critical reliability state r and its renewal time is non-ignored and have a distribution function with mean value $\mu_0(r)$and standard deviation $\sigma_0(r)$ and the cost of the system singular renovation is c_{nig}.

Then, the total operation cost of the repairable system while not ignoring its renovation time during the operation time θ, $\theta \geq 0$, amounts to

$$C_{nig}(\theta) \cong \sum_{b=1}^{v} p_b \sum_{i=1}^{n} c_i(\theta, b) + c_{nig} \bar{\bar{H}}(\theta, r), \quad \theta \geq 0, \tag{6.46}$$

where p_b, $b = 1, 2, \ldots, v$, are transient probabilities defined by (2.22) and $\bar{\bar{H}}(\theta, r)$ is the mean value of the number of renovations of the system operating at the variable conditions during the operation time θ defined by (3.92).

The particular expressions for the mean values $H(\theta, r)$ and $\bar{\bar{H}}(\theta, r)$ for the repairable systems with ignored and non-ignored renovation times existing in the formulae (6.45) and (6.46), respectively defined by (3.58) and (3.92), are determined in Chap. 3 for typical multistate repairable systems, i.e. for multistate series, parallel, "m out of n", consecutive "m out of n: F", series-parallel, parallel-series, series-"m out of k", "m_i out of l_i"-series, series-consecutive "m out of k: F" and consecutive "m_i out of l_i: F"-series systems operating at the variable operating conditions.

After the optimization of the system operation process, the system operation total costs given by (6.44–6.46) assume their optimal values.

The total optimal cost of the non-repairable system during the operation time θ, $\theta \geq 0$, after its operation process optimization is given by

$$\dot{C}(\theta) = \sum_{b=1}^{v} \dot{p}_b \sum_{i=1}^{n} c_i(\theta, b), \quad \theta \geq 0, \tag{6.47}$$

where $\dot{p}_b$, $b = 1, 2, \ldots, v$, are optimal transient probabilities defined by (6.17).

The optimal total operation cost of the repairable system ignoring its renovation time during the operation time θ, $\theta \geq 0$, after its operation process optimization amounts to

$$\dot{C}_{ig}(\theta) \cong \sum_{b=1}^{v} \dot{p}_b \sum_{i=1}^{n} c_i(\theta, b) + c_{ig} \dot{H}(\theta, r), \quad \theta \geq 0, \tag{6.48}$$

where $\dot{p}_b$, $b = 1, 2, \ldots, v$, are optimal transient probabilities defined by (6.17) and $\dot{H}(\theta, r)$ is the mean value of the optimal number of exceedings of the critical reliability state r by the system operating at the variable conditions during the operation time θ defined by (6.29).

The total optimal operation cost of the repairable system while non-ignoring its renovation time during the operation time θ, $\theta \geq 0$, after its operation process optimization amounts to

$$\dot{C}_{nig}(\theta) \cong \sum_{b=1}^{v} \dot{p}_b \sum_{i=1}^{n} c_i(\theta, b) + c_{nig} \dot{\bar{\bar{H}}}(\theta, r), \quad \theta \geq 0, \tag{6.49}$$

where $\dot{p}_b$, $b = 1, 2, \ldots, v$, are optimal transient probabilities defined by (6.17) and $\dot{\bar{\bar{H}}}(\theta, r)$ is the mean value of the optimal number of renovations of the system operating at the variable operating conditions during the operation time θ defined by (6.37).

The particular expressions for the optimal mean values $\dot{H}(\theta, r)$ and $\dot{\bar{\bar{H}}}(\theta, r)$ for the repairable systems with ignored and non-ignored renovation times existing in the formulae (6.48) and (6.49), respectively defined by (6.29) and (6.37), may be obtained by replacing the transient probabilities p_b by their optimal values $\dot{p}_b$ in the expressions for $H(\theta, r)$ and $\bar{\bar{H}}(\theta, r)$ defined by (3.58) and (3.92) which are determined in Chap. 3 for typical multistate repaired systems, i.e. for multistate series, parallel, "m out of n", consecutive "m out of n: F", series-parallel, parallel-series, series-"m out of k", "m_i out of l_i"-series, series-consecutive "m out of k: F" and consecutive "m_i out of l_i: F"-series systems operating at variable operating conditions.

The application of the formulae (6.44–6.46) and (6.47–6.49) allow us to compare the costs of the non-repairable and repairable systems with ignored and non-ignored times of renovations operating at variable operating conditions before and after the optimization of their operation processes.

6.3 Corrective and Preventive Maintenance Policy Optimization of Complex Technical Systems

6.3.1 Corrective and Preventive Maintenance Policy Maximizing Availability of Complex Technical System

We consider the maintenance policy applied to the complex technical multistate system operating at the variable conditions which includes both corrective renovation taken upon the system exceeding the critical reliability state r and preventive renovation taken periodically with the period of time η, $\eta \geq 0$. As before, we denote by $T(r)$, $r = 1, 2, \ldots, z$, the system lifetime in the reliability states subset $\{r, r+1, \ldots, z\}$ and by $\mu_0(r)$ the mean value of the system corrective renovation time upon the system exceeding the critical reliability state r. Additionally, we mark by $\mu_1(r)$ the mean value of the system preventive renovation time upon the system correction after each period of time η independently of the system exceeding the critical reliability state r. In both cases, after renovation, the system becomes "as good as new", i.e. it is in the best reliability state z.

Under these assumptions, the system renovation can be performed either at the moment t, $t < \eta$, or at the moment η and the system mean time between the renovations is given by

$$\mu_U(\eta, r) = P\{T(r) = t \cap t < \eta\} \cdot t + P\{T(r) = t \cap t \geq \eta\} \times \eta.$$

Hence

$$\begin{aligned}
\mu_U(\eta, r) &= \int_0^{\eta} tf(t, r)\mathrm{d}t + P\{T(r) \geq \eta\} \times \eta \\
&= \int_0^{+\infty} tf(t, r)\mathrm{d}t - \int_{\eta}^{+\infty} tf(t, r)\mathrm{d}t + \boldsymbol{R}(\eta, r)\eta \\
&= \int_0^{+\infty} \boldsymbol{R}(t, r)\mathrm{d}t - \int_{\eta}^{+\infty} tf(t, r)\mathrm{d}t + \boldsymbol{R}(\eta, r)\eta \\
&= \int_0^{\eta} \boldsymbol{R}(t, r)\mathrm{d}t + \int_{\eta}^{+\infty} \boldsymbol{R}(t, r)\mathrm{d}t - \int_{\eta}^{+\infty} tf(t, r)\mathrm{d}t + \boldsymbol{R}(\eta, r)\eta,
\end{aligned}$$

where

$$\boldsymbol{R}(t, r) = P\{T(r) > t\}, \quad t \in \langle 0, +\infty),$$

is the rth coordinate of the system reliability function

$$\boldsymbol{R}(t, \cdot) = [\boldsymbol{R}(t, 1), \boldsymbol{R}(t, 2), \ldots, \boldsymbol{R}(t, z)], \quad t \in \langle 0, +\infty),$$

and

$$f(t,r) = -\frac{d}{dt}\boldsymbol{R}(t,r) = -\boldsymbol{R}'(t,r)$$

is its corresponding density function.

From the above, after integrating by parts, we get

$$\mu_U(\eta,r) = \int_0^{\eta} \boldsymbol{R}(t,r)dt + \int_{\eta}^{+\infty} \boldsymbol{R}(t,r)dt + t\boldsymbol{R}(t,r)|_{\eta}^{+\infty} - \int_{\eta}^{+\infty} \boldsymbol{R}(t,r)dt + \boldsymbol{R}(\eta,r)\eta$$

and finally

$$\mu_U(\eta,r) = \int_0^{\eta} \boldsymbol{R}(t,r)dt. \tag{6.50}$$

As the system renovation can be performed either at the moment t, $t<\eta$, with the mean value of the system corrective renovation time $\mu_0(r)$ or at the moment η with the mean value of the system preventive renovation time $\mu_1(r)$, the system mean renovation time is given by

$$\begin{aligned}\mu_D(\eta,r) = P\{T(r) &= t \cap t<\eta\} \times \mu_0(r) + P\{T(r) \\ &= t \cap t \geq \eta\} \times \mu_1(r) \\ &= [1 - \boldsymbol{R}(\eta,r)]\mu_0(r) + \boldsymbol{R}(\eta,r)\mu_1(r).\end{aligned} \tag{6.51}$$

After applying the results (6.50) and (6.51), we get the formula for the availability coefficient of the system

$$A(\eta,r) \cong \frac{\mu_U(\eta,r)}{\mu_U(\eta,r)+\mu_D(\eta,r)} = \frac{\int_0^{\eta} \boldsymbol{R}(t,r)dt}{\int_0^{\eta} \boldsymbol{R}(t,r)dt + [1-\boldsymbol{R}(\eta,r)]\mu_0(r) + \boldsymbol{R}(\eta,r)\mu_1(r)}, \quad \eta \geq 0. \tag{6.52}$$

Our objective now is that of finding the optimal value $\dot{\eta}$ of the preventive maintenance period η that maximizes $A(\eta,r)$ given by (6.52). To solve this problem, we find the derivative of $A(\eta,r)$ with respect to η that takes the form

$$\begin{aligned}\frac{dA(\eta,r)}{d\eta} = &\frac{\boldsymbol{R}(\eta,r)[\int_0^{\eta} \boldsymbol{R}(t,r)dt + [1-\boldsymbol{R}(\eta,r)]\mu_0(r) + \boldsymbol{R}(\eta,r)\mu_1(r)]}{[\int_0^{\eta} \boldsymbol{R}(t,r)dt + [1-\boldsymbol{R}(\eta,r)]\mu_0(r) + \boldsymbol{R}(\eta,r)\mu_1(r)]^2} \\ &- \frac{\int_0^{\eta} \boldsymbol{R}(t,r)dt[\boldsymbol{R}(\eta,r) - \boldsymbol{R}'(\eta,r)\mu_0(r) + \boldsymbol{R}'(\eta,r)\mu_1(r)]}{[\int_0^{\eta} \boldsymbol{R}(t,r)dt + [1-\boldsymbol{R}(\eta,r)]\mu_0(r) + \boldsymbol{R}(\eta,r)\mu_1(r)]^2}\end{aligned}$$

and further we get

$$\frac{dA(\eta,r)}{d\eta} = \frac{-\boldsymbol{R}'(\eta,r)[\mu_1(r)-\mu_0(r)]\int_0^\eta \boldsymbol{R}(t,r)dt + \boldsymbol{R}(\eta,r)\mu_0(r) + \boldsymbol{R}^2(\eta,r)[\mu_1(r)-\mu_0(r)]}{[\int_0^\eta \boldsymbol{R}(t,r)dt + [1-\boldsymbol{R}(\eta,r)]\mu_0(r) + \boldsymbol{R}(\eta,r)\mu_1 r)]^2}. \tag{6.53}$$

Since $\boldsymbol{R}(\eta,r)$ is a non-increasing function of η, $\boldsymbol{R}'(\eta,r) \le 0$ and from (6.53) it follows that in the case when

$$\mu_0(r) \le \mu_1(r)$$

the inequality

$$\frac{dA(\eta,r)}{d\eta} > 0 \quad \text{for } \eta \ge 0$$

holds, which means that $A(\eta,r)$ is an increasing function of η independently of the form of $\boldsymbol{R}(\eta,r)$. Thus, there is no optimal value $\dot{\eta}$ of the preventive maintenance period η that maximizes $A(\eta,r)$ given by (6.52).

If

$$\mu_0(r) > \mu_1(r)$$

and either

$$\frac{dA(\eta,r)}{d\eta} > 0 \quad \text{for } \eta \ge 0$$

or

$$\frac{dA(\eta,r)}{d\eta} < 0 \quad \text{for } \eta \ge 0$$

then also there is no optimal value $\dot{\eta}$ of the preventive maintenance period η that maximizes $A(\eta,r)$ given by (6.52).

Finally, the optimal value $\dot{\eta}$ of the preventive maintenance period η that maximizes $A(\eta,r)$ given by (6.51) can only be found when

$$\mu_0(r) > \mu_1(r)$$

and there exists the solution $\eta \ge 0$ of the equality

$$\frac{dA(\eta,r)}{d\eta} = 0 \tag{6.54}$$

Thus, after joining (6.53) and (6.54), we get the equation

$$\boldsymbol{R}'(\eta,r)[\mu_0(r)-\mu_1(r)]\int_0^\eta \boldsymbol{R}(t,r)\mathrm{d}t-\boldsymbol{R}^2(\eta,r)[\mu_0(r)-\mu_1(r)]=-\boldsymbol{R}(\eta,r)\mu_0(r) \tag{6.55}$$

from which the optimal value $\dot{\eta}$ of η can be determined.

From (6.55), using simple arithmetical operations, we can obtain a slightly simpler equation

$$-\frac{\boldsymbol{R}'(\eta,r)}{\boldsymbol{R}(\eta,r)}\int_0^\eta \boldsymbol{R}(t,r)\mathrm{d}t+\boldsymbol{R}(\eta,r)=\frac{\mu_0(r)}{\mu_0(r)-\mu_1(r)} \tag{6.56}$$

from which the optimal value $\dot{\eta}$ of η can be determined.

Introducing the rate of departure from the system reliability states set $\{r,r+1,\ldots,z\}$

$$\lambda(\eta,r)=-\frac{\boldsymbol{R}'(\eta,r)}{\boldsymbol{R}(\eta,r)} \tag{6.57}$$

from (6.56), we obtain the following expression

$$\lambda(\eta,r)\int_0^\eta \boldsymbol{R}(t,r)\mathrm{d}t+\boldsymbol{R}(\eta,r)=\frac{\mu_0(r)}{\mu_0(r)-\mu_1(r)}. \tag{6.58}$$

Substituting either $\boldsymbol{R}(\eta,r)$ into Eq. 6.56 or $\boldsymbol{R}(\eta,r)$ and $\lambda(\eta,r)$ into Eq. 6.58 we may find the optimal value $\dot{\eta}$ of η. Later substituting this value into the expression (6.52) we may obtain the maximum value $A(\dot{\eta},r)$ of the availability coefficient $A(\eta,r)$ of the system. Unfortunately, as a rule, it is not possible to find the explicit solution of Eqs. 6.56 and 6.58 and they have to be found by the approximate methods.

One of the approximate methods suggested to optimize the complex technical system's corrective and preventive maintenance policy maximizing its availability is the method of secants which we apply by performing the following procedure:

- we check the relationship between the mean value of the system corrective renovation time $\mu_0(r)$ and the mean value of the system preventive renovation time $\mu_1(r)$;
- if the inequality

$$\mu_0(r)\leq\mu_1(r)$$

holds, then there is no optimal value $\dot{\eta}$ of the preventive maintenance period η which maximizes the availability coefficient of the system $A(\eta,r)$ given by (6.52) which is an increasing function of η and we determine its values for the arbitrarily fixed values of the preventive maintenance period

$$\eta_1,\eta_2,\ldots,\eta_K;$$

- if the inequality is

$$\mu_0(r) > \mu_1(r)$$

we are looking for the optimal value $\dot{\eta}$ of the preventive maintenance period η which maximizes the availability coefficient of the system $A(\eta, r)$ given by (6.52) by determining, if it exists, its approximate value from Eq. 6.58 by applying the method of secants in the interval $\langle a,\ b\rangle$ as follows:

- we define the function

$$f(\eta) = \lambda(\eta, r)\int_0^{\eta} \boldsymbol{R}(t, r)\mathrm{d}t + \boldsymbol{R}(\eta, r) - \frac{\mu_0(r)}{\mu_0(r) - \mu_1(r)} \quad \text{for } \eta \geq 0,$$

where $\lambda(\eta, r)$ is given by (6.57),

- we define the interval $\langle a,\ b\rangle$ assuming $a = 0$ and finding b such that

$$f(b) > 0,$$

- we use the recurrent formula

$$\eta_0 = a,$$

$$\eta_{k+1} = \eta_k - \frac{f(\eta_k)}{f(b) - f(\eta_k)}(b - \eta_k) \quad \text{for } k = 0, 1, \ldots, K,$$

where K is such that

$$f(\eta_{K+1}) < \varepsilon$$

and ε is the measure of the method of secants accuracy,

- we fix the optimal value $\dot{\eta}$ of the preventive maintenance period of time η maximizing the system availability assuming

$$\dot{\eta} = \eta_{K+1},$$

and determine the maximum value $A(\dot{\eta}, r)$ of the availability coefficient $A(\eta, r)$ of the system substituting this optimal value $\dot{\eta}$ for η in expression (6.52) in the case when the recurrent procedure is convergent. Otherwise, there is no optimal value $\dot{\eta}$ of the preventive maintenance period η that maximizes the availability coefficient of the system and we find its value for the arbitrarily fixed values of the preventive maintenance period of time $\eta_1, \eta_2, \ldots, \eta_K$.

6.3.2 *Corrective and Preventive Maintenance Policy Minimizing Cost of Renovation of Complex Technical Systems*

We consider the maintenance policy applied to complex technical multistate systems operating in variable conditions which includes both a corrective renovation taken upon the system exceeding the critical reliability state r and a preventive renovation taken at the system's fixed age ς, $\varsigma > 0$. As in Sect. 6.3.1, we denote by $T(r), r = 1, 2, \ldots, z$, the system lifetime in the reliability states subset $\{r, r+1, \ldots, z\}$ and by

$$\boldsymbol{R}(t,r) = P\{T(r) > t\}, \quad t \in \langle 0, +\infty),$$

the rth coordinate of the system reliability function

$$\boldsymbol{R}(t,\ \cdot) = [\boldsymbol{R}(t,1), \boldsymbol{R}(t,2), \ldots, \boldsymbol{R}(t,z)], \quad t \in \langle 0, +\infty),$$

and by

$$f(t,r) = -\frac{\mathrm{d}}{\mathrm{d}t}\boldsymbol{R}(t,r) = -\boldsymbol{R}'(t,r),$$

the density function corresponding to the coordinate $\boldsymbol{R}(t,r)$.

Moreover, we denote by $c_0(r)$ the mean value of the cost of the system corrective renovation upon the system exceeding the critical reliability state r and by $c_1(r)$ the mean value of the cost of the system preventive renovation upon the system correction after reaching by the system the age ς, independently of the system exceeding the critical reliability state r. We assume that the cost $c_0(r)$ is greater than the cost $c_1(r)$ and that the system intensity of departure $\lambda(t,r)$ from the system reliability states $\{r, r+1, \ldots, z\}$ is non decreasing, which are natural assumptions.

Under these assumptions, the system renovation can be performed in one cycle up to the first renovation either at the moment t, $t < \varsigma$, because of exceeding by the system the critical reliability state r with the probability

$$P\{T(r) \le \varsigma\} = 1 - \boldsymbol{R}(\varsigma, r)$$

or at the moment ς because of reaching by the system the age ς with the probability

$$P\{T(r) > \varsigma\} = \boldsymbol{R}(\varsigma, r).$$

Therefore, the expected value of the length of the time of one cycle up to the first renovation is given by

$$L_C(\varsigma, r) = L(\varsigma, r) \cdot [1 - \boldsymbol{R}(\varsigma, r)] + \varsigma \cdot \boldsymbol{R}(\varsigma, r), \tag{6.59}$$

where

$$L(\varsigma, r) = \int_0^{\varsigma} t \frac{f(t,r)}{1 - \boldsymbol{R}(\varsigma, r)} \mathrm{d}t = -\int_0^{\varsigma} t \frac{\boldsymbol{R}'(t,r)}{1 - \boldsymbol{R}(\varsigma, r)} \mathrm{d}t$$

$$= \frac{-\varsigma \boldsymbol{R}(\varsigma, r)] + \int_0^{\varsigma} \boldsymbol{R}(\varsigma, r) \mathrm{d}t}{1 - \boldsymbol{R}(\varsigma, r)} \tag{6.60}$$

is the conditional expected value of the length of the time up to exceeding by the system the critical reliability state r before the system reaches the age ς.

From the assumptions, it also follows that the expected value of the cost of the system renovation during one cycle is

$$C_C(\varsigma, r) = c_0(r) \times [1 - \boldsymbol{R}(\varsigma, r)] + c_1(r) \times \boldsymbol{R}(\varsigma, r). \tag{6.61}$$

Now, using (6.59) and (6.61), we may define the expected cost of the system renovation per unit time as

$$C(\varsigma, r) = \frac{C_C(\varsigma, r)}{L_C(\varsigma, r)} = \frac{c_0(r)[1 - \boldsymbol{R}(\varsigma, r)] + c_1(r)\boldsymbol{R}(\varsigma, r)}{L(\varsigma, r)[1 - \boldsymbol{R}(\varsigma, r)] + \varsigma \boldsymbol{R}(\varsigma, r)}. \tag{6.62}$$

The above formula relates the renovation age ς to the expected renovation cost per unit time.

Taking the derivative of the function $C(\varsigma, r)$ given by (6.62) with respect to ς we get

$$\frac{dC(\varsigma, r)}{d\varsigma} = \frac{[c_0(r) - c_1(r)]\boldsymbol{f}(\varsigma, r)[L(\varsigma, r)[1 - \boldsymbol{R}(\varsigma, r)] + \varsigma \boldsymbol{R}(\varsigma, r)]}{[L(\varsigma, r)[1 - \boldsymbol{R}(\varsigma, r)] + \varsigma \boldsymbol{R}(\varsigma, r)]^2}$$
$$- \frac{[c_0(r) - [c_0(r) - c_1(r)]\boldsymbol{R}(\varsigma, r)][\varsigma \boldsymbol{f}(\varsigma, r) + L(\varsigma, r)\boldsymbol{f}(\varsigma, r) + \boldsymbol{R}(\varsigma, r) - \varsigma \boldsymbol{f}(\varsigma, r)]}{[L(\varsigma, r)[1 - \boldsymbol{R}(\varsigma, r)] + \varsigma \boldsymbol{R}(\varsigma, r)]^2}$$

and then

$$\frac{dC(\varsigma, r)}{d\varsigma} = \frac{[c_0(r) - c_1(r)][\varsigma \boldsymbol{f}(\varsigma, r)\boldsymbol{R}(\varsigma, r) + \boldsymbol{R}^2(\varsigma, r)] - c_0(r)\boldsymbol{R}(\varsigma, r) - c_1(r)\boldsymbol{f}(\varsigma, r)L(\varsigma, r)}{[L(\varsigma, r)[1 - \boldsymbol{R}(\varsigma, r)] + \varsigma \boldsymbol{R}(\varsigma, r)]^2} \tag{6.63}$$

Hence, the equality

$$\frac{dC(\varsigma, r)}{d\varsigma} = 0$$

holds if and only if

$$\boldsymbol{R}^2(\varsigma, r) + \varsigma \boldsymbol{f}(\varsigma, r)\boldsymbol{R}(\varsigma, r) = \frac{c_0(r)\boldsymbol{R}(\varsigma, r) + c_1(r)\boldsymbol{f}(\varsigma, r)L(\varsigma, r)}{c_0(r) - c_1(r)} \tag{6.64}$$

Thus, finally, we get

$$\boldsymbol{R}^2(\varsigma, r) - \varsigma \boldsymbol{R}'(\varsigma, r)\boldsymbol{R}(\varsigma, r) = \frac{c_0(r)\boldsymbol{R}(\varsigma, r) - c_1(r)\boldsymbol{R}'(\varsigma, r)L(\varsigma, r)}{c_0(r) - c_1(r)}, \tag{6.65}$$

where $L(\varsigma, r)$ is given by (6.60).

Substituting $\boldsymbol{R}(\varsigma, r)$ and $\boldsymbol{R}'(\varsigma, r)$ into Eq. 6.65 we receive the expression from which we may find the optimal value $\dot{\varsigma}$ of the system age ς at which the system preventive maintenance minimizing the cost of the system renovation is performed. Afterwards, substituting this value into expression (6.62), we may obtain the minimum value $C(\dot{\varsigma}, r)$ of the expected cost $C(\varsigma, r)$of the system renovation per unit time. Unfortunately, very often, it is not possible to find the explicit solution of Eq. (6.65) and it has to be solved by approximate methods.

One of the approximate methods suggested to optimize the complex technical system corrective and preventive maintenance policy minimizing its cost of renovation is the method of secants that we apply by performing the following procedure:

- we check the relationship between the mean value of the cost of the system corrective renovation $c_0(r)$ and the mean value of the cost of the system preventive renovation $c_1(r)$;
- if the inequality

$$c_0(r) \leq c_1(r)$$

holds, then there is no optimal value $\dot{\varsigma}$ of the system age ς which minimizes the system renovation cost per unit time $C(\varsigma, r)$given by (6.62) and we determine its values for the arbitrarily fixed values of the system age at which the system preventive maintenance is performed

$$\varsigma_1, \varsigma_2, \ldots, \varsigma_K,$$

- if the inequality

$$c_0(r) > c_1(r)$$

we are looking for the optimal value $\dot{\varsigma}$ of the system age ς which minimizes the system renovation cost per unit time $C(\varsigma, r)$given by (6.62) by determining, if it exists, its approximate value from Eq. 6.65 by applying the method of secants in the interval $\langle a,\ b \rangle$ as follows:
- we define the function

$$\psi(\varsigma) = \boldsymbol{R}^2(\varsigma, r) - \varsigma \boldsymbol{R}'(\varsigma, r)\boldsymbol{R}(\varsigma, r) - \frac{c_0(r)\boldsymbol{R}(\varsigma, r) - c_1(r)\boldsymbol{R}'(\varsigma, r)L(\varsigma, r)}{c_0(r) - c_1(r)} \quad \text{for}$$

$$\varsigma > 0,$$

where $L(\varsigma, r)$ is given by (6.60),
- we define the interval $\langle a,\ b \rangle$ assuming $a = 0$ and finding b such that

$$\psi(b) > 0,$$

- we use the recurrent formula

$$\varsigma_0 = a,$$

$$\varsigma_{k+1} = \varsigma_k - \frac{\psi(\eta_k)}{\psi(b) - \psi(\eta_k)}(b - \varsigma_k) \quad \text{for } k = 0, 1, \ldots, K,$$

where K is such that

$$\psi(\varsigma_{K+1}) < \varepsilon$$

and ε is the measure of the method of secants accuracy,

- we fix the optimal value $\dot{\varsigma}$ of the system age ς at which the system preventive maintenance minimizing the cost of the system renovation is performed assuming

$$\dot{\varsigma} = \varsigma_{K+1},$$

and we determine the minimum value $C(\dot{\varsigma}, r)$ of the expected cost $C(\varsigma, r)$ of the system renovation per unit time substituting this optimal value $\dot{\varsigma}$ for ς in to expression (6.62) in the case when the recurrent procedure is convergent. Otherwise, there is no optimal value $\dot{\varsigma}$ of the system age ς which minimizes the system renovation cost and we find its values for the arbitrarily fixed values of the system age at which the system preventive maintenance is performed $\varsigma_1, \varsigma_2, \ldots, \varsigma_K$.

6.4 Applications

6.4.1 Optimization of Operation, Reliability and Availability and Cost Analysis of Exemplary System

The considered exemplary system at the variable operating conditions is analyzed in Sect. 3.6.1 of Chap. 3. Its reliability, renewal and availability characteristics are found there. From these results, it follows that it is possible to improve these characteristics by changing the parameters of its operation process and after that its operation cost analysis can also be performed.

6.4.1.1 Optimization of Exemplary System Operation Process

The objective function defined by (6.1), in this case, as the exemplary system critical state is $r = 2$, takes the form

$$\mu(2) = p_1 \times 483.87 + p_2 \times 694.44 + p_3 \times 383.04 + p_4 \times 253.88. \tag{6.66}$$

Arbitrarily assumed, the lower $\breve{p}_b$ and upper $\hat{p}_b$ bounds of the unknown transient probabilities p_b, $b = 1, 2, 3, 4$, respectively are:

$$\breve{p}_1 = 0.201,\ \breve{p}_2 = 0.03,\ \breve{p}_3 = 0.245,\ \breve{p}_4 = 0.309;$$
$$\hat{p}_1 = 0.351,\ \hat{p}_2 = 0.105,\ \hat{p}_3 = 0.395,\ \hat{p}_4 = 0.459.$$

Therefore, according to (6.2–6.3), we assume the following bound constraints

$$\begin{aligned} &0.201 \le p_1 \le 0.351,\ 0.03 \le p_2 \le 0.105,\\ &0.245 \le p_3 \le 0.395,\ 0.309 \le p_4 \le 0.459. \end{aligned} \tag{6.67}$$

$$\sum_{b=1}^{4} p_b = 1, \tag{6.68}$$

Now, before we find optimal values $\dot{p}_b$ of the transient probabilities p_b, $b = 1, 2, 3, 4$, that maximize the objective function (6.66), w arranges the system conditional lifetime mean values $\mu_b(2)$, $b = 1, 2, 3, 4$, in non-increasing order

$$\mu_2(2) \geq \mu_1(2) \geq \mu_3(2) \geq \mu_4(2).$$

Further, according to (6.6), we substitute

$$x_1 = p_2, \quad x_2 = p_1, \quad x_3 = p_3, \quad x_4 = p_4, \tag{6.69}$$

and

$$\breve{x}_1 = \breve{p}_2 = 0.03, \quad \breve{x}_2 = \breve{p}_1 = 0.201, \quad \breve{x}_3 = \breve{p}_3 = 0.245, \quad \breve{x}_4 = \breve{p}_4 = 0.309; \tag{6.70}$$

$$\hat{x}_1 = \hat{p}_2 = 0.105, \quad \hat{x}_2 = \hat{p}_1 = 0.351, \quad \hat{x}_3 = \hat{p}_3 = 0.395, \quad \hat{x}_4 = \hat{p}_4 = 0.459, \tag{6.71}$$

and we maximize with respect to x_i, $i = 1, 2, 3, 4$, the linear form (6.66) which according to (6.7–6.9) takes the form

$$\mu(2) = x_1 \times 694.44 + x_2 \times 483.87 + x_3 \times 383.04 + x_4 \times 253.88, \tag{6.72}$$

with the following bound constraints

$$\begin{aligned} &0.03 \leq x_1 \leq 0.105, \quad 0.201 \leq x_2 \leq 0.351, \\ &0.245 \leq x_3 \leq 0.395, \quad 0.309 \leq x_4 \leq 0.459. \end{aligned} \tag{6.73}$$

$$\sum_{i=1}^{4} x_i = 1. \tag{6.74}$$

According to (6.11), we calculate

$$\breve{x} = \sum_{i=1}^{4} \breve{x}_i = 0.785, \quad \hat{y} = 1 - \breve{x} = 1 - 0.785 = 0.215 \tag{6.75}$$

and according to (6.12), we determine

$$\begin{aligned} &\breve{x}^{0} = 0, \quad \hat{x}^{0} = 0, \quad \hat{x}^{0} - \breve{x}^{0} = 0, \\ &\breve{x}^{1} = 0.03, \quad \hat{x}^{1} = 0.105, \quad \hat{x}^{1} - \breve{x}^{1} = 0.075, \\ &\breve{x}^{2} = 0.231, \quad \hat{x}^{2} = 0.456, \quad \hat{x}^{2} - \breve{x}^{2} = 0.225, \end{aligned}$$

$$\breve{x}^{3} = 0.476. \quad \widehat{x}^{3} = 0.851, \quad \widehat{x}^{3} - \breve{x}^{3} = 0.375,$$
$$\breve{x}^{4} = 0.785, \quad \widehat{x}^{4} = 1.31, \quad \widehat{x}^{4} - \breve{x}^{4} = 0.525. \tag{6.76}$$

From the above, as according to (6.75), the inequality (6.13) takes the form

$$\widehat{x}^{I} - \breve{x}^{I} < 0.215, \tag{6.77}$$

it follows that the largest value $I \in \{0, 1, 2, 3, 4\}$ such that this inequality holds is $I = 1$.

Therefore, we fix the optimal solution that maximizes the linear function (6.72) according to rule (6.15). Namely, we get

$$\dot{x}_1 = \widehat{x}_1 = 0.105,$$
$$\dot{x}_2 = \hat{y} - \widehat{x}^{1} + \breve{x}^{1} + \breve{x}_2 = 0.215 - 0.105 + 0.03 + 0.201 = 0.341,$$
$$\dot{x}_3 = \breve{x}_3 = 0.245, \ \dot{x}_4 = \breve{x}_4 = 0.309. \tag{6.78}$$

Finally, after making the inverse substitution to (6.69), we get the optimal transient probabilities

$$\dot{p}_2 = \dot{x}_1 = 0.105, \ \dot{p}_1 = \dot{x}_2 = 0.341, \ \dot{p}_3 = \dot{x}_3 = 0.245, \ \dot{p}_4 = \dot{x}_4 = 0.309, \tag{6.79}$$

that maximize the exemplary system mean lifetime $\mu(2)$ in the reliability state subset $\{2, 3\}$ expressed by the linear form (6.66) giving, according to (6.18) and (6.79), its optimal value

$$\dot{\mu}(2) = \dot{p}_1 \times 483.87 + \dot{p}_2 \times 694.44 + \dot{p}_3 \times 383.04 + \dot{p}_4 \times 253.88$$
$$= 0.341 \times 483.87 + 0.105 \times 694.44 + 0.245 \times 383.04 + 0.309$$
$$\times 253.88 \cong 410.21 \tag{6.80}$$

6.4.1.2 Optimal Reliability Characteristics of Exemplary System

Substituting the optimal solution (6.79) into the formula (6.19), we obtain the optimal solution for the mean values of the exemplary system unconditional lifetimes in the reliability state subsets $\{1, 2, 3\}$ and $\{3\}$, that are as follows

$$\dot{\mu}(1) = \dot{p}_1 \times 505 + \dot{p}_2 \times 744.05 + \dot{p}_3 \times 405.56 + \dot{p}_4 \times 271.08$$
$$= 0.341 \times 505 + 0.105 \times 744.05 + 0.245 \times 405.56 + 0.309 \times 271.08$$
$$\cong 433.46, \tag{6.81}$$

$$\begin{aligned}\dot{\mu}(3) &= \dot{p}_1 \times 468.73 + \dot{p}_2 \times 651.04 + \dot{p}_3 \times 370.67 + \dot{p}_4 \times 237.05 \\ &= 0.341 \times 468.73 + 0.105 \times 651.04 + 0.245 \times 370.67 + 0.309 \times 237.05 \\ &\cong 392.26, \end{aligned} \tag{6.82}$$

and according to (6.23), the optimal values of the mean values of the system unconditional lifetimes in the particular reliability states 1, 2 and 3, respectively are

$$\begin{aligned}\dot{\bar{\mu}}(1) = \dot{\mu}(1) - \dot{\mu}(2) = 23.25,\ \dot{\bar{\mu}}(2) = \dot{\mu}(2) - \dot{\mu}(3) = 17.95, \\ \dot{\mu}(3) = \mu(3) = 392.26. \end{aligned} \tag{6.83}$$

Moreover, according to (6.20–6.21), the corresponding optimal unconditional multistate reliability function of the system is of the form

$$\dot{\boldsymbol{R}}(t,\ \cdot) = [1, \dot{\boldsymbol{R}}(t,\ 1),\ \dot{\boldsymbol{R}}(t,\ 2)]\dot{\boldsymbol{R}}(t,\ 3), \quad t \geq 0, \tag{6.84}$$

with the coordinates given by

$$\begin{aligned}\dot{\boldsymbol{R}}(t,\ 1) = 0.341 \times [\boldsymbol{R}(t,1)]^{(1)} + 0.105 \times [\boldsymbol{R}(t,1)]^{(2)} + 0.245 \times [\boldsymbol{R}(t,1)]^{(3)} \\ + 0.309 \times [\boldsymbol{R}(t,1)]^{(4)} \quad \text{for } t \geq 0 \end{aligned} \tag{6.85}$$

$$\begin{aligned}\dot{\boldsymbol{R}}(t,\ 2) = 0.341 \times [\boldsymbol{R}(t,2)]^{(1)} + 0.105 \times [\boldsymbol{R}(t,2)]^{(2)} + 0.245 \times [\boldsymbol{R}(t,2)]^{(3)} \\ + 0.309 \times [\boldsymbol{R}(t,2)]^{(4)} \quad \text{for } t \geq 0, \end{aligned} \tag{6.86}$$

$$\begin{aligned}\dot{\boldsymbol{R}}(t,\ 3) = 0.341 \times [\boldsymbol{R}(t,3)]^{(1)} + 0.105 \times [\boldsymbol{R}(t,3)]^{(2)} + 0.245 \times [\boldsymbol{R}(t,3)]^{(3)} \\ + 0.309 \times [\boldsymbol{R}(t,3)]^{(4)} \quad \text{for } t \geq 0, \end{aligned} \tag{6.87}$$

where $[\boldsymbol{R}(t,1)]^{(b)}$, $[\boldsymbol{R}(t,2)]^{(b)}$, $[\boldsymbol{R}(t,3)]^{(b)}$, $b = 1,2,3,4$, are fixed in Sect. 3.6.1.

The graph of the exemplary system optimal reliability function $\dot{\boldsymbol{R}}(t,\ \cdot)$ given by (6.87) is presented in Fig. 6.1.

Further, by (6.22), the corresponding optimal variances and standard deviations of the system unconditional lifetimes in the system reliability state subsets are

$$\dot{\sigma}^2(1) = 2\int_0^\infty t\,\dot{\boldsymbol{R}}(t,1)\mathrm{d}t - [\dot{\mu}(1)]^2 \cong 110365.45, \quad \dot{\sigma}(1) \cong 332.21, \tag{6.88}$$

$$\dot{\sigma}^2(2) = 2\int_0^\infty t\,\dot{\boldsymbol{R}}(t,2)\mathrm{d}t - [\dot{\mu}(2)]^2 \cong 101665.04, \quad \dot{\sigma}(2) \cong 318.85, \tag{6.89}$$

$$\dot{\sigma}^2(3) = 2\int_0^\infty t\,\dot{\boldsymbol{R}}(t,3)\mathrm{d}t - [\dot{\mu}(3)]^2 \cong 93424.64, \quad \dot{\sigma}(3) \cong 305.65, \tag{6.90}$$

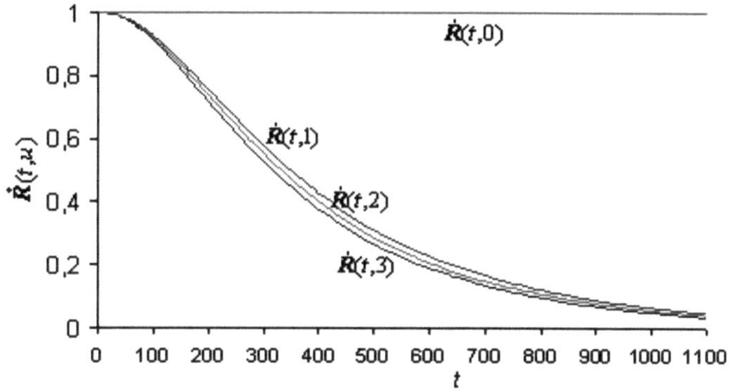

Fig. 6.1 The graph of the exemplary system optimal reliability function $\dot{\boldsymbol{R}}(t, \cdot)$ coordinates

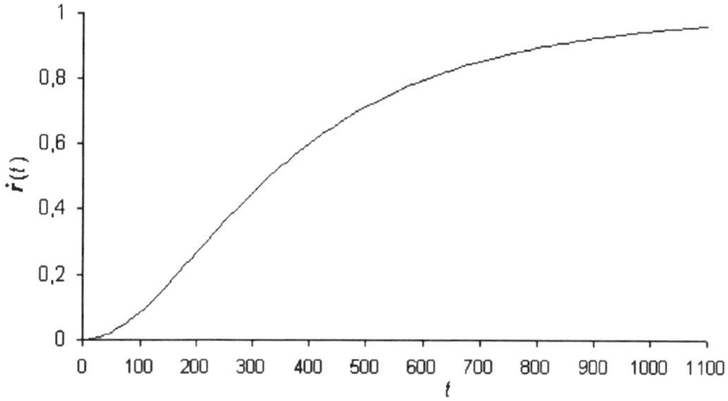

Fig. 6.2 The graph of the exemplary system optimal risk function $\dot{\boldsymbol{r}}(t)$

where $\dot{\boldsymbol{R}}(t,1)$, $\dot{\boldsymbol{R}}(t,2)$, $\dot{\boldsymbol{R}}(t,3)$ are given by (6.85–6.87) and $\dot{\mu}(1)$, $\dot{\mu}(2)$, $\dot{\mu}(3)$, are given by (6.80–6.82).

As the critical reliability state is $r = 2$, the exemplary system optimal system risk function, according to (6.24), is given by (Fig. 6.2)

$$\dot{\boldsymbol{r}}(t) = 1 - \dot{\boldsymbol{R}}(t,\, 2) \quad \text{for } t \geq 0, \tag{6.91}$$

where $\dot{\boldsymbol{R}}(t,2)$ is given by (6.86).

Hence and considering (6.25), the moment when the optimal system risk function exceeds a permitted level, for instance $\delta = 0.05$, is

$$\dot{\tau} = \dot{\boldsymbol{r}}^{-1}(\delta) \cong 80. \tag{6.92}$$

6.4.1.3 Optimal Renewal and Availability Characteristics of Exemplary System

To make the estimation of the renewal and availability characteristics of the exemplary system after its operation process optimization, we distinguish two cases of the system renovation, one with ignored time of renovation and the second with non-ignored time of renovation.

In the case when the exemplary system renovation time is ignored, applying *Proposition 6.1*, we determine the following its optimal characteristics:

(a) the optimal time $\dot{S}_N(2)$ until the *N*th exceeding by the system the reliability critical state 2, for sufficiently large *N*, has approximately normal distribution $N(410.20N, 318.85\sqrt{N})$, i.e.,

$$\dot{F}^{(N)}(t,2) = P(\dot{S}_N(2) < t) \cong F_{N(0,1)}\left(\frac{t - 410.20N}{318.85\sqrt{N}}\right), \quad t \in (-\infty, \infty),$$

(b) the expected value and the variance of the optimal time $\dot{S}_N(2)$ until the *N*th exceeding by the system the reliability critical state 2, for sufficiently large *N*, respectively are

$$E[\dot{S}_N(2)] = 410.20N, \; D[\dot{S}_N(2)] = 101665.32N,$$

(c) the optimal number $\dot{N}(t,2)$ of exceedings by the system of the reliability critical state 2 up to the moment t, $t \geq 0$, for sufficiently large t, has distribution approximately of the form

$$P(\dot{N}(t,2) = N) \cong F_{N(0,1)}\left(\frac{410.20(N+1) - t}{15.74\sqrt{t}}\right) - F_{N(0,1)}\left(\frac{410.20N - t}{15.74\sqrt{t}}\right),$$
$$N = 0, 1, \ldots,$$

(d) the expected value and the variance of the optimal number $N(t,2)$ of exceedings by the system the reliability critical state 2 up to the moment t, $t \geq 0$, for sufficiently large t, respectively are

$$\dot{H}(t,2) = 0.0024t, \quad \dot{D}(t,2) = 0.0015t. \tag{6.93}$$

To make the estimation of the renewal and availability characteristics of the exemplary system in the case when the time of renovation is non-ignored, assuming the mean value of the system renovation time $\mu_0(2) = 10$ and the standard deviation of the system renovation time $\sigma_0(2) = 5$ and applying *Proposition 6.2*, we determine the following optimal characteristics:

(a) the optimal time $\dot{\bar{S}}_N(2)$ until the *N*th exceeding by the system of the reliability critical state 2, for sufficiently large *N*, has approximately normal distribution $N(410.20N + 10(N-1), \sqrt{101665.32N + 25(N-1)}$, i.e.,

$$\dot{\bar{F}}^{(N)}(t,2) = P(\dot{\bar{S}}_N(2) < t) = F_{N(0,1)}\left(\frac{t - 420.20N + 10}{\sqrt{101690.32N - 25}}\right), \quad t \in (-\infty, \infty)$$

(b) the expected value and the variance of the optimal time $\dot{\bar{S}}_N(2)$ until the Nth exceeding by the system of the reliability critical state 2, for sufficiently large N, respectively are

$$E[\dot{\bar{S}}_N(2)] \cong 410.20N + 10(N-1), \quad D[\dot{\bar{S}}(2)] \cong 101665.32N + 25(N-1),$$

(c) the optimal number $\dot{\bar{N}}(t,2)$ of exceeding by the system of the reliability critical state 2 up to the moment t, $t \geq 0$, for sufficiently large t, has approximately distribution of the form

$$P(\dot{\bar{N}}(t,2) = N) \cong F_{N(0,1)}\left(\frac{420.20(N+1) - t - 10}{15.56\sqrt{t+10}}\right) - F_{N(0,1)}\left(\frac{420.20N - t - 10}{15.56\sqrt{t+10}}\right), \quad N = 0, 1, \ldots,$$

(d) the expected value and the variance of the optimal number $\dot{\bar{N}}(t,2)$ of exceedings by the system of the reliability critical state 2 up to the moment t, $t \geq 0$, for sufficiently large t, respectively are

$$\dot{\bar{H}}(t,2) \cong \frac{t+10}{420.20}, \quad \dot{\bar{D}}(t,2) \cong 0.0014(t+10),$$

(e) the optimal time $\dot{\bar{\bar{S}}}_N(2)$ until the Nth system's renovation, for sufficiently large N, has approximately normal distribution $N(420.20N, 318.89\sqrt{N})$, i.e.,

$$\dot{\bar{\bar{F}}}^{(N)}(t,2) = P\left(\dot{\bar{\bar{S}}}_N(2) < t\right) \cong F_{N(0,1)}\left(\frac{t - 420.20N}{318.89\sqrt{N}}\right), \quad t \in (-\infty, \infty),$$
$$N = 0, 1, \ldots,$$

(f) the expected value and the variance of the optimal time $\dot{\bar{\bar{S}}}_N(2)$ until the Nth system's renovation, for sufficiently large N, respectively are

$$E[\dot{\bar{\bar{S}}}_N(2)] \cong 420.20N, \quad D[\dot{\bar{\bar{S}}}^{(N)}(2)] \cong 101690.32N,$$

(g) the optimal number $\dot{\bar{\bar{N}}}(t,2)$ of system's renovations up to the moment t, $t \geq 0$, for sufficiently large t, has approximately distribution of the form

$$P(\dot{\bar{\bar{N}}}(t,2) = N) \cong F_{N(0,1)}\left(\frac{420.20(N+1) - t}{15.56\sqrt{t}}\right) - F_{N(0,1)}\left(\frac{420.20N - t}{15.56\sqrt{t}}\right),$$
$$N = 0, 1, \ldots,$$

(h) the expected value and the variance of the optimal number $\dot{\bar{\bar{N}}}(t,2)$ of system's renovations up to the moment t, $t \geq 0$, for sufficiently large t, respectively are

$$\dot{\bar{\bar{H}}}(t,1) \cong 0.0024t, \quad \dot{\bar{\bar{D}}}(t,1) \cong 0.0014t, \tag{6.94}$$

(i) the optimal steady availability coefficient of the system at the moment t, $t \geq 0$, for sufficiently large t, is

$$\dot{A}(t,2) \cong 0.98, \quad t \geq 0,$$

(j) the optimal steady availability coefficient of the system in the time interval $\langle t, t+\tau)$, $\tau > 0$, $t \geq 0$, for sufficiently large t, is

$$\dot{A}(t,\tau,2) \cong 0.0024 \int_{\tau}^{\infty} \dot{\boldsymbol{R}}(t,2)\mathrm{d}t, \quad t \geq 0, \quad \tau > 0,$$

where $\dot{\boldsymbol{R}}(t,2)$ is given by (6.86).

6.4.1.4 Optimal Sojourn Times at Operation States of Exemplary System and Operation Strategy

Substituting the exemplary operation process optimal transient probabilities at operation states

$$\dot{p}_1 = 0.341, \quad \dot{p}_2 = 0.105, \quad \dot{p}_3 = 0.245, \quad \dot{p}_4 = 0.309,$$

given by (6.79) and the steady probabilities

$$\pi_1 \cong 0.236, \ \pi_2 \cong 0.169, \ \pi_3 \cong 0.234, \pi_4 \cong 0.361.$$

determined by (2.31) in Sect. 2.3.1 into (6.41), we get the following system of equations with the unknown optimal mean values $\dot{M}_b$ of the exemplary system operation process unconditional sojourn times at the operation states we are looking for

$$\begin{aligned} -0.155524\dot{M}_1 + 0.057629\dot{M}_2 + 0.079794\dot{M}_3 + 0.123101\dot{M}_4 &= 0 \\ 0.02478\dot{M}_1 - 0.151255\dot{M}_2 + 0.02457\dot{M}_3 + 0.037905\dot{M}_4 &= 0 \\ 0.05782\dot{M}_1 + 0.041405\dot{M}_2 - 0.17667\dot{M}_3 + 0.088445\dot{M}_4 &= 0 \\ 0.072924\dot{M}_1 + 0.052221\dot{M}_2 + 0.072306\dot{M}_3 - 0.249451\dot{M}_4 &= 0. \end{aligned} \tag{6.95}$$

The determinant of the main matrix of the above homogeneous system of equations is equal to zero and therefore there are non-zero solutions of this system of equations that are ambiguous and dependent on one or more parameters. Thus, we may fix some of them and determine the remaining ones. To show the way of solving this system of equations, we may suppose that we are arbitrarily interested in fixing the value of $\dot{M}_4$ and we put

$$\dot{M}_4 = 400.$$

Substituting the above value into the system of equations (6.95), we get

$$\begin{aligned}
-0.155524\dot{M}_1 + 0.057629\dot{M}_2 + 0.079794\dot{M}_3 &= -49.2404 \\
0.02478\dot{M}_1 - 0.151255\dot{M}_2 + 0.02457\dot{M}_3 &= -15.1620 \\
0.05782\dot{M}_1 + 0.041405\dot{M}_2 - 0.17667\dot{M}_3 &= -35.3780 \\
0.072924\dot{M}_1 + 0.052221\dot{M}_2 + 0.072306\dot{M}_3 &= 99.7804,
\end{aligned}$$

and we solve it with respect to $\dot{M}_1$, $\dot{M}_2$ and $\dot{M}_3$, after omitting its last equation. In this way, we obtained solutions that satisfy (6.95), are

$$\dot{M}_1 \cong 675,\ \dot{M}_2 \cong 290,\ \dot{M}_3 \cong 490,\ \dot{M}_4 = 400. \tag{6.96}$$

It can be seen that these solutions differ much from the values M_1, M_2, M_3 and M_4, estimated in Sect. 2.3.1 and given by (2.27–2.30).

Having these solutions, it is also possible to look for the optimal values $\dot{M}_{bl}$ of the mean values M_{bl} of the exemplary system operation process conditional sojourn times at operation states. Namely, substituting the values $\dot{M}_b$ instead of M_b, the probabilities

$$[p_{bl}] = \begin{bmatrix} 0 & 0.22 & 0.32 & 0.46 \\ 0.20 & 0 & 0.30 & 0.50 \\ 0.12 & 0.16 & 0 & 0.72 \\ 0.48 & 0.22 & 0.30 & 0 \end{bmatrix}$$

of the exemplary system operation process transitions between the operation states given by (2.25) in Sect. 2.3.1 and replacing M_{bl} by $\dot{M}_{bl}$ in (6.42), we get the following system of equations

$$\begin{aligned}
0.22\dot{M}_{12} + 0.32\dot{M}_{13} + 0.46\dot{M}_{14} &= 675 \\
0.20\dot{M}_{21} + 0.30\dot{M}_{23} + 0.50\dot{M}_{24} &= 290 \\
0.12\dot{M}_{31} + 0.16\dot{M}_{32} + 0.72\dot{M}_{34} &= 490 \\
0.48\dot{M}_{41} + 0.22\dot{M}_{42} + 0.30\dot{M}_{14} &= 400
\end{aligned} \tag{6.97}$$

with the unknown optimal values $\dot{M}_{bl}$ we want to find.

As the solutions of the above system of equations are ambiguous, we fix some of them, because of practically important reasons, and we find the remaining ones. For instance:

- we fix in the first equation $\dot{M}_{12} = 200$, $\dot{M}_{13} = 500$ and we find $\dot{M}_{14} \cong 1024$;
- we fix in the second equation $\dot{M}_{21} = 100$, $\dot{M}_{23} = 100$ and we find $\dot{M}_{24} \cong 480$;

- we fix in the third equation $\dot{M}_{31} = 900$, $\dot{M}_{32} = 500$ and we find $\dot{M}_{34} \cong 419$;
- we fix in the fourth equation $\dot{M}_{41} = 300, \dot{M}_{42} = 500$ and we find

$$\dot{M}_{43} \cong 487. \tag{6.98}$$

It can be seen that these solutions differ much from the mean values of the exemplary system conditional sojourn times at the particular operation states before its operation process optimization given by (2.26).

Another very useful and much easier to be applied in practice tool that can help in planning the operation process of the exemplary system are the system operation process optimal mean values of the total sojourn times at the particular operation states during the system operation time which by the same assumption as in Sect. 2.3.1 is equal to $\theta = 1\text{year} = 365$ days. Under this assumption, after applying (6.43), we get the optimal values of the exemplary system operation process total sojourn times at the particular operation states during 1 year

$$\begin{aligned} &\dot{E}[\hat{\theta}_1] = \dot{p}_1\theta = 0.341 \times 365 \cong 124.5,\ \dot{E}[\hat{\theta}_2] = \dot{p}_2\theta = 0.105 \times 365 \cong 38.3, \\ &\dot{E}[\hat{\theta}_3] = \dot{p}_3\theta = 0.245 \times 365 \cong 89.4,\ \dot{E}[\hat{\theta}_4] = \dot{p}_4\theta = 0.309 \times 365 \cong 112.8, \end{aligned} \tag{6.99}$$

which differ much from the values of $E[\hat{\theta}_1]$, $E[\hat{\theta}_2]$, $E[\hat{\theta}_3], E[\hat{\theta}_4]$ determined by (2.33) in Sect. 2.3.1.

In practice, knowledge of the optimal values of $\dot{M}_b$, $\dot{M}_{bl}$ and $\dot{E}[\hat{\theta}_b]$ given respectively by (6.96), (6.98), (6.99), can be very important and helpful for the system operation process planning and improving in order to make the system operation more reliable and safer.

6.4.1.5 Exemplary System Operation Cost Analysis

In Sects. 2.3.1 and 3.6.1, it is fixed that the exemplary system is composed of $n = 14$ components and that the numbers of the system components operating at the various operation states z_b, $b = 1, 2, 3, 4$, are different. Namely, there are operating six system components at the operation states z_1, eight system components at the operation states z_2 and 14 system components at the operation states z_3 and z_4. According to the arbitrary assumption, the approximate mean operation costs of the single basic components of the considered exemplary system that are used during the operation time $\theta = 1$ year, are independent of the operation states z_b, $b = 1, 2, 3, 4$, and they amount to

$$c_i(1, b) = 100 \text{ PLN}, \quad b = 1, 2, 3, 4, \quad i = 1, 2, \ldots, 14,$$

whereas, the cost of each system's single basic component that is not used is equal to 0.

In the case when the exemplary system is repaired after exceeding the critical reliability state $r = 2$ and its renovation time is ignored, we assume that the approximate cost of the system's singular renovation is $c_{ig} = 1000$ PLN. Similarly, in the case when the exemplary system is repaired after exceeding the critical reliability state $r = 2$ and its renewal time is not ignored, we assume that the approximate cost of the system singular renovation is $c_{nig} = 1500$ PLN.

First, under the above assumptions, we analyze the system operation cost before its operation process optimization.

Thus, according to (6.44), the total operation cost of the non-failed exemplary system during the operation time $\theta = 1$ year amounts to

$$C(1) \cong 0.214 \times 100 \times 6 + 0.038 \times 100 \times 8 + 0.293 \times 100 \times 14 + 0.455 \times 100 \times 14 = 1206 \text{ PLN}. \quad (6.100)$$

In the case when the exemplary system is repaired after exceeding the critical reliability state $r = 2$ and its renovation time is ignored, according to (3.163) from Sect. 3.6.1, we know that the mean value of the number of exceedings by the exemplary system of the critical reliability state during the operation time $\theta = 1$ year is

$$H(1,2) \cong 0.0028.$$

Thus, by (6.45). the total operation cost of the repairable exemplary system ignoring its renewal time during the operation time $\theta = 1$ year, amounts to

$$C_{ig}(1) \cong 1206 + 1000 \times 0.0028 = 1206 + 2.8 = 1208.8 \text{ PLN}. \quad (6.101)$$

In the case when the exemplary system is repaired after exceeding the critical reliability state $r = 2$ and its renewal time is not ignored, according to (3.165) from Sect. 3.6.1, we know that the mean value of the number of the system renovations after exceeding the critical reliability state during the operation time $\theta = 1$ year is

$$\bar{\bar{H}}(1,2) \cong 0.0027.$$

Thus, by (6.46), the total operation cost of the renewed exemplary system ignoring its renewal time during the operation time $\theta = 1$ year amounts to

$$C_{nig}(1) \cong 1206 + 1500 \cdot 0.0027 = 1206 + 4.05 = 1210.05 \text{ PLN}. \quad (6.102)$$

Now, we analyze the system operation cost after its operation process optimization.

Proceeding afterwards, according to (6.47), the total operation cost of the non-failed exemplary system during the operation time $\theta = 1$ year is given by

$$\dot{C}(1) \cong 0.341 \times 100 \times 6 + 0.105 \times 100 \times 8 + 0.245 \times 100 \times 14 + 0.309 \times 100 \times 14 = 1064.2 \text{ PLN}. \quad (6.103)$$

In the case when the exemplary system is repaired after exceeding the critical reliability state $r = 2$ and its renewal time is ignored, from (6.93), we know that the mean value of the number of exceedings by the exemplary system of the critical reliability state during the operation time $\theta = 1$ year is

$$\dot{H}(1,2) \cong 0.0024.$$

Thus, according to (6.48), the total operation cost of the repairable exemplary system ignoring its renewal time during the operation time $\theta = 1$ year amounts to

$$\dot{C}_{ig}(1) \cong 1064.2 + 1000 \times 0.0024 = 1064.2 + 2.4 = 1066.6 \text{ PLN}. \quad (6.104)$$

In the case when the exemplary system is repaired after exceeding the critical reliability state $r = 2$ and its renewal time is not ignored, from (6.95), we know that the mean value of the number of system renovations after exceeding the critical reliability state during the operation time $\theta = 1$ year is respectively given by

$$\dot{\bar{\bar{H}}}(1,2) \cong 0.0024.$$

Thus, according to (6.49), the total operation cost of the repairable exemplary system by non-ignoring its renewal time during the operation time $\theta = 1$ year amounts to

$$\dot{C}_{nig}(1) \cong 1064.2 + 1500 \cdot 0.0024 = 1064.2 + 3.6 = 1067.8 \text{ PLN}. \quad (6.105)$$

The comparison of the results (6.100–6.102) with the results (6.103–6.105) justifies the sensibility of the exemplary system operation process optimization.

6.4.2 Optimization of Operation, Reliability and Availability and Cost Analysis of Port Oil Piping Transportation System

The considered port oil piping transportation system at the variable operation conditions is analyzed in Sect. 3.6.2, where its reliability, renewal and availability characteristics are found. From these results, it follows that it is possible to improve these characteristics by changing the parameters of its operation process and after that its operation cost analysis can be performed as well.

6.4.2.1 Optimization of Port Oil Piping Transportation System Operation Process

The objective function defined by (6.1), in this case, as the port oil piping transportation system critical state is $r = 1$, takes the form

$$\mu(1) = p_1 \times 0.364 + p_2 \times 0.805 + p_3 \times 0.307 + p_4 \times 0.207 \\ + p_5 \times 0.307 + p_6 \times 0.207 + p_7 \times 0.364. \tag{6.106}$$

The lower $\breve{p}_b$ and upper $\widehat{p}_b$ bounds of the unknown transient probabilities p_b, $b = 1, 2, \ldots, 7$, coming from experts respectively are:

$$\breve{p}_1 = 0.31, \quad \breve{p}_2 = 0.04, \quad \breve{p}_3 = 0.002, \quad \breve{p}_4 = 0.001, \quad \breve{p}_5 = 0.15,$$
$$\breve{p}_6 = 0.04, \quad \breve{p}_7 = 0.25;$$
$$\widehat{p}_1 = 0.46, \quad \widehat{p}_2 = 0.08, \quad \widehat{p}_3 = 0.006, \quad \widehat{p}_4 = 0.004, \quad \widehat{p}_5 = 0.26,$$
$$\widehat{p}_6 = 0.08, \quad \widehat{p}_7 = 0.40.$$

Therefore, according to (6.2–6.4), we assume the following bound constraints

$$0.31 \le p_1 \le 0.46, \; 0.04 \le p_2 \le 0.08, \\ 0.002 \le p_3 \le 0.006, \; 0.001 \le p_4 \le 0.004, \\ 0.15 \le p_5 \le 0.26, \; 0.04 \le p_6 \le 0.08, \\ 0.25 \le p_7 \le 0.40, \\ \sum_{b=1}^{7} p_b = 1. \tag{6.107}$$

Now, before we find optimal values $\dot{p}_b$ of the transient probabilities p_b, $b = 1, 2, \ldots, 7$, that maximize the objective function (6.1), we arrange the system conditional lifetime mean values $\mu_b(1)$, $b = 1, 2, \ldots, 7$, in non-increasing order

$$\mu_2(1) \ge \mu_1(1) \ge \mu_7(1) \ge \mu_3(1) \ge \mu_5(1) \ge \mu_4(1) \ge \mu_6(1).$$

Next, according to (6.6), we substitute

$$x_1 = p_2, \quad x_2 = p_1, \quad x_3 = p_7, \quad x_4 = p_3, \quad x_5 = p_5, \quad x_6 = p_4, \quad x_7 = p_6, \tag{6.108}$$

and

$$\breve{x}_1 = \breve{p}_2 = 0.04, \quad \breve{x}_2 = \breve{p}_1 = 0.31, \quad \breve{x}_3 = \breve{p}_7 = 0.25, \quad \breve{x}_4 = \breve{p}_3 = 0.002, \\ \breve{x}_5 = \breve{p}_5 = 0.15, \quad \breve{x}_6 = \breve{p}_4 = 0.001, \quad \breve{x}_7 = \breve{p}_6 = 0.04, \tag{6.109}$$

$$\widehat{x}_1 = \widehat{p}_2 = 0.08, \quad \widehat{x}_2 = \widehat{p}_1 = 0.46, \quad \widehat{x}_3 = \widehat{p}_7 = 0.40, \quad \widehat{x}_4 = \widehat{p}_3 = 0.006, \\ \widehat{x}_5 = \widehat{p}_5 = 0.26, \quad \widehat{x}_6 = \widehat{p}_4 = 0.004, \quad \widehat{x}_7 = \widehat{p}_6 = 0.08, \tag{6.110}$$

and we maximize with respect to x_i, $i = 1, 2, \ldots, 7$, the linear form (6.106) which according to (6.7–6.9) takes the form

$$\mu(1) = x_1 \times 0.805 + x_2 \times 0.364 + x_3 \times 0.364 + x_4 \times 0.307 + x_5 \times 0.307 + x_6 \times 0.207 + x_7 \times 0.207, \tag{6.111}$$

with the following bound constraints

$$\begin{aligned} &0.04 \le x_1 \le 0.08, \quad 0.31 \le x_2 \le 0.46, \\ &0.25 \le x_3 \le 0.40, \quad 0.002 \le x_4 \le 0.006, \\ &0.15 \le x_5 \le 0.26, \quad 0.001 \le x_6 \le 0.004, \\ &0.04 \le x_7 \le 0.08, \\ &\sum_{i=1}^{7} x_i = 1. \end{aligned} \tag{6.112}$$

According to (6.11), we calculate

$$\breve{x} = \sum_{i=1}^{7} \breve{x}_i = 0.793, \quad \hat{y} = 1 - \breve{x} = 1 - 0.793 = 0.207 \tag{6.113}$$

and according to (6.12), we find

$$\begin{aligned} &\breve{x}^{0} = 0, \ \widehat{x}^{0} = 0, \ \widehat{x}^{0} - \breve{x}^{0} = 0, \\ &\breve{x}^{1} = 0.04, \ \widehat{x}^{1} = 0.08, \ \widehat{x}^{1} - \breve{x}^{1} = 0.04, \\ &\breve{x}^{2} = 0.35, \ \widehat{x}^{2} = 0.54, \ \widehat{x}^{2} - \breve{x}^{2} = 0.19, \\ &\breve{x}^{3} = 0.60, \ \widehat{x}^{3} = 0.94, \ \widehat{x}^{3} - \breve{x}^{3} = 0.34, \\ &\breve{x}^{7} = 0.793, \ \widehat{x}^{7} = 1.29, \ \widehat{x}^{7} - \breve{x}^{7} = 0.497. \end{aligned} \tag{6.114}$$

From the above, as according to (6.113), the inequality (6.13) takes the form

$$\widehat{x}^{I} - \breve{x}^{I} < 0.207, \tag{6.115}$$

then it follows that the largest value $I \in \{0, 1, \ldots, 7\}$ such that this inequality holds is $I = 2$.

Therefore, we fix the optimal solution that maximizes the linear function (6.111) according to the rule (6.15). Namely, we get

$$\begin{aligned} &\dot{x}_1 = \widehat{x}_1 = 0.08, \ \dot{x}_2 = \widehat{x}_2 = 0.46, \\ &\dot{x}_3 = \hat{y} - \widehat{x}^{2} + \breve{x}^{2} + \breve{x}_3 = 0.207 - 0.54 + 0.35 + 0.25 = 0.267, \\ &\dot{x}_4 = \breve{x}_4 = 0.002, \ \dot{x}_5 = \breve{x}_5 = 0.15, \ \dot{x}_6 = \breve{x}_6 = 0.001, \ \dot{x}_7 = \breve{x}_7 = 0.04. \end{aligned} \tag{6.116}$$

Finally, after making the substitution inverse to (6.108), we get the optimal transient probabilities

$$\dot{p}_2 = \dot{x}_1 = 0.08,\ \dot{p}_1 = \dot{x}_2 = 0.46,\ \dot{p}_7 = \dot{x}_3 = 0.267,\ \dot{p}_3 = \dot{x}_4 = 0.002,$$
$$\dot{p}_5 = \dot{x}_5 = 0.15,\ \dot{p}_4 = \dot{x}_6 = 0.001,\ \dot{p}_6 = \dot{x}_7 = 0.04, \tag{6.117}$$

that maximize the pipeline system mean lifetime $\mu(1)$ in the reliability state subset $\{1,2\}$ expressed by the linear form (6.106) giving, according to (6.18) and (6.117), its optimal value

$$\begin{aligned}\dot{\mu}(1) &= \dot{p}_1 \times 0.364 + \dot{p}_2 \times 0.805 + \dot{p}_3 \times 0.307 + \dot{p}_4 \times 0.207 + \dot{p}_5 \times 0.307\\ &\quad + \dot{p}_6 \times 0.207 + \dot{p}_7 \times 0.364\\ &= 0.46 \times 0.364 + 0.08 \times 0.805 + 0.002 \times 0.307 + 0.001 \times 0.207 + 0.15\\ &\quad \times 0.307 + 0.04 \times 0.207 + 0.267 \times 0.364 \cong 0.384.\end{aligned} \tag{6.118}$$

6.4.2.2 Optimal Reliability Characteristics of Port Oil Piping Transportation System

Further, substituting the optimal solution (6.117) into the formula (6.19), we obtain the optimal solution for the mean value of the port oil piping transportation system unconditional lifetime in the reliability state subset $\{2\}$

$$\begin{aligned}\dot{\mu}(2) &= \dot{p}_1 \times 0.302 + \dot{p}_2 \times 0.665 + \dot{p}_3 \times 0.217 + \dot{p}_4 \times 0.156 + \dot{p}_5 \times 0.217\\ &\quad + \dot{p}_6 \times 0.156 + \dot{p}_7 \times 0.302\\ &= 0.46 \times 0.302 + 0.08 \times 0.665 + 0.002 \times 0.217 + 0.001 \times 0.156 + 0.15\\ &\quad \times 0.217 + 0.04 \times 0.156 + 0.267 \times 0.302 \cong 0.312,\end{aligned} \tag{6.119}$$

and according to (6.23), the optimal values of the mean values of the system unconditional lifetimes in the particular reliability states 1 and 2, respectively are

$$\dot{\bar{\mu}}(1) = \dot{\mu}(1) - \dot{\mu}(2) = 0.072, \quad \dot{\bar{\mu}}(2) = \dot{\mu}(2) = 0.312. \tag{6.120}$$

Moreover, according to (6.20–6.21) the corresponding optimal unconditional multistate reliability function of the pipeline system is of the form

$$\dot{\boldsymbol{R}}(t,\cdot) = [1, \dot{\boldsymbol{R}}(t,1), \quad \dot{\boldsymbol{R}}(t,2)], \tag{6.121}$$

with the coordinates given by

$$\begin{aligned}\dot{\boldsymbol{R}}(t,1) =& 0.46 \times [\boldsymbol{R}(t,1)]^{(1)} + 0.08 \times [\boldsymbol{R}(t,1)]^{(2)}\\ &+ 0.002 \times [\boldsymbol{R}(t,1)]^{(3)} + 0.001 \times [\boldsymbol{R}(t,1)]^{(4)}\\ &+ 0.15 \times [\bar{\boldsymbol{R}}(t,1)]^{(5)} + 0.04 \times [\bar{\boldsymbol{R}}(t,1)]^{(6)}\\ &+ 0.267 \times [\boldsymbol{R}(t,1)]^{(7)} \quad \text{for } t \geq 0,\end{aligned} \tag{6.122}$$

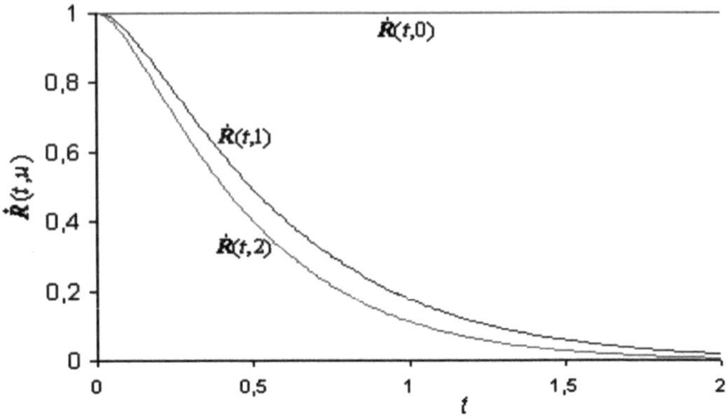

Fig. 6.3 The graph of the port oil piping transportation system optimal reliability function $\dot{\boldsymbol{R}}(t,\cdot)$ coordinates

$$\dot{\boldsymbol{R}}(t,2) = 0.46 \times [\boldsymbol{R}(t,2)]^{(1)} + 0.08 \times [\boldsymbol{R}(t,2)]^{(2)} + 0.002 \times [\boldsymbol{R}(t,2)]^{(3)} + 0.001 \times [\boldsymbol{R}(t,2)]^{(4)} + 0.15 \times [\boldsymbol{R}(t,2)]^{(5)} + 0.04 \times [\boldsymbol{R}(t,2)]^{(6)} + 0.267 \times [\boldsymbol{R}(t,2)]^{(7)} \text{ for } t \geq 0, \tag{6.123}$$

where $[\boldsymbol{R}(t,1)]^{(b)}$, $[\boldsymbol{R}(t,2)]^{(b)}$, $b = 1,2,\ldots,7$, are fixed in Sect. 3.6.2 (Fig. 6.3).

Further, by (6.22) the corresponding optimal variances and standard deviations of the system unconditional lifetime in the system reliability state subsets are

$$\dot{\sigma}^2(1) = 2\int_0^\infty t\,\dot{\boldsymbol{R}}(t,1)\mathrm{d}t - [\dot{\mu}(1)]^2 \cong 0.102, \quad \dot{\sigma}(1) \cong 0.319, \tag{6.124}$$

$$\dot{\sigma}^2(2) = 2\int_0^\infty t\,\dot{\boldsymbol{R}}(t,2)\mathrm{d}t - [\dot{\mu}(2)]^2 \cong 0.068, \quad \dot{\sigma}(2) \cong 0.261, \tag{6.125}$$

where $\dot{\boldsymbol{R}}(t,1)$, $\dot{\boldsymbol{R}}(t,2)$ are given by (6.122–6.123) and $\dot{\mu}(1), \dot{\mu}(2)$, are given by (6.118–6.119).

As the port oil piping transportation system critical safety reliability is $r = 1$, its optimal system risk function, according to (6.24), is given by (Fig. 6.4)

$$\dot{\boldsymbol{r}}(t) = 1 - \dot{\boldsymbol{R}}(t,1) \quad \text{for } t \geq 0, \tag{6.126}$$

where $\dot{\boldsymbol{R}}(t,1)$ is given by (6.122).

Hence, and considering (6.25), the moment when the optimal system risk function exceeds a permitted level, for instance $\delta = 0.05$, is

$$\dot{\tau} = \dot{\boldsymbol{r}}^{-1}(\delta) \cong 0.059 \text{ year}. \tag{6.127}$$

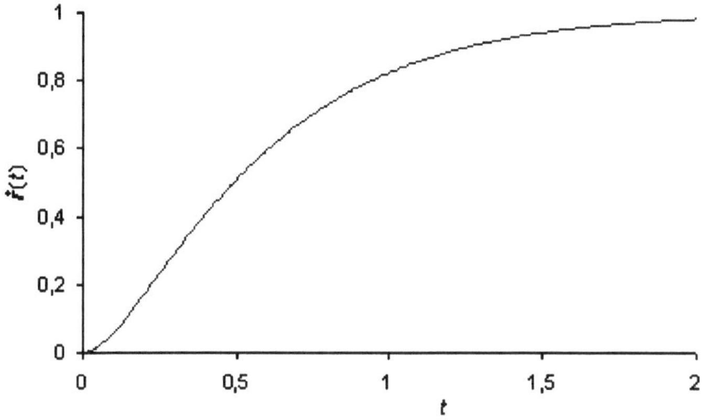

Fig. 6.4 The graph of the port oil piping transportation system optimal risk function $\dot{r}(t)$

6.4.2.3 Optimal Renewal and Availability Characteristics of Port Oil Piping Transportation System

To make the estimation of the renewal and availability characteristics of the port oil piping transportation system after its operation process optimization, we distinguish two cases of the system renovation, one with ignored time of renovation and the other with non-ignored time of renovation.

In the case when the port oil piping transportation system renovation time is ignored, applying *Proposition 6.1*, we determine its following optimal characteristics:

(a) the optimal time $\dot{S}_N(1)$ until the *N*th exceeding by the system the reliability critical state 1, for sufficiently large *N*, has approximately normal distribution $N(0.384N, 0.319\sqrt{N})$, i.e.,

$$\dot{F}^{(N)}(t,1) = P(\dot{S}_N(1) < t) \cong F_{N(0,1)}\left(\frac{t - 0.384N}{0.319\sqrt{N}}\right), \quad t \in (-\infty, \infty),$$

(b) the expected value and the variance of the optimal time $\dot{S}_N(1)$ until the *N*th exceeding by the system the reliability critical state 1, for sufficiently large *N*, respectively are

$$E[\dot{S}_N(1)] = 0.384N, \quad D[\dot{S}_N(1)] = 0.102N,$$

(c) the optimal number $\dot{N}(t,1)$ of exceeding by the system the reliability critical state 1 up to the moment t, $t \geq 0$, for sufficiently large t, has distribution approximately of the form

$$P(\dot{N}(t,1) = N) \cong F_{N(0,1)}\left(\frac{0.384(N+1) - t}{0.515\sqrt{t}}\right) - F_{N(0,1)}\left(\frac{0.384N - t}{0.515\sqrt{t}}\right),$$
$$N = 0, 1, \ldots,$$

(d) the expected value and the variance of the optimal number $N(t,1)$ of exceeding by the system the reliability critical state 1 up to the moment t, $t \geq 0$, for sufficiently large t, respectively are

$$\dot{H}(t,1) = 2.604t, \quad \dot{D}(t,1) = 1.801t. \tag{6.128}$$

To make the estimation of the renewal and availability of the port oil piping transportation system in the case when the time of renovation is non-ignored, assuming the mean value of the system renovation time $\mu_0(1) = 0.005$ year and the standard deviation of the system renovation time $\sigma_0(1) = 0.005$ year and applying *Proposition 6.2*, we determine its following optimal characteristics:

(a) the optimal time $\dot{\bar{S}}_N(1)$ until the Nth exceeding by the system the reliability critical state 1, for sufficiently large N, has approximately normal distribution $N(0.384N + 0.005(N-1), \sqrt{0.102N + 0.000025(N-1)})$, i.e.,

$$\dot{\bar{F}}^{(N)}(t,1) = P(\dot{\bar{S}}_N(1) < t) = F_{N(0,1)}\left(\frac{t - 0.389N + 0.005}{\sqrt{0.102025N - 0.000025}}\right), \quad t \in (-\infty, \infty),$$

(b) the expected value and the variance of the optimal time $\dot{\bar{S}}_N(1)$ until the Nth exceeding by the system the reliability critical state 1, for sufficiently large N, respectively are

$$E[\dot{\bar{S}}_N(1)] \cong 0.384N + 0.005(N-1), \quad D[\dot{\bar{S}}(1)] \cong 0.102N + 0.000025(N-1),$$

(c) the optimal number $\dot{\bar{N}}(t,1)$ of exceeding by the system the reliability critical state 1 up to the moment t, $t \geq 0$, for sufficiently large t, has approximately distribution of the form

$$P(\dot{\bar{N}}(t,1) = N) \cong F_{N(0,1)}\left(\frac{0.389(N+1) - t - 0005}{0.512\sqrt{t + 0.005}}\right) - F_{N(0,1)}\left(\frac{0.389N - t - 0.005}{0.512\sqrt{t + 0.005}}\right), N = 0, 1, \ldots,$$

(d) the expected value and the variance of the optimal number $\dot{\bar{N}}(t,1)$ of exceeding by the system the reliability critical state 1 up to the moment t, $t \geq 0$, for sufficiently large t, respectively are

$$\dot{\bar{H}}(t,1) \cong 2.571(t + 0.005), \quad \dot{\bar{D}}(t,1) \cong 1.733(t + 0.005), \tag{6.129}$$

(e) the optimal time $\dot{\bar{\bar{S}}}_N(1)$ until the Nth system's renovation, for sufficiently large N, has approximately normal distribution $N(0.389N, 0.319\sqrt{N})$, i.e.,

$$\dot{\bar{\bar{F}}}^{(N)}(t,1) = P(\dot{\bar{\bar{S}}}_N(1)<t) \cong F_{N(0,1)}\left(\frac{t-0.389N}{0.319\sqrt{N}}\right), \quad t \in (-\infty,\infty),$$
$$N = 0,1,\ldots,$$

(f) the expected value and the variance of the optimal time $\dot{\bar{\bar{S}}}_N(1)$ until the Nth system's renovation, for sufficiently large N, respectively are

$$E[\dot{\bar{\bar{S}}}_N(1)] \cong 0.389N, \quad D[\dot{\bar{\bar{S}}}^{(N)}(1)] \cong 0.102025N,$$

(g) the optimal number $\dot{\bar{\bar{N}}}(t,1)$ of system's renovations up to the moment t, $t \geq 0$, for sufficiently large t, has approximately distribution of the form

$$P(\dot{\bar{\bar{N}}}(t,1)=N) \cong F_{N(0,1)}\left(\frac{0.389(N+1)-t}{0.512\sqrt{t}}\right) - F_{N(0,1)}\left(\frac{0.389N-t}{0.512\sqrt{t}}\right),$$
$$N = 0,1,\ldots,$$

(h) the expected value and the variance of the optimal number $\dot{\bar{\bar{N}}}(t,1)$ of system's renovations up to the moment t, $t \geq 0$, for sufficiently large t, respectively are

$$\dot{\bar{\bar{H}}}(t,1) \cong 2.571t, \quad \dot{\bar{\bar{D}}}(t,1) \cong 1.733t, \tag{6.130}$$

(i) the optimal steady availability coefficient of the system at the moment t, $t \geq 0$, for sufficiently large t, is

$$\dot{A}(t,1) \cong 0.987, \quad t \geq 0,$$

(j) the optimal steady availability coefficient of the system in the time interval $\langle t, t+\tau)$, $\tau > 0$, $t \geq 0$, for sufficiently large t, is

$$\dot{A}(t,\tau,1) \cong 2.571 \int_{\tau}^{\infty} \dot{\boldsymbol{R}}(t,1)\mathrm{d}t, \quad t \geq 0, \quad \tau > 0,$$

where $\dot{\boldsymbol{R}}(t,1)$is given by (6.122).

6.4.2.4 Optimal Sojourn Times at Operation States of Port Oil Transportation System and Operation Strategy

To obtain the optimal mean sojourn times at the particular operation states maximizing the mean lifetime $\mu(1)$ of the port oil piping transportation system in the reliability states subset $\{1,2\}$, we substitute the optimal transient probabilities $\dot{p}_b$ determined by (6.117) and the steady probabilities π_b determined by (2.37) into the system of Eq. 6.41 and we get its following form

$$
\begin{aligned}
&-0.15552\dot{M}_1+0.01288\dot{M}_2+0.00276\dot{M}_3+0.00322\dot{M}_4+0.13892\dot{M}_5\\
&\quad+0.06716\dot{M}_6+0.10258\dot{M}_7=0\\
&0.02304\dot{M}_1-0.02576\dot{M}_2+0.00048\dot{M}_3+0.00056\dot{M}_4+0.02416\dot{M}_5\\
&\quad+0.01168\dot{M}_6+0.01784\dot{M}_7=0\\
&0.000576\dot{M}_1+0.000056\dot{M}_2-0.005988\dot{M}_3+0.000014\dot{M}_4+0.000604\dot{M}_5\\
&\quad+0.000292\dot{M}_6+0.000446\dot{M}_7=0\\
&0.000288\dot{M}_1+0.000028\dot{M}_2+0.000006\dot{M}_3-0.006993\dot{M}_4+0.000302\dot{M}_5\\
&\quad+0.000146\dot{M}_6+0.000223\dot{M}_7=0\\
&0.0432\dot{M}_1+0.0042\dot{M}_2+0.0009\dot{M}_3+0.00105M_4-0.2567\dot{M}_5+0.0219\dot{M}_6\\
&\quad+0.03345\dot{M}_7=0\\
&0.01152\dot{M}_1+0.00112\dot{M}_2+0.00024\dot{M}_3+0.00028\dot{M}_4+0.01208\dot{M}_5\\
&\quad-0.14016\dot{M}_6+0.00892\dot{M}_7=0\\
&0.076896\dot{M}_1+0.007476\dot{M}_2+0.001602\dot{M}_3+0.001869\dot{M}_4+0.080634\dot{M}_5\\
&\quad+0.038982\dot{M}_6-0.163459\dot{M}_7=0
\end{aligned}
\tag{6.131}
$$

with the unknown optimal mean values $\dot{M}_b$ of the system unconditional sojourn times at the operation states we are looking for.

Since the determinant of the main matrix of the homogeneous system of equations (6.131) is equal to 0, its rank is less than 7 and there are non-zero solutions of this system of equations that are ambiguous and dependent on one or more parameters. Thus, we may fix some of them and determine the remaining ones. In our case, according to (2.36), after considering expert opinion, we conclude that it is sensible to assume

$$\dot{M}_4 \cong 380. \tag{6.132}$$

After this the system of equations (6.131) takes the form

$$
\begin{aligned}
&-0.15552\dot{M}_1+0.01288\dot{M}_2+0.00276\dot{M}_3+0.13892\dot{M}_5+0.06716\dot{M}_6\\
&\quad+0.10258\dot{M}_7=-1.2236\\
&0.02304\dot{M}_1-0.02576\dot{M}_2+0.00048\dot{M}_3+0.02416\dot{M}_5+0.01168\dot{M}_6\\
&\quad+0.01784\dot{M}_7=-0.2128\\
&0.000576\dot{M}_1+0.000056\dot{M}_2-0.005988\dot{M}_3+0.000604\dot{M}_5+0.000292\dot{M}_6\\
&\quad+0.000446\dot{M}_7=-0.00532\\
&0.000288\dot{M}_1+0.000028\dot{M}_2+0.000006\dot{M}_3+0.000302\dot{M}_5+0.000146\dot{M}_6\\
&\quad+0.000223\dot{M}_7=2.65734
\end{aligned}
$$

$$0.0432\dot{M}_1 + 0.0042\dot{M}_2 + 0.0009\dot{M}_3 - 0.2567\dot{M}_5 + 0.0219\dot{M}_6 + 0.03345\dot{M}_7 = -0.399$$
$$0.01152\dot{M}_1 + 0.00112\dot{M}_2 + 0.00024\dot{M}_3 + 0.01208\dot{M}_5 - 0.14016\dot{M}_6 + 0.00892\dot{M}_7 = -0.1064$$
$$0.076896\dot{M}_1 + 0.007476\dot{M}_2 + 0.001602\dot{M}_3 + 0.080634\dot{M}_5 + 0.038982\dot{M}_6 - 0.163459\dot{M}_7 = -0.71022 \quad (6.133)$$

and its solutions are

$$\dot{M}_1 \cong 4248.611, \quad \dot{M}_2 \cong 7600, \quad \dot{M}_3 \cong 886.666, \quad \dot{M}_5 \cong 1321.192, \quad \dot{M}_6 \cong 728.767, \quad \dot{M}_7 \cong 3184.843. \quad (6.134)$$

Hence and considering (6.132), we get the following final solution of Eq. 6.131

$$\dot{M}_1 \cong 4248.611, \quad \dot{M}_2 \cong 7600, \quad \dot{M}_3 \cong 886.666, \quad \dot{M}_4 \cong 380, \quad \dot{M}_5 \cong 1321.192, \quad \dot{M}_6 \cong 728.767, \quad \dot{M}_7 \cong 3184.843. \quad (6.135)$$

Now, substituting into (6.42) the above mean values $\dot{M}_b$ of the system unconditional sojourn times at the particular operation states and the known probabilities p_{bl} of the system operation process transitions between the operation states given in the matrix (2.34), we may look for the optimal values $\dot{M}_{bl}$ of the mean values of the system conditional sojourn times in the particular operation states that maximizes the mean lifetime of the port oil piping transportation system in the reliability state subset $\{1,2\}$. The optimal values $\dot{M}_{bl}$, $b, l = 1, 2, \ldots, 7$, $b \neq l$, should satisfy the following system of equations obtained in this way

$$0.022\dot{M}_{12} + 0.022\dot{M}_{13} + 0.534\dot{M}_{15} + 0.111\dot{M}_{16} + 0.311\dot{M}_{17} = 4248.611$$
$$0.2\dot{M}_{21} + 0.8\dot{M}_{27} = 7600$$
$$1\dot{M}_{31} = 886.666$$
$$1\dot{M}_{47} = 380$$
$$0.488\dot{M}_{51} + 0.023\dot{M}_{52} + 0.023\dot{M}_{54} + 0.233\dot{M}_{56} + 0.233\dot{M}_{57} = 1321.192$$
$$0.095\dot{M}_{61} + 0.667\dot{M}_{65} + 0.238_{67} = 728.767$$
$$0.531\dot{M}_{71} + 0.062\dot{M}_{72} + 0.219\dot{M}_{75} + 0.188\dot{M}_{76} = 3184.843. \quad (6.136)$$

As the solutions of the above system of equations are ambiguous, we may fix some of them, because of practically important reasons, and we find the remaining ones. For instance:

- we fix in the first equation $\dot{M}_{12} = 2000$, $\dot{M}_{13} = 500$, $\dot{M}_{15} = 2000$, $\dot{M}_{16} = 1000$ and we find $\dot{M}_{17} \cong 9693.283$;
- we fix in the second equation $\dot{M}_{21} = 10000$ and we find $\dot{M}_{27} \cong 7000$;

- we find from the third equation $\dot{M}_{31} = 886.666$;
- we find from the fourth equation $\dot{M}_{47} = 380$;
- we fix in the fifth equation $\dot{M}_{51} = 900$, $\dot{M}_{52} = 500$, $\dot{M}_{54} = 300$, $\dot{M}_{56} = 400$,
- and we find $\dot{M}_{57} \cong 3306.4$;
- we fix in the sixth equation $\dot{M}_{61} = 350$, $\dot{M}_{65} = 500$,
- and we find $\dot{M}_{67} \cong 1521.08$;
- we fix in the seventh equation $\dot{M}_{71} = 900$, $\dot{M}_{72} = 500$, $\dot{M}_{75} = 2600$, and we find

$$\dot{M}_{76} \cong 11205.02. \tag{6.137}$$

Another very useful and much easier to be applied in practice tool that can help in planning the operation process of the port oil transportation system are the system operation process optimal mean values of the total sojourn times at the particular operation states during the system operation time θ.

Assuming as in Sect. 2.3.1 the system operation time $\theta = 1\text{year} = 365$ days, after applying (6.43), we get their values

$$\dot{E}[\hat{\theta}_1] = \dot{p}_1\theta = 0.46 \times 365 = 167.9,\ \dot{E}[\hat{\theta}_2] = \dot{p}_2\theta = 0.08 \times 365 = 29.2,$$
$$\dot{E}[\hat{\theta}_3] = \dot{p}_3\theta = 0.002 \times 365 = 0.73,\ \dot{E}[\hat{\theta}_4] = \dot{p}_4\theta = 0.001 \times 365 = 0.365,$$
$$\dot{E}[\hat{\theta}_5] = \dot{p}_5\theta = 0.15 \times 365 = 54.75,\ \dot{E}[\hat{\theta}_6] = \dot{p}_6\theta = 0.04 \times 365 = 14.6,$$
$$\dot{E}[\hat{\theta}_7] = \dot{p}_7\theta = 0.267 \times 365 = 97.46, \tag{6.138}$$

that differ much from the values of $E[\hat{\theta}_i]$, $i = 1, 2, \ldots, 7$, determined by (2.39) in Sect. 2.3.2.

In practice, the knowledge of the optimal values of $\dot{M}_b$, $\dot{M}_{bl}$ and $\dot{E}[\hat{\theta}_b]$ given respectively by (6.135, 6.137, 6.138), can be very important and helpful for the port oil piping transportation system operation process planning and improving in order to make the system operation more reliable and safer.

6.4.2.5 Port Oil Piping Transportation System Operation Cost Analysis

In Sects. 2.3.1 and 3.6.2 it is fixed that the port oil transportation system is composed of $n = 2880$ components and the numbers of the system components operating at the various operation states z_b, $b = 1, 2, \ldots, 7$, are different. Namely, there are operating 1086 system components at the operation states z_1, z_2 and z_7, 1794 system components at the operation states z_3 and z_5, and 2880 system components at the operation states z_4 and z_6.

According to the information from experts, the approximate mean operation cost of the single basic component of the considered pipeline transportation system used during the operation time $\theta = 1$ year, independently of the operation state z_b, $b = 1, 2, \ldots, 7$, amounts to

$$c_i(1,b) = 9.6 \text{ PLN}, \quad b = 1,2,\ldots,7, \quad i = 1,2,\ldots,2880.$$

whereas, the cost of each system singular basic component that is not used is equal to 0.

In the case when the pipeline transportation system is repaired after exceeding the critical reliability state $r = 1$ and its renovation time is ignored, according to expert opinion, we assume that the approximate cost of the system singular renovation is $c_{ig} = 88500$ PLN. Similarly, in the case when the pipeline transportation system is repaired after exceeding the critical reliability state $r = 1$ and its renewal time is not ignored, according to expert opinion, we assume that the approximate cost of the system singular renovation is $c_{nig} = 90000$ PLN.

First, we will analyze the system operation cost before its operation process optimization.

Thus, under the assumptions, the total operation cost of the non-failed pipeline transportation system during the operation time $\theta = 1$ year, according to (6.44), is given by

$$\begin{aligned} C(1) &\cong 0.389 \times 9.6 \times 1086 + 0.062 \times 9.6 \times 1086 + 0.003 \times 9.6 \times 1794 \\ &\quad + 0.002 \times 9.6 \times 2880 \\ &\quad + 0.20 \times 9.6 \times 1794 + 0.058 \times 9.6 \times 2880 + 0.286 \times 9.6 \times 1086 \\ &\cong 12839 \text{ PLN}. \end{aligned} \tag{6.139}$$

In the case when the pipeline transportation system is repaired after exceeding the critical reliability state $r = 1$ and its renovation time is ignored, from (3.233) given in Sect. 3.6.2, we know that the mean value of the number of exceedings by the system the critical reliability state during the operation time $\theta = 1$ year is

$$H(1,1) \cong 2.702.$$

Thus, the total operation cost of the repairable pipeline transportation system ignoring its renewal time during the operation time $\theta = 1$ year, according to (6.45), amounts to

$$C_{ig}(1) \cong 12839 + 88500 \times 2.702 = 12839 + 239127 = 251966 \text{ PLN}. \tag{6.140}$$

In the case when the pipeline transportation system is repaired after exceeding the critical reliability state $r = 1$ and its renovation time is not ignored, from (3.235) given in Sect. 3.6.2, the mean value of the number of system renovations after exceeding by the system the critical reliability state during the operation time $\theta = 1$ year is

$$\bar{\bar{H}}(1,1) \cong 2.667.$$

Thus, the total operation cost of the repairable pipeline transportation system non-ignoring its renovation time during the operation time $\theta = 1$ year, according to (6.46), amounts to

$$C_{nig}(1) \cong 12839 + 90000 \times 2.667 = 12839 + 240030 = 252869 \text{ PLN}. \quad (6.141)$$

Now, we analyze the system operation cost after its operation process optimization.

Proceeding similarly as before, the total operation cost of the non-failed pipeline transportation system during the operation time $\theta = 1$ year, according to (6.47), is given by

$$\begin{aligned}\dot{C}(1) &\cong 0.46 \times 9.6 \times 1086 + 0.08 \times 9.6 \times 1086 + 0.002 \times 9.6 \times 1794 + 0.001 \\ &\quad \times 9.6 \times 2880 \\ &\quad + 0.15 \times 9.6 \times 1794 + 0.04 \times 9.6 \times 2880 + 0.267 \times 9.6 \times 1086 \\ &\cong 12165 \text{ PLN}. \end{aligned} \quad (6.142)$$

In the case when the pipeline transportation system is repaired after exceeding the critical reliability state $r = 1$ and its renewal time is ignored, from (6.128), we know that the mean value of the number of exceedings of the critical reliability state during the operation time $\theta = 1$ year is

$$\dot{H}(1, 1) \cong 2.604.$$

Thus, the total operation cost of the repairable pipeline transportation system ignoring its renewal time during the operation time $\theta = 1$ year, according to (6.48), amounts to

$$\dot{C}_{ig}(1) \cong 12165 + 88500 \times 2.604 = 12165 + 230454 = 242619 \text{ PLN}. \quad (6.143)$$

In the case when the pipeline transportation system is repaired after exceeding the critical reliability state $r = 1$ and its renewal time is not ignored, from (6.130), we know that the mean value of the number of system renovations after exceeding the critical reliability state during the operation time $\theta = 1$ year is respectively given by

$$\dot{\bar{\bar{H}}}(1, 1) \cong 2.571.$$

Thus, the total operation cost of the repairable pipeline transportation system non-ignoring its renewal time during the operation time $\theta = 1$ year, according to (6.49), amounts to

$$\dot{C}_{nig}(1) \cong 12165 + 90000 \times 2.571 = 12165 + 231390 = 243555 \text{ PLN}. \quad (6.144)$$

The comparison of the results (6.139–6.141) with the results (6.142–6.144) justifies the sensibility of the port oil piping transportation system operation process optimization.

6.4.3 Optimization of Operation, Safety and Availability of Maritime Ferry Technical System

The considered ferry technical system at the variable operation conditions is analyzed in Sect. 3.6.3 and its safety characteristics are found. From these results, it follows that it is possible to improve these characteristics by changing the parameters of its operation process.

6.4.3.1 Optimization of Ferry Technical System Operation Process

The objective function given by (6.1), in this case, as the ferry technical system critical safety state is $r = 2$, takes the form

$$\begin{aligned}\mu(2) = {} & p_1 \times 6.45 + p_2 \times 2.43 + p_3 \times 3.90 + p_4 \times 3.80 + p_5 \times 3.80 + p_6 \times 3.24 \\ & + p_7 \times 2.43 + p_8 \times 7.69 + p_9 \times 7.69 + p_{10} \times 2.43 + p_{11} \times 3.37 \\ & + p_{12} \times 3.80 + p_{13} \times 3.80 + p_{14} \times 3.80 + p_{15} \times 3.90 + p_{16} \times 3.37 \\ & + p_{17} \times 2.43 + p_{18} \times 6.45. \end{aligned} \tag{6.145}$$

The lower $\breve{p}_b$ and upper $\hat{p}_b$ bounds of the unknown transient probabilities p_b, $b = 1, 2, \ldots, 18$, from experts, respectively are:

$$\breve{p}_1 = 0.0006,\ \breve{p}_2 = 0.001,\ \breve{p}_3 = 0.018,\ \breve{p}_4 = 0.027,\ \breve{p}_5 = 0.286,\ \breve{p}_6 = 0.018,$$
$$\breve{p}_7 = 0.002,\ \breve{p}_8 = 0.001,\ \breve{p}_9 = 0.001,\ \breve{p}_{10} = 0.001,\ \breve{p}_{11} = 0.002,\ \breve{p}_{12} = 0.013,$$
$$\breve{p}_{13} = 0.286,\ \breve{p}_{14} = 0.025,\ \breve{p}_{15} = 0.018,\ \breve{p}_{16} = 0.002,\ \breve{p}_{17} = 0.002,\ \breve{p}_{18} = 0.001,$$
$$\hat{p}_1 = 0.056,\ \hat{p}_2 = 0.002,\ \hat{p}_3 = 0.027,\ \hat{p}_4 = 0.056,\ \hat{p}_5 = 0.780,\ \hat{p}_6 = 0.024,$$
$$\hat{p}_7 = 0.018,\ \hat{p}_8 = 0.018,\ \hat{p}_9 = 0.056,\ \hat{p}_{10} = 0.003,\ \hat{p}_{11} = 0.004,\ \hat{p}_{12} = 0.024,$$
$$\hat{p}_{13} = 0.780,\ \hat{p}_{14} = 0.043,\ \hat{p}_{15} = 0.024,\ \hat{p}_{16} = 0.004,\ \hat{p}_{17} = 0.007,\ \hat{p}_{18} = 0.018.$$

Therefore, according to (6.2–6.4), we assume the following bound constraints

$$\begin{aligned} & 0.0006 \le p_1 \le 0.056, && 0.001 \le p_2 \le 0.002, && 0.018 \le p_3 \le 0.027, \\ & 0.027 \le p_4 \le 0.056, && 0.286 \le p_5 \le 0.780, && 0.018 \le p_6 \le 0.024, \\ & 0.002 \le p_7 \le 0.018, && 0.001 \le p_8 \le 0.018, && 0.001 \le p_9 \le 0.056, \\ & 0.001 \le p_{10} \le 0.003, && 0.002 \le p_{11} \le 0.004, && 0.013 \le p_{12} \le 0.024, \\ & 0.286 \le p_{13} \le 0.780, && 0.025 \le p_{14} \le 0.043, && 0.018 \le p_{15} \le 0.024, \\ & 0.002 \le p_{16} \le 0.004, && 0.002 \le p_{17} \le 0.007, && 0.001 \le p_{18} \le 0.018, \\ & \sum_{b=1}^{18} p_b = 1. \end{aligned} \tag{6.146}$$

Now, before we find optimal values $\dot{p}_b$ of the transient probabilities p_b, $b = 1, 2, \ldots, 18$, that maximize the objective function (6.145), we arrange the system conditional lifetime mean values $\mu_b(2)$, $b = 1, 2, \ldots, 18$, in non-increasing order

$$\mu_8(2) \geq \mu_9(2) \geq \mu_1(2) \geq \mu_{18}(2) \geq \mu_3(2) \geq \mu_{15}(2) \geq \mu_4(2) \geq \mu_5(2) \geq \mu_{12}(2) \geq \mu_{13}(2) \geq \mu_{14}(2) \geq \mu_{11}(2) \geq \mu_{16}(2) \geq \mu_6(2) \geq \mu_2(2) \geq \mu_7(2) \geq \mu_{10}(2) \geq \mu_{17}(2).$$

Next, according to (6.6), we substitute

$$x_1 = p_8, \quad x_2 = p_9, \quad x_3 = p_1, \quad x_4 = p_{18}, \quad x_5 = p_3, \quad x_6 = p_{15}, \quad x_7 = p_4,$$
$$x_8 = p_5, x_9 = p_{12}, \quad x_{10} = p_{13}, \quad x_{11} = p_{14}, \quad x_{12} = p_{11}, \quad x_{13} = p_{16},$$
$$x_{14} = p_6, \quad x_{15} = p_2, \quad x_{16} = p_7, \quad x_{17} = p_{10}, \quad x_{18} = p_{17}, \tag{6.147}$$

and

$$\breve{x}_1 = \breve{p}_8 = 0.001, \quad \breve{x}_2 = \breve{p}_9 = 0.001, \quad \breve{x}_3 = \breve{p}_1 = 0.0006, \quad \breve{x}_4 = \breve{p}_{18} 0.001,$$
$$\breve{x}_5 = \breve{p}_3 = 0.018, \quad \breve{x}_6 = \breve{p}_{15} 0.018, \quad \breve{x}_7 = \breve{p}_4 = 0.027, \quad \breve{x}_8 = \breve{p}_5 = 0.286,$$
$$\breve{x}_9 = \breve{p}_{12} = 0.013, \quad \breve{x}_{10} = \breve{p}_{13} = 0.286, \quad \breve{x}_{11} = \breve{p}_{14} = 0.025, \quad \breve{x}_{12} = \breve{p}_{11} = 0.02,$$
$$\breve{x}_{13} = \breve{p}_{16} = 0.002, \quad \breve{x}_{14} = \breve{p}_6 = 0.018, \quad \breve{x}_{15} = \breve{p}_2 = 0.001, \quad \breve{x}_{16} = \breve{p}_7 = 0.002,$$
$$\breve{x}_{17} = \breve{p}_{10} = 0.001, \quad \breve{x}_{18} = \breve{p}_{17} = 0.002, \tag{6.148}$$

$$\hat{x}_1 = \hat{p}_8 = 0.018, \quad \hat{x}_2 = \hat{p}_9 = 0.056, \quad \hat{x}_3 = \hat{p}_1 = 0.056, \quad \hat{x}_4 = \hat{p}_{18} = 0.018,$$
$$\hat{x}_5 = \hat{p}_3 = 0.027, \quad \hat{x}_6 = \hat{p}_{15} = 0.024, \quad \hat{x}_7 = \hat{p}_4 = 0.056, \quad \hat{x}_8 = \hat{p}_5 = 0.780,$$
$$\hat{x}_9 = \hat{p}_{12} = 0.024, \quad \hat{x}_{10} = \hat{p}_{13} = 0.780, \quad \hat{x}_{11} = \hat{p}_{14} = 0.043, \quad \hat{x}_{12} = \hat{p}_{11} = 0.004,$$
$$\hat{x}_{13} = \hat{p}_{16} = 0.004, \quad \hat{x}_{14} = \hat{p}_6 = 0.024, \quad \hat{x}_{15} = \hat{p}_2 = 0.002, \quad \hat{x}_{16} = \hat{p}_7 = 0.018,$$
$$\hat{x}_{17} = \hat{p}_{10} = 0.003, \quad \hat{x}_{18} = \hat{p}_{17} = 0.007, \tag{6.149}$$

and we maximize with respect to x_i, $i = 1, 2, \ldots, 18$, the linear form (6.145) which according to (6.7–6.9) takes the form

$$\mu(2) = x_1 \times 7.69 + x_2 \times 7.69 + x_3 \times 6.45 + x_4 \times 6.45 + x_5 \times 3.90 + x_6 \times 3.90 + x_7 \times 3.80 + x_8 \times 3.80 + x_9 \times 3.80 + x_{10} \times 3.80 + x_{11} \times 3.80 + x_{12} \times 3.37 + x_{13} \times 3.37 + x_{14} \times 3.24 + x_{15} \times 2.43 + x_{16} \times 2.43 + x_{17} \times 2.43 + x_{18} \times 2.43, \tag{6.150}$$

with the following bound constraints

$$0.001 \leq x_1 \leq 0.018, \quad 0.001 \leq x_2 \leq 0.056, \quad 0.0006 \leq x_3 \leq 0.056,$$
$$0.001 \leq x_4 \leq 0.018, \quad 0.018 \leq x_5 \leq 0.027, \quad 0.018 \leq x_6 \leq 0.024,$$
$$0.027 \leq x_7 \leq 0.056, \quad 0.286 \leq x_8 \leq 0.780, \quad 0.013 \leq x_9 \leq 0.024,$$
$$0.286 \leq x_{10} \leq 0.780, \quad 0.025 \leq x_{11} \leq 0.043, \quad 0.002 \leq x_{12} \leq 0.004,$$
$$0.002 \leq x_{13} \leq 0.004, \quad 0.018 \leq x_{14} \leq 0.024, \quad 0.001 \leq x_{15} \leq 0.002,$$
$$0.002 \leq x_{16} \leq 0.018, \quad 0.001 \leq x_{17} \leq 0.003, \quad 0.002 \leq x_{18} \leq 0.007.$$

$$\sum_{i=1}^{18} x_i = 1, \tag{6.151}$$

According to (6.11), we calculate

$$\breve{x} = \sum_{i=1}^{18} \breve{x}_i = 0.7046, \quad \hat{y} = 1 - \breve{x}, = 1 - 0.7046 = 0.2954 \tag{6.152}$$

and according to (6.12), we determine

$$\begin{aligned}
&\breve{x}^{0} = 0, \quad \widehat{x}^{0} = 0, \quad \widehat{x}^{0} - \breve{x}^{0} = 0,\\
&\breve{x}^{1} = 0.001, \quad \widehat{x}^{1} = 0.018, \quad \widehat{x}^{1} - \breve{x}^{1} = 0.017,\\
&\breve{x}^{2} = 0.002, \quad \widehat{x}^{2} = 0.074, \quad \widehat{x}^{2} - \breve{x}^{2} = 0.072,\\
&\breve{x}^{3} = 0.0026, \quad \widehat{x}^{3} = 0.13, \quad \widehat{x}^{3} - \breve{x}^{3} = 0.1274,\\
&\breve{x}^{4} = 0.0036, \quad \widehat{x}^{4} = 0.148, \quad \widehat{x}^{4} - \breve{x}^{4} = 0.1444,\\
&\breve{x}^{5} = 0.0216, \quad \widehat{x}^{5} = 0.175, \quad \widehat{x}^{5} - \breve{x}^{5} = 0.1534,\\
&\breve{x}^{6} = 0.0396, \quad \widehat{x}^{6} = 0.199, \quad \widehat{x}^{6} - \breve{x}^{6} = 0.1594,\\
&\breve{x}^{7} = 0.0666, \quad \widehat{x}^{7} = 0.255, \quad \widehat{x}^{7} - \breve{x}^{7} = 0.1884,\\
&\breve{x}^{8} = 0.3526, \quad \widehat{x}^{8} = 1.035, \quad \widehat{x}^{8} - \breve{x}^{8} = 0.6824.
\end{aligned} \tag{6.153}$$

From the above, as according to (6.152), the inequality (6.13) takes the form

$$\widehat{x}^{I} - \breve{x}^{I} < 0.2954, \tag{6.154}$$

it follows that the largest value $I \in \{0, 1, \ldots, 18\}$ such that this inequality holds is $I = 7$.

Therefore, we fix the optimal solution that maximize linear function (6.150) according to the rule (6.15). Namely, we get

$$\begin{aligned}
&\dot{x}_1 = \widehat{x}_1 = 0.018, \quad \dot{x}_2 = \widehat{x}_2 = 0.056, \quad \dot{x}_3 = \widehat{x}_3 = 0.056, \dot{x}_4 = \widehat{x}_4 = 0.018,\\
&\dot{x}_5 = \widehat{x}_5 = 0.027, \quad \dot{x}_6 = \widehat{x}_6 = 0.024, \quad \dot{x}_7 = \widehat{x}_7 = 0.056,\\
&\dot{x}_8 = \hat{y} - \widehat{x}^{7} + \breve{x}^{7} + \breve{x}_8 = 0.2954 - 0.255 + 0.0666 + 0.286 = 0.393,\\
&\dot{x}_9 = \breve{x}_9 = 0.013, \quad \dot{x}_{10} = \breve{x}_{10} = 0.286, \quad \dot{x}_{11} = \breve{x}_{11} = 0.025, \quad \dot{x}_{12} = \breve{x}_{12} = 0.002,\\
&\dot{x}_{13} = \breve{x}_{13} = 0.002, \quad \dot{x}_{14} = \breve{x}_{14} = 0.018, \quad \dot{x}_{15} = \breve{x}_{15} = 0.001, \quad \dot{x}_{16} = \breve{x}_{16}, = 0.002,\\
&\dot{x}_{17} = \breve{x}_{17} = 0.001, \quad \dot{x}_{18} = \breve{x}_{18} = 0.002.
\end{aligned} \tag{6.155}$$

Finally, after making the inverse to (6.147) substitution, we get the optimal transient probabilities

$$\dot{p}_8 = \dot{x}_1 = 0.018, \quad \dot{p}_9 = \dot{x}_2 = 0.056, \quad \dot{p}_1 = \dot{x}_3 = 0.056, \quad \dot{p}_{18} = \dot{x}_4 = 0.018,$$
$$\dot{p}_3 = \dot{x}_5 = 0.027, \quad \dot{p}_{15} = \dot{x}_6 = 0.024, \quad \dot{p}_4 = \dot{x}_7 = 0.056, \quad \dot{p}_5 = \dot{x}_8 = 0.393,$$
$$\dot{p}_{12} = \dot{x}_9 = 0.013, \quad \dot{p}_{13} = \dot{x}_{10} = 0.286, \quad \dot{p}_{14} = \dot{x}_{11} = 0.025, \quad \dot{p}_{11} = \dot{x}_{12} = 0.002,$$
$$\dot{p}_{16} = \dot{x}_{13} = 0.002, \quad \dot{p}_6 = \dot{x}_{14} = 0.018, \quad \dot{p}_2 = \dot{x}_{15} = 0.001, \quad \dot{p}_7 = \dot{x}_{16} = 0.002,$$
$$\dot{p}_{10} = \dot{x}_{17} = 0.001, \quad \dot{p}_{17} = \dot{x}_{18} = 0.002, \tag{6.156}$$

that maximize the system mean lifetime $\mu(2)$ in the safety state subset $\{2,3,4\}$ expressed by the linear form (6.145) giving, according to (6.18) and (6.156), its optimal value

$$\begin{aligned}\dot{\mu}(2) \cong\ & 0.056 \times 6.45 + 0.001 \times 2.43 + 0.027 \times 3.90 + 0.056 \times 3.80 + 0.393 \times 3.80\\ & + 0.018 \times 3.24 + 0.002 \times 2.43 + 0.018 \times 7.69 + 0.056 \times 7.69 + 0.001 \times 2.43\\ & + 0.002 \times 3.37 + 0.013 \times 3.80 + 0.286 \times 3.80 + 0.025 \times 3.80 + 0.024 \times 3.90\\ & + 0.002 \times 3.37 + 0.002 \times 2.43 + 0.018 \times 6.45 = 4.27. \end{aligned} \tag{6.157}$$

6.4.3.2 Optimal Safety Characteristics of Ferry Technical System

Substituting the optimal solution (6.156) into the formulae (6.19), we obtain the optimal solution for the mean values of the system unconditional lifetime in the safety state subset $\{1,2,3,4\}$, $\{3,4\}$ and $\{4\}$, that respectively amount to:

$$\begin{aligned}\dot{\mu}(1) \cong\ & 0.056 \times 8.13 + 0.001 \times 2.86 + 0.027 \times 4.94 + 0.056 \times 4.2 + 0.393 \times 4.2\\ & + 0.018 \times 4.01 + 0.002 \times 2.86 + 0.018 \times 9.71 + 0.056 \times 9.71 + 0.001 \times 2.86\\ & + 0.002 \times 3.91 + 0.013 \times 4.2 + 0.286 \times 4.2 + 0.025 \times 4.2 + 0.024 \times 4.94\\ & + 0.002 \times 3.91 + 0.002 \times 2.86 + 0.018 \times 8.13 = 4.92, \end{aligned} \tag{6.158}$$

$$\begin{aligned}\dot{\mu}(3) \cong\ & 0.056 \times 5.71 + 0.001 \times 2.14 + 0.027 \times 3.44 + 0.056 \times 3.38 + 0.393 \times 3.38\\ & + 0.018 \times 2.88 + 0.002 \times 2.14 + 0.018 \times 6.89 + 0.056 \times 6.89 + 0.001 \times 2.14\\ & + 0.002 \times 3.07 + 0.013 \times 3.38 + 0.286 \times 3.38 + 0.025 \times 3.38 + 0.024 \times 3.44\\ & + 0.002 \times 3.07 + 0.002 \times 2.14 + 0.018 \times 5.71 = 3.79, \end{aligned} \tag{6.159}$$

$$\begin{aligned}\dot{\mu}(4) \cong\ & 0.056 \times 5.00 + 0.001 \times 1.93 + 0.027 \times 3.1 + 0.056 \times 3.05 + 0.552 \times 3.05\\ & + 0.018 \times 2.61 + 0.002 \times 1.93 + 0.018 \times 6.25 + 0.056 \times 6.25 + 0.001 \times 1.93\\ & + 0.002 \times 2.76 + 0.013 \times 3.05 + 0.286 \times 3.05 + 0.025 \times 3.05 + 0.024 \times 3.10\\ & + 0.002 \times 2.76 + 0.002 \times 1.93 + 0.018 \times 5.00 = 3.42, \end{aligned} \tag{6.160}$$

and according to (6.23), the optimal solutions for the mean values of the system unconditional lifetimes in the particular safety states 1, 2, 3 and 4, respectively are

$$\begin{aligned}&\dot{\bar{\mu}}(1) = \dot{\mu}(1) - \dot{\mu}(2) = 0.65, \ \dot{\bar{\mu}}(2) = \dot{\mu}(2) - \dot{\mu}(3) = 0.48,\\ &\dot{\bar{\mu}}(3) = \dot{\mu}(3) - \dot{\mu}(4) = 0.37, \ \dot{\bar{\mu}}(4) = \dot{\mu}(4) = 3.42,\end{aligned} \tag{6.161}$$

Moreover, according to (6.20–6.21), the corresponding optimal unconditional multistate safety function of the system is of the form

$$\dot{s}(t,\cdot) = [1, \dot{s}(t,1), \dot{s}(t,2), \dot{s}(t,3), \dot{s}(t,4)], \quad t \geq 0, \tag{6.162}$$

with the coordinates given by

$$\begin{aligned}\dot{s}(t,1) = {} & 0.056 \times [s(t,1)]^{(1)} + 0.001 \times [s(t,1)]^{(2)} + 0.027 \times [s(t,1)]^{(3)} \\ & + 0.056 \times [s(t,1)]^{(4)} + 0.393 \times [s(t,1)]^{(5)} + 0.018 \times [s(t,1)]^{(6)} \\ & + 0.002 \times [s(t,1)]^{(7)} + 0.018 \times [s(t,1)]^{(8)} + 0.056 \times [s(t,1)]^{(9)} \\ & + 0.001 \times [s(t,1)]^{(10)} + 0.002 \times [s(t,1)]^{(11)} + 0.013 \times [s(t,1)]^{(12)} \\ & + 0.286 \times [s(t,1)]^{(13)} + 0.025 \times [s(t,1)]^{(14)} + 0.024 \times [s(t,1)]^{(15)} \\ & + 0.002 \times [s(t,1)]^{(16)} + 0.002 \times [\bar{s}(t,1)]^{(17)} + 0.018 \times [s(t,1)]^{(18)},\end{aligned} \tag{6.163}$$

$$\begin{aligned}\dot{s}(t,2) = {} & 0.056 \times [s(t,2)]^{(1)} + 0.001 \times [s(t,2)]^{(2)} + 0.027 \times [s(t,2)]^{(3)} \\ & + 0.056 \times [s(t,2)]^{(4)} + 0.393 \times [s(t,2)]^{(5)} + 0.018 \times [s(t,2)]^{(6)} \\ & + 0.002 \times [s(t,2)]^{(7)} + 0.018 \times [s(t,2)]^{(8)} + 0.056 \times [s(t,2)]^{(9)} \\ & + 0.001 \times [s(t,2)]^{(10)} + 0.002 \times [s(t,2)]^{(11)} + 0.013 \times [s(t,2)]^{(12)} \\ & + 0.286 \times [s(t,2)]^{(13)} + 0.025 \times [s(t,2)]^{(14)} + 0.024 \times [s(t,2)]^{(15)} \\ & + 0.002 \times [s(t,2)]^{(16)} + 0.002 \times [\bar{s}(t,2)]^{(17)} + 0.018 \times [s(t,2)]^{(18)},\end{aligned} \tag{6.164}$$

$$\begin{aligned}\dot{s}(t,3) = {} & 0.056 \times [s(t,3)]^{(1)} + 0.001 \times [s(t,3)]^{(2)} + 0.027 \times [s(t,3)]^{(3)} \\ & + 0.056 \times [s(t,3)]^{(4)} + 0.393 \times [s(t,3)]^{(5)} + 0.018 \times [s(t,3)]^{(6)} \\ & + 0.002 \times [s(t,3)]^{(7)} + 0.018 \times [s(t,3)]^{(8)} + 0.056 \times [s(t,3)]^{(9)} \\ & + 0.001 \times [s(t,3)]^{(10)} + 0.002 \times [s(t,3)]^{(11)} + 0.013 \times [s(t,3)]^{(12)} \\ & + 0.286 \times [s(t,3)]^{(13)} + 0.025 \times [s(t,3)]^{(14)} + 0.024 \times [s(t,3)]^{(15)} \\ & + 0.002 \times [s(t,3)]^{(16)} + 0.002 \times [\bar{s}(t,3)]^{(17)} + 0.018 \times [s(t,3)]^{(18)},\end{aligned} \tag{6.165}$$

$$\begin{aligned}\dot{s}(t,4) = {} & 0.056 \times [s(t,4)]^{(1)} + 0.001 \times [s(t,4)]^{(2)} + 0.027 \times [s(t,4)]^{(3)} \\ & + 0.056 \times [s(t,4)]^{(4)} + 0.393 \times [s(t,4)]^{(5)} + 0.018 \times [s(t,4)]^{(6)} \\ & + 0.002 \times [s(t,4)]^{(7)} + 0.018 \times [s(t,4)]^{(8)} + 0.056 \times [s(t,4)]^{(9)} \\ & + 0.001 \times [s(t,4)]^{(10)} + 0.002 \times [s(t,4)]^{(11)} + 0.013 \times [s(t,4)]^{(12)} \\ & + 0.286 \times [s(t,4)]^{(13)} + 0.025 \times [s(t,4)]^{(14)} + 0.024 \times [s(t,4)]^{(15)} \\ & + 0.002 \times [s(t,4)]^{(16)} + 0.002 \times [\bar{s}(t,4)]^{(17)} + 0.018 \\ & \times [s(t,4)]^{(18)} \quad \text{for } t \geq 0,\end{aligned} \tag{6.166}$$

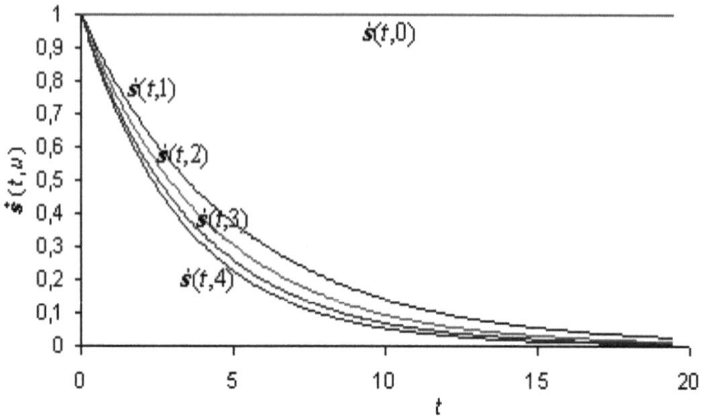

Fig. 6.5 The graph of the ferry technical system optimal safety function $\dot{s}(t, \times)$ coordinates

where $s(t,u)]^{(b)}$, $u = 1,2,3,4$, $b = 1,2,\ldots,18$, are fixed in Sect. 3.6.3 (Fig. 6.5).

Further, according to (6.22), the corresponding optimal variances and standard deviations of the system unconditional lifetimes in the system safety state subsets are

$$\dot{\sigma}^2(1) = 2\int_0^\infty t\,\dot{s}(t,1)\mathrm{d}t - [\dot{\mu}(1)]^2 \cong 21.16, \quad \dot{\sigma}(1) \cong 4.60, \tag{6.167}$$

$$\dot{\sigma}^2(2) = 2\int_0^\infty t\,\dot{s}(t,2)\mathrm{d}t - [\dot{\mu}(2)]^2 \cong 13.54, \quad \dot{\sigma}(2) = 3.68, \tag{6.168}$$

$$\dot{\sigma}^2(3) = 2\int_0^\infty t\,\dot{s}(t,3)\mathrm{d}t - [\dot{\mu}(3)]^2 \cong 11.69, \quad \dot{\sigma}(3) \cong 3.42, \tag{6.169}$$

$$\dot{\sigma}^2(4) = 2\int_0^\infty t\,\dot{s}(t,4)\mathrm{d}t - [\dot{\mu}(4)]^2 \cong 9.06, \quad \dot{\sigma}(4) = 3.01, \tag{6.170}$$

where $\dot{s}(t,1)$, $\dot{s}(t,2)$, $\dot{s}(t,3)$, $\dot{s}(t,4)$, are given by (6.163–6.166) and $\dot{\mu}(1), \dot{\mu}(2), \dot{\mu}(3), \dot{\mu}(4)$,are given by (6.157–6.160).

As the critical safety state is $r = 2$, the system risk function, according to (6.24) is given by (Fig. 6.6)

$$\dot{r}(t) = 1 - \dot{s}(t,2) \quad \text{for } t \geq 0, \tag{6.171}$$

where $\dot{s}(t,2)$ is given by (6.164).

Hence, considering (6.25) the moment when the optimal system risk function exceeds a permitted level, for instance $\delta = 0.05$, is

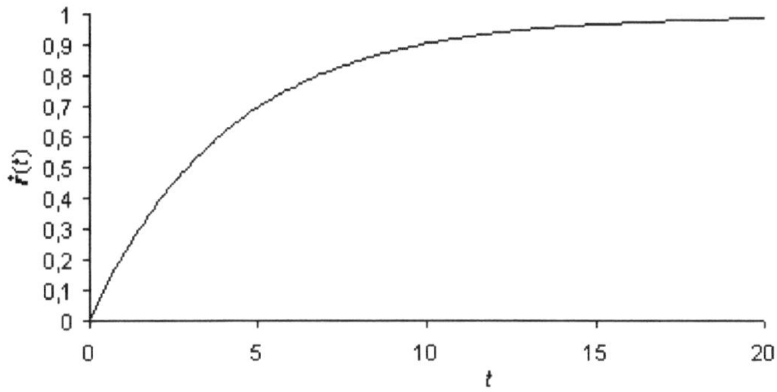

Fig. 6.6 The graph of the ferry technical system optimal risk function $\dot{\boldsymbol{r}}(t)$

$$\dot{\tau} = \dot{\boldsymbol{r}}^{-1}(\delta) \cong 0.22 \text{ year.} \tag{6.172}$$

6.4.3.3 Optimal Renewal and Availability Characteristics of Ferry Technical System

To determine the optimal renewal and availability characteristics of the ferry technical system after its operation process optimization, we use the results of the system safety characteristics optimization performed in Sects. 6.4.3.1–6.4.3.2 and the results of Sect. 6.2.3.

In the case when the ferry technical system renovation time is ignored, considering the optimal values $\dot{\mu}(2)$ determined by (6.157) and $\dot{\sigma}(2)$ determined by (6.168) and applying *Proposition 6.1* from Sect. 6.2.3, we determine its following optimal characteristics:

(a) the optimal time $\dot{S}_N(2)$ until the Nth exceeding by the system the safety critical state 2, for sufficiently large N, has approximately normal distribution $N(4.27N, 3.68\sqrt{N})$, i.e.,

$$\dot{F}^{(N)}(t,2) = P(\dot{S}_N(2) < t) \cong F_{N(0,1)}\left(\frac{t - 4.27N}{3.68\sqrt{N}}\right), \quad t \in (-\infty, \infty),$$

(b) the expected value and the variance of the optimal time $\dot{S}_N(2)$ until the Nth exceeding by the system the safety critical state 2, for sufficiently large N, respectively are

$$E[\dot{S}_N(2)] = 4.27N, \quad D[\dot{S}_N(2)] = 13.542N,$$

(c) the optimal number $\dot{N}(t,2)$ of exceeding by the system the safety critical state 2 up to the moment t, $t \geq 0$, for sufficiently large t, has distribution approximately of the form

$$P(\dot{N}(t,2)=N) \cong F_{N(0,1)}\left(\frac{4.27(N+1)-t}{1.781\sqrt{t}}\right) - F_{N(0,1)}\left(\frac{4.27N-t}{1.781\sqrt{t}}\right),$$
$$N=0,1,\ldots,$$

(d) the expected value and the variance of the optimal number $N(t,2)$ of exceeding by the system the safety critical state 2 up to the moment t, $t \geq 0$, for sufficiently large t, respectively are

$$\dot{H}(t,2)=0.234t, \quad \dot{D}(t,2)=0.174t.$$

To make the estimation of the renewal and availability of the ferry technical system in the case when the time of renovation is non-ignored, considering the optimal values $\dot{\mu}(2)$ determined by (6.157) and $\dot{\sigma}(2)$ determined by (6.168), assuming the mean value of the system renovation time $\mu_0(2)=0.019$ year and the standard deviation of the system renovation time $\sigma_0(2)=0.0095$ year and applying *Proposition 6.2* from Sect. 6.2.3, we determine its following optimal characteristics:

(a) the optimal time $\dot{\breve{S}}_N(2)$ until the Nth exceeding by the system the safety critical state 2, for sufficiently large N, has approximately normal distribution $N(4.27N+0.019(N-1), \sqrt{13.542N+0.000093(N-1)})$, i.e.,

$$\dot{\breve{F}}^{(N)}(t,2)=P(\dot{\breve{S}}_N(2)<t)=F_{N(0,1)}\left(\frac{t-4.289N+0.019}{\sqrt{13.542093N-0.000093}}\right),$$
$$t \in (-\infty,\infty),$$

(b) the expected value and the variance of the optimal time $\dot{\breve{S}}_N(2)$ until the Nth exceeding by the system the safety critical state 2, for sufficiently large N, respectively are

$$E[\dot{\breve{S}}_N(2)] \cong 4.27N+0.019(N-1), \quad D[\dot{\breve{S}}(2)] \cong 13.542N+0.000093(N-1),$$

(c) the optimal number $\dot{\breve{N}}(t,2)$ of exceeding by the system the safety critical state 2 up to the moment t, $t \geq 0$, for sufficiently large t, has approximately distribution of the form

$$P(\dot{\breve{N}}(t,2)=N) \cong F_{N(0,1)}\left(\frac{4.289(N+1)-t-0.019}{1.778\sqrt{t+0.019}}\right)$$
$$- F_{N(0,1)}\left(\frac{4.289N-t-0.019}{1.778\sqrt{t+0.019}}\right), \quad N=0,1,\ldots,$$

(d) the expected value and the variance of the optimal number $\dot{\breve{N}}(t,2)$ of exceeding by the system the safety critical state 2 up to the moment t, $t \geq 0$, for sufficiently large t, respectively are

$\dot{\breve{H}}(t,2) \cong 0.233(t+0.019), \quad \dot{\breve{D}}(t,2) \cong 0.1716(t+0.019),$

(e) the optimal time $\dot{\breve{\breve{S}}}_N(2)$ until the Nth system's renovation, for sufficiently large N, has approximately normal distribution $N(4.289N, 3.680\sqrt{N})$, i.e.,

$$\dot{\bar{\bar{F}}}^{(N)}(t,2) = P(\dot{\bar{\bar{S}}}_N(2) < t) \cong F_{N(0,1)}\left(\frac{t - 4.89N}{3.680\sqrt{N}}\right), \quad t \in (-\infty, \infty), \quad N = 0, 1, \ldots,$$

(f) the expected value and the variance of the optimal time $\dot{\bar{\bar{S}}}_N(2)$ until the Nth system's renovation, for sufficiently large N, respectively are

$$E\left[\dot{\bar{\bar{S}}}_N(2)\right] \cong 4.289N, \quad D\left[\dot{\bar{\bar{S}}}^{(N)}(2)\right] \cong 13.542093N,$$

(g) the optimal number $\dot{\bar{\bar{N}}}(t,2)$ of system's renovations up to the moment t, $t \geq 0$, for sufficiently large t, has approximately distribution of the form

$$P(\dot{\bar{\bar{N}}}(t,2) = N) \cong F_{N(0,1)}\left(\frac{4.289(N+1) - t}{1.778\sqrt{t}}\right) - F_{N(0,1)}\left(\frac{4.289N - t}{1.778\sqrt{t}}\right),$$
$$N = 0, 1, \ldots,$$

(h) the expected value and the variance of the optimal number $\dot{\bar{\bar{N}}}(t,2)$ of system's renovations up to the moment t, $t \geq 0$, for sufficiently large t, respectively are

$$\dot{\bar{\bar{H}}}(t,2) \cong 0.233t, \quad \dot{\bar{\bar{D}}}(t,2) \cong 0.1716t,$$

(i) the optimal steady availability coefficient of the system at the moment t, $t \geq 0$, for sufficiently large t, is

$$\dot{A}(t,2) \cong 0.996, \quad t \geq 0,$$

(j) the optimal steady availability coefficient of the system in the time interval $\langle t, t+\tau)$, $\tau > 0$, $t \geq 0$, for sufficiently large t, is

$$\dot{A}(t,\tau,2) \cong 0.233 \int_{\tau}^{\infty} \dot{s}(t,2)\mathrm{d}t, \quad t \geq 0, \quad \tau > 0,$$

where $\dot{s}(t,2)$ is given by (6.164).

6.4.3.4 Optimal Sojourn Times at Operation States of Ferry Technical System and Operation Strategy

To obtain the optimal mean sojourn times at the particular operation states maximizing the mean lifetime $\mu(2)$ of the ferry technical system in the safety state subset $\{2,3,4\}$, we substitute the optimal transient probabilities $\dot{p}_b$ determined by (6.156) and the steady probabilities π_b determined by (2.47) into the system of Eq. 6.41 and get its following form

$$-0.944\dot{M}_1 + 0.056(\dot{M}_2 + \dot{M}_3 + \dot{M}_4 + \dot{M}_5 + \dot{M}_6 + \dot{M}_7 + \dot{M}_8 + \dot{M}_9 + \dot{M}_{10}$$
$$+ \dot{M}_{11} + \dot{M}_{12} + \dot{M}_{13} + \dot{M}_{14} + \dot{M}_{15} + \dot{M}_{16} + \dot{M}_{17} + \dot{M}_{18}) = 0$$

$$-0.999\dot{M}_2 + 0.001(\dot{M}_1 + \dot{M}_3 + \dot{M}_4 + \dot{M}_5 + \dot{M}_6 + \dot{M}_7 + \dot{M}_8 + \dot{M}_9 + \dot{M}_{10}$$
$$+ \dot{M}_{11} + \dot{M}_{12} + \dot{M}_{13} + \dot{M}_{14} + \dot{M}_{15} + \dot{M}_{16} + \dot{M}_{17} + \dot{M}_{18}) = 0$$

$$-0.973\dot{M}_3 + 0.027(\dot{M}_1 + \dot{M}_2 + \dot{M}_4 + \dot{M}_5 + \dot{M}_6 + \dot{M}_7 + \dot{M}_8 + \dot{M}_9 + \dot{M}_{10}$$
$$+ \dot{M}_{11} + \dot{M}_{12} + \dot{M}_{13} + \dot{M}_{14} + \dot{M}_{15} + \dot{M}_{16} + \dot{M}_{17} + \dot{M}_{18}) = 0$$

$$-0.944\dot{M}_4 + 0.056(\dot{M}_1 + \dot{M}_2 + \dot{M}_3 + \dot{M}_5 + \dot{M}_6 + \dot{M}_7 + \dot{M}_8 + \dot{M}_9 + \dot{M}_{10}$$
$$+ \dot{M}_{11} + \dot{M}_{12} + \dot{M}_{13} + \dot{M}_{14} + \dot{M}_{15} + \dot{M}_{16} + \dot{M}_{17} + \dot{M}_{18}) = 0$$

$$-0.607\dot{M}_5 + 0.393(\dot{M}_1 + \dot{M}_2 + \dot{M}_3 + \dot{M}_4 + \dot{M}_6 + \dot{M}_7 + \dot{M}_8 + \dot{M}_9 + \dot{M}_{10}$$
$$+ \dot{M}_{11} + \dot{M}_{12} + \dot{M}_{13} + \dot{M}_{14} + \dot{M}_{15} + \dot{M}_{16} + \dot{M}_{17} + \dot{M}_{18}) = 0$$

$$-0.982\dot{M}_6 + 0.018(\dot{M}_1 + \dot{M}_2 + \dot{M}_3 + \dot{M}_4 + \dot{M}_5 + \dot{M}_7 + \dot{M}_8 + \dot{M}_9 + \dot{M}_{10}$$
$$+ \dot{M}_{11} + \dot{M}_{12} + \dot{M}_{13} + \dot{M}_{14} + \dot{M}_{15} + \dot{M}_{16} + \dot{M}_{17} + \dot{M}_{18}) = 0$$

$$-0.998\dot{M}_7 + 0.002(\dot{M}_1 + \dot{M}_2 + \dot{M}_3 + \dot{M}_4 + \dot{M}_5 + \dot{M}_6 + \dot{M}_8 + \dot{M}_9 + \dot{M}_{10}$$
$$+ \dot{M}_{11} + \dot{M}_{12} + \dot{M}_{13} + \dot{M}_{14} + \dot{M}_{15} + \dot{M}_{16} + \dot{M}_{17} + \dot{M}_{18}) = 0$$

$$-0.982\dot{M}_8 + 0.018(\dot{M}_1 + \dot{M}_2 + \dot{M}_3 + \dot{M}_4 + \dot{M}_5 + \dot{M}_6 + \dot{M}_7 + \dot{M}_9 + \dot{M}_{10}$$
$$+ \dot{M}_{11} + \dot{M}_{12} + \dot{M}_{13} + \dot{M}_{14} + \dot{M}_{15} + \dot{M}_{16} + \dot{M}_{17} + \dot{M}_{18}) = 0$$

$$-0.944\dot{M}_9 + 0.056(\dot{M}_1 + \dot{M}_2 + \dot{M}_3 + \dot{M}_4 + \dot{M}_5 + \dot{M}_6 + \dot{M}_7 + \dot{M}_8 + \dot{M}_{10}$$
$$+ \dot{M}_{11} + \dot{M}_{12} + \dot{M}_{13} + \dot{M}_{14} + \dot{M}_{15} + \dot{M}_{16} + \dot{M}_{17} + \dot{M}_{18}) = 0$$

$$-0.999\dot{M}_{10} + 0.001(\dot{M}_1 + \dot{M}_2 + \dot{M}_3 + \dot{M}_4 + \dot{M}_5 + \dot{M}_6 + \dot{M}_7 + \dot{M}_8 + \dot{M}_9$$
$$+ \dot{M}_{11} + \dot{M}_{12} + \dot{M}_{13} + \dot{M}_{14} + \dot{M}_{15} + \dot{M}_{16} + \dot{M}_{17} + \dot{M}_{18}) = 0$$

$$-0.998 + \dot{M}_{11} + 0.002(\dot{M}_1 + \dot{M}_2 + \dot{M}_3 + \dot{M}_4 + \dot{M}_5 + \dot{M}_6 + \dot{M}_7 + \dot{M}_8 + \dot{M}_9$$
$$+ \dot{M}_{10} + \dot{M}_{12} + \dot{M}_{13} + \dot{M}_{14} + \dot{M}_{15} + \dot{M}_{16} + \dot{M}_{17} + \dot{M}_{18}) = 0$$

$$-0.987\dot{M}_{12} + 0.013(\dot{M}_1 + \dot{M}_2 + \dot{M}_3 + \dot{M}_4 + \dot{M}_5 + \dot{M}_6 + \dot{M}_7 + \dot{M}_8 + \dot{M}_9$$
$$+ \dot{M}_{10} + + \dot{M}_{11} + \dot{M}_{13} + \dot{M}_{14} + \dot{M}_{15} + \dot{M}_{16} + \dot{M}_{17} + \dot{M}_{18}) = 0$$

$$-0.714\dot{M}_{13} + 0.286(\dot{M}_1 + \dot{M}_2 + \dot{M}_3 + \dot{M}_4 + \dot{M}_5 + \dot{M}_6 + \dot{M}_7 + \dot{M}_8 + \dot{M}_9$$
$$+ \dot{M}_{10} + + \dot{M}_{11} + \dot{M}_{12} + \dot{M}_{14} + \dot{M}_{15} + \dot{M}_{16} + \dot{M}_{17} + \dot{M}_{18}) = 0$$

$$-0.975\dot{M}_{14} + 0.025(\dot{M}_1 + \dot{M}_2 + \dot{M}_3 + \dot{M}_4 + \dot{M}_5 + \dot{M}_6 + \dot{M}_7 + \dot{M}_8 + \dot{M}_9$$
$$+ \dot{M}_{10} + \dot{M}_{11} + \dot{M}_{12} + \dot{M}_{13} + \dot{M}_{15} + \dot{M}_{16} + \dot{M}_{17} + \dot{M}_{18}) = 0$$

$$-0.976\dot{M}_{15} + 0.024(\dot{M}_1 + \dot{M}_2 + \dot{M}_3 + \dot{M}_4 + \dot{M}_5 + \dot{M}_6 + \dot{M}_7 + \dot{M}_8 + \dot{M}_9$$
$$+ \dot{M}_{10} + \dot{M}_{11} + \dot{M}_{12} + \dot{M}_{13} + \dot{M}_{14} + \dot{M}_{16} + \dot{M}_{17} + \dot{M}_{18}) = 0$$

$$-0.998\dot{M}_{16} + 0.002(\dot{M}_1 + \dot{M}_2 + \dot{M}_3 + \dot{M}_4 + \dot{M}_5 + \dot{M}_6 + \dot{M}_7 + \dot{M}_8 + \dot{M}_9$$
$$+ \dot{M}_{10} + \dot{M}_{11} + \dot{M}_{12} + \dot{M}_{13} + \dot{M}_{14} + \dot{M}_{15} + \dot{M}_{17} + \dot{M}_{18}) = 0$$

$$- 0.998\dot{M}_{17} + 0.002(\dot{M}_1 + \dot{M}_2 + \dot{M}_3 + \dot{M}_4 + \dot{M}_5 + \dot{M}_6 + \dot{M}_7 + \dot{M}_8 + \dot{M}_9$$
$$+ \dot{M}_{10} + \dot{M}_{11} + \dot{M}_{12} + \dot{M}_{13} + \dot{M}_{14} + \dot{M}_{15} + \dot{M}_{16} + \dot{M}_{18}) = 0$$
$$- 0.982\dot{M}_{18} + 0.018(\dot{M}_1 + \dot{M}_2 + \dot{M}_3 + \dot{M}_4 + \dot{M}_5 + \dot{M}_6 + \dot{M}_7 + \dot{M}_8 + \dot{M}_9$$
$$+ \dot{M}_{10} + \dot{M}_{11} + \dot{M}_{12} + \dot{M}_{13} + \dot{M}_{14} + \dot{M}_{15} + \dot{M}_{16} + \dot{M}_{17}) = 0 \quad (6.173)$$

From the above we get

$$\dot{M}_1 = 0.056\dot{M}, \quad \dot{M}_2 = 0.001\dot{M}, \quad \dot{M}_3 = 0.027\dot{M}, \quad \dot{M}_4 = 0.056\dot{M},$$
$$\dot{M}_5 = 0.393\dot{M}, \quad \dot{M}_6 = 0.018\dot{M}, \quad \dot{M}_7 = 0.002\dot{M}, \quad \dot{M}_8 = 0.018\dot{M},$$
$$\dot{M}_9 = 0.056\dot{M}, \quad \dot{M}_{10} = 0.001\dot{M} \quad \dot{M}_{11} = 0.002\dot{M}, \quad \dot{M}_{12} = 0.013\dot{M},$$
$$\dot{M}_{13} = 0.286\dot{M}, \quad \dot{M}_{14} = 0.025\dot{M}, \quad \dot{M}_{15} = 0.024\dot{M} \quad \dot{M}_{16} = 0.002\dot{M},$$
$$\dot{M}_{17} = 0.002\dot{M}, \quad \dot{M}_{18} = 0.018\dot{M}, \quad (6.174)$$

where $\dot{M} = \sum_{b=1}^{18} \dot{M}_b$.

Assuming that $\dot{M}$ is equal to the total time of one voyage given in the ferry timetable, i.e. $\dot{M} = 1350$ min, we get the following solution of the system of equations

$$\dot{M}_1 = 75.60, \quad \dot{M}_2 = 1.35, \quad \dot{M}_3 = 36.45, \quad \dot{M}_4 = 75.60, \quad \dot{M}_5 = 530.55,$$
$$\dot{M}_6 = 24.30, \quad \dot{M}_7 = 2.70, \quad \dot{M}_8 = 24.30, \quad \dot{M}_9 = 75.60, \quad \dot{M}_{10} = 1.35,$$
$$\dot{M}_{11} = 2.70, \quad \dot{M}_{12} = 17.55, \quad \dot{M}_{13} = 386.10, \quad \dot{M}_{14} = 33.75,$$
$$\dot{M}_{15} = 32.40, \quad \dot{M}_{16} = 2.70, \quad \dot{M}_{17} = 2.70, \quad \dot{M}_{18} = 24.30. \quad (6.175)$$

Substituting into (6.42) the above mean values $\dot{M}_b$ of the system unconditional sojourn times at the particular operation states and the known probabilities p_{bl} of the system operation process transitions between the operation states given in the matrix (2.40), we may look for the optimal values $\dot{M}_{bl}$ of the mean values of the system conditional sojourn times at the particular operation states that maximize the mean lifetime $\mu(2)$ of the ferry technical system in the safety states subset $\{2, 3, 4\}$. The optimal values $\dot{M}_{bl}$, $b, l = 1, 2, \ldots, 18$, $b \neq l$, should satisfy the following system of equations obtained in this way

$$\dot{M}_{12} = 75.60, \quad \dot{M}_{23} = 1.35, \quad \dot{M}_{34} = 36.45, \quad \dot{M}_{45} = 75.60, \quad \dot{M}_{56} = 530.55,$$
$$\dot{M}_{67} = 24.30, \quad \dot{M}_{78} = 2.70, \quad \dot{M}_{89} = 24.30, \quad \dot{M}_{910} = 75.60, \quad \dot{M}_{1011} = 1.35,$$
$$\dot{M}_{1112} = 2.70, \quad \dot{M}_{1213} = 17.55, \quad \dot{M}_{1314} = 386.10, \quad \dot{M}_{1415} = 33.75,$$
$$\dot{M}_{1516} = 32.40, \quad \dot{M}_{1617} = 2.70, \quad \dot{M}_{1718} = 2.70, \quad \dot{M}_{181} = 24.30. \quad (6.176)$$

Another very useful and much easier to be applied in practice tool that can help in planning the operation process of the ferry technical systems is the system operation process optimal mean values of the total sojourn times at the particular operation states during the system operation time θ.

Assuming as in Sect. 2.3.1 the system operation time $\theta = 1$ year $= 365$ days, after applying (6.43), we get their values

$$\begin{aligned}
&\dot{E}[\hat{\theta}_1] = \dot{p}_1\theta = 0.056 \times 365 = 20.44, \quad \dot{E}[\hat{\theta}_2] = \dot{p}_2\theta = 0.001 \times 365 = 0.365,\\
&\dot{E}[\hat{\theta}_3] = \dot{p}_3\theta = 0.027 \times 365 = 9.855, \quad \dot{E}[\hat{\theta}_4] = \dot{p}_4\theta = 0.056 = 20.44,\\
&\dot{E}[\hat{\theta}_5] = \dot{p}_5\theta = 0.393 \times 365 = 143.445, \quad \dot{E}[\hat{\theta}_6] = \dot{p}_6\theta = 0.018 \times 365 = 6.57,\\
&\dot{E}[\hat{\theta}_7] = \dot{p}_7\theta = 0.002 \times 365 = 0.73, \quad \dot{E}[\hat{\theta}_8] = \dot{p}_8\theta = 0.018 \times 365 = 6.57,\\
&\dot{E}[\hat{\theta}_9] = \dot{p}_9\theta = 0.056 \times 365 = 20.44, \quad \dot{E}[\hat{\theta}_{10}] = \dot{p}_{10}\theta = 0.001 \times 365 = 0.365,\\
&\dot{E}[\hat{\theta}_{11}] = \dot{p}_{11}\theta = 0.002 \times 365 = 0.73, \quad \dot{E}[\hat{\theta}_{12}] = \dot{p}_{12}\theta = 0.013 \times 365 = 4.745,\\
&\dot{E}[\hat{\theta}_{13}] = \dot{p}_{13}\theta = 0.286 \times 365 = 104.39, \quad \dot{E}[\hat{\theta}_{14}] = \dot{p}_{14}\theta = 0.025 \times 365 = 9.125,\\
&\dot{E}[\hat{\theta}_{15}] = \dot{p}_{15}\theta = 0.024 \times 365 = 8.76, \quad \dot{E}[\hat{\theta}_{16}] = \dot{p}_{16}\theta = 0.002 \times 365 = 0.73,\\
&\dot{E}[\hat{\theta}_{17}] = \dot{p}_{17}\theta = 0.002 \times 365 = 0.73, \quad \dot{E}[\hat{\theta}_{18}] = \dot{p}_{18}\theta = 0.018 \times 365 = 6.57,
\end{aligned} \tag{6.177}$$

that differ from the values of $E[\hat{\theta}_i]$, $i = 1, 2, \ldots, 18$, determined by (2.49) in Sect. 2.3.3.

In practice, the knowledge of the optimal values of $\dot{M}_b$, $\dot{M}_{bl}$, and $\dot{E}[\hat{\theta}_b]$ given respectively by (6.175–6.177), can be very important and helpful for the system operation process planning and improving in order to make the system operation more reliable and safer.

6.4.4 *Corrective and Preventive Maintenance Policy of Exemplary Technical System*

6.4.4.1 Corrective and Preventive Maintenance Policy Maximizing Exemplary Technical System Availability

To optimize the exemplary system corrective and preventive maintenance policy maximizing its availability, we use its following reliability and renewal parameters:

- the system and components critical reliability state $r = 2$,
- the 2nd coordinate of the system unconditional reliability function $\boldsymbol{R}(t, \cdot)$

$$\begin{aligned}
\boldsymbol{R}(t, 2) = {}& 0.214 \times [\boldsymbol{R}(t,2)]^{(1)} + 0.038 \times [\boldsymbol{R}(t,2)]^{(2)} + 0.293 \times [\boldsymbol{R}(t,2)]^{(3)}\\
& + 0.455 \times [\boldsymbol{R}(t,2)]^{(4)} \quad \text{for } t \geq 0,
\end{aligned} \tag{6.178}$$

where $[\boldsymbol{R}(t,2)]^{(b)}$, $b = 1,2,3,4$, are respectively given by (3.108), (3.115), (3.130), (3.145).

- the derivative of the 2nd coordinate of the system unconditional reliability function $\boldsymbol{R}(t,\ \cdot)$

$$\boldsymbol{R}'(t,\ 2) = 0.214 \times [\boldsymbol{R}'(t,2)]^{(1)} + 0.038 \times [\boldsymbol{R}'(t,2)]^{(2)} + 0.293 \times [\boldsymbol{R}'(t,2)]^{(3)} + 0.455 \times [\boldsymbol{R}'(t,2)]^{(4)} \quad \text{for } t \geq 0, \tag{6.179}$$

- the mean value of the system corrective renovation time $\mu_0(2) = 10$,
- the mean value of the system preventive renovation time $\mu_1(2) = 5$.

Moreover, to apply the algorithm proposed in Sect. 6.3.1, we fix:

- the measure of the method of secants accuracy $\varepsilon = 0.001$,
- the number of values of the system preventive maintenance period η for which we find the values of the availability coefficient of the exemplary system in cases when there is no optimal value $\kappa = 20$,
- the values of the system preventive maintenance period η for which we find the values of the availability coefficient of the system in cases when there is no optimal value

$$\eta_i = (i-1)\ 0.2\mu(2) = (i-1)0.2 \times 358, \quad i = 1,2,\ldots,20. \tag{6.180}$$

where $\mu(2) \cong 358$, is determined by (3.157).

Since

$$\mu_0(2) = 10 > \mu_1(2) = 5,$$

we are looking for the optimal value $\dot{\eta}$ of the preventive maintenance period η that maximizes the availability coefficient of the system $A(\eta, r)$ given by (6.52) in Sect. 6.3.1 by determining, if it exists, its approximate value from Eq. 6.58 by applying the method of secants in the interval $\langle a,\ b\rangle$ as follows:

- we define, obtained after the transformation of Eq. 6.58, the function

$$f(\eta) = \lambda(\eta,2)\int_0^{\eta} \boldsymbol{R}(t,2)\mathrm{d}t + \boldsymbol{R}(\eta,2) - \frac{\mu_0(2)}{\mu_0(2) - \mu_1(2)}$$

$$= \lambda(\eta,2)\int_0^{\eta} \boldsymbol{R}(t,2)\mathrm{d}t + \boldsymbol{R}(\eta,2) - 2 \text{ for } \quad \eta \geq 0,$$

where $\boldsymbol{R}(t,\ 2)$ is given by (6.178) and

$$\lambda(\eta,2) = -\frac{\boldsymbol{R}'(\eta,2)}{\boldsymbol{R}(\eta,2)},$$

Table 6.1 The values of the availability coefficient of the exemplary system

η	$A(\eta,2)$	η	$A(\eta,2)$
0.0	0.0	716.0	0.97203
71.6	0.9304	787.6	0.97220
143.2	0.9574	859.2	0.97233
214.8	0.9650	930.8	0.97243
286.4	0.9681	1002.4	0.97250
358.0	0.9696	1074.0	0.97256
429.6	0.97054	1145.6	0.97262
501.2	0.97110	1217.2	0.97265
572.8	0.97152	1288.8	0.97268
644.4	0.97180	1360.4	0.97271

- we define the interval $\langle a,\ b\rangle$ assuming $a = 0$ and finding b such that $f(b) > 0$,
- we use the recurrent formula

$$\eta_0 = a,$$

$$\eta_{k+1} = \eta_k - \frac{f(\eta_k)}{f(b) - f(\eta_k)}(b - \eta_k) \text{ for } k = 0, 1, \ldots, K,$$

where K is such that

$$f(\eta_{K+1}) < \varepsilon$$

and $\varepsilon = 0.001$ is the measure of the method of secants accuracy,
- we fix the optimal value $\dot{\eta}$ of the preventive maintenance period η assuming

$$\dot{\eta} = \eta_{K+1}.$$

As a result of computer calculations, we recognize that there is no optimal value $\dot{\eta}$ of the exemplary system preventive maintenance period η that maximizes the value of its availability coefficient. The values of the system preventive maintenance period η defined by (6.180) and the corresponding values of the availability coefficient of the exemplary system are given in Table 6.1.

6.4.4.2 Corrective and Preventive Maintenance Policy Minimizing Exemplary Technical System Renovation Cost

To optimize the exemplary system corrective and preventive maintenance policy minimizing its cost of renovation, we use its following reliability and operation cost parameters:

- the system and components critical reliability state $r = 2$,
- the 2nd coordinate $\boldsymbol{R}(t,\ 2)$ of the system unconditional reliability function $\boldsymbol{R}(t,\ \cdot)$ given by (3.178),

- the derivative $\boldsymbol{R}'(t, 2)$ of the 2nd coordinate of the system unconditional reliability function $\boldsymbol{R}(t, \cdot)$ given by (3.179),
- the mean value of the cost of the exemplary system corrective renovation $c_0(1) = 1000$ PLN,
- the mean value of the cost of the exemplary system preventive renovation $c_1(1) = 800$ PLN,
- the measure of the method of secants accuracy $\varepsilon = 0.001$,
- the number of values of the system age ς for which we find the values of the system renovation cost in cases when there is no optimal value $\kappa = 20$,
- the values of the system age ς for which we find the values of the system renovation cost in cases when there is no optimal value

$$\varsigma_i = i\,0.2\mu(2) = i\,0.2 \times 358, \quad i = 1, 2, \ldots, 20 \tag{6.181}$$

where $\mu(2) \cong 358$, is determined by (3.157).

After fixing the above system reliability and operation cost input parameters, we use the procedure described in Sect. 6.3.2.

Since

$c_0(2) = 1000 > c_1(2) = 800$,

we are looking for the optimal value $\dot{\varsigma}$ of the system age ς at which the system preventive renovation is performed that minimizes the system renovation cost per unit time $C(\varsigma, 2)$ given by (6.62) in Sect. 6.3.2 by determining, if it exists, its approximate value from Eq. 6.65 by applying the method of secants in the interval $\langle a,\ b\rangle$ as follows:

- we define, obtained after the transformation of Eq. 6.65, the function

$$\psi(\varsigma) = \boldsymbol{R}^2(\varsigma, 2) - \varsigma \boldsymbol{R}'(\varsigma, 2)\boldsymbol{R}(\varsigma, 2) - \frac{c_0(2)\boldsymbol{R}(\varsigma, 2) - c_1(2)\boldsymbol{R}'(\varsigma, 2)L(\varsigma, 2)}{c_0(2) - c_1(2)}$$

for $\varsigma > 0$,

where $\boldsymbol{R}(t, 2)$ is given by (6.178) and

$$L(\varsigma, 2) = \frac{-\varsigma \boldsymbol{R}(\varsigma, 2)] + \int_0^{\varsigma} \boldsymbol{R}(\varsigma, 2)\mathrm{d}t}{1 - \boldsymbol{R}(\varsigma, 2)},$$

- we define the interval $\langle a,\ b\rangle$ assuming $a = 0$ and finding b such that

$$\psi(b) > 0,$$

- we use the recurrent formula

$$\varsigma_0 = a,$$
$$\varsigma_{k+1} = \varsigma_k - \frac{\psi(\eta_k)}{\psi(b) - \psi(\eta_k)}(b - \varsigma_k) \quad \text{for } k = 0, 1, \ldots, K,$$

where K is such that

Table 6.2 The values of the renovation cost of the port oil pipeline transportation system

ς	$C(\varsigma, 1)$	ς	$C(\varsigma, 1)$
71.6	11.5295	787.6	2.92419
143.2	6.27350	859.2	2.89513
214.8	4.68207	930.8	2.87310
286.4	3.96252	1002.4	2.85626
358.0	3.57631	1074.0	2.84328
429.6	3.34259	1145.6	2.83309
501.2	3.19124	1217.2	2.82536
572.8	3.08762	1288.8	2.81921
644.4	3.01558	1360.4	2.81438
716.0	2.96299	1432.0	2.81056

$$\psi(\varsigma_{K+1}) < \varepsilon$$

and ε is the measure of the method of secants accuracy,

- we fix the optimal value $\dot{\varsigma}$ of the system age ς assuming

$$\dot{\varsigma} = \varsigma_{K+1},$$

As a result of computer calculations, we recognize that there is no optimal value $\dot{\varsigma}$ of the exemplary system age ς at which the system preventive renovation is performed that minimizes the system renovation cost. The exemplary values of the system age ς at which the system preventive renovation is performed defined by (6.181) and the corresponding values of the renovation cost of the exemplary system are given in Table 6.2.

6.4.5 *Corrective and Preventive Maintenance Policy of Port Oil Transportation System*

6.4.5.1 Corrective and Preventive Maintenance Policy Maximizing Port Oil Transportation System Availability

To optimize the port oil pipeline transportation system corrective and preventive maintenance policy maximizing its availability, we use its following reliability and renewal parameters:

- the oil pipeline system and its components' critical reliability state $r = 1$,
- the 1st coordinate of the oil pipeline system unconditional reliability function $\boldsymbol{R}(t, \cdot)$

$$\begin{aligned}\boldsymbol{R}(t,1) = {} & 0.395 \times [\boldsymbol{R}(t,1)]^{(1)} + 0.060 \times [\boldsymbol{R}(t,1)]^{(2)} + 0.003 \times [\boldsymbol{R}(t,1)]^{(3)} \\ & + 0.002 \times [\boldsymbol{R}(t,1)]^{(4)} + 0.2 \times [\boldsymbol{R}(t,1)]^{(5)} + 0.058 \times [\boldsymbol{R}(t,1)]^{(6)} \\ & + 0.282 \times [\boldsymbol{R}(t,1)]^{(7)} \quad \text{for } t \geq 0, \qquad (6.182)\end{aligned}$$

where $[\boldsymbol{R}(t,1)]^{(b)}$, $b = 1, 2, \ldots, 7$, are respectively given by (3.167), (3.173), (3.185), (3.200), (3.206), (3.212), (3.218).

- the derivative of the 1st coordinate of the oil pipeline system unconditional reliability function $\boldsymbol{R}(t, \cdot)$

$$\begin{aligned}\boldsymbol{R}'(t,1) = 0.395 \times [\boldsymbol{R}'(t,1)]^{(1)} + 0.060 \times [\boldsymbol{R}'(t,1)]^{(2)} + 0.003 \times [\boldsymbol{R}'(t,1)]^{(3)} \\ + 0.002 \times [\boldsymbol{R}'(t,1)]^{(4)} + 0.2 \times [\boldsymbol{R}'(t,1)]^{(5)} + 0.058 \times [\boldsymbol{R}'(t,1)]^{(6)} \\ + 0.282 \times [\boldsymbol{R}'(t,1)]^{(7)} \quad \text{for } t \geq 0, \end{aligned} \tag{6.183}$$

- the mean value of the oil pipeline system corrective renovation time $\mu_0(1) = 0.005$,
- the mean value of the oil pipeline system preventive renovation time $\mu_1(1) = 0.003$.

Moreover, to apply the algorithm proposed in Sect. 6.3.1, we fix:

- the measure of the method of secants accuracy $\varepsilon = 0.001$,
- the number of values of the system preventive maintenance period η for which we find the values of the availability coefficient of the oil pipeline system in cases when there is no optimal value $\kappa = 20$,
- the values of the system preventive maintenance period η for which we find the values of the availability coefficient of the oil pipeline system in cases when there is no optimal value

$$\eta_i = (i-1)\, 0.2\mu(1) = (i-1)\, 0.2 \times 0.370, \quad i = 1, 2, \ldots, 20, \tag{6.184}$$

where $\mu(1) \cong 0.370$, is determined by (3.226).

After fixing the above reliability and renewal input parameters, we use the procedure described in Sect. 6.3.1.

Since

$$\mu_0(1) = 0.005 > \mu_1(1) = 0.003,$$

we are looking for the optimal value $\dot{\eta}$ of the preventive maintenance period η that maximizes the availability coefficient of the system $A(\eta, 1)$ given by (6.52) in Sect. 6.3.1 by determining, if it exists, its approximate value from Eq. 6.58 by applying the method of secants in the interval $\langle a, b \rangle$ as follows:

- we define, obtained after the transformation of Eq. 6.58, the function

$$\begin{aligned} f(\eta) &= \lambda(\eta,1) \int_0^\eta \boldsymbol{R}(t,1)\mathrm{d}t + \boldsymbol{R}(\eta,1) - \frac{\mu_0(1)}{\mu_0(1) - \mu_1(1)} \\ &= \lambda(\eta,1) \int_0^\eta \boldsymbol{R}(t,1)\mathrm{d}t + \boldsymbol{R}(\eta,1) - 2.5 \quad \text{for } \eta \geq 0, \end{aligned}$$

Table 6.3 The values of the availability coefficient of the port oil transportation system

η	$A(\eta, 1)$	η	$A(\eta, 1)$
0.0	0.0	0.74	0.98619
0.074	0.95865	0.814	0.98632
0.148	0.97566	0.888	0.98641
0.222	0.98081	0.962	0.98647
0.296	0.98306	1.036	0.98653
0.370	0.98428	1.110	0.98657
0.444	0.98503	1.184	0.98660
0.518	0.98550	1.258	0.98663
0.592	0.98581	1.332	0.98665
0.666	0.98603	1.406	0.98667

where $\boldsymbol{R}(t, 1)$is given by (6.182) and

$$\lambda(\eta, 1) = -\frac{\boldsymbol{R}'(\eta, 1)}{\boldsymbol{R}(\eta, 1)},$$

- we define the interval $\langle a, b \rangle$ assuming $a = 0$ and finding b such that

$$f(b) > 0,$$

- we use the recurrent formula

$$\eta_0 = a,$$
$$\eta_{k+1} = \eta_k - \frac{f(\eta_k)}{f(b) - f(\eta_k)}(b - \eta_k) \quad \text{for } k = 0, 1, \ldots, K,$$

where K is such that

$$f(\eta_{K+1}) < \varepsilon$$

and $\varepsilon = 0.001$ is the measure of the method of secants accuracy,

- we fix the optimal value $\dot{\eta}$ of the preventive maintenance period η assuming

$$\dot{\eta} = \eta_{K+1}.$$

As a result of the computer calculations, we recognize that there is no optimal value $\dot{\eta}$ of the oil pipeline transportation system preventive maintenance period η that maximizes the value of its availability coefficient. The exemplary values of the system preventive maintenance period η defined by (6.184) and the values of the availability coefficient of the exemplary system are given in Table 6.3.

6.4.5.2 Corrective and Preventive Maintenance Policy Minimizing Port Oil Transportation System Renovation Cost

To optimize the port oil pipeline transportation system corrective and preventive maintenance policy minimizing its cost of renovation, we use its following reliability and operation cost parameters:

- the oil pipeline system and components critical reliability state is $r = 1$,
- the 1st coordinate $\boldsymbol{R}(t, 1)$ of the system unconditional reliability function $\boldsymbol{R}(t, \cdot)$ given by (6.182),
- the derivative $\boldsymbol{R}'(t, 1)$ of the 1st coordinate of the system unconditional reliability function $\boldsymbol{R}(t, \cdot)$ given by (6.183),
- the mean value of the cost of the oil pipeline system corrective renovation $c_0(1) = 90000$ PLN,
- the mean value of the cost of the oil pipeline system preventive renovation $c_1(1) = 80000$ PLN,
- the measure of the method of secants accuracy $\varepsilon = 0.001$,
- the number of values of the system age ς for which we find the values of the system renovation cost in the cases when there is no its optimal value $\kappa = 20$,
- the values of the system age ς for which we find the values of the system renovation cost in cases when there is no optimal value

$$\varsigma_i = i\,0.2\mu(1) = i\,0.2 \times 0.370, \quad i = 1, 2, \ldots, 20 \tag{6.185}$$

where $\mu(1) \cong 0.370$ is determined by (3.226).

After fixing the above system reliability and operation cost input parameters, we use the procedure described in Sect. 6.3.2.

Since

$$c_0(2) = 90000 > c_1(2) = 80000,$$

we are looking for the optimal value $\dot{\varsigma}$ of the system age ς at which the system preventive renovation is performed that minimizes the system renovation cost per unit time $C(\varsigma, 1)$ given by (6.62) in Sect. 6.3.2 by determining, if it exists, its approximate value from Eq. 6.65 by applying the method of secants in the interval $\langle a,\ b\rangle$ as follows:

- we define, obtained after the transformation of Eq. 6.65, the function

$$\psi(\varsigma) = \boldsymbol{R}^2(\varsigma, 1) - \varsigma \boldsymbol{R}'(\varsigma, 1)\boldsymbol{R}(\varsigma, 1) - \frac{c_0(1)\boldsymbol{R}(\varsigma, 1) - c_1(1)\boldsymbol{R}'(\varsigma, 1)L(\varsigma, 1)}{c_0(1) - c_1(1)}$$

for $\varsigma > 0$,

where $\boldsymbol{R}(t,\ 1)$ is given by (6.182) and

$$L(\varsigma, 1) = \frac{-\varsigma \boldsymbol{R}(\varsigma, 1)] + \int_0^{\varsigma} \boldsymbol{R}(\varsigma, 1)dt}{1 - \boldsymbol{R}(\varsigma, 1)},$$

Table 6.4 The values of the renovation cost of the port oil piping transportation system

ς	$C(\varsigma, 1)$	ς	$C(\varsigma, 1)$
0.074	12181207	0.814	290550
0.148	2721949	0.888	279928
0.222	1250203	0.962	272036
0.296	785079	1.036	265898
0.37	574767	1.110	261223
0.444	462143	1.184	257615
0.518	395952	1.258	254704
0.592	353887	1.332	252445
0.666	325210	1.406	250608
0.74	305135	1.48	249121

- we define the interval $\langle a,\ b\rangle$ assuming $a = 0$ and finding b such that

$$\psi(b) > 0,$$

- we use the recurrent formula

$$\varsigma_0 = a,$$
$$\varsigma_{k+1} = \varsigma_k - \frac{\psi(\eta_k)}{\psi(b) - \psi(\eta_k)}(b - \varsigma_k) \quad \text{for } k = 0, 1, \ldots, K,$$

where K is such that

$$\psi(\varsigma_{K+1}) < \varepsilon$$

and ε is the measure of the method of secants accuracy,
- we fix the optimal value $\dot{\varsigma}$ of the system age ς assuming $\dot{\varsigma} = \varsigma_{K+1}$.

As a result of the computer calculations, we recognize that there is no optimal value $\dot{\varsigma}$ of the port oil pipeline system age ς at which the system preventive renovation is performed that minimizes the system renovation cost. The exemplary values of the system age ς at which the system preventive renovation is performed defined by (6.185) and the values of the renovation cost of the oil pipeline system are given in Table 6.4.

6.4.6 *Corrective and Preventive Maintenance Policy of Ferry Technical System*

6.4.6.1 Corrective and Preventive Maintenance Policy Maximizing Ferry Technical System Availability

To optimize the maritime ferry technical system corrective and preventive maintenance policy maximizing its availability, we use its following safety and renewal parameters:

- the ferry technical system and components critical safety state $r = 2$,
- the 2nd coordinate of the ferry technical system unconditional safety function $\mathbf{s}(t, \cdot)$

$$\begin{aligned} s(t,2) = {} & 0.038 \times [s(t,2)]^{(1)} + 0.002 \times [s(t,2)]^{(2)} + 0.026 \times [s(t,2)]^{(3)} \\ & + 0.036 \times [s(t,2)]^{(4)} + 0.363 \times [s(t,2)]^{(5)} + 0.026 \times [s(t,2)]^{(6)} \\ & + 0.005 \times [s(t,2)]^{(7)} + 0.016 \times [s(t,2)]^{(8)} + 0.037 \times [s(t,2)]^{(9)} \\ & + 0.002 \times [s(t,2)]^{(10)} + 0.003 \times [s(t,2)]^{(11)} + 0.016 \times [s(t,2)]^{(12)} \\ & + 0.351 \times [s(t,2)]^{(13)} + 0.034 \times [s(t,2)]^{(14)} + 0.024 \times [s(t,2)]^{(15)} \\ & + 0.003 \times [s(t,2)]^{(16)} + 0.005 \times [\bar{s}(t,2)]^{(17)} + 0.013 \times [s(t,2)]^{(18)} \\ & \text{for } t \geq 0, \end{aligned} \tag{6.186}$$

where $[s(t,)]^{(b)}$, $b = 1, 2, \ldots, 18$, are respectively given by (3.248), (3.271), (3.289), (3.312), (3.320), (3.343), (3.351), (3.369), (3.377), (3.385), (3.403), (3.411), (3.420), (3.427), (3.435), (3.443), (3.451), (3.459).

- the derivative of the 2nd coordinate of the ferry technical system unconditional safety function $s(t, \cdot)$

$$\begin{aligned} s'(t,2) = {} & 0.038 \times [s'(t,2)]^{(1)} + 0.002 \times [s'(t,2)]^{(2)} + 0.026 \times [s'(t,2)]^{(3)} \\ & + 0.036 \times [s'(t,2)]^{(4)} + 0.363 \times [s'(t,2)]^{(5)} + 0.026 \times [s'(t,2)]^{(6)} \\ & + 0.005 \times [s'(t,2)]^{(7)} + 0.016 \times [s'(t,2)]^{(8)} + 0.037 \times [s'(t,2)]^{(9)} \\ & + 0.002 \times [s'(t,2)]^{(10)} + 0.003 \times [s'(t,2)]^{(11)} + 0.016 \times [s'(t,2)]^{(12)} \\ & + 0.351 \times [s'(t,2)]^{(13)} + 0.034 \times [s'(t,2)]^{(14)} + 0.024 \times [s'(t,2)]^{(15)} \\ & + 0.003 \times [s'(t,2)]^{(16)} + 0.005 \times [\bar{s}'(t,2)]^{(17)} + 0.013 \times [s'(t,2)]^{(18)} \\ & \text{for } \quad t \geq 0, \end{aligned}$$

- the mean value of the ferry technical system corrective renovation time $\mu_0(2) = 0.019$,
- the mean value of the ferry technical system preventive renovation time $\mu_1(2) = 0.016$.

Moreover, to apply the algorithm proposed in Sect. 6.3.1, we fix:

- the measure of the method of secants accuracy $\varepsilon = 0.001$,
- the number of values of the system preventive maintenance period η for which we find the values of the availability coefficient of the ferry technical system in the cases when there is no optimal value $\kappa = 20$,
- the values of the system preventive maintenance period η for which we find the values of the availability coefficient of the ferry technical system in cases when there is no optimal value

$$\eta_i = (i-1)\ 0.2\mu(2) = (i-1)\ 0.2 \times 4.11, \quad i = 1, 2, \ldots, 20. \tag{6.187}$$

where $\mu(2) \cong 4.11$is determined by (3.472).

After fixing the above safety and renewal input parameters, we use the procedure described in Sect. 6.3.1.

Since

$$\mu_0(2) = 0.019 > \mu_1(2) = 0.016,$$

we are looking for the optimal value $\dot{\eta}$ of the preventive maintenance period η that maximizes the availability coefficient of the system $A(\eta, 2)$ given by (6.52) in Sect. 6.3.1 by determining, if it exists, its approximate value from Eq. 6.58 by applying the method of secants in the interval $\langle a, b \rangle$ as follows:

- we define, obtained after the transformation of the Eq. 6.58, the function

$$f(\eta) = \lambda(\eta, 2) \int_0^\eta \mathbf{s}(t, 2)\mathrm{d}t\, \mathbf{s}(\eta, 2) - \frac{\mu_0(2)}{\mu_0(2) - \mu_1(2)}$$

$$= \lambda(\eta, 2) \int_0^\eta \mathbf{s}(t, 2)\mathrm{d}t \; + \mathbf{s}(\eta, 2) - 6.33 \text{ for } \eta \geq 0,$$

where $\mathbf{s}(t, 2)$ is given by (6.186) and

$$\lambda(\eta, 2) = -\frac{\mathbf{s}'(\eta, 2)}{\mathbf{s}(\eta, 2)},$$

- we define the interval $\langle a, b \rangle$ assuming $a = 0$ and finding b such that

$$f(b) > 0,$$

- we use the recurrent formula

$$\eta_0 = a,$$

$$\eta_{k+1} = \eta_k - \frac{f(\eta_k)}{f(b) - f(\eta_k)}(b - \eta_k) \quad \text{for } k = 0, 1, \ldots, K,$$

where K is such that

$$f(\eta_{K+1}) < \varepsilon$$

and $\varepsilon = 0.001$ is the measure of the method of secants accuracy,
- we fix the optimal value $\dot{\eta}$ of the preventive maintenance period η assuming

$$\dot{\eta} = \eta_{K+1}.$$

As a result of computer calculations, we recognize that there is no optimal value $\dot{\eta}$ of the ferry technical system preventive maintenance period η that maximizes the

Table 6.5 The values of the availability coefficient of the ferry technical system

η	$A(\eta, 2)$	η	$A(\eta, 2)$
0	0.0	8.220	0.99476
0.822	0.97829	9.042	0.99489
1.644	0.98762	9.864	0.994981
2.466	0.99071	10.686	0.99506
3.288	0.99225	1.508	0.99511
4.110	0.99313	12.330	0.99516
4.932	0.99371	13.152	0.99520
5.754	0.99411	13.974	0.99523
6.576	0.99440	14.796	0.99526
7.398	0.99461	15.618	0.99528

value of its availability coefficient. The exemplary values of the system preventive maintenance period η defined by (6.187) and the values of the availability coefficient of the ferry technical system are given in Table 6.5.

6.5 Summary

In this chapter tools useful in reliability, availability and safety optimization and operation cost analysis of a very wide class of real technical systems operating at varying conditions that have an influence on changing their reliability and safety structures and their components' reliability and safety characteristics are presented. The results achieved are interesting for reliability practitioners from the maritime transport industry and from other industrial sectors as well. The testing of the tools performed in the maritime industry for reliability, risk, availability, safety and cost optimization of the port oil piping transportation system and the maritime ferry technical system justifies their practical utility.

References

1. Helvacioglu S, Insel M (2008) Expert system applications in marine technologies. Ocean Eng 35:1067–1074
2. Klabjan D, Adelman D (2006) Existence of optimal policies for semi-Markov decision processes using duality for infinite linear programming. SIAM J Contr Optim 44(6):2104–2122
3. Kolowrocki K, Soszynska J (2009) Reliability, risk and availability based optimization of complex technical systems operation processes. Part 1. Theoretical backgrounds. Electron J Reliab Risk Anal Theory Appl 2(44):141–152
4. Kolowrocki K, Soszynska J (2009) Reliability, risk and availability based optimization of complex technical systems operation processes. Part 2. Application in port transportation. Electron J Reliab Risk Anal Theory Appl 2(4):153–167
5. Kolowrocki K, Soszynska J (2010) Reliability, availability and safety of complex technical systems: modelling–identification–prediction–optimization. Summer safety & reliability seminars. J Pol Saf Reliab Assoc 4(1):133–158

6. Kolowrocki K, Soszynska J (2010) Reliability modeling of a port oil transportation system's operation processes. Int J Perform Eng 6(1):77–87
7. Kolowrocki K, Soszynska J (2010) Safety and risk optimization of a ferry technical system Summer Safety & Reliability Seminars. J Pol Saf Reliab Assoc 4(1):159–172
8. Kuo W, Prasad VR (2000) An annotated overview of system-reliability optimization. IEEE Trans Reliab 49(2):176–187
9. Kuo W, Zuo M J (2003) Optimal reliability modeling: principles and applications. Wiley, Hoboken
10. Levitin G, Lisnianski A (2000) Optimisation of imperfect preventive maintenance for multistate systems. Reliab Eng Sys Saf 67:193–203
11. Levitin G, Lisnianski A (2003) Optimal replacement scheduling in multi-state series-parallel systems. Qual Reliab Eng Int 16:157–162
12. Lisnianski A, Levitin G (2003) Multi-State System Reliability Assessment Optimisation and Applications. World Scientific Publishing Co. Pte. Ltd, Singapore
13. Suich RC, Patterson RL (1991) k-out-of-n:G system: some cost considerations. IEEE Trans Reliab 40(3):259–264
14. Tang H, Yin BQ, Xi HS (2007) Error bounds of optimization algorithms for semi-Markov decision processes. Int J Sys Sci 38(9):725–736
15. Merrick JRW, van Dorp R (2006) Speaking the truth in maritime risk assessment. Risk Anal 26(1):223–237
16. Vercellis S (2009) Data mining and optimization for decision making. John Wiley & Sons Ltd, Indianapolis
17. Zio E (2006) An introduction to the basics of reliability and risk analysis. World Scientific Publishing Co Pte. Ltd, Singapore

Chapter 7
Additional Applications

7.1 Introduction

The applications of some theoretical results of the book are not illustrated in the previous chapters. The direct applications of the results included in the propositions of Chaps. 1 and 4 to much simpler systems than that considered in the appliqué sections of these chapters like series, parallel [3–7, 10] and consecutive "*m* out of *n*: F" [1, 9] systems operating at the variable conditions are not illustrated. The application of the asymptotic approach [2, 5, 10] to the large system reliability analysis [11–13] also is not illustrated in the previous chapters. The complementary applications of the results presented in the book to these systems, reliability and safety analysis are included in this chapter. Moreover, the application of the results of the book to the reliability, renewal and availability identification, prediction and optimization of the container gantry crane [8] is illustrated to give the reader the method of using these tools for an overall reliability or safety analysis of any real complex technical system.

7.2 Operation, Reliability, Availability, Safety and Cost Analysis of Complex Technical Systems

7.2.1 Reliability and Availability of Series System

We consider a series system composed of components E_i, $i = 1, 2, 3, 4$, operating at two operation states z_1 and z_2, i.e. $v = 2$, with the same reliability structure presented in Fig. 7.1.

We arbitrarily assume that the transient probabilities of the system at the particular operation states z_1 and z_2 respectively are

K. Kołowrocki and J. Soszyńska-Budny, *Reliability and Safety of Complex Technical Systems and Processes*, Springer Series in Reliability Engineering,
DOI: 10.1007/978-0-85729-694-8_7,

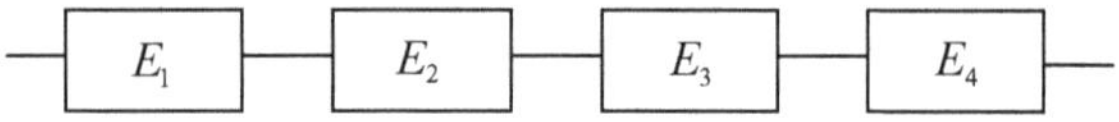

Fig. 7.1 The scheme of the series system reliability structure at the operation states z_1 and z_2

$$p_1 = 0.25, \quad p_2 = 0.75.$$

Moreover, we distinguish four reliability states of the system components 0, 1, 2, 3, i.e. $z = 3$. We fix that the critical reliability state is $r = 2$ and we assume that the components' reliability characteristics are different at the operation states z_1 and z_2. Consequently, we define the four-state conditional reliability functions of the system components E_i, $i = 1, 2, 3, 4$, at the operation state z_1 in the form of the vector

$$[R_i(t,\cdot)]^{(1)} = [1,\ [R_i(t,1)]^{(1)},\ [R_i(t,2)]^{(1)},\ [R_i(t,3)]^{(1)}], \quad i = 1,2,3,4,$$

with the exponential coordinates

$$[R_i(t,u)]^{(1)} = \exp[-2ut],\ u = 1,2,3, \quad i = 1,2,3,4.$$

and the four-state conditional reliability functions of the system components E_i, $i = 1, 2, 3, 4$, at the operation state z_2 in the form of the vector

$$[R_i(t,\cdot)]^{(2)} = [1,\ [R_i(t,1)]^{(2)},\ [R_i(t,2)]^{(2)},\ [R_i(t,3)]^{(2)}],\ i = 1,2,3,4,$$

with the exponential coordinates

$$[R_i(t,u)]^{(2)} = \exp[-ut],\ u = 1,2,3,\ i = 1,2,3,4.$$

After direct application of the formula (3.18) stated in *Proposition 3.1*, we get the unconditional system reliability function

$$\boldsymbol{R}(t,\cdot) = [1,\ \boldsymbol{R}(t,1),\ \boldsymbol{R}(t,2),\ \boldsymbol{R}(t,3)],\ t \geq 0, \tag{7.1}$$

where

$$\begin{aligned}\boldsymbol{R}(t,u) = \bar{\boldsymbol{R}}_4(t,u) &= \sum_{b=1}^{2} p_b \exp\left[-\sum_{i=1}^{4}[\lambda_i(u)]^{(b)}t\right]\\ &= p_1\exp[-4\cdot 2ut] + p_2\exp[-4\cdot ut] = 0.25\exp[-8ut]\\ &\quad + 0.75\exp[-4ut],\ u = 1,2,3,\end{aligned}$$

and particularly (Fig. 7.2)

$$\boldsymbol{R}(t,1) = 0.25\exp[-8t] + 0.75\exp[-4t], \tag{7.2}$$

$$\boldsymbol{R}(t,2) = 0.25\exp[-16t] + 0.75\exp[-8t], \tag{7.3}$$

$$\boldsymbol{R}(t,3) = 0.25\exp[-24t] + 0.75\exp[-12t]. \tag{7.4}$$

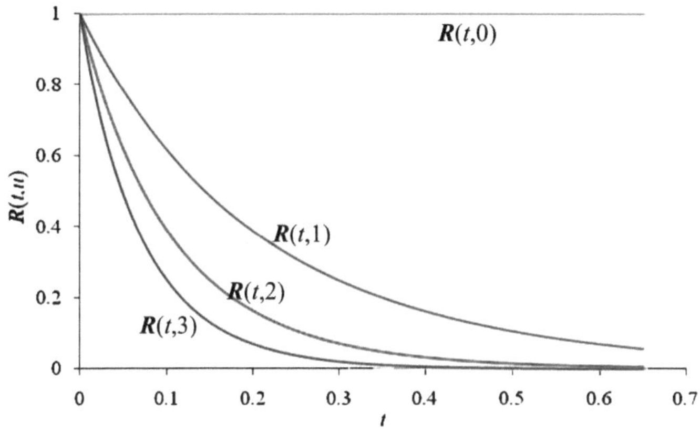

Fig. 7.2 The graph of the series system reliability function ***R***(t, •) coordinates

The expected values and standard deviations of the system unconditional lifetimes in the reliability state subsets {1, 2, 3}, {2, 3}, {3}, calculated from the results given by (7.2–7.4), according to (3.7–3.9), respectively are:

$$\mu(1) = 0.25\frac{1}{8} + 0.75\frac{1}{4} \cong 0.219,\ \sigma(1) \cong 0.232, \tag{7.5}$$

$$\mu(2) = 0.25\frac{1}{16} + 0.75\frac{1}{8} \cong 0.109,\ \sigma(2) \cong 0.116, \tag{7.6}$$

$$\mu(3) = 0.25\frac{1}{24} + 0.75\frac{1}{12} \cong 0.073,\ \sigma(3) \cong 0.077. \tag{7.7}$$

Consequently, considering (3.10) and (7.5–7.7), the mean values of the system unconditional lifetimes in the particular reliability states 1, 2, 3, respectively are:

$$\bar{\mu}(1) = \mu(1) - \mu(2) = 0.110,\ \bar{\mu}(2) = \mu(2) - \mu(3) = 0.036,$$
$$\bar{\mu}(3) = \mu(3) = 0.073. \tag{7.8}$$

Since the critical reliability state is $r = 2$, the system risk function, according to (3.11), is given by

$$\boldsymbol{r}(t) = 1 - \boldsymbol{R}(t,2) = 1 - 0.25\exp[-16t] - 0.75\exp[-8t] \quad \text{for } t \geq 0. \tag{7.9}$$

Hence, by (3.12), the moment when the system risk function exceeds a permitted level, for instance $\delta = 0.05$, is (Fig. 7.3)

$$\tau = \boldsymbol{r}^{-1}(\delta) \cong 0.00515. \tag{7.10}$$

Further, assuming that the system is repaired after the exceeding its critical reliability state $r = 2$ and that the time of the system renovation is ignored and applying *Proposition 3.4*, we obtain the following results:

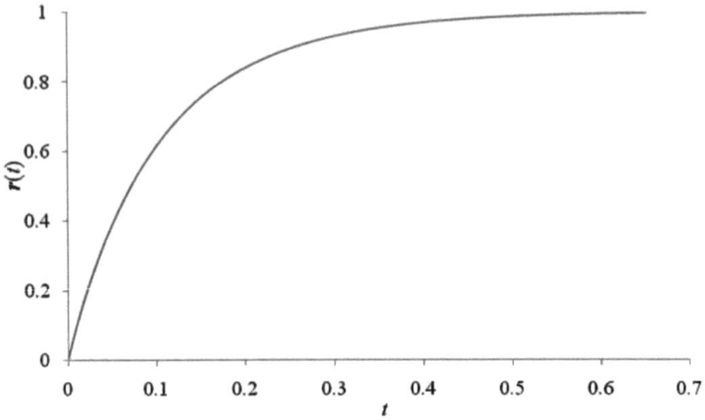

Fig. 7.3 The graph of the series system risk function $\boldsymbol{r}(t)$

(a) the time $S_N(2)$ until the Nth exceeding by the system the reliability critical state $r = 2$, for sufficiently large N, has approximately normal distribution $N(0.109\,N, 0.116\sqrt{N})$, i.e.,

$$F^{(N)}(t,2) = P(S_N(2) < t) \cong F_{N(0,1)}\left(\frac{t - 0.109\,N}{0.116\sqrt{N}}\right), \quad t \in (-\infty, \infty);$$

(b) the expected value and the variance of the time $S_N(2)$ until the Nth exceeding by the system the reliability critical state $r = 2$ are respectively given by

$$E[S_N(2)] \cong 0.109\,N, \; D[S_N(2)] \cong 0.0135\,N;$$

(c) the number $N(t, 2)$ of exceedings by the system the reliability critical state $r = 2$ up to the moment t, $t \geq 0$, for sufficiently large t, approximately has the distribution of the form

$$P(N(t,2) = N) \cong F_{N(0,1)}\left(\frac{0.109(N+1) - t}{0.351\sqrt{t}}\right) - F_{N(0,1)}\left(\frac{0.109\,N - t}{0.351\sqrt{t}}\right),$$
$$N = 0, 1, \ldots;$$

(d) the expected value and the variance of the number $N(t, 2)$ of exceedings by the system the reliability critical state $r = 2$ up to the moment t, $t \geq 0$, for sufficiently large t, approximately are respectively given by

$$H(t,2) \cong 9.174t, \; D(t,2) \cong 10.424t.$$

Assuming that the system is repaired after exceeding its reliability critical state $r = 2$ and that the time of the system renovation is non-ignored and it has the mean value $\mu_0(2) = 0.05$ and the standard deviation $\sigma_0(2) = 0.02$ and applying *Proposition 3.5*, we obtain the following results:

(a) the time $\bar{S}_N(2)$ until the Nth exceeding by the system the reliability critical state $r = 2$, for sufficiently large N, has approximately normal distribution $N(0.109\,N + 0.05(N-1), \sqrt{0.0135\,N + 0.0004(N-1)})$, i.e.,

$$\bar{F}^{(N)}(t,2) = P(\bar{S}_N(2) < t) \cong F_{N(0,1)}\left(\frac{t - 0.159\,N + 0.05}{\sqrt{0.0139\,N - 0.0004}}\right), \quad t \in (-\infty, \infty);$$

(b) the expected value and the variance of the time $\bar{S}_N(2)$ until the Nth exceeding by the system the reliability critical state $r = 2$, for sufficiently large N, are respectively given by

$$E[\bar{S}_N(2)] \cong 0.109\,N + 0.05(N-1),\ D[\bar{S}_N(2)] \cong 0.0135\,N + 0.0004(N-1);$$

(c) the number $\bar{N}(t,2)$ of exceedings by the system the reliability critical state $r = 2$ up to the moment t, $t \geq 0$, for sufficiently large t, has approximately distribution of the form

$$\begin{aligned} P(\bar{N}(t,2) = N) &\cong F_{N(0,1)}\left(\frac{0.159(N+1) - t - 0.05}{0.296\sqrt{t + 0.05}}\right) \\ &\quad - F_{N(0,1)}\left(\frac{0.159\,N - t - 0.05}{0.296\sqrt{t + 0.05}}\right), \\ N &= 0, 1, \ldots; \end{aligned}$$

(d) the expected value and the variance of the number $\bar{N}(t,2)$ of exceedings by the system the reliability critical state $r = 2$ up to the moment t, $t \geq 0$, for sufficiently large t, are respectively given by

$$\bar{H}(t,2) \cong \frac{t + 0.05}{0.159},\ \bar{D}(t,2) \cong 3.458(t + 0.05);$$

(e) the time $\bar{\bar{S}}_N(2)$ until the Nth system's renovation, for sufficiently large N, has approximately normal distribution $N(0.159\,N, 0.118\sqrt{N})$, i.e.,

$$\bar{\bar{F}}^{(N)}(t,2) = P(\bar{\bar{S}}_N(2) < t) \cong F_{N(0,1)}\left(\frac{t - 0.159\,N}{0.118\sqrt{N}}\right),\ t \in (-\infty, \infty);$$

(f) the expected value and the variance of the time $\bar{\bar{S}}_N(2)$ until the Nth system's renovation, for sufficiently large N, are respectively given by

$$E[\bar{\bar{S}}_N(2)] \cong 0.159\,N,\ D[\bar{\bar{S}}_N(2)] \cong 0.0139\,N;$$

(g) the number $\bar{\bar{N}}(t,2)$ of the system's renovations up to the moment t, $t \geq 0$, for sufficiently large t, has approximately distribution of the form

$$P(\bar{\bar{N}}(t,2) = N) \cong F_{N(0,1)}\left(\frac{0.159(N+1) - t}{0.296\sqrt{t}}\right) - F_{N(0,1)}\left(\frac{0.159N - t}{0.296\sqrt{t}}\right),$$
$$N = 0, 1, \ldots;$$

(h) the expected value and the variance of the number $\bar{\bar{N}}(t,2)$ of system's renovations up to the moment t, $t \geq 0$, for sufficiently large t, are respectively given by

$$\bar{\bar{H}}(t,2) \cong 6.289t, \quad \bar{\bar{D}}(t,2) \cong 3.458t;$$

(i) the steady availability coefficient of the system at the moment t, $t \geq 0$, for sufficiently large t, is given by

$$A(t,2) \cong 0.686, \; t \geq 0;$$

(j) the steady availability coefficient of the system in the time interval $\langle t, t+\tau)$, $\tau\rangle 0$, for sufficiently large t, is given by

$$A(t,\tau,2) \cong 6.289 \int_{\tau}^{\infty} \boldsymbol{R}(t,2)dt, \quad t \geq 0, \; \tau > 0,$$

where $\boldsymbol{R}$(t, 2) is given by (7.3).

7.2.2 Reliability and Availability of Parallel System

We consider a parallel system composed of components E_i, $i = 1, 2, 3, 4$, operating at two operation states z_1 and z_2, i.e. $v = 2$. We assume that the system reliability structure and the system components' reliability characteristics are changing at the various operation states. At the operation state z_1 the system is composed of two components E_i, $i = 1, 2$, with the reliability structure presented in Fig. 7.4.

At the operation state z_2 the system is composed of four components E_i, $i = 1, 2, 3, 4$, with the reliability structure illustrated in Fig. 7.5.

We arbitrarily assume that the transient probabilities of the system at the particular operation states z_1 and z_2 respectively are

$$p_1 = 0.25, \quad p_2 = 0.75.$$

Moreover, we distinguish three reliability states of the system components 0, 1, 2, i.e. $z = 2$, and we fix that the critical reliability state is $r = 1$. Consequently, we define the three-state reliability functions of the system components E_i, $i = 1, 2$, at the operation state z_1 in the form of the vector

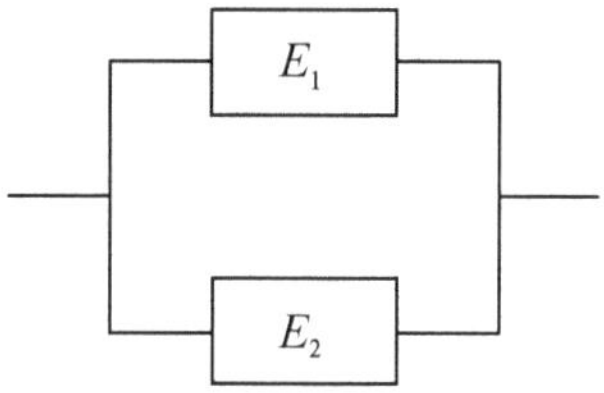

Fig. 7.4 The scheme of the parallel system reliability structure at the operation state z_1

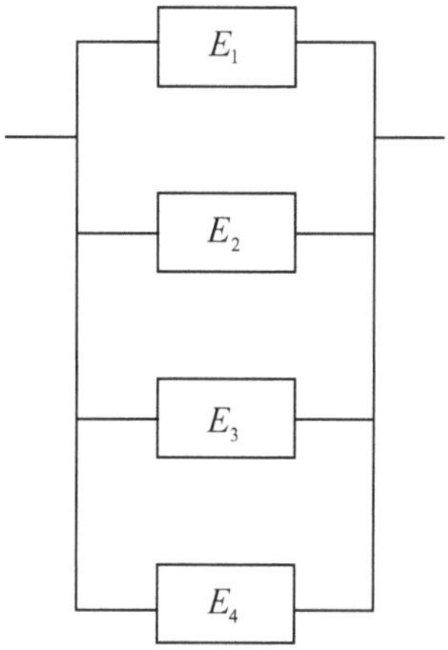

Fig. 7.5 The scheme of the parallel system reliability structure at the operation state z_2

$$[R_i(t,\ \cdot)]^{(1)} = [1,\ [R_i(t,\ 1)]^{(1)},\ [R_i(t,\ 2)]^{(1)}], \quad i = 1, 2,$$

with the exponential coordinates

$$[R_i(t,\ u)]^{(1)} = \exp[-2ut], \quad u = 1, 2, \quad i = 1, 2,$$

and the three-state reliability functions of the system components E_i, $i = 1, 2$, at the operation state z_2 in the form of the vector

$$[R_i(t,\ \cdot)]^{(2)} = [1,\ [R_i(t,\ 1)]^{(2)},\ [R_i(t,\ 2)]^{(2)}], \quad i = 1, 2, 3, 4,$$

with the exponential coordinates

$$[R_i(t,\ u)]^{(2)} = \exp[-ut], \quad u = 1, 2, \quad i = 1, 2, 3, 4.$$

After considering *Remark 3.1* given at the end of Sect. 3.2 and direct application of the formula (3.20) stated in *Proposition 3.1*, we get the system reliability function

$$\boldsymbol{R}(t,\ \cdot) = [1,\ \boldsymbol{R}(t, 1),\ \boldsymbol{R}(t, 2)],\ t \geq 0, \tag{7.11}$$

where

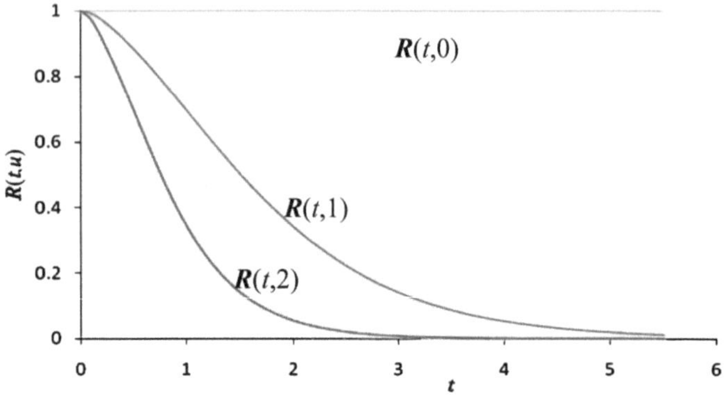

Fig. 7.6 The graph of the parallel system reliability function $\boldsymbol{R}(t,\cdot)$ coordinates

$$
\begin{aligned}
\boldsymbol{R}(t,u) &= 1 - \sum_{b=1}^{2} p_b \prod_{i=1}^{n^{(b)}} [1 - \exp[-[\lambda_i(u)]^{(b)} t]] \\
&= 1 - p_1[1 - \exp[-2ut]]^2 - p_2[1 - \exp[-ut]]^4 \\
&= 1 - 0.25[1 - 2\exp[-2ut] + \exp[-4ut]] \\
&\quad - 0.75[1 - 4\exp[-ut] + 6\exp[-2ut] - 4\exp[-3ut] + \exp[-4ut]] \\
&= 0.25[2\exp[-2ut] - \exp[-4ut]] + 0.75[4\exp[-ut] - 6\exp[-2ut] \\
&\quad + 4\exp[-3ut] - \exp[-4ut]] \\
&= 3\exp[-ut] - 4\exp[-2ut] + 3\exp[-3ut] - \exp[-4ut], \quad u = 1, 2,
\end{aligned}
$$

and particularly (Fig. 7.6)

$$\boldsymbol{R}(t,1) = 3\exp[-t] - 4\exp[-2t] + 3\exp[-3t] - \exp[-4t], \tag{7.12}$$

$$\boldsymbol{R}(t,2) = 3\exp[-2t] - 4\exp[-4t] + 3\exp[-6t] - \exp[-8t]. \tag{7.13}$$

The expected values and standard deviations of the system unconditional lifetimes in the reliability state subsets {1,2}, {2}, calculated from the results given by (7.12–7.13), according to (3.7–3.9), respectively are:

$$\mu(1) = 3\frac{1}{1} - 4\frac{1}{2} + 3\frac{1}{3} - 1\frac{1}{4} \cong 1.750,\ \sigma(1) \cong 1.070, \tag{7.14}$$

$$\mu(2) = 3\frac{1}{2} - 4\frac{1}{4} + 3\frac{1}{6} - 1\frac{1}{8} \cong 0.875,\ \sigma(2) \cong 0.608, \tag{7.15}$$

and further, considering (3.10) and (7.14–7.15), the mean values of the unconditional lifetimes in the particular reliability states 1, 2, respectively are:

$$\bar{\mu}(1) = \mu(1) - \mu(2) = 0.875, \quad \bar{\mu}(2) = \mu(2) = 0.875. \tag{7.16}$$

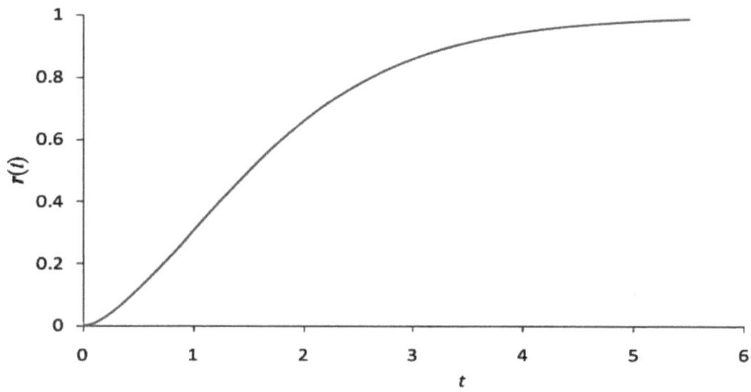

Fig. 7.7 The graph of the parallel system risk function $\boldsymbol{r}(t)$

Since the critical reliability state is $r = 1$, the system risk function, according to (3.11), is given by (Fig. 7.7)

$$\boldsymbol{r}(t) = 1 - \boldsymbol{R}(t,1)$$
$$= 1 - [3\exp[-t] - 4\exp[-2t] + 3\exp[-3t] - \exp[-4t]] \quad \text{for}\, t \geq 0. \qquad (7.17)$$

Hence, by (3.12), the moment when the system risk function exceeds a permitted level, for instance $\delta = 0.05$, is

$$\tau = \boldsymbol{r}^{-1}(\delta) \cong 0.285. \qquad (7.18)$$

Further, assuming that the system is repaired after the exceeding its reliability critical state $r = 1$ and that the time of the system renovation is ignored and applying *Proposition 3.4*, we obtain the following results:

(a) the time $S_N(1)$ until the Nth exceeding by the system the reliability critical state $r = 1$, for sufficiently large N, has approximately normal distribution $N(1.750\,N, 1.070\sqrt{N})$, i.e.,

$$F^{(N)}(t,1) = P(S_N(1) < t) \cong F_{N(0,1)}\left(\frac{t - 1.750\,N}{1.070\sqrt{N}}\right),\ t \in (-\infty, \infty);$$

(b) the expected value and the variance of the time S_N (1) until the Nth exceeding by the system the reliability critical state $r = 1$ are respectively given by

$$E[S_N(1)] \cong 1.750\,N, \quad D[S_N(1)] \cong 1.1449\,N;$$

(c) the number $N(t, 1)$ of exceedings by the system the reliability critical state $r = 1$ up to the moment t, $t \geq 0$, for sufficiently large t, approximately has the distribution of the form

$$P(N(t,1) = N) \cong F_{N(0,1)}\left(\frac{1.750(N+1) - t}{0.809\sqrt{t}}\right) - F_{N(0,1)}\left(\frac{1.750\,N - t}{0.809\sqrt{t}}\right),$$
$$N = 0, 1, \ldots;$$

(d) the expected value and the variance of the number $N(t, 1)$ of exceedings by the system the reliability critical state $r = 1$ up to the moment t, $t \geq 0$, for sufficiently large t, approximately are respectively given by

$$H(t,1) \cong 0.571t, \; D(t,1) \cong 0.214t.$$

Assuming that the system is repaired after the exceeding of its reliability critical state $r = 1$ and that the time of the system renovation is not ignored and it has the mean value $\mu_0(1) = 0.10$ and the standard deviation $\sigma_0(1) = 0.05$ and applying *Proposition 3.5*, we obtain the following results:

(a) the time $\bar{S}_N(1)$ until the Nth exceeding by the system the reliability critical state $r = 1$, for sufficiently large N, has approximately normal distribution $N(1.750N + 0.1(N-1), \sqrt{1.1449N + 0.0025(N-1)})$, i.e.,

$$\bar{F}^{(N)}(t,1) = P(\bar{S}_N(1) < t) \cong F_{N(0,1)}\left(\frac{t - 1.850N + 0.1}{\sqrt{1.1474N - 0.0025}}\right), \quad t \in (-\infty, \infty);$$

(b) the expected value and the variance of the time $\bar{S}_N(1)$ until the Nth exceeding by the system the reliability critical state $r = 1$, for sufficiently large N, are respectively given by

$$E[\bar{S}_N(1)] \cong 1.750N + 0.10(N-1), \quad D[\bar{S}_N(1)] \cong 1.1449N + 0.0025(N-1);$$

(c) the number $\bar{N}(t,1)$ of exceeding by the system the reliability critical state $r = 1$ up to the moment t, $t \geq 0$, for sufficiently large t, has approximately distribution of the form

$$\begin{aligned} P(\bar{N}(t,1) = N) \\ &\cong F_{N(0,1)}\left(\frac{1.850(N+1) - t - 0.10}{0.786\sqrt{t + 0.10}}\right) \\ &\quad - F_{N(0,1)}\left(\frac{1.850N - t - 0.10}{0.786\sqrt{t + 0.10}}\right), \\ N = 0, 1, \ldots; \end{aligned}$$

(d) the expected value and the variance of the number $\bar{N}(t,1)$ of exceeding by the system the reliability critical state $r = 1$ up to the moment t, $t \geq 0$, for sufficiently large t, are respectively given by

$$\bar{H}(t,1) \cong \frac{t + 0.10}{1.85}, \quad \bar{D}(t,1) \cong 0.181(t + 0.10);$$

(e) the time $\bar{\bar{S}}_N(1)$ until the Nth system's renovation, for sufficiently large N, has approximately normal distribution $N(1.850N, 1.071\sqrt{N})$, i.e.,

$$\bar{\bar{F}}^{(N)}(t,1) = P(\bar{\bar{S}}_N(1) < t) \cong F_{N(0,1)}\left(\frac{t - 1.850N}{1.071\sqrt{N}}\right), \quad t \in (-\infty, \infty);$$

(f) the expected value and the variance of the time $\bar{\bar{S}}_N(1)$ until the Nth system's renovation, for sufficiently large N, are respectively given by

$$E[\bar{\bar{S}}_N(1)] \cong 1.850N, \; D[\bar{\bar{S}}_N(1)] \cong 1.1474N;$$

(g) the number $\bar{\bar{N}}(t,1)$ of the system's renovations up to the moment t, $t \geq 0$, for sufficiently large t, has approximately distribution of the form

$$P(\bar{\bar{N}}(t,1) = N) \cong F_{N(0,1)}\left(\frac{1.850(N+1) - t}{0.786\sqrt{t}}\right) - F_{N(0,1)}\left(\frac{1.850N - t}{0.786\sqrt{t}}\right),$$
$$N = 0, 1, \ldots;$$

(h) the expected value and the variance of the number $\bar{\bar{N}}(t,1)$ of system's renovations up to the moment t, $t \geq 0$, for sufficiently large t, are respectively given by

$$\bar{\bar{H}}(t,1) \cong 0.541t, \; \bar{\bar{D}}(t,1) \cong 0.181t;$$

(i) the steady availability coefficient of the system at the moment t, $t \geq 0$, for sufficiently large t, is given by

$$A(t,1) \cong 0.946, \; t \geq 0;$$

(j) the steady availability coefficient of the system in the time interval $\langle t, t+\tau)$, $\tau\rangle 0$, for sufficiently large t, is given by

$$A(t,\tau,1) \cong 0.541 \int_{\tau}^{\infty} \boldsymbol{R}(t,1)\mathrm{d}t, \; t \geq 0, \; \tau > 0,$$

where $\boldsymbol{R}$(t, 1) is given by (7.12).

7.2.3 Reliability of Large Parallel-Series System

We consider a parallel-series system composed of components E_{ij}, $i = 1, 2, 3$, $j = 1, 2, \ldots, 36$, operating at three operation states z_1, z_2 and z_3, i.e. $v = 3$. We assume that the system reliability structure and the system components' reliability characteristics are changing at the various operation states.

At the operation state z_1 the system is composed of one parallel subsystem composed of components E_{ij}, $i = 1, j = 1, 2, \ldots, 36$, with the reliability structure presented in Fig. 7.8.

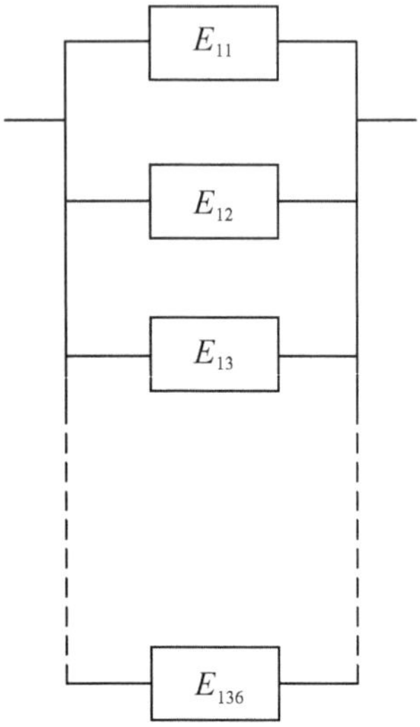

Fig. 7.8 The scheme of the parallel-series system reliability structure at the operation state z_1

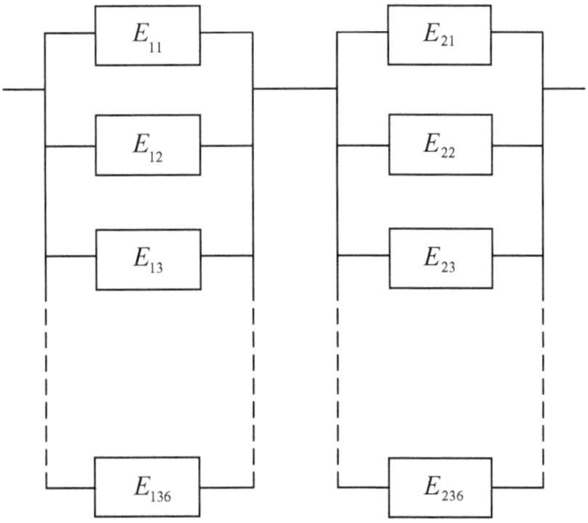

Fig. 7.9 The scheme of the parallel-series system reliability structure at the operation state z_2

At the operation state z_2 the system is composed of two parallel subsystems linked in series and composed of components E_{ij}, $i = 1, 2$, $j = 1, 2, \ldots, 36$, with the reliability structure presented in Fig. 7.9.

At the operation state z_3, the system is composed of three parallel subsystems linked in series and composed of components E_{ij}, $i = 1, 2, 3$, $j = 1, 2, \ldots, 36$, with the reliability structure presented in Fig. 7.10.

We arbitrarily assume that the transient probabilities of the system at particular operation states z_1, z_2 and z_3 respectively are

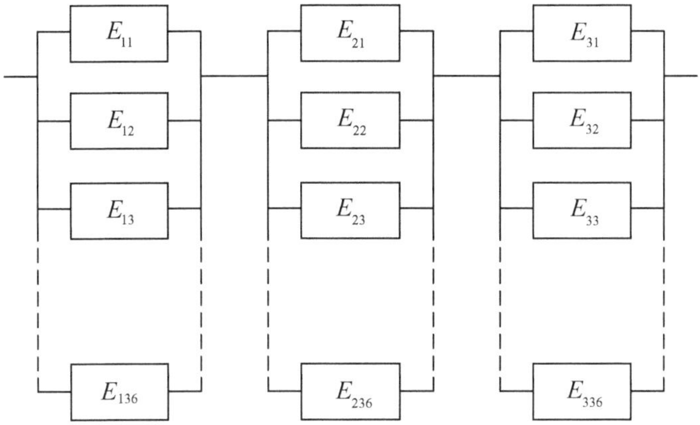

Fig. 7.10 The scheme of the parallel-series system structure at the operation state z_3

$$p_1 = 0.2, \quad p_2 = 0.4, \quad p_3 = 0.4.$$

Moreover, we distinguish three reliability states of the system components 0, 1, 2, i.e. $z = 2$, and we fix that the critical reliability state is $r = 1$. Consequently, we define the three-state reliability functions of the system components E_{ij}, $i = 1$, $j = 1, 2, \ldots, 36$, at the operation state z_1 in the form of the vector

$$[R_{ij}(t, \cdot)]^{(1)} = [1, [R_{ij}(t, 1)]^{(1)}, [R_{ij}(t, 2)]^{(1)}], \quad i = 1, \quad j = 1, 2, \ldots, 36,$$

with the exponential coordinates

$$[R_{ij}(t, u)]^{(1)} = \exp[-ut], u = 1, 2, i = 1, j = 1, 2, \ldots, 36,$$

and the three-state reliability functions of the system components E_{ij}, $i = 1, 2$, $j = 1, 2, \ldots, 36$, at the operation state z_2 in the form of the vector

$$[R_{ij}(t, \cdot)]^{(2)} = [1, [R_{ij}(t, 1)]^{(2)}, [R_{ij}(t, 2)]^{(2)}], i = 1, 2, j = 1, 2, \ldots, 36,$$

with the exponential coordinates

$$[R_{ij}(t, u)]^{(2)} = \exp[-2ut], \quad u = 1, 2, \quad i = 1, 2, \quad j = 1, 2, \ldots, 36,$$

and the three-state reliability functions of the system components E_{ij}, $i = 1, 2, 3$, $j = 1, 2, \ldots, 36$, at the operation state z_3 in the form of the vector

$$[R_{ij}(t, \cdot)]^{(3)} = [1, [R_{ij}(t, 1)]^{(3)}, [R_{ij}(t, 2)]^{(3)}], \quad i = 1, 2, 3, \quad j = 1, 2, \ldots, 36,$$

with the exponential co-ordinates

$$[R_{ij}(t,\,u)]^{(3)} = \exp[-3ut],\; u = 1,2,3\; i = 1,2,3,\; j = 1,2,\ldots,36.$$

Since the shape parameters of the considered parallel-series system are:

- $k_n^{(1)} \to k^{(1)} = 1,\; l^{(1)} = 36$, at the operation state z_1,
- $k_n^{(2)} \to k^{(2)} = 2,\; l^{(2)} = 36$, at the operation state z_2,
- $k_n^{(3)} \to k^{(3)} = 3,\; l^{(3)} = 36$, at the operation state z_3,

the application of the second part of *Proposition 3.3* is suitable. Thus, applying directly the formulae (3.53–3.54), we get the system reliability function

$$\boldsymbol{R}(t,\,\cdot) = [1,\; \boldsymbol{R}(t,1),\, \boldsymbol{R}(t,2)],\; t \ge 0, \tag{7.19}$$

where

$$\begin{aligned}
\boldsymbol{R}(t,\,u) &\cong \sum_{b=1}^{3} p_b[1 - \exp[-l_n^{(b)} \exp[-[\lambda(u)]^{(b)} t]]]^{k(b)} \\
&= p_1[1 - \exp[-l_n^{(1)} \exp[-[\lambda(u)]^{(1)} t]]]^{k(1)} + p_2[1 - \exp[-l_n^{(2)} \exp[-[\lambda(u)]^{(2)} t]]]^{k(2)} \\
&\quad + p_3[1 - \exp[-l_n^{(3)} \exp[-[\lambda(u)]^{(3)} t]]]^{k(3)} \\
&= 0.2[1 - \exp[-36\exp[-ut]]]^1 + 0.4[1 - \exp[-36\exp[-2ut]]]^2 \\
&\quad + 0.4[1 - \exp[-36\exp[-3ut]]]^3 \\
&= 1 - 0.2\exp[-36\exp[-ut]] - 0.8\exp[-36\exp[-2ut]] + 0.4\exp[-72\exp[-2ut]] \\
&\quad - 1.2\exp[-36\exp[-3ut]] + 1.2\exp[-72\exp[-3ut]] - 0.4\exp[-108\exp[-3ut]] \\
&\quad \text{for}\; t \ge 0,\, u = 1,2,
\end{aligned}$$

and particularly

$$\begin{aligned}
\boldsymbol{R}(t,1) \cong\; & 1 - 0.2\exp[-36\exp[-t]] - 0.8\exp[-36\exp[-2t]] + 0.4\exp[-72\exp[-2t]] \\
& - 1.2\exp[-36\exp[-3t]] + 1.2\exp[-72\exp[-3t]] \\
& - 0.4\exp[-108\exp[-3t]],\; t \ge 0,
\end{aligned} \tag{7.20}$$

$$\begin{aligned}
\boldsymbol{R}(t,2) \cong\; & 1 - 0.2\exp[-36\exp[-2t]] - 0.8\exp[-36\exp[-4t]] + 0.4\exp[-72\exp[-4t]] \\
& - 1.2\exp[-36\exp[-6t]] + 1.2\exp[-72\exp[-6t]] \\
& - 0.4\exp[-108\exp[-6t]],\; \text{for}\; t \ge 0.
\end{aligned} \tag{7.21}$$

The graphs of the approximate coordinates of the parallel-series system reliability function are presented in Fig. 7.11.

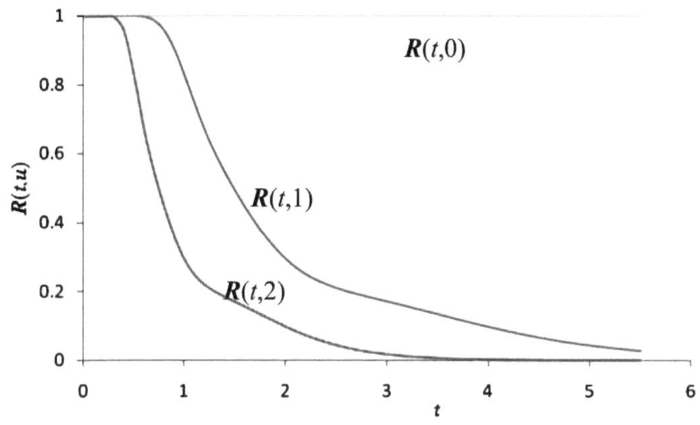

Fig. 7.11 The graph of the parallel-series system reliability function $\boldsymbol{R}(t, \cdot)$ coordinates

The expected values and standard deviations of the system unconditional lifetimes in the reliability state subsets {1, 2}, {2}, calculated from the results given by (7.20–7.21), according to (3.7–3.9), respectively are:

$$\mu(1) \cong 1.939, \ \sigma(1) \cong 0.924, \tag{7.22}$$

$$\mu(2) \cong 0.969, \ \sigma(2) \cong 0.462, \tag{7.23}$$

and further, considering (3.10) and (7.20–7.21), the mean values of the unconditional lifetimes in the particular reliability states 1, 2, respectively are:

$$\bar{\mu}(1) = \mu(1) - \mu(2) = 0.970, \quad \bar{\mu}(2) = \mu(2) = 0.969. \tag{7.24}$$

Since the critical reliability state is $r = 1$, the system risk function, according to (3.11), is given by (Fig. 7.12)

$$\begin{aligned} \boldsymbol{r}(t) &= 1 - \boldsymbol{R}(t, 1) \\ &\cong 0.2\exp[-36\exp[-t]] + 0.8\exp[-36\exp[-2t]] - 0.4\exp[-72\exp[-2t]] \\ &\quad + 1.2\exp[-36\exp[-3t]] - 1.2\exp[-72\exp[-3t]] \\ &\quad + 0.4\exp[-108\exp[-3t]], \ t \geq 0, \text{for}\, t \geq 0. \end{aligned} \tag{7.25}$$

Hence, by (3.12), the moment when the system risk function exceeds a permitted level, for instance $\delta = 0.05$, is

$$\tau = \boldsymbol{r}^{-1}(\delta) \cong 0.812. \tag{7.26}$$

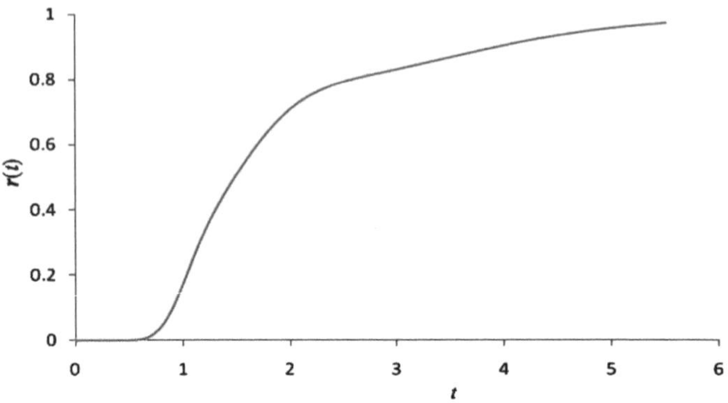

Fig. 7.12 The graph of the parallel-series system risk function $r(t)$

7.2.4 Safety of Steel Cover

We consider the safety of the steel cover composed of $n = 24$ arranged identical sheets $E_1, E_2, \ldots, E_{24}$. We assume that $z = 4$, i.e. the steel cover and the sheets it is composed of may be in one of the safety states from the safety state set $\{0, 1, 2, 3, 4\}$. The cover is out of the safety state subset $\{u, u+1, \ldots, 4\}$, $u = 1, 2, 3, 4$, if at least $m = 2$ of its neighboring sheets is out of this safety state subset. Suppose that the considered steel cover critical safety state is $r = 2$, we conclude that the steel cover is failed, i.e. it is out of the safety state subset $\{2, 3, 4\}$, if at least 2 of its neighboring sheets from 24 sheets are out of the safety state subset $\{2, 3, 4\}$. Thus, the considered steel cover is the five-state consecutive "2 out of 24: F" system, and according to formulae (1.32–1.33) from Sect. 1.2, after replacing in them the reliability notations $\boldsymbol{R}(t, \cdot)$, $\boldsymbol{CR}^2_{24}(t, \cdot)$by the safety notations $\boldsymbol{s}(t, \cdot)$, $\boldsymbol{Cs}^2_{24}(t, \cdot)$, we get the following expression for the steel cover safety function,

$$\boldsymbol{Cs}^2_{24}(t, \cdot) = [1, \boldsymbol{Cs}^2_{24}(t, 1), \boldsymbol{Cs}^2_{24}(t, 2), \boldsymbol{Cs}^2_{24}(t, 3), \boldsymbol{Cs}^2_{24}(t, 4)], \tag{7.27}$$

where

$$\begin{aligned} \boldsymbol{Cs}^2_{24}(t, u) &= s(t,u)\boldsymbol{Cs}^2_{23}(t, u) + s(t,u)F(t,u)\boldsymbol{Cs}^2_{22}(t, u) \\ &\text{for } t \in \langle 0, \infty),\ u = 1, 2, 3, 4, \end{aligned} \tag{7.28}$$

and

$$Cs_n^2(t,u) = \begin{cases} 1 & \text{for } n<2, \\ 1 - \prod_{i=1}^{n} F_i(t,u) & \text{for } n = 2, \\ s_n(t,u)\boldsymbol{Cs}_{n-1}^2(t,u) + \sum_{i=1}^{1} s_{n-i}(t,u)\boldsymbol{Cs}_{n-i-1}^2(t,u) & \\ \cdot \prod_{j=n-i+1}^{n} F_j(t,u) & \text{for } n > 2 \end{cases} \quad (7.29)$$

for $t \geq 0,\ \ u = 1,2,\ldots,z,\ n = 2,3,\ldots,24.$

In the particular case, when the lifetimes $T_i(u)$, $u = 1,2,3,4$, of the sheets E_i, $i = 1,2,\ldots,24$, in the safety state subsets have the exponential safety functions given by the vector

$$s(t,\ \cdot) = [1,\ s(t,1),\ s(t,2),\ s(t,3),\ s(t,4)], t \in <0,\ \infty),$$

with the coordinates

$$s(t,1) = \exp[-0.01t],\ s(t,2) = \exp[-0.02t],\ s(t,3) = \exp[-0.05t],$$
$$s(t,4) = \exp[-0.10t] \quad \text{for } t \geq 0,$$

considering (7.28–7.29), we get the following recurrent formulae for the cover safety function coordinates:

- $\boldsymbol{Cs}_{24}^2(t,\ 1)$ is determined by the formulae

$$\boldsymbol{Cs}_2^2(t,\ 1) = 1 - [1 - \exp[-0.01t]]^2 \text{for } t \in \langle 0,\infty),$$
$$\boldsymbol{Cs}_n^2(t,\ 1) = \exp[-0.01t]\boldsymbol{Cs}_{n-1}^2(t,\ 1) + \exp[-0.01t][1 - \exp[-0.01t]]\boldsymbol{Cs}_{n-2}^2(t,\ 1)$$
$$\text{for } t \in \langle 0,\infty), \quad n = 3,4,\ldots,24, \quad (7.30)$$

- $\boldsymbol{Cs}_{24}^2(t,\ 2)$ is determined by the formulae

$$\boldsymbol{Cs}_2^2(t,\ 2) = 1 - [1 - \exp[-0.02t]]^2 \text{ for } t \in \langle 0,\infty),$$
$$\boldsymbol{Cs}_n^2(t,\ 2) = \exp[-0.02t]\boldsymbol{Cs}_{n-1}^2(t,\ 2) + \exp[-0.02t][1 - \exp[-0.02t]]\boldsymbol{Cs}_{n-2}^2(t,\ 2)$$
$$\text{for } t \in \langle 0,\infty),\ n = 3,4,\ldots,24, \quad (7.31)$$

- $\boldsymbol{Cs}_{24}^2(t,\ 3)$ is determined by the formulae

$$\boldsymbol{Cs}_2^2(t,\ 3) = 1 - [1 - \exp[-0.05t]]^2 \text{ for } t \in <0,\infty),$$
$$\boldsymbol{Cs}_n^2(t,\ 3) = \exp[-0.05t]\boldsymbol{Cs}_{n-1}^2(t,\ 3) + \exp[-0.05t][1 - \exp[-0.05t]]\boldsymbol{Cs}_{n-2}^2(t,\ 3)$$
$$\text{for } t \in <0,\infty),\ n = 3,4,\ldots,24, \quad (7.32)$$

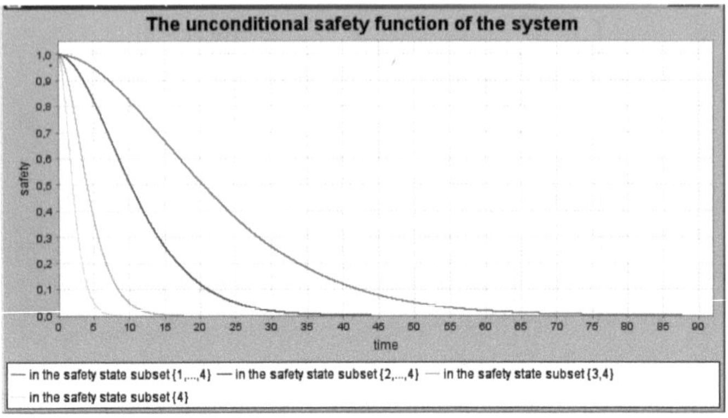

Fig. 7.13 The graphs of the steel cover five-state safety function coordinates

- $\boldsymbol{Cs}^2_{24}(t, 4)$ is determined by the formulae (Fig. 7.13)

$$\boldsymbol{Cs}^2_2(t, 4) = 1 - [1 - \exp[-0.10t]]^2 \text{for } t \in <0, \infty),$$
$$\boldsymbol{Cs}^2_n(t, 4) = \exp[-0.10t]\boldsymbol{Cs}^2_{n-1}(t, 4) + \exp[-0.10t][1 - \exp[-0.10t]]\boldsymbol{Cs}^2_{n-2}(t, 4)$$
$$\text{for } t \in <0, \infty),\ n = 3, 4, \ldots, 24. \tag{7.33}$$

The expected values and standard deviations of the system unconditional lifetimes in the reliability state subsets {1, 2, 3, 4}, {2, 3, 4}, {3, 4}and {4}, calculated from the results given by (7.30–7.32), according to (3.7–3.9), and using the computer programme respectively are:

$$\mu(1) = 22.969, \quad \sigma(1) \cong 14.086, \tag{7.34}$$

$$\mu(2) = 11.485, \quad \sigma(2) \cong 7.043, \tag{7.35}$$

$$\mu(3) = 4.594, \quad \sigma(3) \cong 2.817, \tag{7.36}$$

$$\mu(4) = 2.297, \quad \sigma(4) \cong 1.409, \tag{7.37}$$

and further, considering (3.10) and (7.34–7.37), the mean values of the unconditional lifetimes in the particular reliability states 1, 2, 3, 4 respectively are:

$$\begin{aligned} \bar{\mu}(1) = \mu(1) - \mu(2) \cong 11.484,\ \bar{\mu}(2) = \mu(2) - \mu(3) \cong 6.891, \\ \bar{\mu}(3) = \mu(3) - \mu(4) \cong 2.297, \quad \bar{\mu}(4) = \mu(4) \cong 2.297. \end{aligned} \tag{7.38}$$

Since the critical reliability state is $r = 2$, the system risk function, according to (3.11) and (7.31), is given by (Fig. 7.14)

$$\begin{aligned} \boldsymbol{r}(t) = 1 - \boldsymbol{Cs}^2_{24}(t, 2) = 1 - \exp[-0.02t]\boldsymbol{Cs}^2_{23}(t, 2) \\ - \exp[-0.02t][1 - \exp[-0.02t]]\boldsymbol{Cs}^2_{23}(t, 2) \text{ for } t \in <0, \infty). \end{aligned} \tag{7.39}$$

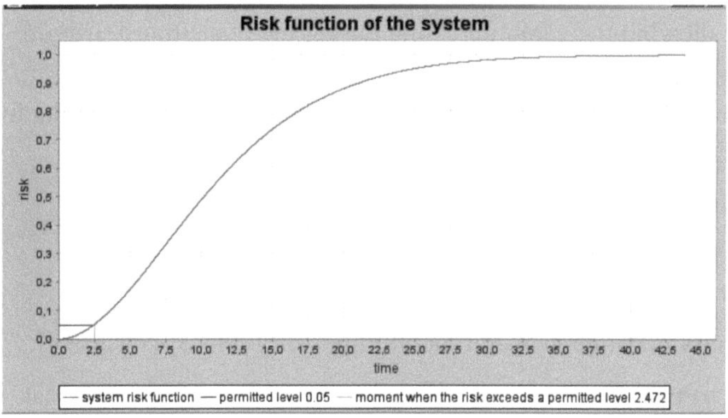

Fig. 7.14 The graph of the steel cover risk function

Hence, by (3.12), the moment when the system risk function exceeds a permitted level, for instance $\delta = 0.05$, is

$$\tau = \boldsymbol{r}^{-1}(\delta) \cong 2.5. \tag{7.40}$$

7.2.5 Identification, Prediction and Optimization of Container Gantry Crane Operation, Reliability and Availability

7.2.5.1 Container Gantry Crane System Analysis

We analyze the reliability of the container gantry crane operating at the container terminal placed at the seashore. The considered container terminal is engaged in trans-shipment of containers. The loading of containers is carried out by using the gantry cranes called Ship-To-Shore (STS).

We consider the STS container gantry crane that is composed of 5 basic subsystems S_1, S_2, S_3, S_4 and S_5 having an essential influence on its reliability. The subsystems are as follows:

S_1 — the power supply subsystem,
S_2 — the control and monitoring subsystem,
S_3 — the arm getting up and getting down subsystem,
S_4 — the transferring subsystem,
S_5 — the loading and unloading subsystem.

The gantry crane power supply subsystem S_1 consists of:

- a high voltage cable delivering the energy from the substation to the gantry crane $E_1^{(1)}$,

- a drum allowing the cable unreeling during the crane transferring $E_2^{(1)}$,
- an inner crane power supply cable $E_3^{(1)}$,
- a device transmitting the energy from the high voltage cable to the inner crane cable $E_4^{(1)}$,
- main and supporting voltage transformers $E_5^{(1)}$,
- a low voltage power supply cable $E_6^{(1)}$,
- relaying and protective electrical components $E_7^{(1)}$.

The gantry crane control and monitoring subsystem S_2 consists of:

- a crane software controller precisely analyzing the situation and takes suitable actions in order to assure correct work of the crane $E_1^{(2)}$,
- a measuring and diagnostic device sending signals about the crane state to the software controller $E_2^{(2)}$,
- a transmitter of signals from the controller to elements executing the set of commands $E_3^{(2)}$,
- devices carrying out the controller's orders (a permission to work, a blockade of work, etc.) $E_4^{(2)}$,
- control panels (an engine room, an operator's cabin, a crane arm cabin) $E_5^{(2)}$,
- control and steering cables' connections $E_6^{(2)}$.

The gantry crane arm getting up and getting down subsystem S_3 consists of:

- a propulsion unit (an engine, a rope drum, a transmission gear, a clutch, breaks, a rope) $E_1^{(3)}$,
- a set of rollers and multi-wheels $E_2^{(3)}$,
- a crane arm (joints, hooks fastening the arm) $E_3^{(3)}$.

The gantry crane transferring subsystem S_4 consists of:

- a driving unit (an engine, a clutch, breaks, a transmission gear, gantry crane wheels) $E_1^{(4)}$.

The gantry crane loading and unloading subsystem S_5 consists of the winch unit $E_1^{(5)}$ composed of:

- a propulsion unit (an engine, a clutch, breaks, a transmission gear, ropes),
- a winch head (which a container grab is connected to),
- a container's grab,
- a container's grab stabilizing unit

and the cart unit $E_2^{(5)}$ composed of:

- a propulsion unit (an engine, a clutch, breaks, a transmission gear, cart wheels, ropes),

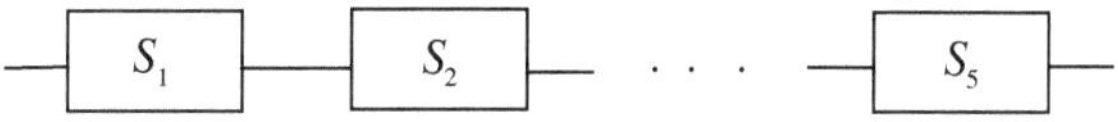

Fig. 7.15 General scheme of gantry crane reliability structure

- rails which cart is moving on during the operation,
- a crane cart.

The subsystems S_1, S_2, S_3, S_4, S_5 form a general series gantry crane reliability structure presented in Fig. 7.15.

7.2.5.2 Container Gantry Crane Operation Process Modelling

The container gantry crane reliability structure and the subsystems and components reliability depend on the changing in time operation states.

Taking into account expert opinion on the varying in time operation process of the considered container gantry crane we fix the number of system operation process states $\nu = 6$ and we distinguish the following as its six operation states:

- operation state z_1 — the crane standby with the power supply on and the control system off,
- operation state z_2 — the crane preparing either to start or finish the work with the crane arm angle position of 90°,
- operation state z_3 — the crane preparing either to start or finish the work with the crane arm angle position of 0°,
- operation state z_4 — the crane transferring either to or from the loading and unloading area with the crane arm angle position of 90°,
- operation state z_5 — the crane transferring either to or from the loading and unloading area with the crane arm angle position of 0°,
- operation state z_6 — loading and unloading of the containers with the crane arm angle position of 0°.

Moreover, we fix that there are possible transitions between all system operation states. Thus, according to Sect. 2.2, the unknown parameters of the container gantry crane operation process semi-Markov model are:

- the vector of probabilities $[p_b(0)]_{1\times 6}$ of the container gantry crane operation process stay at the particular operation states at the initial moment $t = 0$,
- the matrix of probabilities $[p_{bl}]_{6\times 6}$ of the container gantry crane operation process transitions between the operation states
- the matrix of conditional distribution functions $[H_{bl}(t)]_{6\times 6}$ of the container gantry crane operation process conditional sojourn times θ_{bl} at the operation states
- the mean values M_{bl} of the conditional sojourn times $\theta_{bl}, b, l = 1, 2, \ldots, 6, b \neq l$.

To identify all these parameters of the container gantry crane operation process the statistical data coming from this process realizations are needed. The statistical data collected by the system operators are described in the following section.

7.2.5.3 Container Gantry Crane Operation Process Identification

Data Collection for Estimating Unknown Parameters of Container Gantry Crane Operation Process

The statistical data necessary collected by experts to evaluate the probabilities of the container gantry crane operation process straying at the particular operation states at the initial moment $t = 0$, defined by (2.1), are:

- the container gantry crane operation process observation/experiment time

$$\Theta = 29\,\text{days};$$

- the number of container gantry crane operation process realizations

$$n(0) = 25;$$

- the vector of realizations

$$n_1(0) = 1,\ n_2(0) = 18,\ n_3(0) = 1,\ n_4(0) = 0,\ n_5(0) = 2,\ n_6(0) = 3,$$

of the number of strayings of the system operation process respectively at the operation states $z_1,\ z_2,\ \ldots,\ z_6$, at the initial moments $t = 0$ of all $n(0) = 25$ observed realizations of the system operation process

$$[n_b(0)] = [1,\ 18,\ 1,\ 0,\ 2,\ 3],$$

where

$$n_1(0) + n_2(0) + n_3(0) + n_4(0) + n_5(0) + n_6(0) = 25.$$

The collected statistical data necessary to evaluate the probabilities p_{bl} of the container gantry crane transitions between the system operation states, defined by (2.2), are:

- the matrix of realizations

$$\begin{aligned}
n_{11} = 0,\ n_{12} = 77,\ n_{13} = 40,\ n_{14} = 1,\ n_{15} = 0,\ n_{16} = 1,\\
n_{21} = 62,\ n_{22} = 0,\ n_{23} = 44,\ n_{24} = 11,\ n_{25} = 0,\ n_{26} = 1,\\
n_{31} = 40,\ n_{32} = 42,\ n_{33} = 0,\ n_{34} = 0,\ n_{35} = 45,\ n_{36} = 253,\\
n_{41} = 5,\ n_{42} = 7,\ n_{43} = 0,\ n_{44} = 0,\ n_{45} = 0,\ n_{46} = 0,\\
n_{51} = 1,\ n_{52} = 0,\ n_{53} = 42,\ n_{54} = 0\, n_{55} = 0,\ n_{56} = 148,\\
n_{61} = 5,\ n_{62} = 0,\ n_{63} = 225,\ n_{64} = 0,\ n_{65} = 146,\ n_{66} = 0;
\end{aligned}$$

of the number of container gantry crane operation process transitions from state z_b into state z_l during the experiment time $\Theta = 29$ days

$$[n_{bl}] = \begin{bmatrix} 0 & 77 & 40 & 1 & 0 & 1 \\ 62 & 0 & 44 & 11 & 0 & 1 \\ 40 & 42 & 0 & 0 & 45 & 253 \\ 5 & 7 & 0 & 0 & 0 & 0 \\ 1 & 0 & 42 & 0 & 0 & 148 \\ 5 & 0 & 255 & 0 & 146 & 0 \end{bmatrix};$$

- the vector of realizations

$$n_1 = n_{11} + n_{12} + n_{13} + n_{14} + n_{15} + n_{16} = 119,$$
$$n_2 = n_{21} + n_{22} + n_{23} + n_{24} + n_{25} + n_{26} = 118,$$
$$n_3 = n_{31} + n_{32} + n_{33} + n_{34} + n_{35} + n_{36} = 380,$$
$$n_4 = n_{41} + n_{42} + n_{43} + n_{44} + n_{45} + n_{46} = 12,$$
$$n_5 = n_{51} + n_{52} + n_{53} + n_{54} + n_{55} + n_{56} = 191,$$
$$n_6 = n_{61} + n_{62} + n_{63} + n_{64} + n_{65} + n_{66} = 406,$$

(the sums of the number of matrix $[n_{bl}]$ rows) of the number of container gantry crane operation process departures from the operation state z_b during the experiment time $\Theta = 29$ days

$$[n_b] = [119,\ 118,\ 380,\ 12,\ 191,\ 406].$$

The collected statistical data necessary to evaluate the unknown parameters of the distributions of the conditional sojourn times of the container gantry crane operation process in the particular operation states, defined by (2.3), are as follows:

- the numbers n_{bl}, $b, l = 1, 2, \ldots, 6$, $b \neq l$, of the realizations θ_{bl}^k, $k = 1,2, \ldots, n_{bl}$, of the conditional sojourn times θ_{bl} of the container gantry crane operation process at the operation state z_b when the next transition is to the operation state z_l during the observation time;
- the realizations θ_{bl}^k, $k = 1,2, \ldots, n_{bl}$, of the conditional sojourn times θ_{bl} of the container gantry crane operation process at the operation state z_b when the next transition is to the operation state z_l during the observation time Θ= 29 days.

For instance, the collected statistical data for the sojourn time θ_{12} are as follows:

- the number of realizations $n_{12} = 77$,
- the realizations θ_{12}^k, $k = 1, 2, \ldots, 77$: 1480, 159, 356, 422, 369, 235, 47, 30, 7, 1304, 997, 1375, 2, 2, 12, 2301, 411, 8, 1, 1, 1, 1, 1, 1, 1, 1, 435, 631, 1, 304, 37, 8, 1, 1, 1, 1, 1, 1, 1, 1, 1, 435, 631, 1, 304, 37, 1625, 286, 25, 1157, 11, 225, 22,

5, 862, 127, 64, 978, 93, 1126, 82, 337, 144, 777, 401, 270, 1669, 60, 329, 16, 75, 663, 2203, 409, 219, 469, 2898.

Estimating Parameters of Container Gantry Crane Operation Process

On the basis of the statistical data presented in "Data Collection for Estimating Unknown Parameters of Container Gantry Crane Operation Process", using respectively the formulae (4.1–4.3) and (4.4–6.6), we evaluate

- the vector

$$[p(0)] = [0.04,\ 0.72,\ 0.04,\ 0,\ 0.08,\ 0.08]$$

  of the probabilities $p_b(0)$, $b = 1, 2, \ldots, 6$, of the container gantry crane operation process stay at the particular states z_b at the initial moment $t = 0$,
- the matrix

$$[p_{bl}] = \begin{bmatrix} 0 & 0.648 & 0.336 & 0.008 & 0 & 0.008 \\ 0.525 & 0 & 0.373 & 0.093 & 0 & 0.009 \\ 0.105 & 0.111 & 0 & 0 & 0.118 & 0.666 \\ 0.417 & 0.583 & 0 & 0 & 0 & 0 \\ 0.005 & 0 & 0.220 & 0 & 0 & 0.775 \\ 0.012 & 0 & 0.628 & 0 & 0.360 & 0 \end{bmatrix}, \qquad (7.41)$$

  of the probabilities p_{bl}, $b, l = 1, 2, \ldots, 6$, of the container gantry crane operation process transitions from the operation state z_b to the operation state z_l.

Estimating Parameters of Distributions of Container Gantry Crane Conditional Sojourn Times at Operation States

On the basis of the statistical data partly presented in "Data Collection for Estimating Unknown Parameters of Container Gantry Crane Operation Process", using the procedure and the formulae given in Sect. 4.2.3, it is possible to determine the empirical characteristics of the realizations of the conditional sojourn time of the container gantry crane operation process at the particular operation states. To illustrate this procedure and the formulae application, we perform it for the conditional sojourn time θ_{12}, using their realizations presented in "Data Collection for Estimating Unknown Parameters of Container Gantry Crane Operation Process".

The results for the conditional sojourn time θ_{12} are:

- defined by (4.7) the realization of the mean value $\bar{\theta}_{12}$ of the conditional sojourn time θ_{12} of the container gantry crane operation process at the operation state z_1 when the next transition is to the operation state z_2

$$\bar{\theta}_{12} = \frac{1}{77}\sum_{k=1}^{77}\theta_{12}^{k} = 389.4,$$

- the number $\bar{r}_{12}$ of the disjoint intervals $I_j = \langle a_{bl}^j, b_{bl}^j), j = 1, 2, \ldots, \bar{r}_{12}$ that include the realizations θ_{12}^k, $k = 1, 2, \ldots, 77$, of the conditional sojourn time θ_{12} at the operation state z_1 when the next transition is to the operation state z_2

$$\bar{r}_{12} \cong \sqrt{77} \cong 9,$$

- the length d_{12} of the intervals $I_j = \langle a_{bl}^j, b_{bl}^j)$, $j = 1, 2, \ldots, 9$, which after considering

$$\bar{R}_{12} = \max_{1 \le k \le 77} \theta_{12}^k - \min_{1 \le k \le 77} \theta_{12}^k = 2898 - 1 = 2897$$

is

$$d_{12} = \frac{\bar{R}_{12}}{\bar{r}_{12} - 1} = \frac{2897}{8} \cong 362,$$

- the ends a_{bl}^j, b_{bl}^j, of the intervals $I_j = \langle a_{bl}^j, b_{bl}^j)$, $j = 1, 2, \ldots, 9$, which after considering

$$\min_{1 \le k \le 77} \theta_{12}^k - \frac{d_{12}}{2} = 1 - \frac{362}{2} = -180,$$

are

$$a_{12}^1 = \max\{-180, 0\} = 0,\ b_{12}^1 = a_{12}^1 + 362 = 0 + 362 = 362,$$
$$a_{12}^2 = b_{12}^1 = 362,\ b_{12}^2 = a_{12}^1 + 2 \times 362 = 0 + 724 = 724,$$
$$a_{12}^3 = b_{12}^2 = 724,\ b_{12}^3 = a_{12}^1 + 3 \times 362 = 0 + 1086 = 1086,$$
$$a_{12}^4 = b_{12}^3 = 1086,\ b_{12}^4 = a_{12}^1 + 4 \times 362 = 0 + 1448 = 1448,$$
$$a_{12}^5 = b_{12}^4 = 1448,\ b_{12}^5 = a_{12}^1 + 5 \times 362 = 0 + 1810 = 1810,$$
$$a_{12}^6 = b_{12}^5 = 1810,\ b_{12}^6 = a_{12}^1 + 6 \times 362 = 0 + 2172 = 2172,$$
$$a_{12}^7 = b_{12}^6 = 2172,\ b_{12}^7 = a_{12}^1 + 7 \times 362 = 0 + 2534 = 2534,$$
$$a_{12}^8 = b_{12}^7 = 2534,\ b_{12}^8 = a_{12}^1 + 8 \times 362 = 0 + 2896 = 2896,$$
$$a_{12}^9 = b_{12}^8 = 2896,\ b_{12}^9 = a_{12}^1 + 9 \times 362 = 0 + 3258 = 3258,$$

- the numbers n_{12}^j of the realizations θ_{12}^k in particular intervals $I_j = \langle a_{bl}^j, b_{bl}^j)$, $j = 1, 2, \ldots, 9$,

$$n_{12}^1 = 52,\ n_{12}^2 = 11,\ n_{12}^3 = 4,\ n_{12}^4 = 4,\ n_{12}^5 = 3,\ n_{12}^6 = 0,\ n_{12}^7 = 2,\ n_{12}^8 = 0,\ n_{12}^9 = 1.$$

Table 7.1 The realization of the histogram of the conditional sojourn time θ_{12}

Histogram of the conditional sojourn time θ_{12}									
$I_j = \langle a_{12}^j, b_{12}^j)$	0–362	362–724	724–1086	1086–1448	1448–1810	1810–2172	2172–2534	2534–2896	2896–3258
n_{12}^j	52	11	4	4	3	0	2	0	1
$\bar{h}_{12}(t) = n_{12}^j/n_{12}$	52/77	11/77	4/77	4/77	3/77	0/77	2/77	0/77	1/77

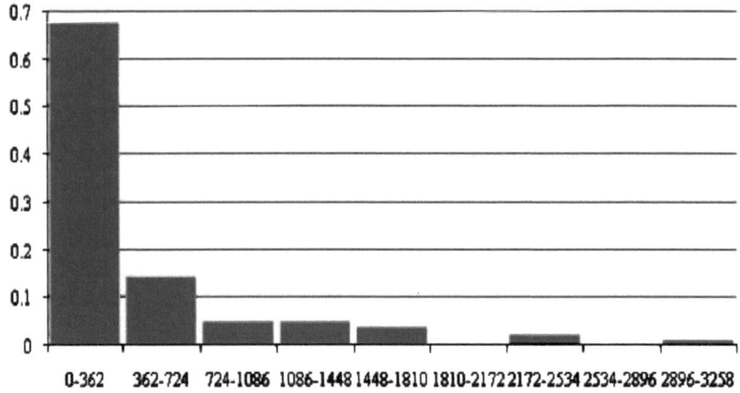

Fig. 7.16 The graph of the histogram of the conditional sojourn time θ_{12}

Identifying Distribution Functions of Container Gantry Crane Conditional Sojourn Times at Operation States

Using the procedure given in Sect. 4.2.4 and the results partly presented in "Estimating Parameters of Distributions of Container Gantry Crane Conditional Sojourn Times at Operation States", we verify the hypotheses on the distributions of the system conditional sojourn times θ_{bl}, $b, l = 1, 2, \ldots, 6$, $b \neq l$, at the particular operation states. To make the procedure familiar to the reader, we perform it for the conditional sojourn time for $\theta_{12,}$ preliminary analyzed in "Estimating Parameters of Distributions of Container Gantry Crane Conditional Sojourn Times at Operation States".

The realization $\bar{h}_{12}(t)$ of the histogram of the conditional sojourn time θ_{12} defined by (4.45) is presented in Table 7.1 and illustrated in Fig. 7.16.

After analyzing and comparing the realization $\bar{h}_{12}(t)$ of the histogram with the graphs of the density functions $h_{bl}(t)$ of the previously distinguished in Sect. 2.2 distributions, we formulate the null hypothesis H_0 in the following form:

H_0: The system conditional sojourn time θ_{12} at the operation state z_1 when the next transition is to the operation state z_2, has the Weibull's distribution with the density function defined by (2.10) in the following form

$$h_{12}(t) = \begin{cases} 0, & t > x_{12} \\ \alpha_{12}\beta_{12}(t - x_{12})^{\beta_{12}-1} \exp[-\alpha_{12}(t - x_{12})^{\beta_{12}}], & t \geq x_{12}. \end{cases} \quad (7.42)$$

We estimate the unknown parameters of the density function of the hypothetical Weibull's distribution using the formulae (4.14) that take the following forms

$$x_{12} = a_{12}^1 = 0,$$
$$\alpha_{12} = \frac{77}{\sum_{k=1}^{77} (\theta_{bl}^k)^{\beta_{12}}}, \quad \alpha_{12} = \frac{\frac{77}{\beta_{12}} + \sum_{k=1}^{77} \ln(\theta_{12}^k - x_{12})}{\sum_{k=1}^{77} (\theta_{bl}^k)^{\beta_{bl}} \ln(\theta_{12}^k - x_{12})}. \quad (7.43)$$

The approximate solution of the above system of equations (7.43) are $\alpha_{12} = 0.097$ and $\beta_{12} = 0.448$.

Substituting the above results into (7.42), we completely define the hypothetical density function in the form

$$h_{12}(t) = \begin{cases} 0, & t > 0 \\ 0.097 \times 0.448 t^{0.448-1} \exp[-0.097 t^{0.448}], & t < 0. \end{cases} \quad (7.44)$$

Hence, the hypothetical distribution function $H_{12}(t)$ of the conditional sojourn time θ_{12}, after taking the integral of the hypothetical density function $h_{12}(t)$ given by (7.44), takes the following form

$$H_{12}(t) = \int_0^t h_{12}(t)dt = \begin{cases} 0, & t < 0 \\ 1 - \exp[-0.097 t^{0.448}], & t \geq 0. \end{cases} \quad (7.45)$$

Next, we join the intervals defined in the realization of the histogram $h_{12}(t)$ that have the numbers n_{12}^j, of realizations less than 4 into new intervals performing the following steps:

- we fix the new number of intervals $\bar{\bar{r}}_{12} = 5$,
- we determine the new intervals

$$\bar{I}_1 = \langle 0,\ 362), \bar{I}_2 = \langle 362,\ 724), \bar{I}_3 = \langle 724,\ 1086), \bar{I}_4 = \langle 1086,\ 1448),$$
$$\bar{I}_5 = \langle 1448,\ 3258),$$

- we fix the number of realizations in the new intervals

$$\bar{n}_{12}^1 = 52,\ \bar{n}_{12}^2 = 11,\ \bar{n}_{12}^3 = 4,\ \bar{n}_{12}^4 = 4,\ \bar{n}_{12}^5 = 6,$$

- we calculate, using (4.46) and (7.45), the hypothetical probabilities that the variable θ_{12} takes values from the new intervals

$$p_1 = P(\theta_{12} \in \bar{I}_1) = P(0 \leq \theta_{12} < 362) = H_{12}(362) - H_{12}(0)$$
$$= (1 - \exp[-0.097 \cdot 362^{0.448}]) - 0 = 1 - -0.257 = 0.734,$$

$$\begin{aligned} p_2 &= P(\theta_{12} \in \bar{I}_2) = P(362 \le \theta_{12} < 724) = H_{12}(724) - H_{12}(362) \\ &= (1 - \exp[-0.097 \cdot 724^{0.448}]) - (1 - \exp[-0.097 \cdot 362^{0.448}]) \\ &= -0.157 + 0.257 = 0.1, \end{aligned}$$

$$\begin{aligned} p_3 &= P(\theta_{12} \in \bar{I}_3) = P(724 \le \theta_{13} < 1086) = H_{12}(1086) - H_{12}(724) \\ &= (1 - \exp[-0.097 \cdot 1086^{0.448}]) - (1 - \exp[-0.097 \cdot 724^{0.448}]) \\ &= -0.108 + 0.157 = 0.049, \end{aligned}$$

$$\begin{aligned} p_4 &= P(\theta_{12} \in \bar{I}_4) = P(1086 \le \theta_{13} < 1448) = H_{12}(1448) - H_{12}(1086) \\ &= (1 - \exp[-0.097 \cdot 1448^{0.448}]) - (1 - \exp[-0.097 \cdot 1086^{0.448}]) \\ &= -0.08 + 0.108 = 0.028, \end{aligned}$$

$$\begin{aligned} p_5 &= P(\theta_{12} \in \bar{I}_5) = P(1448 \le \theta_{13} <; \infty) = 1 - [p_1 + p_2 + p_3 + p_4] \\ &= 1 - [0.743 + 0.1 + 0.049 + 0.028] \\ &= 1 - 0.92 = 0.08 \end{aligned}$$

- we calculate, using (4.47), the realization of the χ^2(chi-square)-Pearson's statistics

$$\begin{aligned} u_{77} &= \sum_{j=1}^{5} \frac{(\bar{n}_{12}^j - n_{12}p_j)^2}{n_{12}p_j} = \frac{(52 - 77 \times 0.743)^2}{77 \times 0.743} + \frac{(11 - 77 \times 0.1)^2}{77 \times 0.1} + \frac{(4 - 77 \times 0.049)^2}{77 \times 0.049} \\ &+ \frac{(4 - 77 \times 0.028)^2}{77 \times 0.028} + \frac{(6 - 77 \times 0.08)^2}{77 \times 0.08} \\ &\cong 0.47 + 1.41 + 0.06 + 0.86 + 0.004 = 2.804, \end{aligned} \tag{7.46}$$

- we assume the significance level $\alpha = 0.05$,
- we fix the number of degrees of freedom

$$\bar{\bar{r}}_{12} - l - 1 = 5 - 2 - 1 = 2,$$

- we read from the Tables of the χ^2-Pearson's distribution the value u_α for the fixed values of the significance level $\alpha = 0.05$ and the number of degrees of freedom $\bar{\bar{r}}_{12} - l - 1 = 2$, such that, according to (4.48), the following equality holds

$$P(U_{77} > u_\alpha) = \alpha = 0.05$$

which amounts to $u_\alpha = 5.99$ and we determine the critical domain in the form of the interval $(5.99, +\infty)$ and the acceptance domain in the form of the interval $<0, 5.99>$ (Fig. 7.17).
- we compare the obtained in (7.46) value $u_{77} = 2.804$ of the realization of the statistics U_{77} with that read from the Tables, critical value $u_\alpha = 5.99$ of the chi-square random variable and since the value $u_{24} = 2.804$ does not belong to the critical domain, i.e.

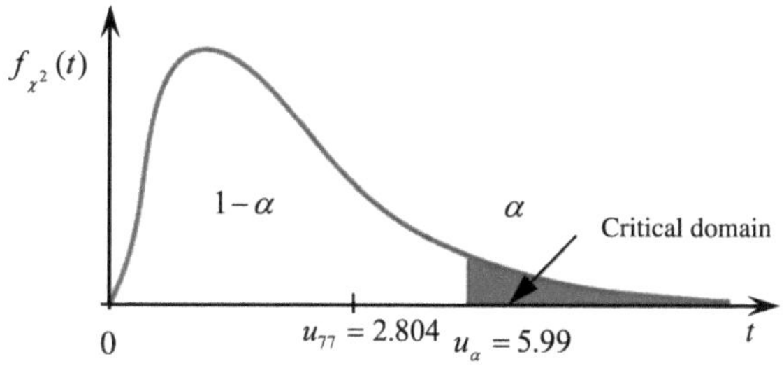

Fig. 7.17 The graphical interpretation of the critical interval and the acceptance interval for the chi-square goodness-of-fit test

$$u_{77} = 2.804 \le u_\alpha = 5.99,$$

we do not reject the hypothesis H_0.

For the remaining cases, when the realizations of conditional sojourn times θ_{bl} are more than 30, proceeding afterwards in an analogous way as in the case of the conditional sojourn time θ_{12}, we can get the following results:

- the conditional sojourn time θ_{13} has a chimney distribution with the density function

$$h_{13}(t) = \begin{cases} 0, & t<0 \\ 0.0218, & 0 \le t < 39 \\ 0.00076, & 39 \le t < 234 \\ 0, & t \ge 234. \end{cases}$$

- the conditional sojourn time θ_{21} has a chimney distribution with the density function

$$h_{21}(t) = \begin{cases} 0, & t<0 \\ 0.066923, & 0 \le t < 13 \\ 0.003769, & 13 \le t < 104 \\ 0, & t \ge 104. \end{cases}$$

- the conditional sojourn time θ_{23} has a chimney distribution with the density function

$$h_{23}(t) = \begin{cases} 0, & t<0 \\ 0.09253, & 0 \le t < 9.33 \\ 0.002436, & 9.33 \le t < 65.33 \\ 0, & t \ge 65.33, \end{cases}$$

- the conditional sojourn time θ_{31} has an exponential distribution with the density function

$$h_{31}(t) = \begin{cases} 0, & t<0 \\ 0.182 \exp[-0.182t], & t \ge 0, \end{cases}$$

- the conditional sojourn time θ_{32} has a chimney distribution with the density function

$$h_{32}(t) = \begin{cases} 0, & t<0 \\ 0.20677, & 0 \le t < 3.8 \\ 0.011278, & 3.8 \le t < 22.8 \\ 0, & t \ge 22.8, \end{cases}$$

- the conditional sojourn time θ_{35} has an exponential distribution with the density function

$$h_{35}(t) = \begin{cases} 0, & t<0 \\ 0.147\exp[-0.147t], & t \ge 0, \end{cases}$$

- the conditional sojourn time θ_{36} has a chimney distribution with the density function

$$h_{36}(t) = \begin{cases} 0, & t<0 \\ 0.099, & 0 \le t < 7 \\ 0.0148081, & 7 \le t < 112 \\ 0, & t \ge 112, \end{cases}$$

- the conditional sojourn time θ_{53} has a chimney distribution with the density function

$$h_{53}(t) = \begin{cases} 0, & t<0.3 \\ 0.25510, & 0.3 \le t < 3.1 \\ 0.05102, & 3.1 \le t < 8.7 \\ 0, & t \ge 8.7, \end{cases}$$

- the conditional sojourn time θ_{56} has a chimney distribution with the density function

$$h_{56}(t) = \begin{cases} 0, & t<0 \\ 0.02610, & 0 \le t < 37.27 \\ 0.00000659, & 37.27 \le t < 447.27 \\ 0, & t \ge 447.27, \end{cases}$$

- the conditional sojourn time θ_{63} has Weibull's distribution with the density function

$$h_{63}(t) = \begin{cases} 0, & t<0 \\ 0.061 \cdot 0.904 t^{0.904-1}\exp[-0.061 t^{0.904}], & t \ge 0, \end{cases}$$

- the conditional sojourn time θ_{65} has Weibull's distribution with the density function

$$h_{65}(t) = \begin{cases} 0, & t<0 \\ 0.081 \times 0.854 t^{0.854-1} \exp[-0.081 t^{0.854}], & t \geq 0. \end{cases}$$

For the distributions identified in this section, by application either the general formulae for the mean value given by (2.12) or the particular formulae (2.13–2.19), the mean values $M_{bl} = E[\theta_{bl}]$, $b, l = 1, 2, \ldots, 6$, $b \neq l$, of the container gantry crane operation process conditional sojourn times at the particular operation states can be determined and they amount to:

$$\begin{gathered} M_{12} = 456.978,\ M_{13} = 36.860,\ M_{21} = 7.887,\ M_{23} = 9.121,\ M_{31} = 5.5, \\ M_{32} = 4.343,\ M_{35} = 6.822,\ M_{36} = 7.857,\ M_{53} = 2.899,\ M_{56} = 24.681, \\ M_{63} = 23.117,\ M_{65} = 20.512. \end{gathered} \quad (7.47)$$

In the remaining cases, when the number of realizations of the sojourn times are less than 30 and the distributions cannot be identified, using formula (4.7), it is possible to find the approximate empirical values of the mean values $M_{bl} = E[\theta_{bl}]$ of the conditional sojourn times at the particular operation states that are as follow:

$$\begin{gathered} M_{14} = 50,\ M_{16} = 3,\ M_{24} = 1.545,\ M_{26} = 16, \\ M_{41} = 2,\ M_{42} = 2.143,\ M_{51} = 10,\ M_{61} = 22.6. \end{gathered} \quad (7.48)$$

These approximate mean values given by (7.48) and previously fixed mean values given by (7.47) are used in Sect. 7.2.5.4 for the prediction of the characteristics of the container gantry crane operation process.

As there are no realizations of conditional sojourn times

$$\theta_{15},\ \theta_{25},\ \theta_{34},\ \theta_{43},\ \theta_{45},\ \theta_{46},\ \theta_{52},\ \theta_{54},\ \theta_{62},\ \theta_{64},$$

it is impossible to estimate their empirical conditional mean values

$$M_{15},\ M_{25},\ M_{34},\ M_{43},\ M_{45},\ M_{46},\ M_{52},\ M_{54},\ M_{62},\ M_{64}.$$

7.2.5.4 Container Gantry Crane Operation Process Prediction

After considering the results (7.47–7.48) and applying the formulae (2.21), we conclude that the unconditional mean sojourn times of the container gantry crane operation process at the particular operation states are given by:

$$\begin{aligned} M_1 = E[\theta_1] &= p_{12}M_{12} + p_{13}M_{13} + p_{14}M_{14} + p_{16}M_{16} \\ &= 0.648 \times 456.978 + 0.336 \times 36.860 + 0.008 \times 50 + 0.008 \times 3 \cong 308.93, \end{aligned} \quad (7.49)$$

$$M_2 = E[\theta_2] = p_{21}M_{21} + p_{23}M_{23} + p_{24}M_{24} + p_{26}M_{26}$$
$$= 0.525 \times 7.887 + 0.373 \times 9.121 + 0.093 \times 1.545 + 0.009 \times 16 \cong 7.83, \quad (7.50)$$

$$M_3 = E[\theta_3] = p_{31}M_{31} + p_{32}M_{32} + p_{35}M_{35} + p_{36}M_{36}$$
$$= 0.105 \times 5.5 + 0.111 \times 4.343 + 0.118 \times 6.822 + 0.666 \times 7.857 \cong 7.09, \quad (7.51)$$

$$M_4 = E[\theta_4] = p_{41}M_{41} + p_{42}M_{42} = 0.417 \times 2 + 0.583 \times 2.143 \cong 2.08, \quad (7.52)$$

$$M_5 = E[\theta_5] = p_{51}M_{51} + p_{53}M_{53} + p_{56}M_{56}$$
$$= 0.005 \times 10 + 0.220 \times 2.899 + 0.775 \times 24.681 \cong 19.82, \quad (7.53)$$

$$M_6 = E[\theta_6] = p_{61}M_{61} + p_{63}M_{63} + p_{65}M_{65}$$
$$= 0.012 \times 22.6 + 0.628 \times 23.117 + 0.360 \times 20.512 \cong 22.17. \quad (7.54)$$

Since, according to (2.23), from the system of equations

$$\begin{cases} [\pi_1, \pi_2, \pi_3, \pi_4, \pi_5, \pi_6] = [\pi_1, \pi_2, \pi_3, \pi_4, \pi_5, \pi_6][p_{bl}]_{6\times 6} \\ \pi_1 + \pi_2 + \pi_3 + \pi_4 + \pi_5 + \pi_6 = 1, \end{cases}$$

we get

$$\pi_1 = 0.0951,\ \pi_2 = 0.1020,\ \pi_3 = 0.3100,\ \pi_4 = 0.0102,$$
$$\pi_5 = 0.1547,\ \pi_6 = 0.3280. \quad (7.55)$$

Then, the limit values of the transient probabilities $p_b(t)$ of the gantry crane operation process at the operation states z_b, according to (2.22), are given by

$$p_1 = 0.6874,\ p_2 = 0.0187,\ p_3 = 0.0515,\ p_4 = 0.0005,$$
$$p_5 = 0.0717,\ p_6 = 0.1702. \quad (7.56)$$

Hence, the expected values of the total sojourn times $\hat{\theta}_b$, $b = 1, 2, \ldots, 6$, of the container gantry crane operation process at the particular operation states z_b, $b = 1, 2, \ldots, 6$, during the fixed operation time

$$\theta = 1 \text{ year} = 365 \text{ days},$$

after applying (2.24), amount to:

$$E[\hat{\theta}_1] = 0.6874 \text{ year} \cong 251 \text{ days},\ E[\hat{\theta}_2] = 0.0187 \text{ year} \cong 7 \text{ days},$$
$$E[\hat{\theta}_3] = 0.0515 \text{ year} \cong 19 \text{ day},\ E[\hat{\theta}_4] = 0.0005 \text{ year} \cong 0.2 \text{ day}, \quad (7.57)$$
$$E[\hat{\theta}_5] = 0.0717 \text{ year} \cong 26 \text{ days},\ E[\hat{\theta}_6] = 0.1702 \text{ year} \cong 62 \text{ days}.$$

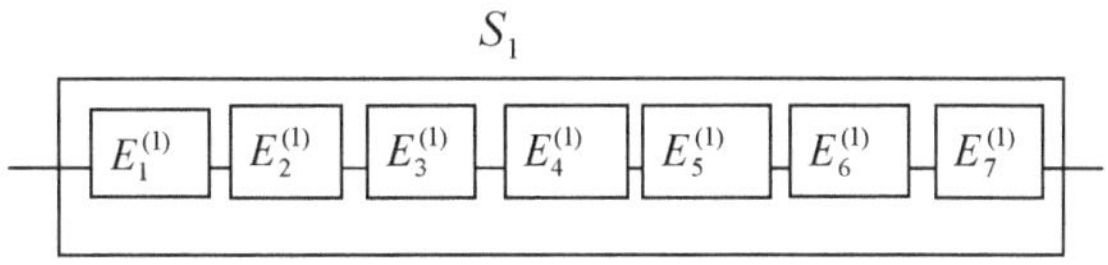

Fig. 7.18 The scheme of the container gantry crane at operation state z_1

7.2.5.5 Container Gantry Crane Components Reliability Modeling

After discussion with experts, taking into account the effectiveness of the operation of the container gantry crane, we fix that the system and its components have four reliability states 0, 1, 2, 3, i.e. $z = 3$. And consequently, at all operation states z_b, $b = 1, 2, \ldots, 6$, we distinguish the following reliability states of the system and its components:

- a reliability state 3 — the gantry operation is fully effective,
- a reliability state 2 — the gantry operation is less effective because of ageing,
- a reliability state 1 — the gantry operation is less effective because of ageing and more dangerous,
- a reliability state 0 — the gantry is destroyed.

We assume that there are possible transitions between the components, reliability states only from better to worse and we fix that the system and components, critical reliability state is $r = 2$. Moreover, we assume that the changes of the operation states of the container gantry crane operation process have an influence on its reliability structure and its multistate components reliability as well.

The container gantry crane operation process influence on the system reliability structure is expressed as follows.

At the system operation state z_1, the container gantry crane is composed of the subsystem S_1 which is a series system composed of $n = 7$ components $E_i^{(1)}$, $i = 1, 2, .., 7$ (subsystems) with the structure shown in Fig. 7.18.

At the system operation states z_2 and z_3, the container gantry crane is composed of subsystems S_1, S_2 and S_3 forming a series structure shown in Fig. 7.19. Subsystem S_1 is a series system composed of $n = 7$ components $E_i^{(1)}$, $i = 1, 2, .., 7$, subsystem S_2 is a series system composed of $n = 6$ components $E_i^{(2)}$, $i = 1, 2, .., 6$, and subsystem S_3 is a series system composed of $n = 3$ components $E_i^{(3)}$, $i = 1, 2, 3$.

At the system operation states z_4 and z_5, the container gantry crane is composed of subsystems S_1, S_2, S_3 and S_4 forming a series structure shown in Fig. 7.20. Subsystem S_1 is a series system composed of $n = 7$ components $E_i^{(1)}$, $i = 1, 2, .., 7$, subsystem S_2 is a series system composed of $n = 6$ components $E_i^{(2)}$, $i = 1, 2, .., 6$, subsystem S_3 is a series system composed of $n = 3$ components $E_i^{(3)}$, $i = 1, 2, 3$, and subsystem S_4 consists of a component $E_1^{(4)}$.

At the system operation state z_6, the container gantry crane is composed of subsystems S_1, S_2, S_3 and S_5 forming a series structure shown in Fig. 7.21.

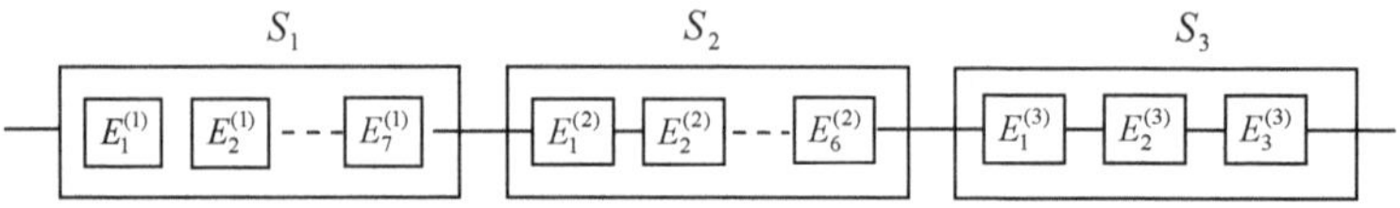

Fig. 7.19 The scheme of the container gantry crane at operation states z_2 and z_3

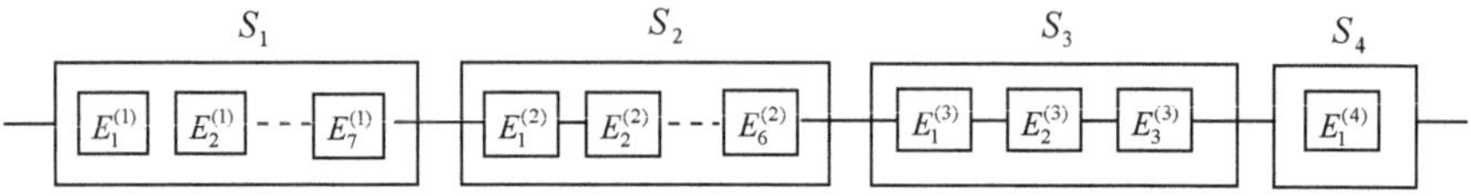

Fig. 7.20 The scheme of the container gantry crane at operation states z_4 and z_5

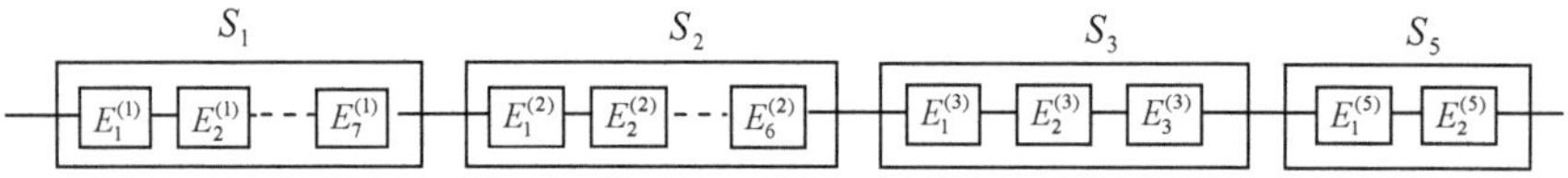

Fig. 7.21 The scheme of the container gantry crane at operation state z_6

Subsystem S_1 is a series system composed of $n = 7$ components $E_i^{(1)}, i = 1, 2, .., 7$, subsystem S_2 is a series system composed of $n = 6$ components $E_i^{(2)}, i = 1, 2, .., 6$, subsystem S_3 is a series system composed of $n = 3$ components $E_i^{(3)}, i = 1, 2, 3$ and subsystem S_5 is a series system composed of $n = 2$ components $E_i^{(5)}, i = 1, 2$.

The container gantry crane operation process, influence on its components reliability is expressed by the assumption that, its subsystems S_υ, $\upsilon = 1, 2, \ldots 5$, are composed of four-state, i.e. $z = 3$, components $E_i^{(\upsilon)}$, $\upsilon = 1, 2, \ldots, 5$ having the conditional four-state reliability functions

$$[R_i^{(\upsilon)}(t, \cdot)]^{(b)} = [1, [R_i^{(\upsilon)}(t, 1)]^{(b)}, [R_i^{(\upsilon)}(t, 2)]^{(b)}, [R_i^{(\upsilon)}(t, 3)]^{(b)}], b = 1, 2, \ldots, 6, \quad (7.58)$$

with the exponential co-ordinates

$$[R_i^{(\upsilon)}(t, 1)]^{(b)} = \exp[-[\lambda_i^{(\upsilon)}(1)]^{(b)} t], [R_i^{(\upsilon)}(t, 2)]^{(b)} = \exp[-[\lambda_i^{(\upsilon)}(2)]^{(b)} t],$$
$$[R_i^{(\upsilon)}(t, 3)]^{(b)} = \exp[-[\lambda_i^{(\upsilon)}(3)]^{(b)} t], t \geq 0, b = 1, 2, \ldots, 6, \upsilon = 1, 2, \ldots, 5, \quad (7.59)$$

different at the various operation states z_b, $b = 1, 2, \ldots, 6$, where

$$[\lambda_i^{(\upsilon)}(1)]^{(b)}, [\lambda_i^{(\upsilon)}(2)]^{(b)}, [\lambda_i^{(\upsilon)}(3)]^{(b)}, b = 1, 2, .., 6, \upsilon = 1, 2, \ldots, 5,$$

are the subsystems components, unknown intensities of departures respectively from the reliability state subsets {1, 2, 3}, {2, 3}, {3}.

7.2.5.6 Container Gantry Crane Components Reliability Identification

Data Collections Coming from Experts

To estimate existing in the formulae (7.58–7.59) the subsystem components, unknown intensities $[\lambda_{ij}^{(\upsilon)}(1)]^{(b)}$, $[\lambda_{ij}^{(\upsilon)}(2)]^{(b)}$, $[\lambda_{i}^{(\upsilon)}(3)]^{(b)}$,$b = 1,2,..,6$, $\upsilon = 1,2,\ldots,5$, of departure respectively from the reliability state subsets $\{1, 2, 3\}$, $\{2, 3\}$, $\{3\}$ we have data from experts.

The approximate realizations

$$[\hat{\mu}_i^{(v)}(u)]^{(b)},\ \upsilon = 1,2,\ldots,5,\ u = 1,2,3,\ b = 1,2,..,6,$$

of the mean values

$$[\hat{\mu}_i^{(v)}(u)]^{(b)} = E[[T_i^{(\upsilon)}(u)]^{(b)}],\ \upsilon = 1,2,\ldots,5,\ u = 1,2,3,\ b = 1,2,..,6,$$

of the conditional lifetimes $[T_i^{(\upsilon)}(u)]^{(b)}$, $\upsilon = 1,2,\ldots,5$, $u = 1, 2, 3$, $b = 1,2,..,6$, in reliability state subsets $\{u, u+1,\ldots,3\}$, $u = 1, 2, 3$, of the component $E_i^{(\upsilon)}$of the container gantry crane subsystems S_υ, $\upsilon = 1,2,\ldots,5$, at the particular operation states z_b, $b = 1,2,\ldots,6$, estimated on the basis of expert opinions are collected.

For instance, in Table 7.2, there are data from experts concerning the components $E_i^{(1)}$, $i = 1,2,\ldots,7$, of subsystem S_1.

Evaluating Container Gantry Crane Components Intensities of Departures from Reliability State Subsets

To evaluate the approximate values $[\hat{\lambda}_i^{(\upsilon)}(1)]^{(b)}$, $[\hat{\lambda}_i^{(\upsilon)}(2)]^{(b)}$, $[\hat{\lambda}_i^{(\upsilon)}(3)]^{(b)}$, of subsystems S_υ, $\upsilon = 1,2,..,5$ components, unknown intensities $[\lambda_i^{(\upsilon)}(1)]^{(b)}$, $[\lambda_i^{(\upsilon)}(2)]^{(b)}$and $[\lambda_i^{(\upsilon)}(3)]^{(b)}$ of departure respectively from the reliability states subsets $\{1, 2, 3\}$, $\{2, 3\}$, $\{3\}$ while the container gantry crane is operating at the operation state z_b, $b = 1,2,\ldots,6$, existing in (7.59), we can use statistical data from experts partly presented in "Data Collections Coming from Experts". The statistical data collected by experts operating the gantry crane and resulting from (5.22) formula

$$[\lambda_i^{(\upsilon)}(u)]^{(b)} = \frac{1}{[\hat{\mu}_i^{(\upsilon)}(u)]^{(b)}},\qquad \upsilon = 1,2,..,5,\quad u = 1,2,3,\quad b = 1,2,\ldots,6, \tag{7.60}$$

application yield the approximate values of the system components, unknown intensities of departures.

Table 7.2 The approximate mean values $[\hat{\mu}_i^{(1)}(u)]^{(b)}$ of subsystem S_1 components, conditional lifetimes $[T_i^{(1)}(u)]^{(b)}$ at operation states z_b

Component of subsystem S_1		$E_1^{(1)}$	$E_2^{(1)}$	$E_3^{(1)}$	$E_4^{(1)}$	$E_5^{(1)}$	$E_6^{(1)}$	$E_7^{(1)}$
Operation state z_b	Reliability state subset $\{u, u+1, \ldots, 3\}$	Mean value $[\hat{\mu}_i^{(1)}(u)]^{(b)}$ (*in years*)						
z_1	{1,2,3}	50	25	25	50	47	55	30
	{2,3}	30	20	20	30	39	35	25
	{3}	20	15	15	20	23	25	20
z_2	{1,2,3}	50	25	25	50	45	55	30
	{2,3}	30	20	20	30	37	35	25
	{3}	20	15	15	20	21	25	20
z_3	{1,2,3}	50	25	25	50	45	55	28
	{2,3}	30	20	20	30	37	35	23
	{3}	20	15	15	20	21	25	18
z_4	{1,2,3}	35	20	25	50	45	53	28
	{2,3}	25	15	20	30	37	33	23
	{3}	15	10	15	20	21	23	18
z_5	{1,2,3}	35	20	25	50	45	53	28
	{2,3}	25	15	20	30	37	33	23
	{3}	15	10	15	20	21	23	18
z_6	{1,2,3}	50	25	25	50	42	50	27
	{2,3}	30	20	20	30	34	30	22
	{3}	20	15	15	20	18	20	17

For instance, substituting into (7.60) the values of the mean lifetimes

$$[\hat{\mu}_1^{(1)}(1)]^{(1)} = 50, \quad [\hat{\mu}_1^{(1)}(2)]^{(1)} = 30, \quad [\hat{\mu}_1^{(1)}(3)]^{(1)} = 20,$$

taken from the third, fourth and fifth rows of the third column of Table 7.2, we obtain the approximate evaluations of the unknown intensities of departure of the component $E_1^{(1)}$ of subsystem S_1 from the reliability states subset {1, 2, 3}, {2, 3}, {3} while the container gantry crane is operating at the operation state z_1 that respectively amount to:

$$[\lambda_1^{(1)}(1)]^{(1)} = \frac{1}{[\hat{\mu}_1^{(1)}(1)]^{(1)}} = \frac{1}{50} = 0.020, \; [\lambda_1^{(1)}(2)]^{(1)} = \frac{1}{[\hat{\mu}_1^{(1)}(2)]^{(1)}} = \frac{1}{30} \cong 0.033,$$
$$[\lambda_1^{(1)}(3)]^{(1)} = \frac{1}{[\hat{\mu}_1^{(1)}(3)]^{(1)}} = \frac{1}{20} = 0.050. \tag{7.61}$$

The evaluations of all unknown intensities of departure from the reliability states subset {1, 2, 3}, {2, 3}, {3} of components of the container gantry crane operating at various operation states, obtained in this way, allows to determine the system components' reliability functions.

Substituting the evaluation given by (7.61) and the remaining evaluations of the intensities of the subsystems components, departures from the reliability state subsets respectively into the formulae (7.59), we get the exponential coordinates of the container gantry crane components, four-state reliability functions defined by (7.58) which after arbitrary acceptance in "Identifying Container Gantry Crane Components Exponential Reliability Functions" are used in Sect. 7.2.5.7 for the prediction of the system reliability characteristics.

Identifying Container Gantry Crane Components, Exponential Reliability Functions

As there are no data collected from the container gantry crane components reliability states changing processes, it is not possible to verify the hypotheses on the exponential forms of the container gantry crane components, conditional reliability functions. Consequently, we arbitrarily assume that these reliability functions are exponential and using the results partly presented in the previous section and the relationships (7.58–7.59) given in Sect. 7.2.5.5 we fix their forms.

For instance, using evaluations (7.61) and expressions (7.58–7.59), we conclude that at the system operation state z_1 subsystem S_1 component $E_1^{(1)}$ has the conditional reliability function

$$[R_1^{(1)}(t,\cdot)]^{(1)} = [1, [R_1^{(1)}(t,1)]^{(1)}, [R_1^{(1)}(t,2)]^{(1)}, [R_1^{(1)}(t,3)]^{(1)}], \; t \geq 0,$$

with the exponential coordinates

$$[R_1^{(1)}(t,1)]^{(1)} = \exp[-0.020t], \; [R_1^{(1)}(t,2)]^{(1)} = \exp[-0.033t],$$
$$[R_1^{(1)}(t,3)]^{(1)} = \exp[-0.050t].$$

In this way, arbitrarily fixed the exponential four-state reliability functions of the container gantry crane components are used in Sect. 7.2.5.7 for the prediction of the system reliability characteristics.

7.2.5.7 Container Gantry Crane Reliability Prediction

Considering the results of the system components reliability modeling from Sect. 7.2.5.5 concerned with the fixed system reliability structures and their shape parameters and with the assumed exponential models of the reliability functions of the system components at various operation states and the results of the evaluations of the system components, intensities of departures from the reliability state subsets partly presented in Sect. 7.2.5.6, we perform the prediction of the container gantry crane reliability characteristics.

Thus, as we fixed in Sect. 7.2.5.5 that at the system operation state z_1, the container gantry crane is identical with subsystem S_1, which is a four-state series system with its structure shape parameter $n = 7$ and according to (1.22–1.23), its four-state reliability function is given by the vector

$$[\boldsymbol{R}(t,\cdot)]^{(1)} = [1,\ [\boldsymbol{R}(t,1)]^{(1)},\ [\boldsymbol{R}(t,2)]^{(1)},\ [\boldsymbol{R}(t,3)]^{(1)}],\ t \geq 0,$$

with the coordinates

$$[\boldsymbol{R}(t,1)]^{(1)} = \bar{\boldsymbol{R}}_7(t,1) = \prod_{i=i}^{7} [R_i^{(1)}(t,1)]^{(1)},$$

$$[\boldsymbol{R}(t,2)]^{(1)} = \bar{\boldsymbol{R}}_7(t,1) = \prod_{i=i}^{7} [R_i^{(1)}(t,2)]^{(1)},$$

$$[\boldsymbol{R}(t,3)]^{(1)} = \bar{\boldsymbol{R}}_7(t,1) = \prod_{i=i}^{7} [R_i^{(1)}(t,3)]^{(1)}.$$

After substituting in the above expressions for the coordinates, the suitable evaluations of the container gantry crane components intensities of departures from the reliability state subsets partly found in Sect. 7.2.5.6, we get:

$$[\boldsymbol{R}(t,1)]^{(1)} = \exp[-0.020t]\exp[-0.040t]\exp[-0.040t]\exp[-0.020t] \times \exp[-0.020t]\exp[-0.018t]\exp[-0.033t] = \exp[-0.191t], \tag{7.62}$$

$$[\boldsymbol{R}(t,2)]^{(1)} = \exp[-0.033t]\exp[-0.050t]\exp[-0.050t]\exp[-0.033t] \times \exp[-0.030t]\exp[-0.028t]\exp[-0.040t] = \exp[-0.264t], \tag{7.63}$$

$$[\boldsymbol{R}(t,3)]^{(1)} = \exp[-0.050t]\exp[-0.066t]\exp[-0.066t]\exp[-0.050t] \times \exp[-0.040t]\exp[-0.040t]\exp[-0.050t] = \exp[-0.362t]. \tag{7.64}$$

The expected values of the container gantry crane conditional lifetimes in the reliability state subsets {1, 2, 3}, {2, 3}, {3} at the operation state z_1, calculated from the results given by (7.62–7.64), according to (3.8), respectively are:

$$\mu_1(1) \cong 5.24,\ \mu_1(2) \cong 3.79,\ \mu_1(3) \cong 2.76 \text{ years.} \tag{7.65}$$

At the system operation state z_2, the container gantry crane is composed of subsystems S_1, S_2 and S_3 forming a series structure.

At this operation state, subsystem S_1 is a four-state series system with its structure shape parameter $n = 7$ and according to (1.22–1.23), its four-state reliability function is given by the vector

$$[\boldsymbol{R}^{(1)}(t,\cdot)]^{(2)} = [1,\ [\boldsymbol{R}^{(1)}(t,1)]^{(2)},\ [\boldsymbol{R}^{(1)}(t,2)]^{(2)},\ [\boldsymbol{R}^{(1)}(t,3)]^{(2)}],\ t \geq 0,$$

with the coordinates

$$[\boldsymbol{R}^{(1)}(t,1)]^{(2)} = \bar{\boldsymbol{R}}_7(t,1) = \prod_{i=i}^{7} [R_i^{(1)}(t,1)]^{(2)},$$

$$[\boldsymbol{R}^{(1)}(t,2)]^{(2)} = \bar{\boldsymbol{R}}_7(t,1) = \prod_{i=i}^{7} [R_i^{(1)}(t,2)]^{(2)},$$

$$[\boldsymbol{R}^{(1)}(t,3)]^{(2)} = \bar{\boldsymbol{R}}_7(t,1) = \prod_{i=i}^{7} [R_i^{(1)}(t,3)]^{(2)}.$$

After substituting in the above expressions for the coordinates, the suitable evaluations of the container gantry crane components intensities of departures from the reliability state subsets partly found in Sect. 7.2.5.6, we get:

$$[\boldsymbol{R}^{(1)}(t,1)]^{(2)} = \exp[-0.020t]\exp[-0.040t]\exp[-0.040t]\exp[-0.020t] \times \exp[-0.022t]\exp[-0.018t]\exp[-0.033t] = \exp[-0.193t], \tag{7.66}$$

$$[\boldsymbol{R}^{(1)}(t,2)]^{(2)} = \exp[-0.033t]\exp[-0.050t]\exp[-0.050t]\exp[-0.033t] \times \exp[-0.027t]\exp[-0.028t]\exp[-0.040t] = \exp[-0.261t], \tag{7.67}$$

$$[\boldsymbol{R}^{(1)}(t,3)]^{(2)} = \exp[-0.050t]\exp[-0.066t]\exp[-0.066t]\exp[-0.050t] \times \exp[-0.048t]\exp[-0.040t]\exp[-0.050t] = \exp[-0.370t]. \tag{7.68}$$

Subsystem S_2 at the operation state z_2, is a four-state series system with its structure shape parameter $n = 6$ and according to (1.22–1.23), its four-state reliability function is given by the vector

$$[\boldsymbol{R}^{(2)}(t,\cdot)]^{(2)} = [1,\ [\boldsymbol{R}^{(2)}(t,1)]^{(2)},\ [\boldsymbol{R}^{(2)}(t,2)]^{(2)},\ [\boldsymbol{R}^{(2)}(t,3)]^{(2)}],\ t \geq 0,$$

with the coordinates

$$[\boldsymbol{R}^{(2)}(t,1)]^{(2)} = \bar{\boldsymbol{R}}_6(t,1) = \prod_{i=i}^{6} [R_i^{(2)}(t,1)]^{(2)},$$

$$[\boldsymbol{R}^{(2)}(t,2)]^{(2)} = \bar{\boldsymbol{R}}_6(t,1) = \prod_{i=i}^{6} [R_i^{(2)}(t,2)]^{(2)},$$

$$[\boldsymbol{R}^{(2)}(t,3)]^{(2)} = \bar{\boldsymbol{R}}_6(t,1) = \prod_{i=i}^{6} [R_i^{(2)}(t,3)]^{(2)}.$$

After substituting in the above expressions for the coordinates, the suitable evaluations of the container gantry crane components intensities of departures from the reliability state subsets, we get:

$$[\boldsymbol{R}^{(2)}(t,1)]^{(2)} = \exp[-0.053t]\exp[-0.048t]\exp[-0.048t]\exp[-0.048t]\exp[-0.020t] \times \exp[-0.018t] = \exp[-0.235t], \tag{7.69}$$

$$[\boldsymbol{R}^{(2)}(t,2)]^{(2)} = \exp[-0.059t]\exp[-0.053t]\exp[-0.053t]\exp[-0.053t]\exp[-0.025t] \times \exp[-0.029t] = \exp[-0.272t], \tag{7.70}$$

$$[\boldsymbol{R}^{(2)}(t,3)]^{(2)} = \exp[-0.066t]\exp[-0.059t]\exp[-0.059t]\exp[-0.059t]\exp[-0.033t] \times \exp[-0.040t] = \exp[-0.316t]. \tag{7.71}$$

Subsystem S_3 at the operation state z_2, is a four-state series system with its structure shape parameter $n = 3$ and according to (1.22–1.23), its four-state reliability function is given by the vector

$$[\boldsymbol{R}^{(3)}(t, \cdot)]^{(2)} = [1, \ [\boldsymbol{R}^{(3)}(t,1)]^{(2)}, \ [\boldsymbol{R}^{(3)}(t,2)]^{(2)}, \ [\boldsymbol{R}^{(3)}(t,3)]^{(2)}], \quad t \geq 0,$$

with the coordinates

$$[\boldsymbol{R}^{(3)}(t,1)]^{(2)} = \bar{\boldsymbol{R}}_3(t,1) = \prod_{i=i}^{3} [R_i^{(3)}(t,1)]^{(2)}, [\boldsymbol{R}^{(3)}(t,2)]^{(2)}$$
$$= \bar{\boldsymbol{R}}_3(t,1) = \prod_{i=i}^{3} [R_i^{(3)}(t,2)]^{(2)},$$
$$[\boldsymbol{R}^{(3)}(t,3)]^{(2)} = \bar{\boldsymbol{R}}_3(t,1) = \prod_{i=i}^{3} [R_i^{(3)}(t,3)]^{(2)}.$$

After substituting in the above expressions for the coordinates, the suitable evaluations of the container gantry crane components intensities of departures from the reliability state subsets, we get:

$$[\boldsymbol{R}^{(3)}(t,1)]^{(2)} = \exp[-0.025t]\exp[-0.033t]\exp[-0.033t] = \exp[-0.091t], \quad (7.72)$$

$$[\boldsymbol{R}^{(3)}(t,2)]^{(2)} = \exp[-0.040t]\exp[-0.040t]\exp[-0.040t] = \exp[-0.120t], \quad (7.73)$$

$$[\boldsymbol{R}^{(3)}(t,3)]^{(2)} = \exp[-0.066t]\exp[-0.066t]\exp[-0.066t] = \exp[-0.198t]. \quad (7.74)$$

Considering that the container gantry crane at the operation state z_2 is a four-state series system composed of subsystems S_1, S_2 and S_3, after applying (1.22–1.23), its conditional four-state reliability function is given by the vector

$$[\boldsymbol{R}(t, \cdot)]^{(2)} = [1, \ [\boldsymbol{R}(t,1)]^{(2)}, \ [\boldsymbol{R}(t,2)]^{(2)}, \ [\boldsymbol{R}(t,3)]^{(2)}], \ t \geq 0,$$

with the coordinates

$$[\boldsymbol{R}(t, 1)]^{(2)} = \bar{\boldsymbol{R}}_3(t, 1) = [\boldsymbol{R}^{(1)}(t, 1)]^{(2)}[\boldsymbol{R}^{(2)}(t, 1)]^{(2)}[\boldsymbol{R}^{(3)}(t, 1)]^{(2)},$$

$$[\boldsymbol{R}(t, 2)]^{(2)} = \bar{\boldsymbol{R}}_3(t, 2) = [\boldsymbol{R}^{(1)}(t, 2)]^{(2)}[\boldsymbol{R}^{(2)}(t, 2)]^{(2)}[\boldsymbol{R}^{(3)}(t, 2)]^{(2)},$$

$$[\boldsymbol{R}(t, 3)]^{(2)} = \bar{\boldsymbol{R}}_3(t, 2) = [\boldsymbol{R}^{(1)}(t, 3)]^{(2)}[\boldsymbol{R}^{(2)}(t, 3)]^{(2)}[\boldsymbol{R}^{(3)}(t, 3)]^{(2)}.$$

After substituting in the above expressions for the coordinates the results (7.66–7.68), (7.69–7.71), and (7.72–7.74), we get:

$$[\boldsymbol{R}(t, 1)]^{(2)} = \exp[-0.193t]\exp[-0.235t]\exp[-0.091t] = \exp[-0.519t], \quad (7.75)$$

$$[\boldsymbol{R}(t, 2)]^{(2)} = \exp[-0.261t]\exp[-0.272t]\exp[-0.120t] = \exp[-0.653t], \quad (7.76)$$

$$[\boldsymbol{R}(t, 3)]^{(2)} = \exp[-0.370t]\exp[-0.316t]\exp[-0.198t] = \exp[-0.884t]. \quad (7.77)$$

The expected values of the container gantry crane conditional lifetimes in the reliability state subsets {1, 2, 3}, {2, 3}, {3} at the operation state z_2, calculated from the results given by (7.75–7.77), according to (3.8), respectively are:

$$\mu_2(1) \cong 1.93,\ \mu_2(2) \cong 1.53,\ \mu_2(3) \cong 1.13 \text{ year}. \tag{7.78}$$

After proceeding in the analogous way in the system reliability analysis and evaluation at the remaining operation states z_3, z_4, z_5 and z_6, we may determine the system conditional reliability function presented below.

At the operation state z_3, the container gantry crane conditional reliability function of the system is given by the vector

$$[\boldsymbol{R}(t,\ \cdot)]^{(3)} = [1,\ [\boldsymbol{R}(t,1)]^{(3)},\ [\boldsymbol{R}(t,2)]^{(3)},\ [\boldsymbol{R}(t,3)]^{(3)}],\ t \geq 0,$$

with the coordinates

$$[\boldsymbol{R}(t,\ 1)]^{(3)} = \exp[-0.196t]\exp[-0.235t]\exp[-0.091t] = \exp[-0.522t], \tag{7.79}$$

$$[\boldsymbol{R}(t,\ 2)]^{(3)} = \exp[-0.264t]\exp[-0.272t]\exp[-0.120t] = \exp[-0.656t], \tag{7.80}$$

$$[\boldsymbol{R}(t,\ 3)]^{(3)} = \exp[-0.375t]\exp[-0.316t]\exp[-0.198t] = \exp[-0.889t]. \tag{7.81}$$

The expected values of the container gantry crane conditional lifetimes in the reliability state subsets {1, 2, 3}, {2, 3}, {3} at the operation state z_3, calculated from the results given by (7.79–7.81), according to (3.8), respectively are:

$$\mu_3(1) \cong 1.91,\ \mu_3(2) \cong 1.52,\ \mu_3(3) \cong 1.12 \text{ year}. \tag{7.82}$$

At the operation state z_4, the container gantry crane conditional reliability function of the system is given by the vector

$$[\boldsymbol{R}(t,\ \cdot)]^{(4)} = [1,\ [\boldsymbol{R}(t,1)]^{(4)},\ [\boldsymbol{R}(t,2)]^{(4)},\ [\boldsymbol{R}(t,3)]^{(4)}],\ t \geq 0,$$

with the coordinates

$$\begin{aligned}[\boldsymbol{R}(t,\ 1)]^{(4)} &= \exp[-0.216t]\exp[-0.241t]\exp[-0.061t]\exp[-0.029t] \\ &= \exp[-0.547t],\end{aligned} \tag{7.83}$$

$$\begin{aligned}[\boldsymbol{R}(t,\ 2)]^{(4)} &= \exp[-0.289t]\exp[-0.278t]\exp[-0.091t]\exp[-0.04t] \\ &= \exp[-0.698t],\end{aligned} \tag{7.84}$$

$$\begin{aligned}[\boldsymbol{R}(t,\ 3)]^{(4)} &= \exp[-0.428t]\exp[-0.328t]\exp[-0.133t]\exp[-0.066t] \\ &= \exp[-0.955t].\end{aligned} \tag{7.85}$$

The expected values and of the container gantry crane conditional lifetimes in the reliability state subsets {1, 2, 3}, {2, 3}, {3} at the operation state z_4, calculated from the results given by (7.83–7.85), according to (3.8), respectively are:

$$\mu_4(1) \cong 1.83,\ \mu_4(2) \cong 1.43,\ \mu_4(3) \cong 1.05 \text{ year}. \tag{7.86}$$

At the operation state z_5, the container gantry crane conditional reliability function of the system is given by the vector

$$[\boldsymbol{R}(t,\ \cdot)]^{(5)} = [1,\ [\boldsymbol{R}(t,1)]^{(5)},\ [\boldsymbol{R}(t,2)]^{(5)},\ [\boldsymbol{R}(t,3)]^{(5)}],\ t \geq 0,$$

with the coordinates

$$\begin{aligned}[\boldsymbol{R}(t,\ 1)]^{(5)} &= \exp[-0.216t]\exp[-0.241t]\exp[-0.061t]\exp[-0.025t] \\ &= \exp[-0.543t],\end{aligned} \tag{7.87}$$

$$\begin{aligned}[\boldsymbol{R}(t,\ 2)]^{(5)} &= \exp[-0.289t]\exp[-0.278t]\exp[-0.091t]\exp[-0.029t] \\ &= \exp[-0.687t],\end{aligned} \tag{7.88}$$

$$\begin{aligned}[\boldsymbol{R}(t,\ 3)]^{(5)} &= \exp[-0.428t]\exp[-0.328t]\exp[-0.133t]\exp[-0.050t] \\ &= \exp[-0.939t].\end{aligned} \tag{7.89}$$

The expected values of the container gantry crane conditional lifetimes in the reliability state subsets {1, 2, 3}, {2, 3}, {3} at the operation state z_5, calculated from the results given by (7.87–7.89), according to (3.8), respectively are:

$$\mu_5(1) \cong 1.84,\ \mu_5(2) \cong 1.46,\ \mu_5(3) \cong 1.06 \text{ year}. \tag{7.90}$$

At the operation state z_6, the container gantry crane conditional reliability function of the system is given by the vector

$$[\boldsymbol{R}(t,\ \cdot)]^{(6)} = [1,\ [\boldsymbol{R}(t,1)]^{(6)},\ [\boldsymbol{R}(t,2)]^{(6)},\ [\boldsymbol{R}(t,3)]^{(6)}],\quad t \geq 0,$$

with the coordinates

$$\begin{aligned}[\boldsymbol{R}(t,\ 1)]^{(6)} &= \exp[-0.201t]\exp[-0.250t]\exp[-0.087t]\exp[-0.080t] \\ &= \exp[-0.618t],\end{aligned} \tag{7.91}$$

$$\begin{aligned}[\boldsymbol{R}(t,\ 2)]^{(6)} &= \exp[-0.273t]\exp[-0.29t]\exp[-0.12t]\exp[-0.1t] \\ &= \exp[-0.783t],\end{aligned} \tag{7.92}$$

$$\begin{aligned}[\boldsymbol{R}(t,\ 3)]^{(6)} &= \exp[-0.396t]\exp[-0.337t]\exp[-0.16t]\exp[-0.132t] \\ &= \exp[-1.025t].\end{aligned} \tag{7.93}$$

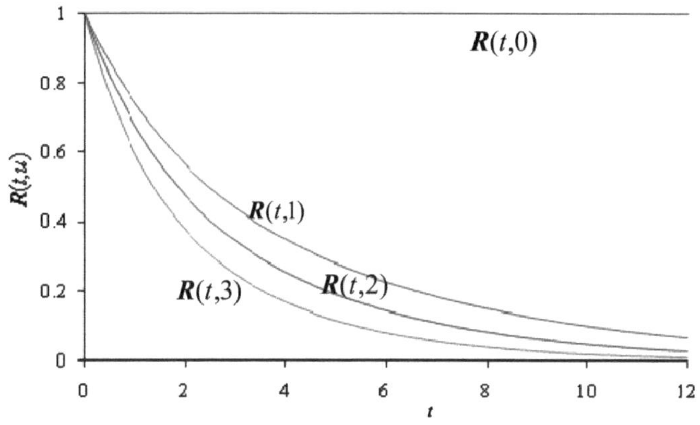

Fig. 7.22 The graph of the container gantry crane reliability function $\boldsymbol{R}(t, \cdot)$ coordinates

The expected values of the container gantry crane conditional lifetimes in the reliability state subsets {1, 2, 3}, {2, 3}, {3} at the operation state z_6, calculated from the results (7.91–7.93), according to (3.8), respectively are:

$$\mu_6(1) \cong 1.62, \quad \mu_6(2) \cong 1.28, \quad \mu_6(3) \cong 0.98 \text{ year.} \tag{7.94}$$

In the case when the operation time is large enough the unconditional four-state reliability function of the container gantry crane is given by the vector

$$\boldsymbol{R}(t, \cdot) = [1, \boldsymbol{R}(t, 1), \boldsymbol{R}(t, 2), \boldsymbol{R}(t, 3)], \quad t \geq 0, \tag{7.95}$$

where according to (3.5–3.6) and considering (7.56), the vector coordinates are given respectively by

$$\boldsymbol{R}(t, 1) = 0.6874 \cdot [\boldsymbol{R}(t,1)]^{(1)} + 0.0187 \cdot [\boldsymbol{R}(t,1)]^{(2)} + 0.0515 \cdot [\boldsymbol{R}(t,1)]^{(3)} + 0.0005 \cdot [\boldsymbol{R}(t,1)]^{(4)} + 0.0717 \cdot [\boldsymbol{R}(t,1)]^{(5)} + 0.1702 \cdot [\boldsymbol{R}(t,1)]^{(6)} \quad \text{for } t \geq 0, \tag{7.96}$$

$$\boldsymbol{R}(t, 2) = 0.6874 \cdot [\boldsymbol{R}(t,2)]^{(1)} + 0.0187 \cdot [\boldsymbol{R}(t,2)]^{(2)} + 0.0515 \cdot [\boldsymbol{R}(t,2)]^{(3)} + 0.0005 \cdot [\boldsymbol{R}(t,2)]^{(4)} + 0.0717 \cdot [\boldsymbol{R}(t,2)]^{(5)} + 0.1702 \cdot [\boldsymbol{R}(t,2)]^{(6)} \quad \text{for } t \geq 0, \tag{7.97}$$

$$\boldsymbol{R}(t, 3) = 0.6874 \cdot [\boldsymbol{R}(t,3)]^{(1)} + 0.0187 \cdot [\boldsymbol{R}(t,3)]^{(2)} + 0.0515 \cdot [\boldsymbol{R}(t,3)]^{(3)} + 0.0005 \cdot [\boldsymbol{R}(t,3)]^{(4)} + 0.0717 \cdot [\boldsymbol{R}(t,3)]^{(5)} + 0.1702 \cdot [\boldsymbol{R}(t,3)]^{(6)} \quad \text{for } t \geq 0, \tag{7.98}$$

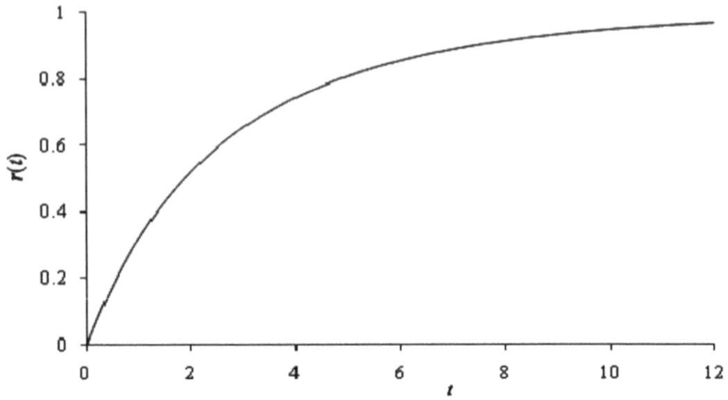

Fig. 7.23 The graph of the container gantry crane risk function $\boldsymbol{r}(t)$

and the coordinates $[\boldsymbol{R}(t,1)]^{(1)}$, $[\boldsymbol{R}(t,1)]^{(2)}$, $[\boldsymbol{R}(t,1)]^{(3)}$, $[\boldsymbol{R}(t,1)]^{(4)}$, $[\boldsymbol{R}(t,1)]^{(5)}$, $[\boldsymbol{R}(t,1)]^{(6)}$, are given by (7.62), (7.75), (7.79), (7.83), (7.87), (7.91), and $[\boldsymbol{R}(t,2)]^{(1)}$, $[\boldsymbol{R}(t,2)]^{(2)}$, $[\boldsymbol{R}(t,2)]^{(3)}$, $[\boldsymbol{R}(t,2)]^{(4)}$, $[\boldsymbol{R}(t,2)]^{(5)}$, $[\boldsymbol{R}(t,2)]^{(6)}$, are given by (7.63), (7.76), (7.80), (7.84), (7.88), (7.92), and $[\boldsymbol{R}(t,3)]^{(1)}$, $[\boldsymbol{R}(t,3)]^{(2)}$, $[\boldsymbol{R}(t,3)]^{(3)}$, $[\boldsymbol{R}(t,3)]^{(4)}$, $[\boldsymbol{R}(t,3)]^{(5)}$, $[\boldsymbol{R}(t,3)]^{(6)}$,are given by (7.64), (7.77), (7.81), (7.85), (7.89), (7.93).

The graphs of the coordinates of the container gantry crane reliability function are presented in Fig. 7.22.

The expected values and standard deviations of the container gantry crane unconditional lifetimes in the reliability state subsets {1, 2, 3}, {2, 3}, {3} calculated from the results given by (7.96–7.98), according to (3.7–3.9) and considering (7.56), (7.65), (7.78), (7.82), (7.86), (7.90), (7.94), respectively are:

$$\mu(1)=0.6874\times5.24+0.0187\times1.93+0.0515\times1.91+0.0005\times1.83+0.0717\times1.84$$
$$+0.1702\times1.62\cong4.14\text{years},$$
$$\sigma(1)\cong4.71\text{years}, \quad (7.99)$$

$$\mu(2)=0.6874\times3.79+0.0187\times1.53+0.0515\times1.52+0.0005\times1.43+0.0717\times1.46$$
$$+0.1702\times1.28\cong3.04\text{years}, \quad (7.100)$$

$$\sigma(2)\cong3.43\text{years}, \quad (7.101)$$

$$\mu(3)=0.6874\times2.76+0.0187\times1.13+0.0515\times1.12+0.0005\times1.05$$
$$+0.0717\times1.06+0.1702\times0.98\cong2.22\text{ years},$$
$$\sigma(3)\cong2.50\text{ years}, \quad (7.102)$$

Further, considering (3.10) and (7.99), (7.100) and (7.102), the mean values of the unconditional lifetimes in the particular reliability states 1, 2, 3 respectively are:

$$\bar{\mu}(1) = \mu(1) - \mu(2) = 1.10,\ \bar{\mu}(2) = \mu(2) - \mu(3) = 0.82,$$
$$\bar{\mu}(3) = \mu(3) = 2.22 \text{ years.} \quad (7.103)$$

Since the critical reliability state is $r = 2$, the system risk function, according to (3.11), is given by

$$\boldsymbol{r}(t) = 1 - \boldsymbol{R}(t, 2), \quad (7.104)$$

where $\boldsymbol{R}$(t, 2) is given by (7.97).

Hence, the moment when the system risk function exceeds a permitted level, for instance $\delta = 0.05$, from (3.12), is

$$\tau = \boldsymbol{r}^{-1}(\delta) \cong 0.126 \text{ year.} \quad (7.105)$$

The graph of the risk function $\boldsymbol{r}$(t) of the container gantry crane operating at the variable conditions is given in Fig. 7.23.

7.2.5.8 Container Gantry Crane Renewal and Availability Prediction

Using the results of the container gantry crane reliability prediction given by (7.100–7.101) and the results of the classical renew theory presented in Sect. 3.4, we may predict the renewal and availability characteristics of this system in the case when it is repairable and its time of renovation is either ignored or non-ignored.

First, assuming that the container gantry crane is repaired after the exceeding of its reliability critical state $r = 2$ and that the time of the system renovation is ignored and applying *Proposition 3.4*, we obtain the following results:

(a) the time $S_N(2)$ until the Nth exceeding by the system the reliability critical state $r = 2$, for sufficiently large N, has approximately normal distribution $N(3.04\,N, 3.43\sqrt{N})$, i.e.,

$$F^{(N)}(t,2) = P(S_N(2) < t) \cong F_{N(0,1)}\left(\frac{t - 3.04\,N}{3.43\sqrt{N}}\right), \quad t \in (-\infty, \infty);$$

(b) the expected value and the variance of the time $S_N(2)$ until the Nth exceeding by the system the reliability critical state $r = 2$ are respectively given by

$$E[S_N(2)] \cong 3.04\,N,\ D[S_N(2)] \cong 11.76\,N;$$

(c) the number $N(t, 2)$ of exceedings by the system the reliability critical state $r = 2$ up to the moment t, $t \geq 0$, for sufficiently large t, approximately has the distribution of the form

$$P(N(t,2) = N) \cong F_{N(0,1)}\left(\frac{3.04(N+1)-t}{1.967\sqrt{t}}\right) - F_{N(0,1)}\left(\frac{3.04N-t}{1.967\sqrt{t}}\right),$$
$$N = 0, 1, \ldots;$$

(d) the expected value and the variance of the number $N(t, 2)$ of exceedings by the system the reliability critical state $r = 2$ up to the moment t, $t \geq 0$, for sufficiently large t, approximately are respectively given by

$$H(t,2) = 0.3289t,\ D(t,2) = 0.419t.$$

Further, assuming that the container gantry crane is repaired after the exceeding of its reliability critical state $r = 2$, the time of the system renovation is not ignored and it has the mean value $\mu_0(2) = 0.0027$, the standard deviation $\sigma_0(2) = 0.0014$ and applying Proposition 3.5, we obtain the following results:

(a) the time $\bar{S}_N(2)$ until the Nth exceeding by the system the reliability critical state $r = 2$, for sufficiently large N, has approximately normal distribution $N(3.04\,N + 0.0027(N-1), \sqrt{11.76\,N - 0.0000019(N-1)})$, i.e.,

$$\bar{F}^{(N)}(t,2) = P(\bar{S}_N(2) < t) \cong F_{N(0,1)}\left(\frac{t - 3.0427\,N + 0.0027}{\sqrt{11.760002\,N - 0.0000019}}\right),$$
$$t \in (-\infty, \infty);$$

(b) the expected value and the variance of the time $\bar{S}_N(2)$ until the Nth exceeding by the system the reliability critical state $r = 2$, for sufficiently large N, are respectively given by

$$E[\bar{S}_N(2)] \cong 3.04\,N + 0.0027(N-1), \quad D[\bar{S}_N(2)] \cong 11.76\,N + 0.0000019(N-1);$$

(c) the number $\bar{N}(t,2)$ of exceeding by the system the reliability critical state $r = 2$ up to the moment t, $t \geq 0$, for sufficiently large t, has approximately distribution of the form

$$P(\bar{N}(t,2) = N) \cong F_{N(0,1)}\left(\frac{3.0427(N+1) - t - 0.0027}{1.966\sqrt{t + 0.0027}}\right)$$
$$- F_{N(0,1)}\left(\frac{3.0427\,N - t - 0.0027}{1.966\sqrt{t + 0.005}}\right),\ N = 0, 1, \ldots;$$

(d) the expected value and the variance of the number $\bar{N}(t,2)$ of exceedings by the system the reliability critical state $r = 2$ up to the moment t, $t \geq 0$, for sufficiently large t, are respectively given by

$$\bar{H}(t,2) \cong 0.329(t + 0.0027), \quad \bar{D}(t,2) \cong 0.417(t + 0.0027);$$

(e) the time $\bar{\bar{S}}_N(2)$ until the Nth system's renovation, for sufficiently large N, has approximately normal distribution $N(3.0427\,N, 3.429\sqrt{N})$, i.e.,

$$\bar{\bar{F}}^{(N)}(t,2) = P(\bar{\bar{S}}_N(2) < t) \cong F_{N(0,1)}\left(\frac{t - 3.0427\,N}{3.429\sqrt{N}}\right),\ t \in (-\infty, \infty);$$

(f) the expected value and the variance of the time $\bar{\bar{S}}_N(2)$ until the Nth system's renovation, for sufficiently large N, are respectively given by

$$E[\bar{\bar{S}}_N(2)] \cong 3.0427\,N,\ D[\bar{\bar{S}}_N(2)] \cong 11.76002\,N;$$

(g) the number $\bar{\bar{N}}(t,2)$ of the system's renovations up to the moment t, $t \geq 0$, for sufficiently large t, has approximately distribution of the form

$$P(\bar{\bar{N}}(t,2) = N) \cong F_{N(0,1)}\left(\frac{3.0427(N+1) - t}{1.966\sqrt{t}}\right) - F_{N(0,1)}\left(\frac{3.0427\,N - t}{1.966\sqrt{t}}\right),\ N = 0, 1, \ldots;$$

(h) the expected value and the variance of the number $\bar{\bar{N}}(t,2)$ of system's renovations up to the moment t, $t \geq 0$, for sufficiently large t, are respectively given by

$$\bar{\bar{H}}(t,2) \cong 0.3286t, \quad \bar{\bar{D}}(t,2) \cong 0.417t;$$

(i) the steady availability coefficient of the system at the moment t, $t \geq 0$, for sufficiently large t, is given by

$$A(t,2) \cong 0.9989, \quad t \geq 0;$$

(j) the steady availability coefficient of the system in the time interval $<t, t+\tau)$, $\tau > 0$, for sufficiently large t, is given by

$$A(t,\tau,2) \cong 0.329 \int_{\tau}^{\infty} \boldsymbol{R}(t,2)dt, \quad t \geq 0, \quad \tau > 0,$$

where $\boldsymbol{R}$(t, 2) is given by (7.97).

7.2.5.9 Container Gantry Crane Operation Process Optimization

Optimal Transient Probabilities of Container Gantry Crane Operation Process at Operation States

Considering Eqs. 7.95–7.98, it is natural to assume that the container gantry crane operation process has a significant influence on the system reliability. This influence is also clearly expressed in the formulae (7.99), (7.100) and (7.102) for the mean values of the container gantry crane unconditional lifetimes in the reliability state subsets that can be used for this system operation process optimization performed in accordance with the procedure proposed in Sect. 6.2.1.

The objective function defined by (6.1), in this case as the container gantry crane critical reliability state is $r = 2$, takes the form

$$\mu(2) = p_1 \times 3.79 + p_2 \times 1.53 + p_3 \times 1.52 + p_4 \times 1.43 + p_5 \times 1.46 + p_6 \times 1.28. \tag{7.106}$$

The lower $\breve{p}_b$ and upper $\widehat{p}_b$ bounds of the unknown transient probabilities p_b, $b = 1, 2, \ldots, 6$, defined by (6.5), coming from experts respectively are:

$$\breve{p}_1 = 0.50,\ \breve{p}_2 = 0.01,\ \breve{p}_3 = 0.03,\ \breve{p}_4 = 0.0004,\ \breve{p}_5 = 0.05,\ \breve{p}_6 = 0.09;$$
$$\widehat{p}_1 = 0.90,\ \widehat{p}_2 = 0.03,\ \widehat{p}_3 = 0.07,\ \widehat{p}_4 = 0.0007,\ \widehat{p}_5 = 0.09,\ \widehat{p}_6 = 0.30.$$

Therefore, according to (6.2–6.3), we assume the following bound constraints

$$\begin{aligned} &0.50 \le p_1 \le 0.90,\ 0.01 \le p_2 \le 0.03,\ 0.03 \le p_3 \le 0.07,\\ &0.0004 \le p_4 \le 0.0007,\ 0.05 \le p_5 \le 0.09,\ 0.09 \le p_6 \le 0.30,\\ &\sum_{b=1}^{6} p_b = 1. \end{aligned} \tag{7.107}$$

Now, in order to find the optimal values $\dot{p}_b$ of the transient probabilities p_b, $b = 1, 2, \ldots, 6$, that maximize the objective function (7.106), we arrange the system conditional lifetimes mean values $\mu_b(2)$, $b = 1, 2, \ldots, 6$, in non-increasing order

$$\mu_1(2) \ge \mu_2(2) \ge \mu_3(2) \ge \mu_5(2) \ge \mu_4(2) \ge \mu_6(2).$$

Next, according to (6.6), we substitute

$$x_1 = p_1,\ x_2 = p_2,\ x_3 = p_3,\ x_4 = p_5,\ x_5 = p_4,\ x_6 = p_6, \tag{7.108}$$

and

$$\begin{aligned} &\breve{x}_1 = \breve{p}_1 = 0.50,\ \breve{x}_2 = \breve{p}_2 = 0.01,\ \breve{x}_3 = \breve{p}_3 = 0.03,\ \breve{x}_4 = \breve{p}_5 = 0.05,\\ &\breve{x}_5 = \breve{p}_4 = 0.0004,\ \breve{x}_6 = \breve{p}_6 = 0.09,\\ &\widehat{x}_1 = \widehat{p}_1 = 0.90,\ \widehat{x}_2 = \widehat{p}_2 = 0.03,\ \widehat{x}_3 = \widehat{p}_3 = 0.07,\ \widehat{x}_4 = \widehat{p}_5 = 0.09,\\ &\widehat{x}_5 = \widehat{p}_4 = 0.0007,\ \widehat{x}_6 = \widehat{p}_6 = 0.30, \end{aligned} \tag{7.109}$$

and we maximize with respect to x_i, $i = 1, 2, \ldots, 6$, the linear form (7.106) which according to (6.7–6.10) takes the form

$$\mu(2) = x_1 \times 3.79 + x_2 \times 1.53 + x_3 \times 1.52 + x_4 \times 1.46 + x_5 \times 1.43 + x_6 \times 1.28, \tag{7.110}$$

with the following bound constraints

$0.50 \le x_1 \le 0.90,\ 0.01 \le x_2 \le 0.03,\ 0.03 \le x_3 \le 0.07,\ 0.05 \le x_4 \le 0.09,$
$0.0004 \le x_5 \le 0.0007,\ 0.09 \le x_6 \le 0.30,$

$$\sum_{i=1}^{6} x_i = 1. \tag{7.111}$$

According to (6.11), we calculate

$$\breve{x} = \sum_{i=1}^{6} \breve{x}_i = 0.6804, \quad \hat{y} = 1 - \breve{x} = 1 - 0.6804 = 0.3196 \tag{7.112}$$

and according to (6.12), we find

$$\begin{aligned} &\breve{x}^{0} = 0,\ \hat{x}^{0} = 0,\ \hat{x}^{0} - \breve{x}^{0} = 0, \\ &\breve{x}^{1} = 0.50,\ \hat{x}^{1} = 0.90,\ \hat{x}^{1} - \breve{x}^{1} = 0.40, \\ &\breve{x}^{2} = 0.51,\ \hat{x}^{2} = 0.93,\ \hat{x}^{2} - \breve{x}^{2} = 0.42, \\ &\breve{x}^{6} = 0.6804,\ \hat{x}^{6} = 1.3907,\ \hat{x}^{6} - \breve{x}^{6} = 0.7103. \end{aligned} \tag{7.113}$$

By (7.113), the inequality (6.13) takes the form

$$\hat{x}^{I} - \breve{x}^{I} < 0.3196. \tag{7.114}$$

Thus, from the above and from (7.113), it follows that the largest value $I \in \{0, 1, \ldots, 6\}$ such that the inequality (7.114) is satisfied, is $I = 0$.

Therefore, we fix the optimal solution that maximizes linear function (7.110) according to the rule (6.14). Namely, we get

$$\begin{aligned} &\dot{x}_1 = \hat{y} + \breve{x}_1 = 0.3196 + 0.5 = 0.8196, \\ &\dot{x}_2 = \breve{x}_2 = 0.01,\ \dot{x}_3 = \breve{x}_3 = 0.03,\ \dot{x}_4 = \breve{x}_4 = 0.05, \\ &\dot{x}_5 = \breve{x}_5 = 0.0004,\ \dot{x}_6 = \breve{x}_6 = 0.09. \end{aligned} \tag{7.115}$$

Finally, after making the substitution inverse to (7.108), we get the optimal transient probabilities

$$\begin{aligned} &\dot{p}_1 = \dot{x}_1 = 0.8196,\ \dot{p}_2 = \dot{x}_1 = 0.01,\ \dot{p}_3 = \dot{x}_3 = 0.03,\ \dot{p}_4 = \dot{x}_5 = 0.0004, \\ &\dot{p}_5 = \dot{x}_4 = 0.05,\ \dot{p}_6 = \dot{x}_6 = 0.09, \end{aligned} \tag{7.116}$$

that maximize the system mean lifetime in the reliability state subset {2, 3} expressed by the linear form (7.106) giving, according (6.18) and (7.116), its optimal value

$$\dot{\mu}(2) = \dot{p}_1 \times 3.79 + \dot{p}_2 \times 1.53 + \dot{p}_3 \times 1.52 + \dot{p}_4 \times 1.43 + \dot{p}_5 \times 1.46 + \dot{p}_6 \times 1.28$$
$$= 0.8196 \times 3.79 + 0.01 \times 1.53 + 0.03 \times 1.52 + 0.0004 \times 1.43 + 0.05 \times 1.46$$
$$+ 0.09 \times 1.28 \cong 3.36. \tag{7.117}$$

Optimal Sojourn Times of Container Gantry Crane Operation Process at Operation States

Having the values of the optimal transient probabilities determined by (7.116), it is possible to find the optimal conditional and unconditional mean values of the sojourn times of the container gantry crane operation process at the operation states and the optimal mean values of the total unconditional sojourn times of the container gantry crane operation process at the operation states during the fixed operation time as well.

Substituting the optimal transient probabilities at operation states

$$\dot{p}_1 = 0.8196,\ \dot{p}_2 = 0.01,\ \dot{p}_3 = 0.03,\ \dot{p}_4 = 0.0004,\ \dot{p}_5 = 0.05,\ \dot{p}_6 = 0.09,$$

determined in (7.116) and the steady probabilities

$$\pi_1 \cong 0.0951,\ \pi_2 \cong 0.1020,\ \pi_3 \cong 0.3100,\ \pi_4 \cong 0.0102,\ \pi_5 \cong 0.1547,$$
$$\pi_6 \cong 0.3280.$$

determined by (7.55) into (6.41), we get the following system of equations

$$-0.017156\dot{M}_1 + 0.0835992\dot{M}_2 + 0.254076\dot{M}_3 + 0.0083599\dot{M}_4 + 0.1267921\dot{M}_5$$
$$+ 0.2688288\dot{M}_6 = 0$$
$$0.000951\dot{M}_1 - 0.10098\dot{M}_2 + 0.0031\dot{M}_3 + 0.000102\dot{M}_4 + 0.001547\dot{M}_5$$
$$+ 0.00328\dot{M}_6 = 0$$
$$0.002853\dot{M}_1 + 0.00306\dot{M}_2 - 0.3007\dot{M}_3 + 0.000306\dot{M}_4 + 0.004641\dot{M}_5$$
$$+ 0.00984\dot{M}_6 = 0$$
$$0.000038\dot{M}_1 + 0.0000408\dot{M}_2 + 0.000124\dot{M}_3 - 0.0101959\dot{M}_4 + 0.0000618\dot{M}_5$$
$$+ 0.0001312\dot{M}_6 = 0$$
$$0.004755\dot{M}_1 + 0.0051\dot{M}_2 + 0.0155\dot{M}_3 + 0.00051M_4 - 0.146965\dot{M}_5$$
$$+ 0.0164\dot{M}_6 = 0$$
$$0.008559\dot{M}_1 + 0.00918\dot{M}_2 + 0.0279\dot{M}_3 + 0.000918\dot{M}_4 + 0.013923\dot{M}_5$$
$$- 0.29848\dot{M}_6 = 0 \tag{7.118}$$

with the unknown optimal mean values $\dot{M}_b$ of the system unconditional sojourn times in the operation states we are looking for.

Since the determinant of the main matrix of the homogeneous system of Eqs. 7.118 is equal to 0, its rank is less than 6 and there are non-zero solutions of this system of equations that are ambiguous and dependent on one or more parameters. Thus, we may fix some of them and determine the remaining ones. In our case, according to (7.52), after considering expert opinion, we conclude that it is sensible to assume

$$\dot{M}_4 \cong 2. \tag{7.119}$$

Consequently, from (7.118), we get the system of equations

$$\begin{aligned}
&-0.017156\dot{M}_1 + 0.0835992\dot{M}_2 + 0.254076\dot{M}_3 + 0.1267921\dot{M}_5 \\
&+ 0.2688288\dot{M}_6 = -0.0167198 \\
&0.000951\dot{M}_1 - 0.10098\dot{M}_2 + 0.0031\dot{M}_3 + 0.001547\dot{M}_5 \\
&+ 0.00328\dot{M}_6 = -0.000204 \\
&0.002853\dot{M}_1 + 0.00306\dot{M}_2 - 0.3007\dot{M}_3 + 0.004641\dot{M}_5 \\
&+ 0.00984\dot{M}_6 = -0.000612 \\
&0.000038\dot{M}_1 + 0.0000408\dot{M}_2 + 0.000124\dot{M}_3 + 0.0000618\dot{M}_5 \\
&+ 0.0001312\dot{M}_6 = 0.0203918 \\
&0.004755\dot{M}_1 + 0.0051\dot{M}_2 + 0.0155\dot{M}_3 - 0.146965\dot{M}_5 \\
&+ 0.0164\dot{M}_6 = -0.00102 \\
&0.008559\dot{M}_1 + 0.00918\dot{M}_2 + 0.0279\dot{M}_3 \\
&+ 0.013923\dot{M}_5 - 0.29848\dot{M}_6 = -0.001836
\end{aligned}$$

and we solve it with respect to $\dot{M}_1$, $\dot{M}_2$, $\dot{M}_3$, $\dot{M}_5$ and $\dot{M}_6$. The solutions of the system of Eqs. 7.118 obtained in this way, are

$$\begin{aligned}
&\dot{M}_1 \cong 439.9400,\ \dot{M}_2 \cong 5.0046,\ \dot{M}_3 \cong 4.9401,\ \dot{M}_4 \cong 2,\ \dot{M}_5 \cong 16.4988, \\
&\dot{M}_6 \cong 14.0069.
\end{aligned} \tag{7.120}$$

It can be seen that these solutions differ much from the values M_1, M_2, M_3, M_4, M_5 and M_6 estimated in Sect. 7.2.5.4 and given by (7.49–7.54).

Having these solutions, it is also possible to look for the optimal values $\dot{M}_{bl}$ of the mean values M_{bl} of the conditional sojourn times at the operation states. Namely, substituting the probabilities of the system operation process transitions between the operation states, determined by (7.41) and the optimal mean values $\dot{M}_b$ given by (7.120) into (6.42), we get the following system of equations

$$0.648\dot{M}_{12} + 0.336\dot{M}_{13} + 0.008\dot{M}_{14} + 0.008\dot{M}_{16} = 439.9400$$
$$0.525\dot{M}_{21} + 0.373\dot{M}_{23} + 0.093\dot{M}_{24} + 0.009\dot{M}_{26} = 5.0046$$
$$0.105\dot{M}_{31} + 0.111\dot{M}_{32} + 0.118\dot{M}_{35} + 0.666\dot{M}_{36} = 4.9401$$
$$0.417\dot{M}_{41} + 0.583\dot{M}_{42} = 2$$
$$0.005\dot{M}_{51} + 0.220\dot{M}_{53} + 0.775\dot{M}_{56} + = 16.4988$$
$$0.012\dot{M}_{61} + 0.628\dot{M}_{63} + 0.360\dot{M}_{65} = 14.0069.$$

with the unknown optimal values $\dot{M}_{bl}$ we want to find.

As the solutions of the above system of equations are ambiguous, we arbitrarily fix some of them, for practically important reasons, and we find the remaining ones. In this case we proceed as follows:

- we fix in the first equation $\dot{M}_{13} = 40$, $\dot{M}_{14} = 50$, $\dot{M}_{16} = 4$ and we find $\dot{M}_{12} \cong 657.5$;
- we fix in the second equation $\dot{M}_{23} = 9$, $\dot{M}_{24} = 2, \dot{M}_{26} = 16$and we find $\dot{M}_{21} \cong 2.51$;
- we fix in the third equation $\dot{M}_{31} = 6, \dot{M}_{32} = 4, \dot{M}_{35} = 7$ and we find $\dot{M}_{36} \cong 4.56$;
- we fix in the fourth equation $\dot{M}_{41} = 2$ and we find $\dot{M}_{42} \cong 2$;
- we fix in the fifth equation $\dot{M}_{51} = 10, \dot{M}_{53} = 3$ and we find $\dot{M}_{56} \cong 20.37$;
- we fix in the sixth equation $\dot{M}_{61} = 23, \dot{M}_{65} = 21$ and we find

$$\dot{M}_{63} \cong 9.83. \tag{7.121}$$

Other very useful and much easier to be applied in practice tools that can help in planning the operation process of the ferry technical system are the system operation process optimal mean values of the total sojourn times at the particular operation states during the fixed system operation time θ.

Assuming as in Sect. 7.2.5.4, the system operation time $\theta = 1\text{year} = 365$ days, after applying (6.43), we get their values

$$\dot{E}[\hat{\theta}_1] = \dot{p}_1\theta = 0.8196 \times 365 \cong 299.15,\ \dot{E}[\hat{\theta}_2] = \dot{p}_2\theta = 0.01 \times 365 = 3.65,$$
$$\dot{E}[\hat{\theta}_3] = \dot{p}_3\theta = 0.03 \times 365 = 10.95,\ \dot{E}[\hat{\theta}_4] = \dot{p}_4\theta = 0.0004 \times 365 \cong 0.15,$$
$$\dot{E}[\hat{\theta}_5] = \dot{p}_5\theta = 0.05 \times 365 = 18.25,\ \dot{E}[\hat{\theta}_6] = \dot{p}_6\theta = 0.09 \times 365 = 32.85. \tag{7.122}$$

that differ from the values of $E[\hat{\theta}_i]$, $i = 1, 2, \ldots, 6$, determined by (7.57).

In practice, the knowledge of the optimal values of $\dot{M}_b$, $\dot{M}_{bl}$ and $\dot{E}[\hat{\theta}_b]$ given respectively by (7.120–7.122), can be very important and helpful for the container gantry crane operation process planning and improving in order to make the system operation more reliable and safer.

7.2.5.10 Container Gantry Crane Reliability Optimization

To make the optimization of the reliability of the container gantry crane we need the optimal values $\dot{p}_b$, $b = 1, 2, \ldots, 6$, of the transient probabilities p_b, $b = 1, 2, \ldots, 6$, in particular operation states determined by (7.116). Substituting these optimal solutions into the formula (6.19), we obtain the optimal mean values of the container gantry crane unconditional lifetimes in the reliability state subset $\{1, 2, 3\}$ and $\{3\}$ that respectively are

$$\begin{aligned}\dot{\mu}(1) &= \dot{p}_1 \times 5.24 + \dot{p}_2 \times 1.93 + \dot{p}_3 \times 1.91 + \dot{p}_4 \times 1.83 + \dot{p}_5 \times 1.84 + \dot{p}_6 \times 1.62 \\ &= 0.8196 \times 5.24 + 0.01 \times 1.93 + 0.03 \times 1.91 + 0.0004 \times 1.83 + 0.05 \times 1.84 \\ &\quad + 0.09 \times 1.62 \cong 4.61.\end{aligned} \tag{7.123}$$

$$\begin{aligned}\dot{\mu}(3) &= \dot{p}_1 \cdot 2.76 + \dot{p}_2 \cdot 1.13 + \dot{p}_3 \cdot 1.12 + \dot{p}_4 \cdot 1.05 + \dot{p}_5 \cdot 1.06 + \dot{p}_6 \cdot 0.98 \\ &= 0.8196 \cdot 2.76 + 0.01 \cdot 1.13 + 0.03 \cdot 1.12 + 0.0004 \cdot 1.05 + 0.05 \cdot 1.06 \\ &\quad + 0.09 \cdot 0.98 \cong 2.45.\end{aligned} \tag{7.124}$$

According to (6.23), the optimal solutions for the mean values of the container gantry crane unconditional lifetimes in the particular reliability states 1, 2 and 3, respectively are

$$\begin{aligned}&\dot{\bar{\mu}}(1) = \dot{\mu}(1) - \dot{\mu}(2) = 1.25,\ \dot{\bar{\mu}}(2) = \dot{\mu}(2) - \dot{\mu}(3) = 0.91, \\ &\dot{\bar{\mu}}(3) = \dot{\mu}(3) = 2.45.\end{aligned} \tag{7.125}$$

Moreover, according to (6.20–6.21), the corresponding optimal unconditional multistate reliability function of the container gantry crane is given by the vector

$$\dot{\boldsymbol{R}}(t, \cdot) = [1, \dot{\boldsymbol{R}}(t, 1), \dot{\boldsymbol{R}}(t, 2), \dot{\boldsymbol{R}}(t, 3)], \tag{7.126}$$

with the coordinates

$$\begin{aligned}\dot{\boldsymbol{R}}(t, 1) = {} & 0.8196 \times [\boldsymbol{R}(t, 1)]^{(1)} + 0.01 \times [\boldsymbol{R}(t, 1)]^{(2)} + 0.03 \times [\boldsymbol{R}(t, 1)]^{(3)} \\ & + 0.0004 \times [\boldsymbol{R}(t, 1)]^{(4)} + 0.05 \times [\boldsymbol{R}(t, 1)]^{(5)} + 0.09 \times [\boldsymbol{R}(t, 1)]^{(6)} \text{ for } t \geq 0,\end{aligned} \tag{7.127}$$

$$\begin{aligned}\dot{\boldsymbol{R}}(t, 2) = {} & 0.8196 \times [\boldsymbol{R}(t, 2)]^{(1)} + 0.01 \times [\boldsymbol{R}(t, 2)]^{(2)} + 0.03 \times [\boldsymbol{R}(t, 2)]^{(3)} \\ & + 0.0004 \times [\boldsymbol{R}(t, 2)]^{(4)} + 0.05 \times [\boldsymbol{R}(t, 2)]^{(5)} + 0.09 \times [\boldsymbol{R}(t, 2)]^{(6)} \text{ for } t \geq 0,\end{aligned} \tag{7.128}$$

$$\begin{aligned}\dot{\boldsymbol{R}}(t, 3) = {} & 0.8196 \times [\boldsymbol{R}(t, 3)]^{(1)} + 0.01 \times [\boldsymbol{R}(t, 3)]^{(2)} + 0.03 \times [\boldsymbol{R}(t, 3)]^{(3)} \\ & + 0.0004 \times [\boldsymbol{R}(t, 3)]^{(4)} + 0.05 \times [\boldsymbol{R}(t, 3)]^{(5)} + 0.09 \times [\boldsymbol{R}(t, 3)]^{(6)} \text{ for } t \geq 0,\end{aligned} \tag{7.129}$$

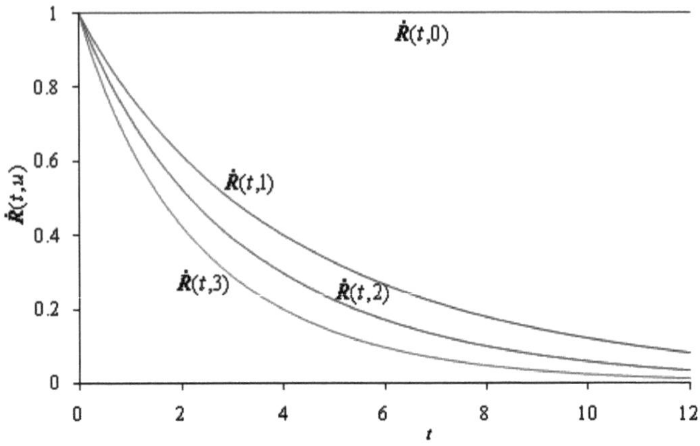

Fig. 7.24 The graph of the container gantry crane optimal reliability function $\boldsymbol{R}(t, \cdot)$ coordinates

where $[\boldsymbol{R}(t,1)]^{(b)}$, $[\boldsymbol{R}(t,2)]^{(b)}$, $[\boldsymbol{R}(t,3)]^{(b)}$, $b = 1,2,\ldots,6$, are fixed in Sect. 7.2.5.7, respectively by (7.62–7.64), (7.75–7.77), (7.79–7.81), (7.83–7.85), (7.87–7.89), (7.91–7.93).

The graphs of the coordinates of the container gantry crane optimal reliability function are given in Fig. 7.24.

Further, by (6.22), the corresponding optimal variances and standard deviations of the container gantry crane unconditional lifetime in the reliability state subsets are

$$\dot{\sigma}^2(1) = 2\int_0^\infty t\dot{\boldsymbol{R}}(t,1)\mathrm{d}t - [\dot{\mu}(1)]^2 \cong 24.758,\ \dot{\sigma}(1) \cong 4.9,$$

$$\dot{\sigma}^2(2) = 2\int_0^\infty t\dot{\boldsymbol{R}}(t,2)\mathrm{d}t - [\dot{\mu}(2)]^2 \cong 12.923,\ \dot{\sigma}(2) \cong 3.59, \tag{7.130}$$

$$\dot{\sigma}^2(3) = 2\int_0^\infty t\dot{\boldsymbol{R}}(t,2)\mathrm{d}t - [\dot{\mu}(3)]^2 \cong 6.893,\ \dot{\sigma}(3) \cong 2.63,$$

where $\dot{\boldsymbol{R}}(t,1)$, $\dot{\boldsymbol{R}}(t,2)$, $\dot{\boldsymbol{R}}(t,3)$are given by (7.127–7.129) and $\dot{\mu}(1)$, $\dot{\mu}(2)$, $\dot{\mu}(3)$ are given by (7.117) and (7.123), (7.124).

Since the critical reliability state is $r = 2$, the optimal system risk function, according to (6.24), is given by (Fig. 7.25)

$$\dot{r}(t) = 1 - \dot{\boldsymbol{R}}(t,\ 2)\ \text{for}\ t \geq 0, \tag{7.131}$$

where $\dot{\boldsymbol{R}}(t,2)$ is given by (7.128).

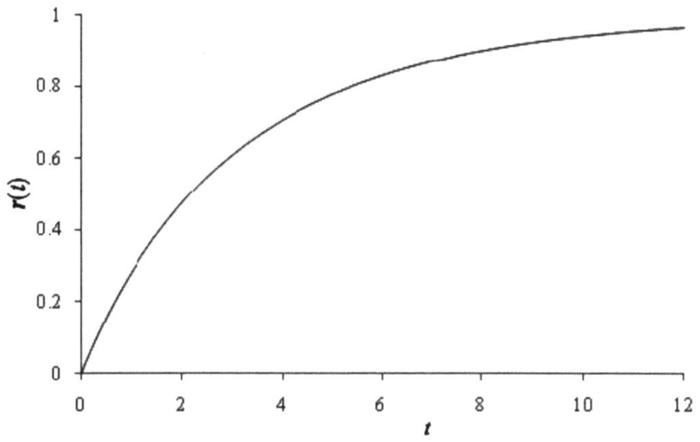

Fig. 7.25 The graph of the container gantry crane optimal risk function $\dot{\boldsymbol{r}}(t)$

Hence, considering (6.25), the moment when the optimal system risk function exceeds a permitted level, for instance $\delta = 0.05$, is

$$\dot{\tau} = \dot{\boldsymbol{r}}^{-1}(\delta) \cong 0.149\,\text{year}. \tag{7.132}$$

7.2.5.11 Container Gantry Crane Renewal and Availability Optimization

To determine the optimal renewal and availability characteristics of the container gantry crane after its operation process optimization, we use the results of the system reliability characteristics optimization performed in Sects. 7.2.5.9 and 7.2.5.10 and the results of Sect. 6.2.3.

In the case when the container gantry crane renovation time is ignored, considering the optimal values $\dot{\mu}(2) \cong 3.36$ determined by (7.117) and $\dot{\sigma}(2) \cong 3.59$ determined by (7.130) and applying *Proposition 6.1*, we determine its following optimal characteristics:

(a) the optimal time $\dot{S}_N(2)$ until the Nth exceeding by the system the reliability critical state 2, for sufficiently large N, has approximately normal distribution $N(3.36\,N, 3.59\sqrt{N})$, i.e.,

$$\dot{F}^{(N)}(t,2) = P(\dot{S}_N(2) < t) \cong F_{N(0,1)}\left(\frac{t - 3.36\,N}{3.59\sqrt{N}}\right), \quad t \in (-\infty, \infty),$$

(b) the expected value and the variance of the optimal time $\dot{S}_N(2)$ until the Nth exceeding by the system the reliability critical state 2, for sufficiently large N, respectively are

$$E[\dot{S}_N(2)] = 3.36\,N, \quad D[\dot{S}_N(2)] = 12.888\,N,$$

(c) the optimal number $\dot{N}(t,2)$ of exceedings by the system the reliability critical state 2 up to the moment t, $t \geq 0$, for sufficiently large t, has distribution approximately of the form

$$P(\dot{N}(t,2) = N) \cong F_{N(0,1)}\left(\frac{3.36(N+1) - t}{1.959\sqrt{t}}\right) - F_{N(0,1)}\left(\frac{3.36N - t}{1.959\sqrt{t}}\right), N = 0, 1, \ldots,$$

(d) the expected value and the variance of the optimal number $N(t, 2)$ of exceedings by the system the reliability critical state 2 up to the moment t, $t \geq 0$, for sufficiently large t, respectively are

$$\dot{H}(t,2) = 0.298t, \; \dot{D}(t,2) = 0.339t.$$

To make the estimation of the renewal and availability of the container gantry crane in the case when the time of renovation is non-ignored, considering the optimal values $\dot{\mu}(2) \cong 3.36$ determined by (7.117) and $\dot{\sigma}(2) \cong 3.59$ determined by (7.130), assuming the mean value of the system renovation time $\mu_0(2) = 0.0027$ year and the standard deviation of the system renovation time $\sigma_0(2) = 0.0014$ year and applying *Proposition 6.2*, we determine its following optimal characteristics:

(a) the optimal time $\dot{\bar{S}}_N(2)$ until the Nth exceeding by the system the reliability critical state 2, for sufficiently large N, has approximately normal distribution $N(3.36N + 0.0027(N-1), \sqrt{12.888N + 0.0000019(N-1)})$, i.e.,

$$\dot{\bar{F}}^{(N)}(t,2) = P(\dot{\bar{S}}_N(2) < t) = F_{N(0,1)}\left(\frac{t - 3.3627N + 0.0027}{\sqrt{12.888002N - 0.0000019}}\right), \quad t \in (-\infty, \infty);$$

(b) the expected value and the variance of the optimal time $\dot{\bar{S}}_N(2)$ until the Nth exceeding by the system the reliability critical state 2, for sufficiently large N, respectively are

$$E[\dot{\bar{S}}_N(2)] \cong 3.36N + 0.0027(N-1), \; D[\dot{\bar{S}}(2)] \cong 12.888N + 0.0000019(N-1),$$

(c) the optimal number $\dot{\bar{N}}(t,2)$ of exceedings by the system the reliability critical state 2 up to the moment t, $t \geq 0$, for sufficiently large t, has approximately distribution of the form

$$P(\dot{\bar{N}}(t,2) = N) \cong F_{N(0,1)}\left(\frac{3.3627(N+1) - t - 0.0027}{1.98\sqrt{t + 0.0027}}\right) - F_{N(0,1)}\left(\frac{3.3627N - t - 0.0027}{1.98\sqrt{t + 0.0027}}\right), N = 0, 1, \ldots,$$

(d) the expected value and the variance of the optimal number $\dot{\bar{N}}(t,2)$ of exceedings by the system the reliability critical state 2 up to the moment t, $t \geq 0$, for sufficiently large t, respectively are

$$\dot{\bar{H}}(t,2) \cong 0.297(t+0.0027),\ \dot{\bar{D}}(t,2) \cong 0.3389(t+0.0027),$$

(e) the optimal time $\dot{\bar{\bar{S}}}_N(2)$ until the Nth system's renovation, for sufficiently large N, has approximately normal distribution $N(3.3627\,N, 3.5899\sqrt{N})$, i.e.,

$$\dot{\bar{\bar{F}}}^{(N)}(t,2) = P(\dot{\bar{\bar{S}}}_N(2) < t) \cong F_{N(0,1)}\left(\frac{t - 3.3627\,N}{3.5899\sqrt{N}}\right), \quad t \in (-\infty, \infty),$$

(f) the expected value and the variance of the optimal time $\dot{\bar{\bar{S}}}_N(2)$ until the Nth system's renovation, for sufficiently large N, respectively are

$$E[\dot{\bar{\bar{S}}}_N(2)] \cong 3.3627\,N, \quad D[\dot{\bar{\bar{S}}}^{(N)}(2)] \cong 12.888002\,N,$$

(g) the optimal number $\dot{\bar{\bar{N}}}(t,2)$ of system's renovations up to the moment t, $t \geq 0$, for sufficiently large t, has approximately distribution of the form.

$$P(\dot{\bar{\bar{N}}}(t,2) = N) \cong F_{N(0,1)}\left(\frac{3.3627(N+1) - t}{1.98\sqrt{t}}\right) - F_{N(0,1)}\left(\frac{3.3627\,N - t}{1.98\sqrt{t}}\right),$$
$$N = 0, 1, \ldots,$$

(h) the expected value and the variance of the optimal number $\dot{\bar{\bar{N}}}(t,2)$ of system's renovations up to the moment t, $t \geq 0$, for sufficiently large t, respectively are

$$\dot{\bar{\bar{H}}}(t,2) \cong 0.297t, \quad \dot{\bar{\bar{D}}}(t,2) \cong 0.3389t,$$

(i) the optimal steady availability coefficient of the system at the moment t, $t \geq 0$, for sufficiently large t, is

$$\dot{A}(t,2) \cong 0.9992, \quad t \geq 0,$$

(j) the optimal steady availability coefficient of the system in the time interval $<t, t+\tau)$, $\tau > 0$,$t \geq 0$, for sufficiently large t, is

$$\dot{A}(t,\tau,2) \cong 0.297 \int_{\tau}^{\infty} \dot{\boldsymbol{R}}(t,2)\mathrm{d}t, \quad t \geq 0, \quad \tau > 0,$$

where $\dot{\boldsymbol{R}}(t,2)$ is given by (7.128).

7.2.5.12 Container Gantry Crane Corrective and Preventive Maintenance Policy Optimization

Maintenance Policy Maximizing System Availability

To optimize the container gantry crane corrective and preventive maintenance policy maximizing its availability, we use its following reliability and renewal parameters:

- the container gantry crane and its components critical reliability state $r = 2$,
- the 2nd coordinate of the container gantry crane unconditional reliability function $\boldsymbol{R}(t,\ \cdot)$

$$\boldsymbol{R}(t,2) = 0.6874 \times [\boldsymbol{R}(t,2)]^{(1)} + 0.0187 \times [\boldsymbol{R}(t,2)]^{(2)} + 0.0515 \times [\boldsymbol{R}(t,2)]^{(3)} + 0.0005 \times [\boldsymbol{R}(t,2)]^{(4)} + 0.0717 \times [\boldsymbol{R}(t,2)]^{(5)} + 0.1702 \times [\boldsymbol{R}(t,2)]^{(6)} \quad \text{for} \quad t \geq 0, \tag{7.133}$$

where $[\boldsymbol{R}(t,2)]^{(b)}$, $b = 1,2,\ldots,6$, are respectively given by (7.63), (7.76), (7.80), (7.84), (7.88), (7.92),
- the derivative of the 2-nd coordinate of the container gantry crane unconditional reliability function $\boldsymbol{R}(t,\ \cdot)$

$$\boldsymbol{R}'(t,2) = 0.6874 \times [\boldsymbol{R}'(t,2)]^{(1)} + 0.0187 \times [\boldsymbol{R}'(t,2)]^{(2)} + 0.0515 \times [\boldsymbol{R}'(t,2)]^{(3)} + 0.0005 \times [\boldsymbol{R}'(t,2)]^{(4)} + 0.0717 \times [\boldsymbol{R}'(t,2)]^{(5)} + 0.1702 \times [\boldsymbol{R}'(t,2)]^{(6)} \quad \text{for}\, t \geq 0$$

- the mean value of the container gantry crane unconditional corrective renovation time $\mu_0(1) = 0.0027$,
- the mean value of the container gantry crane unconditional preventive renovation time $\mu_1(1) = 0.0025$.

Moreover, to apply the algorithm proposed in Sect. 6.3.1, we fix:

- the measure of the method of secants accuracy $\varepsilon = 0.001$,
- the number of values of the system preventive maintenance period η for which we find the values of the availability coefficient of the container gantry crane in cases when there is no optimal value $\kappa = 20$,
- the values of the system preventive maintenance period η for which we find the values of the availability coefficient of the container gantry crane in cases when there is no optimal value

$$\eta_i = (i-1)0.2\mu(2) = (i-1)0.2 \cdot 3.04, \quad i = 1,2,\ldots,20. \tag{7.134}$$

where $\mu(2) \cong 3.04$, is given by (7.100).

After fixing the above reliability and renewal input parameters, we use the procedure described in Sect. 6.3.1.Since

$$\mu_0(2) = 0.0027 > \mu_1(2) = 0.0025$$

we are looking for the optimal value $\dot{\eta}$ of the preventive maintenance period η that maximizes the availability coefficient of the system $A(\eta, 1)$ given by (6.52) by determining, if it exists, its approximate value from Eq. 6.58 by applying the method of secants in the interval $\langle a,\ b\rangle$ as follows:

- we define, obtained after the transformation of the Eq. 6.58, the function

$$f(\eta) = \lambda(\eta, 2)\int_0^\eta \boldsymbol{R}(t,2)\mathrm{d}t + \boldsymbol{R}(\eta,2) - \frac{\mu_0(2)}{\mu_0(2) - \mu_1(2)}$$

$$= \lambda(\eta, 2)\int_0^\eta \boldsymbol{R}(t,2)\mathrm{d}t + \boldsymbol{R}(\eta,2) - 13.5 \quad \text{for } \eta \geq 0,$$

where $\boldsymbol{R}(t,\ 2)$ is given by (7.133) and

$$\lambda(\eta, 2) = -\frac{\boldsymbol{R}'(\eta, 2)}{\boldsymbol{R}(\eta, 2)},$$

- we define the interval $\langle\ a,\ b\ \rangle$ assuming $a = 0$ and finding b such that

$$f(b) > 0,$$

- we use the recurrent formula

$$\eta_0 = a,$$

$$\eta_{k+1} = \eta_k - \frac{f(\eta_k)}{f(b) - f(\eta_k)}(b - \eta_k) \quad \text{for } k = 0, 1, \ldots, K,$$

where K is such that

$$f(\eta_{K+1}) < \varepsilon$$

and $\varepsilon = 0.001$ is the measure of the method of secants accuracy,

- we fix the optimal value $\dot{\eta}$ of the preventive maintenance period η assuming

$$\dot{\eta} = \eta_{K+1}.$$

As a result of computer calculations, we recognize that there is no optimal value $\dot{\eta}$ of the container gantry crane preventive maintenance period η that maximizes the value of its availability coefficient. The exemplary values of the system preventive maintenance period η defined by (7.134) and the values of the availability coefficient of the container gantry crane are given in Table 7.3.

Table 7.3 The values of the availability coefficient of the container gantry crane

η	$A(\eta, 2)$	η	$A(\eta, 2)$
0.0	0.0	6.080	0.99894
0.608	0.99532	6.688	0.99897
1.216	0.99735	7.296	0.99899
1.824	0.99802	7.904	0.99901
2.432	0.99836	8.512	0.99903
3.04	0.99856	9.120	0.99904
3.648	0.99869	9.728	0.99905
4.256	0.99878	10.336	0.99906
4.864	0.99885	0.944	0.99907
5.472	0.99890	11.552	0.99908

7.2.5.13 Container Gantry Crane Operation and Reliability New Strategy

Parameters and Characteristics of Container Gantry Crane Operation Process Before and After Its Optimization

From Sect. 7.2.5.3, we have the values of the following container gantry crane operation process parameters before its optimization:

- the conditional mean sojourn times of the container gantry crane at the particular operation states

$$\begin{aligned}
&M_{12} = 456.978,\ M_{13} = 36.860,\ M_{14} = 50,\ M_{16} = 3,\\
&M_{21} = 7.887,\ M_{23} = 9.121,\ M_{24} = 1.545,\ M_{26} = 16,\\
&M_{31} = 5.5,\ M_{32} = 4.343,\ M_{35} = 6.822,\ M_{36} = 7.857,\\
&M_{41} = 2,\ M_{42} = 2.143,\ M_{51} = 10,\ M_{53} = 2.899,\ M_{56} = 24.681,\\
&M_{61} = 22.6,\ M_{63} = 23.117,\ M_{65} = 20.512.
\end{aligned} \tag{7.135}$$

From Sect. 7.2.5.9, we have the values of the following container gantry crane operation process parameters after its optimization:

- the optimal conditional mean sojourn times of the container gantry crane at the particular operation states

$$\begin{aligned}
&\dot{M}_{12} \cong 657.5,\ \dot{M}_{13} = 40,\ \dot{M}_{14} = 50,\ \dot{M}_{16} = 4,\\
&\dot{M}_{21} \cong 2.51,\ \dot{M}_{23} = 9,\ \dot{M}_{24} = 2,\ \dot{M}_{26} = 16,\\
&\dot{M}_{31} = 6,\ \dot{M}_{32} = 4,\ \dot{M}_{35} = 7,\ \dot{M}_{36} \cong 4.56,\\
&\dot{M}_{41} = 2,\ \dot{M}_{42} \cong 2,\ \dot{M}_{51} = 10,\ \dot{M}_{53} = 3,\ \dot{M}_{56} \cong 20,\\
&\dot{M}_{61} = 23,\ \dot{M}_{65} = 21,\ \dot{M}_{63} \cong 9.83.
\end{aligned} \tag{7.136}$$

From Sect. 7.2.5.4, we have the values of the following container gantry crane operation process characteristics before its optimization:

- the unconditional mean sojourn times of the container gantry crane at the particular operation states

$$M_1 = 308.93,\ M_2 = 7.83,\ M_3 = 7.09,\ M_4 = 2.08,\ M_5 \cong 19.82,\ M_6 \cong 22.17; \quad (7.137)$$

- the transient probabilities of the container gantry crane operation process at the operational states

$$p_1 = 0.6874,\ p_2 = 0.0187,\ p_3 = 0.0515,\ p_4 = 0.0005,\ p_5 = 0.0717,$$
$$p_6 = 0.1702; \quad (7.138)$$

- the total sojourn times of the container gantry crane operation process in particular operation states during the operation time $\theta = 1$ year $= 365$ days

$$E[\hat{\theta}_1] = 251\,\text{days},\ E[\hat{\theta}_2] = 7\,\text{days},\ E[\hat{\theta}_3] = 19\,\text{day},$$
$$E[\hat{\theta}_4] = 0.2\,\text{day},\ E[\hat{\theta}_5] = 26\,\text{days},\ E[\hat{\theta}_6] = 62\,\text{days}. \quad (7.139)$$

From Sect. 7.2.5.9, we have the values of the following container gantry crane operation process characteristics after its optimization:

- the optimal unconditional mean sojourn times of the container gantry crane in the particular operation states

$$\dot{M}_1 \cong 439.94,\ \dot{M}_2 \cong 5.00,\ \dot{M}_3 \cong 439.94,\ \dot{M}_4 \cong 2,\ \dot{M}_5 \cong 16.50,$$
$$\dot{M}_6 \cong 14.01; \quad (7.140)$$

- the optimal transient probabilities of the container gantry crane operation process at the operational states

$$\dot{p}_1 = 0.8196,\ \dot{p}_2 = 0.01,\ \dot{p}_3 0.03,\ \dot{p}_4 = 0.0004,\ \dot{p}_5 = 0.05,\ \dot{p}_6 = 0.09; \quad (7.141)$$

- the optimal total sojourn times of the container gantry crane operation process in particular operation states during the operation time $\theta = 1$ year $= 365$ days

$$\dot{E}[\hat{\theta}_1] = 299.15\ \text{days},\ \dot{E}[\hat{\theta}_2] = 3.65\ \text{days},\ \dot{E}[\hat{\theta}_3] = 10.95\text{days},$$
$$\dot{E}[\hat{\theta}_4] = 0.15\ \text{days},\ \dot{E}[\hat{\theta}_5] = 18.25\ \text{days},\ \dot{E}[\hat{\theta}_6] = 32.85\text{days}. \quad (7.142)$$

Characteristics of Container Gantry Crane Reliability Before and After Operation Process Optimization

From Sect. 7.2.5.7, we have the values of the following container gantry crane reliability characteristics before its operation process optimization:

- the expected values of the container gantry crane unconditional lifetimes respectively in the reliability state subsets {1, 2, 3},{2, 3}, {3}

$$\mu(1) = 4.14\,\text{years},\ \mu(2) = 3.04\,\text{years},\ \mu(3) = 2.22\,\text{years}; \tag{7.143}$$

- the mean values of the unconditional lifetimes respectively in the particular reliability states 1, 2, 3

$$\bar{\mu}(1) = 1.10, \quad \bar{\mu}(2) = 0.82, \quad \bar{\mu}(3) = 2.22\ \text{years}; \tag{7.144}$$

- the moment when the system risk function exceeds a permitted level

$$\tau \cong 0.126\ \text{year}. \tag{7.145}$$

From Sects. 7.2.5.9 and 7.2.5.10, we have the values of the following container gantry crane reliability parameters and characteristics after its operation process optimization:

- the optimal expected values of the container gantry crane unconditional lifetimes respectively in the reliability state subsets {1, 2, 3}, {2, 3}, {3}

$$\dot{\mu}(1) = 4.61\ \text{years},\ \dot{\mu}(2) = 3.36\ \text{years},\ \dot{\mu}(3) = 2.45\,\text{years}; \tag{7.146}$$

- the optimal mean values of the unconditional lifetimes respectively in the particular reliability states 1, 2, 3

$$\dot{\bar{\mu}}(1) = 1.25,\ \dot{\bar{\mu}}(2) = 0.91,\ \dot{\bar{\mu}}(3) = 2.45\ \text{years}; \tag{7.147}$$

- the optimal moment when the system risk function exceeds a permitted level

$$\dot{\tau} \cong 0.149\,\text{year}. \tag{7.148}$$

Characteristics of Container Gantry Crane Renewal and Availability Before and After Operation Process Optimization

From Sect. 7.2.5.8, we have the values of the following container gantry crane renewal and availability characteristics before its operation process optimization:

- the expected value of the number of exceedings by the system with ignored time of renovation the reliability critical state during 1 year = 365 days

$$H(365, 2) = 120.05; \tag{7.149}$$

- the expected value of the number of renovations of the system with non-ignored time of renovation during 1 year = 365 days

$$\bar{\bar{H}}(365, 2) \cong 119.94; \tag{7.150}$$

- the steady availability coefficient of the system

$$A(t, 2) \cong 0.9989. \tag{7.151}$$

From Sect. 7.2.5.11, we have the values of the following container gantry crane renewal and availability characteristics after its operation process optimization:

- the optimal expected value of the number of exceeding by the system with ignored time of renovation the reliability critical state during 1 year = 365 days

$$\dot{H}(365, 2) = 108.77; \tag{7.152}$$

- the optimal expected value of the number of renovations of the system with non-ignored time of renovation during 1 year = 365 days

$$\dot{\bar{\bar{H}}}(365, 2) \cong 108.41; \tag{7.153}$$

- the optimal steady availability coefficient of the system

$$\dot{A}(t, 2) \cong 0.9992. \tag{7.154}$$

Results of Container Gantry Crane Corrective and Preventive Maintenance Policy Optimization

In Sect. 7.2.5.12, we fixed that there is no optimal value $\dot{\eta}$ of the container gantry crane preventive maintenance period η that maximizes the value of its availability coefficient. The exemplary values of the system preventive maintenance period η and the values of the availability coefficient of the container gantry crane that are given in Table 7.3 justify that the container gantry crane coefficient is an increasing function of the preventive period.

Analysis of Container Gantry Crane Operation and Reliability Characteristics and Maintenance Optimization Results

The comparison of the values of the selected container gantry crane characteristics before the system operation process optimization given by (7.143–7.145) for the system reliability and by (7.149–7.151) for the system renewal and availability with their values after the system operation process optimization respectively given by (7.146–7.148) and by (7.152–7.154) justifies the sensibility of the performed system operation process optimization.

The results of the system corrective and preventive optimization does not give the value of the preventive maintenance period that maximizes the system availability coefficient. The value of the system availability coefficient increases while the preventive maintenance period increases.

Suggestions on New Strategy of Container Gantry Crane Operation Process Organizing

From the analysis performed of the results of the container gantry crane operation process optimization it can be suggested to organize the system operation process in a way that causes the replacing (or the approaching/convergence to) the conditional mean sojourn times M_{bl} of the system at the particular operation states before the optimization given by (7.135) by their optimal values $\dot{M}_{bl}$ after the optimization given by (7.136). The possibility of fulfilling this suggestion of the operation process parameters changing is not easy and has to be checked in practice.

It seems to be practically easier to, change the operation process characteristics which results in replacing (or the approaching/convergence to) the unconditional mean sojourn times M_b of the container gantry crane at the particular operation states before the optimization given by (7.137) by their optimal values $\dot{M}_b$ after the optimization given by (7.140).

The easiest way of the system operation process reorganizing is that leading to the replacing (or the approaching/convergence to) the total sojourn times $E[\hat{\theta}_b]$ of the container gantry crane operation process at the particular operation states during the operation time $\theta = 1$ year before the optimization given by (7.139) by their optimal values $\dot{E}[\hat{\theta}_b]$ after the optimization given by (7.142).

Suggestions on New Strategy of Container Gantry Crane Maintenance Policy

From the analysis of the results of the system corrective and preventive optimization it can only be suggested that in order to keep high availability of the system, the value of the preventive maintenance period of time that maximizes its availability coefficient should not be too small.

Other Suggestions

The container gantry crane operating at the varying in time operation conditions reliability and availability evaluation and prediction are approximate because of non sufficiently exact input data. Although the container gantry crane operation process parameters are fixed with a high accuracy, their changing reliability structures are fixed on a high level of generality and the reliability parameters of its components are not sufficiently exact because of the lack of statistical data necessary for their estimation. Therefore, the achieved results may only be considered as an illustration of the possibilities of applications of the proposed methods and procedures to this container gantry crane operation, reliability and availability analysis, prediction and optimization. However, the obtained evaluation may be a very useful example in the container gantry cranes and other complex technical systems, reliability and availability characteristics, identification and prediction, especially during the design and when planning and improving their operation processes safety and effectiveness.

7.3 Summary

The examples of applications, complementary to the appliqué parts of the previous chapters, are presented. They illustrate the possibilities of the theoretical results, applications in the book that are different from that presented in the previous chapters. Studying them makes much wider the range of real complex technical systems, whose reliability and safety identification, prediction and optimization can be performed with the methods and procedures presented in the book.

References

1. Guze S, Kołowrocki K (2008) Reliability analysis of multistate ageing consecutive "k out of n: F" systems. Int J Mat Struct Reliab 6(1):47–60
2. Kołowrocki K (2004) Reliability of large systems. Elsevier, Amsterdam
3. Kołowrocki K (2007) Reliability modelling of complex systems–Part 1. Int J Gnedenko e-Forum Reliab: Theory Appl 2(3–4):116–127
4. Kołowrocki K (2007) Reliability modelling of complex systems–Part 2. Int J Gnedenko e-Forum Reliab: Theory Appl 2(3–4):128–139
5. Kołowrocki K (2008) Reliability of large systems. In: Encyclopedia of quantitative risk analysis and assessment, vol 4. Wiley, New York, pp 1466–1471
6. Kołowrocki K, Soszyńska J (2006) Reliability and availability of complex systems. Qual Reliab Eng Int 22(1):79–99
7. Kolowrocki K, Soszynska J (2010) Reliability, availability and safety of complex technical systems: modelling–identification–prediction–optimization. Summer Safety & Reliability Seminars. J Pol Saf Reliab Assoc 4(1):133–158
8. Kolowrocki K, Soszynska J (2010) Preliminary statistical identification and orediction of the container gantry crane operation process. Summer Safety & Reliability Seminars. J Pol Saf and Reliab Assoc 4(1):173–181
9. Kossow A, Preuss W (1995) Reliability of linear consecutively-connected systems with multistate components. IEEE Trans Reliab 44:518–522
10. Soszyńska J (2004) Reliability of large parallel systems in variable operation conditions. Fac of Navig Res Works, vol 16, Gdynia, pp 168–180
11. Soszyńska J (2006) Reliability of large series-parallel system in variable operation conditions. Int J Autom Comput 3(2):199–206

12. Soszyńska J (2007) Systems reliability analysis in variable operation conditions. PhD thesis, Gdynia Maritime University-System Research Institute Warsaw (in Polish)
13. Soszyńska J (2008) Asymptotic approach to reliability evaluation of large "m out of l"–series system in variable operation conditions. Summer Safety & Reliability Seminars. J Pol Saf Reliab Assoc 2(2):323–346

Chapter 8
Summary

In this book the comprehensive approach to the analysis, identification, evaluation, prediction and optimization of complex technical systems, operation, reliability, availability and safety is presented. The results obtained may play the role of an easy-to-use guide necessary in reliability and safety evaluations of real complex technical systems, as well as during their operation and when they are designed. To make the results and the proposed methods and algorithms easy and useful tools for reliability practitioners their usage is illustrated by various examples of their application to the identification of the unknown parameters and to evaluation and optimization of the real system reliability and safety characteristics. The general analytical reliability, availability and safety models are the basis for the procedures and algorithms that allow to find the main and practically important reliability and safety and renewal and availability characteristics of the complex technical multistate systems. These models together with the linear pogramming are very useful in the complex technical systems operation, reliability, availability and safety improvement, optimization and their operation cost analysis. The methods and the procedures of identification of operation processes and the reliability and safety models of the complex technical systems presented in the book may be successfully applied to the real complex technical systems operation, reliability and safety unknown parameters estimation. All the tools are useful in the reliability, availability and safety optimization and operation cost analysis of a very wide class of real technical systems operating at the varying conditions that have an influence on changing their reliability and safety structures and their components, reliability and safety characteristics.

In addition to the general solutions, theoretical results are illustrated by practical examples of their application in reliability and safety evaluation and optimization of complex technical systems. The identification of the real technical port and maritime transportation systems, unknown parameters are obtained on the basis of enough precise systems and components operation data and on non-precise component reliability and safety data and therefore, first of all, they should

K. Kołowrocki and J. Soszyńska-Budny, *Reliability and Safety of Complex Technical Systems and Processes*, Springer Series in Reliability Engineering, DOI: 10.1007/978-0-85729-694-8_8,

be treated as an illustration of a wide possibility of applications of the proposed approach in system reliability analysis. Reliability and safety data come from experts and are concerned with component's mean lifetimes and their arbitrary assumed exponential multistate reliability or safety functions. These evaluations, despite not being precise may be a very useful, simple and quick tool in approximate reliability evaluation, especially during the design of large systems, and when planning and improving their safety and effectiveness operation processes. Nonetheless, the approximate optimisation of the operation processes and reliability and safety characteristics of the considered real complex technical systems also with respect to their operation costs may be performed with the application of the methods and procedures presented and used in the reorganization of the system operation and its new strategy planning. Thus, the results that can be achieved by the proposed tools application may be very interesting for reliability and safety practitioners from the port and maritime transport industry and from other industrial sectors as well.

The results presented in the book suggest that it seems reasonable to continue the investigations focusing on:

- methods and procedures of increasing the reliability and safety of complex multistate systems,
- methods and procedures of reliability and safety optimization of complex technical multistate systems related to their operation costs,
- availability and maintenance optimization of complex technical systems,
- elaboration of universal practical tools in the form of the reliability and safety decision support system including the methods and procedures and the computer program packages addressed to the operators of the complex technical systems, allowing them to evaluate and optimize automatically these systems, operation, reliability, availability and safety,
- elaboration of the offer of the addressed to industry training courses in the scope of operation, safety, reliability and availability of complex technical systems based on the methods and procedures and corresponding computer program packages included in the reliability and safety decision support system.

Appendix A: Table of Standard Normal Distribution—Laplace's Function

The values of Laplace's function

The values of a function
$\Phi(t) = P(0 \le U \le t), t \ge 0$

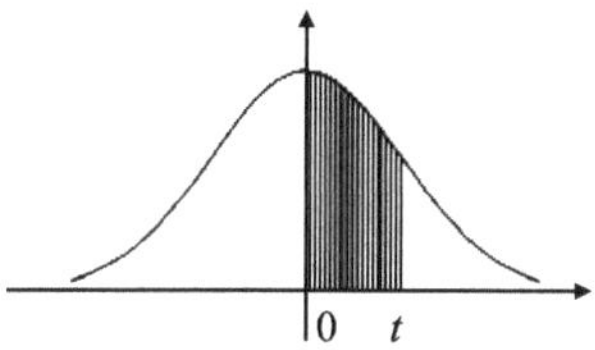

t	0	1	2	3	4	5	6	7	8	9
0.0	0.0000	0.0040	0.0080	0.0120	0.0160	0.0199	0.0239	0.0279	0.0319	0.0359
0.1	0.0398	0.0438	0.0478	0.0517	0.0557	0.0596	0.0636	0.0675	0.0714	0.0753
0.2	0.0793	0.0832	0.0871	0.0910	0.0948	0.0987	0.1026	0.1064	0.1103	0.1141
0.3	0.1179	0.1217	0.1255	0.1293	0.1331	0.1368	0.1406	0.1443	0.1480	0.1517
0.4	0.1554	0.1591	0.1628	0.1664	0.1700	0.1736	0.1772	0.1808	0.1844	0.1879
0.5	**0.1915**	**0.1950**	**0.1985**	**0.2019**	**0.2054**	**0.2088**	**0.2123**	**0.2157**	**0.2190**	**0.2224**
0.6	0.2257	0.2291	0.2324	0.2357	0.2389	0.2422	0.2454	0.2486	0.2517	0.2549
0.7	0.2580	0.2611	0.2642	0.2673	0.2704	0.2734	0.2764	0.2794	0.2823	0.2752
0.8	0.2781	0.2910	0.2939	0.2967	0.2995	0.3023	0.3051	0.3078	0.3106	0.3133
0.9	0.3159	0.3186	0.3212	0.3238	0.3264	0.3489	0.3315	0.3340	0.3365	0.3389
1.0	**0.3413**	**0.3438**	**0.3461**	**0.3485**	**0.3508**	**0.3531**	**0.3554**	**0.3577**	**0.3599**	**0.3621**
1.1	0.3643	0.3665	0.3686	0.3708	0.3729	0.3749	0.3770	0.3790	0.3810	0.3830
1.2	0.3849	0.3869	0.3888	0.3907	0.3925	0.3945	0.3962	0.3980	0.3997	0.4015
1.3	0.4032	0.4049	0.4066	0.4082	0.4099	0.4115	0.4131	0.4147	0.4162	0.4177
1.4	0.4192	0.4270	0.4222	0.4236	0.4251	0.4265	0.4279	0.4292	0.4306	0.4319
1.5	**0.4332**	**0.4345**	**0.4357**	**0.4370**	**0.4382**	**0.4394**	**0.4406**	**0.4418**	**0.4429**	**0.4441**
1.6	0.4452	0.4463	0.4474	0.4484	0.4495	0.4505	0.4515	0.4525	0.4535	0.4545
1.7	0.4554	0.4564	0.4573	0.4582	0.4591	0.4599	0.4608	0.4616	0.4625	0.4633
1.8	0.4641	0.4649	0.4656	0.4664	0.4671	0.4678	0.4686	0.4693	0.4699	0.4706
1.9	0.4713	0.4719	0.4726	0.4732	0.4738	0.4744	0.4750	0.4756	0.4761	0.4767

(continued)

K. Kołowrocki and J. Soszyńska-Budny, *Reliability and Safety of Complex Technical Systems and Processes*, Springer Series in Reliability Engineering, DOI: 10.1007/978-0-85729-694-8,

(continued)

The values of a function
$\Phi(t) = P(0 \le U \le t), t \ge 0$

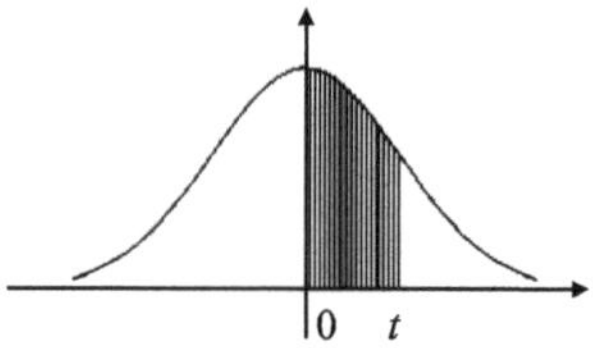

t	0	1	2	3	4	5	6	7	8	9
2.0	**0.4772**	**0.4778**	**0.4783**	**0.4788**	**0.4793**	**0.4798**	**0.4803**	**0.4808**	**0.4812**	**0.4817**
2.1	0.4821	0.4826	0.4830	0.4834	0.4838	0.4842	0.4846	0.4850	0.4854	0.4857
2.2	0.4861	0.4864	0.4868	0.4871	0.4875	0.4878	0.4881	0.4884	0.4887	0.4890
2.3	0.4892	0.4896	0.4898	0.4901	0.4904	0.4906	0.4909	0.4911	0.4913	0.4916
2.4	0.4918	0.4920	0.4922	0.4925	0.4927	0.4929	0.4931	0.4932	0.4934	0.4936
2.5	**0.4938**	**0.4940**	**0.4941**	**0.4943**	**0.4945**	**0.4946**	**0.4948**	**0.4949**	**0.4951**	**0.4952**
2.6	0.4953	0.4955	0.4956	0.4957	0.4959	0.4960	0.4961	0.4962	0.4963	0.4964
2.7	0.4965	0.4966	0.4967	0.4968	0.4969	0.4970	0.4971	0.4972	0.4973	0.4974
2.8	0.4974	0.4975	0.4976	0.4977	0.4977	0.4978	0.4979	0.4979	0.4980	0.4981
2.9	0.4981	0.4982	0.4982	0.4983	0.4984	0.4984	0.4985	0.4985	0.4986	0.4986
3.0	**0.4987**	**0.4987**	**0.4987**	**0.4988**	**0.4988**	**0.4989**	**0.4989**	**0.4989**	**0.4990**	**0.4990**
3.1	0.4990	0.4991	0.4991	0.4991	0.4992	0.4992	0.4992	0.4992	0.4993	0.4993
3.2	0.4993	0.4993	0.4994	0.4994	0.4994	0.4994	0.4994	0.4995	0.4995	0.4995
3.3	0.4995	0.4995	0.4995	0.4996	0.4996	0.4996	0.4996	0.4996	0.4996	0.4997
3.4	0.4997	0.4997	0.4997	0.4997	0.4997	0.4997	0.4997	0.4997	0.4997	0.4998
3.5	**0.4998**									

The relationship between the standard normal distribution $F_{N(0,1)}(t)$ and the Laplace's function $\Phi(t)$

$$F_{N(0,1)}(t) = P(T \le t) = \frac{1}{\sqrt{2\pi}} \int_{-\infty}^{t} e^{-x^2} dx = 0.5 + \text{sign}(|t|) \frac{1}{\sqrt{2\pi}} \int_{0}^{t} e^{-x^2} dx$$

$$= 0.5 + \text{sign}(|t|) P(0 \le T \le t) = 0.5 + \text{sign}(t)\Phi(t) \quad \text{for } t \in (-\infty, \infty)$$

Appendix B: Table of Chi-Square Distribution

The χ^2 with n degrees of freedom distribution of a variable U

The values of u such that $P(U > u_\alpha) = \alpha$

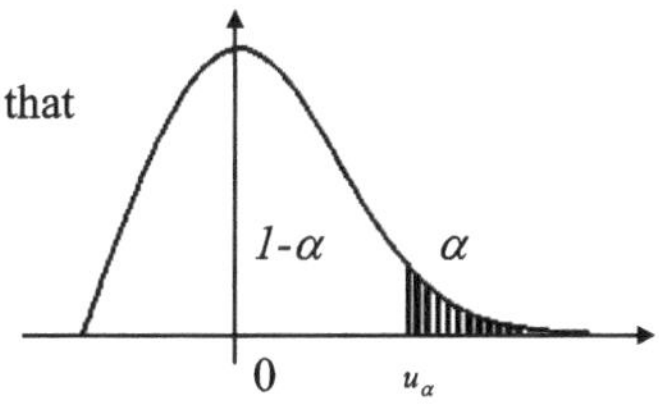

α / n	0.01	0.02	0.05	0.10	0.20	0.30	0.70	0.80	0.90	0.95	0.98	0.99
1	6.63	5.41	3.84	2.71	1.64	1.07	0.15	0.06	0.02	0.00	0.00	0.00
2	9.21	7.82	5.99	4.60	3.22	2.41	0.71	0.45	0.21	0.10	0.04	0.02
3	11.34	9.35	7.81	6.25	4.64	3.66	1.42	1.00	0.58	0.35	0.18	0.11
4	13.28	11.67	9.49	7.78	5.99	4.88	2.19	1.65	1.06	0.71	0.43	0.30
5	**15.09**	**13.39**	**11.07**	**9.24**	**7.29**	**6.06**	**3.00**	**2.34**	**1.61**	**1.14**	**0.75**	**0.55**
6	16.81	15.03	12.59	10.68	8.56	7.23	3.83	3.07	2.20	1.63	1.13	0.87
7	18.47	16.62	14.07	12.02	9.80	8.38	4.67	3.82	2.83	2.17	1.56	1.24
8	20.09	18.17	15.51	13.36	11.03	9.52	5.53	4.59	3.49	2.73	2.03	1.65
9	21.67	19.68	16.92	14.68	12.24	10.66	6.39	5.38	4.17	3.32	2.53	2.09
10	**23.21**	**21.16**	**18.31**	**15.99**	**13.44**	**11.78**	**7.27**	**6.18**	**4.86**	**3.94**	**3.06**	**2.56**
11	24.72	22.62	19.67	17.27	14.63	12.90	8.15	6.99	5.58	4.57	3.61	3.08
12	26.22	24.05	21.03	18.55	15.81	14.01	9.03	7.81	6.30	5.23	4.18	3.57
13	27.69	25.47	22.36	19.81	16.98	15.12	9.93	8.63	7.04	5.89	4.76	4.11
14	29.14	26.87	23.68	21.06	18.15	16.22	10.82	9.47	7.79	6.57	5.37	4.66
15	**30.58**	**28.26**	**25.00**	**22.31**	**19.31**	**17.32**	**11.72**	**10.31**	**8.55**	**7.26**	**5.98**	**5.23**
16	32.00	29.63	26.30	23.54	20.46	18.42	12.62	11.15	9.31	7.96	6.61	5.81
17	33.41	30.99	27.59	24.77	21.61	19.51	13.53	12.00	10.08	8.67	7.25	6.41
18	34.80	32.35	28.87	25.99	22.76	20.60	14.44	12.86	10.86	9.39	7.91	7.01
19	36.19	33.69	30.15	27.20	23.90	21.69	15.35	13.72	11.65	10.12	8.57	7.63
20	**37.57**	**35.02**	**31.41**	**28.41**	**25.04**	**22.77**	**16.27**	**14.59**	**12.44**	**10.85**	**9.24**	**8.26**
21	38.93	36.34	32.67	29.61	26.17	23.86	17.18	15.45	13.24	11.59	9.91	8.90
22	40.29	37.66	33.92	30.81	27.30	24.94	18.10	16.31	14.04	12.34	10.60	9.54

(continued)

K. Kołowrocki and J. Soszyńska-Budny, *Reliability and Safety of Complex Technical Systems and Processes*, Springer Series in Reliability Engineering, DOI: 10.1007/978-0-85729-694-8, © Springer-Verlag London Limited 2011

(continued)

The values of u such that $P(U > u_\alpha) = \alpha$

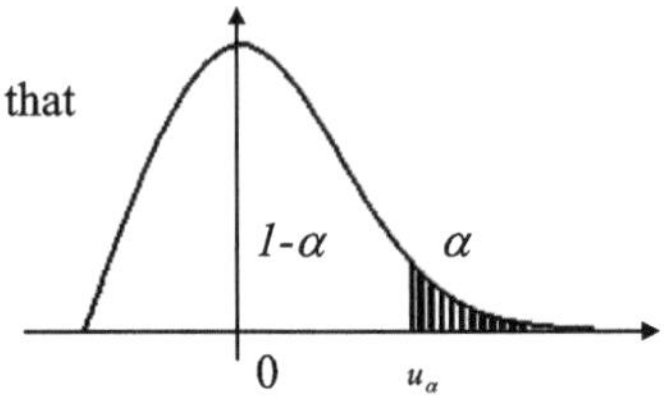

α / n	0.01	0.02	0.05	0.10	0.20	0.30	0.70	0.80	0.90	0.95	0.98	0.99
23	41.64	38.97	35.17	32.01	28.43	26.02	19.02	17.19	14.85	13.09	11.29	10.20
24	42.98	40.27	36.41	33.20	29.55	27.10	19.94	18.06	15.66	13.85	11.99	10.86
25	**44.31**	**41.57**	**37.66**	**34.38**	**30.67**	**28.17**	**20.87**	**18.94**	**16.47**	**14.61**	**12.70**	**11.52**
26	45.64	42.86	38.88	35.56	31.79	29.25	21.79	19.82	17.29	15.38	13.41	12.20
27	46.96	44.14	40.11	36.74	32.91	30.32	22.72	20.70	18.11	15.15	14.12	12.88
28	48.28	45.42	41.34	37.92	34.03	31.39	23.65	21.59	18.84	16.93	14.85	13.56
29	49.60	46.69	42.56	39.09	35.14	32.46	24.58	22.47	19.77	17.71	14.57	14.26
30	**50.89**	**47.96**	**43.77**	**40.26**	**36.25**	**33.53**	**25.51**	**23.36**	**20.60**	**18.49**	**16.31**	**14.95**

Appendix C: Table of λ-Kolmogorov Distribution

The values of $Q(\lambda)$ function

λ	$Q(\lambda)$	λ	$Q(\lambda)$	λ	$Q(\lambda)$	λ	$Q(\lambda)$
The values of a function $Q(\lambda) = \sum_{k=-\infty}^{\infty} (-1)^k \exp[-2k^2\lambda^2]$							
0.32	0.0000	0.82	0.4880	1.32	0.9387	1.82	0.9973
0.33	0.0001	0.83	0.5038	1.33	0.9418	1.83	0.9975
0.34	0.0002	0.84	0.5194	1.34	0.9449	1.84	0.9977
0.35	**0.0003**	**0.85**	**0.5347**	**1.35**	**0.9478**	**1.85**	**0.9979**
0.36	0.0005	0.86	0.5497	1.36	0.9505	1.86	0.9980
0.37	0.0008	0.87	0.5645	1.37	0.9531	1.87	0.9981
0.38	0.0013	0.88	0.5791	1.38	0.9556	1.88	0.9983
0.39	0.0019	0.89	0.5933	1.39	0.9580	1.89	0.9984
0.40	**0.0028**	**0.90**	**0.6073**	**1.40**	**0.9603**	**1.90**	**0.9985**
0.41	0.0040	0.91	0.6209	1.41	0.9625	1.91	0.9986
0.42	0.0055	0.92	0.6343	1.42	0.9646	1.92	0.9987
0.43	0.0074	0.93	0.6473	1.43	0.9665	1.93	0.9988
0.44	0.0097	0.94	0.6601	1.44	0.9684	1.94	0.9989
0.45	**0.0126**	**0.95**	**0.6725**	**1.45**	**0.9702**	**1.95**	**0.9990**
0.46	0.0160	0.96	0.6846	1.46	0.9718	1.96	0.9991
0.47	0.0200	0.97	0.6964	1.47	0.9734	1.97	0.9991
0.48	0.0247	0.98	0.7079	1.48	0.9750	1.98	0.9992
0.49	0.0300	0.99	0.7191	1.49	0.9764	1.99	0.9993
0.50	**0.0361**	**1.00**	**0.7300**	**1.50**	**0.9778**	**2.00**	**0.9993**
0.51	0.0428	1.01	0.7406	1.51	0.9791	2.01	0.9994
0.52	0.0503	1.02	0.7508	1.52	0.9803	2.02	0.9994
0.53	0.0585	1.03	0.7608	1.53	0.9815	2.03	0.9995
0.54	0.0675	1.04	0.7704	1.54	0.9826	2.04	0.9995
0.55	**0.0772**	**1.05**	**0.7798**	**1.55**	**0.9836**	**2.05**	**0.9996**
0.56	0.0876	1.06	0.7889	1.56	0.9846	2.06	0.9996
0.57	0.0987	1.07	0.7976	1.57	0.9855	2.07	0.9996
0.58	0.1104	1.08	0.8061	1.58	0.9864	2.08	0.9996

(continued)

K. Kołowrocki and J. Soszyńska-Budny, *Reliability and Safety of Complex Technical Systems and Processes*, Springer Series in Reliability Engineering,
DOI: 10.1007/978-0-85729-694-8,

(continued)

λ	$Q(\lambda)$	λ	$Q(\lambda)$	λ	$Q(\lambda)$	λ	$Q(\lambda)$
The values of a function $Q(\lambda) = \sum_{k=-\infty}^{\infty} (-1)^k \exp[-2k^2\lambda^2]$							
0.59	0.1228	1.09	0.8143	1.59	0.9873	2.09	0.9997
0.60	**0.1357**	**1.10**	**0.8223**	**1.60**	**0.9880**	**2.10**	**0.9997**
0.61	0.1492	1.11	0.8299	1.61	0.9888	2.11	0.9997
0.62	0.1632	1.12	0.8374	1.62	0.9895	2.12	0.9997
0.63	0.1778	1.13	0.8445	1.63	0.9902	2.13	0.9998
0.64	0.1927	1.14	0.8514	1.64	0.9908	2.14	0.9998
0.65	**0.2080**	**1.15**	**0.8580**	**1.65**	**0.9914**	**2.15**	**0.9998**
0.66	0.2236	1.16	0.8644	1.66	0.9919	2.16	0.9998
0.67	0.2396	1.17	0.8706	1.67	0.9924	2.17	0.9998
0.68	0.2558	1.18	0.8765	1.68	0.9929	2.18	0.9999
0.69	0.2722	1.19	0.8823	1.69	0.9934	2.19	0.9999
0.70	**0.2888**	**1.20**	**0.8877**	**1.70**	**0.9938**	**2.20**	**0.9999**
0.71	0.3055	1.21	0.8930	1.71	0.9942	2.21	0.9999
0.72	0.3223	1.22	0.8981	1.72	0.9946	2.22	0.9999
0.73	0.3391	1.23	0.9030	1.73	0.9950	2.23	0.9999
0.74	0.3560	1.24	0.9076	1.74	0.9953	2.24	0.9999
0.75	**0.3728**	**1.25**	**0.9121**	**1.75**	**0.9956**	**2.25**	0.9999
0.76	0.3896	1.26	0.9164	1.76	0.9959	2.26	0.9999
0.77	0.4064	1.27	0.9206	1.77	0.9962	2.27	0.9999
0.78	0.4230	1.28	0.9245	1.78	0.9965	2.28	0.9999
0.79	0.4395	1.29	0.9283	1.79	0.9967	2.29	0.9999
0.80	**0.4559**	**1.30**	**0.9319**	**1.80**	**0.9969**	**2.30**	**0.9999**
0.81	0.4720	1.31	0.9354	1.81	0.9971	2.31	1.0000

Appendix D: Table of Gamma Function

The values of a gamma function

p	$\Gamma(p)$
1.00	1.00000
1.05	0.97350
1.10	0.95135
1.15	0.93304
1.20	0.91817
1.25	0.90640
1.30	0.89747
1.35	0.89115
1.40	0.88726
1.45	0.88566
1.50	0.88623
1.55	0.88887
1.60	0.89352
1.65	0.90012
1.70	0.90864
1.75	0.91906
1.80	0.93138
1.85	0.94561
1.90	0.96177
1.95	0.97988
2.00	1.00000

The gamma function properties

$$\Gamma(1) = 1,$$

$$\Gamma(p + 1) = p\Gamma(p), \; p > 0,$$

$$\Gamma(n + 1) = n!, \; n \in N.$$

K. Kołowrocki and J. Soszyńska-Budny, *Reliability and Safety of Complex Technical Systems and Processes*, Springer Series in Reliability Engineering,
DOI: 10.1007/978-0-85729-694-8, © Springer-Verlag London Limited 2011

MIX
Papier aus verantwortungsvollen Quellen
Paper from responsible sources
FSC® C105338

If you have any concerns about our products,
you can contact us on
ProductSafety@springernature.com

In case Publisher is established outside the EU,
the EU authorized representative is:
Springer Nature Customer Service Center GmbH
Europaplatz 3, 69115 Heidelberg, Germany

Printed by Libri Plureos GmbH
in Hamburg, Germany